Transform Methods for
SOLVING PARTIAL DIFFERENTIAL EQUATIONS

Dean G. Duffy

Department of Mathematics
United States Naval Academy

CRC Press

Boca Raton   Ann Arbor   London   Tokyo

Library of Congress Cataloging-in-Publication Data

Catalog record is available from the Library of Congress

This book contains information obtained from authentic and highly regarded sources. Reprinted material is quoted with permission, and sources are indicated. A wide variety of references are listed. Reasonable efforts have been made to publish reliable data and information, but the author and the publisher cannot assume responsibility for the validity of all materials or for the consequences of their use.

Neither this book nor any part may be reproduced or transmitted in any form or by any means, electronic or mechanical, including photocopying, microfilming, and recording, or by any information storage and retrieval system, without prior permission in writing from the publisher.

CRC Press, Inc.'s consent does not extend to copying for general distribution, for promotion, for creating new works, or for resale. Specific permission must be obtained in writing from CRC Press for such copying.

Direct all inquiries to CRC Press, Inc., 2000 Corporate Blvd., N.W., Boca Raton, Florida 33431.

© 1994 by CRC Press, Inc.

No claim to original U.S. Government works
International Standard Book Number 0-8493-7374-3
Printed in the United States of America   2 3 4 5 6 7 8 9 0
Printed on acid-free paper

Contents

Dedicated to the Brigade of Midshipmen

Acknowledgments

I would like to thank the many midshipmen who have taken Engineering Mathematics II from me. They have been willing or unwilling guinea pigs in testing out many of the ideas and problems in this book. I would especially like to thank Prof. Tom Mahar for his many useful and often humorous suggestions for improving this book. Finally, I would like to express my appreciation to all those authors and publishers who allowed me the use of their material from the scientific and engineering literature.

Introduction

Purpose. Slightly over one hundred years ago, Oliver Heaviside introduced to the world his version of Laplace transforms. Poorly understood at that time it remained for T. J. I'a. Bromwich and others to win over the skeptical mathematical community. Today Fourier and Laplace transforms form the cornerstone of engineering analysis. However, that cannot be said of the use of operational methods in solving partial differential equations. The reason has been the incredible success enjoyed by separation of variables in solving most elementary problems associated with the heat and wave equations.

Despite these successes, however, separation of variables and the associated eigenfunction expansions are very limited. For example, the introduction of time dependent boundary conditions, especially if they are discontinuous, renders these techniques useless. For this reason, in certain specialities such as seismology, applied mechanics and antenna theory, the use of operational calculus to solve the governing equations has become commonplace. The purpose of this book is to make these techniques available to the general engineering and science community through a book that can be used in the classroom or for reference.

Prerequisites. The book assumes the usual undergraduate sequence of mathematics in engineering or the sciences: the traditional calculus and differential equations. A course in complex variables and Fourier and Laplace transforms is also essential. Finally some knowledge of Bessel functions is desirable to completely understand the book.

Audience. This book may be used as either a textbook or a reference book for applied physicists, geophysicists, civil, mechanical or electrical engineers and applied mathematicians. The material presented in Chapter 2 has been used by me as supplementary material in an operational calculus course taken by sophomore and juniors at the U. S. Naval Academy.

Chapter Overview. The purpose of Chapter 1 is two-fold. The first four sections serve as a refresher on the background material: linear ordinary differential equations, transform methods and complex variables. The amount of time spent with this material depends upon the background of the class. At least one class period should be spent on each section. The last two sections cover multivalued complex functions. This section can be omitted if you only plan to teach the second chapter. Otherwise, several class periods will be necessary to master this material because most students have never seen it. Due to the complexity of the problems, it is suggested that take-home problems are given to test the student's knowledge.

Chapters 2 and 3 are the meat-and-potatoes of the book. We subdivide the material according to whether we invert a single-valued or multivalued transform. Each chapter is then subdivided into the first two sections which deal with simply the mechanics of how to invert the transform and the last two sections which deal with actually applying the transform methods to solving partial differential equations. Undergraduates with a strong mathematics background should be able to handle Chapter 2 while Chapter 3 is really graduate-level material. The constant theme is the repeated application of the residue theorem to invert Fourier and Laplace transforms.

In Chapter 4 we solve partial differential equations by repeated applications of transform methods. We are now in advanced topics and this material is really only suitable for graduate students. The first section is the straightforward, brute-force application of Laplace and Fourier transforms in solving partial differential equations where we hope that we can invert both transforms to find the solution. Sections 2 and 3 are devoted to the very clever inversion techniques of Cagniard and de Hoop. For too long this interesting work has been restricted to the seismic, acoustic and electrodynamic communities. Finally Section 4 gives some of my own research which reflects the latest work in the area of joint transform methods.

In Chapter 5 we treat the classic Wiener-Hopf problem. This is very difficult material because of the very complicated analysis that is usually involved. I have tried to break the chapter into two parts; the first two sections deals with infinite scatterers while the the third deals with finite ones. Of the three sections, the first section is most amenable for a course because it occurs most often and the factorization is straightforward.

Features. This is an unabashedly applied book because the intended audience is problem solvers in engineering and the applied sciences. However, references are given to other books that do cover any unproven point, should the reader be interested. Also I have tried to give some human touch to this field by including references to the original works and photographs of some of the leading figures.

It is always difficult to write a book that satisfies both the student and the researcher. The student usually wants all of the gory details while the researcher wants the answer now. I have tried to accommodate both by centering most of the text around 72 examples of increasing difficulty. There are plenty of details for the student but the researcher may quickly leaf through the examples to find the material that interests him.

As anyone who has taken a course knows, the only way that you know a subject is by working the problems. For that reason I have included 179 well-crafted problems, most of which were taken from the scientific and engineering literature. Because many of these problems are difficult, I have included detailed solutions to most (110) of them. The student is asked however to refrain from looking at the solution before he has really tried to solve it on his own. No pain; no gain. The remaining problems have intermediate results so that the student has confidence that he is on the right track. The researcher also might look at these problems because his problem might already have been solved.

Finally, an important aspect of this book is the 1600 references that can serve as further grist for the student or point the researcher toward a solution of his problem. Of course, we must strike a balance between having a book of references and leaving out some interesting papers. The criteria for inclusion were three-fold. First, the paper had to have used the technique and not merely chanted the magic words; quoting results was unacceptable. Second, the papers had to compute both the forward and inverse transforms. The use of asymptotic or numerical methods to invert the transform excluded the reference. Finally, we required some details of the process.

Chapter 1
The Fundamentals

Many physical processes in nature evolve with time and are wave-like in structure. For this reason, the use of Laplace and Fourier transforms has proven a powerful analytic technique for solving partial differential equations that occur in engineering and the sciences. The purpose of this book is to illustrate their use.

We have designed this first chapter to bring everyone to a common starting point. Sections 1.1–1.4 provide a review for those who are a little rusty on their Laplace and Fourier transforms, ordinary differential equations and complex variables. We limit it to those topics that we will use later. Finally, sections 1.5 and 1.6 give an in-depth examination of multivalued complex functions because most books on complex variables treat these functions in a rather perfunctory manner.

1.1 LAPLACE TRANSFORMS

Let $f(t)$ be a function defined for $t \geq 0$. We define its *Laplace transform*[1] by the integral operation

$$\mathcal{L}[f(t)] = F(s) = \int_0^\infty f(t)e^{-st}dt \qquad (1.1.1)$$

---

[1] The standard reference for Laplace transforms is Doetsch, G., 1950: *Handbuch der Laplace-Transformation. Band 1. Theorie der Laplace-Transformation.* Basel: Birkhäuser Verlag, 581 pp.; Doetsch, G., 1955:

1

provided that the integral converges.

An important property of Laplace transforms involves their application to derivatives. Let $f(t)$ be continuous and have a sectionally continuous derivative $f'(t)$. Then

$$\mathcal{L}[f'(t)] = \int_0^\infty f'(t)e^{-st}dt \tag{1.1.2}$$

$$= e^{-st}f(t)\big|_0^\infty + s\int_0^\infty f(t)e^{-st}dt \tag{1.1.3}$$

$$= sF(s) - f(0). \tag{1.1.4}$$

Repeated applications of (1.1.4) on higher derivatives yield

$$\mathcal{L}[f^{(n)}(t)] = s^n F(s) - s^{n-1}f(0) - \cdots - sf^{(n-2)}(0) - f^{(n-1)}(0). \tag{1.1.5}$$

Of the numerous functions that possess Laplace transforms, two discontinuous ones are particularly useful: the Heaviside and delta functions. The *Heaviside step function* is defined as

$$H(t) = \begin{cases} 1 & \text{if } t > 0, \\ 0 & \text{if } t < 0. \end{cases} \tag{1.1.6}$$

From this definition,

$$\mathcal{L}[H(t)] = \int_0^\infty e^{-st}dt = \frac{1}{s}, \qquad s > 0. \tag{1.1.7}$$

The *unit impulse* or *Dirac delta function* is formally defined by

$$\delta(t) = \begin{cases} \infty & \text{if } t = 0, \\ 0 & \text{if } t \neq 0, \end{cases} \qquad \int_0^\infty \delta(t)\,dt = 1. \tag{1.1.8}$$

However, for many purposes we can represent it by

$$\delta(t) = \lim_{\epsilon \to 0} \begin{cases} 1/\epsilon & \text{if } 0 < t < \epsilon, \\ 0 & \text{if } t > \epsilon. \end{cases} \tag{1.1.9}$$

Using this definition, the Laplace transform of the delta function is

$$\mathcal{L}[\delta(t)] = \int_0^\infty \delta(t)e^{-st}dt = \lim_{\epsilon \to 0} \frac{1}{\epsilon}\int_0^\epsilon e^{-st}dt \tag{1.1.10}$$

$$= \lim_{\epsilon \to 0} \frac{-1}{\epsilon s}\left(e^{-\epsilon s} - 1\right) \tag{1.1.11}$$

$$= \lim_{\epsilon \to 0}\left[\frac{-1}{\epsilon s}\left(1 - \epsilon s + \frac{\epsilon^2 s^2}{2} - \cdots - 1\right)\right] = 1. \tag{1.1.12}$$

Handbuch der Laplace-Transformation. Band 2. Anwendungen der Laplace-Transformation. 1. Abteilung. Basel: Birkhäuser Verlag, 433 pp. and Doetsch, G., 1956: *Handbuch der Laplace-Transformation. Band 2. Anwendungen der Laplace-Transformation. 2. Abteilung.* Basel: Birkhäuser Verlag, 298 pp.

Another useful class of functions is $f(t-a)H(t-a)$ where $a > 0$. Now

$$\mathcal{L}[f(t-a)H(t-a)] = \int_0^\infty f(t-a)H(t-a)e^{-st}dt \qquad (1.1.13)$$

$$= \int_a^\infty f(t-a)e^{-st}dt \qquad (1.1.14)$$

$$= e^{-as}\int_0^\infty f(x)e^{-sx}dx \qquad (1.1.15)$$

$$= e^{-as}F(s) \qquad (1.1.16)$$

if $x = t - a$. Thus, the *second shifting theorem* states that

$$\mathcal{L}[f(t-a)H(t-a)] = e^{-as}F(s). \qquad (1.1.17)$$

• **Example 1.1.1**

Use the second shifting theorem to find the Laplace transform of

$$f(t) = \sin(t)[H(t) - H(t-a)]. \qquad (1.1.18)$$

Because

$$f(t) = \sin(t)H(t) + \sin(t-\pi)H(t-\pi), \qquad (1.1.19)$$

the second shifting theorem gives

$$F(s) = \frac{1}{s^2+1} + \frac{1}{s^2+1}e^{-s\pi}. \qquad (1.1.20)$$

Let us now consider the transform $\mathcal{L}\left[\int_0^t f(\tau)\,d\tau\right]$. From the definition of the Laplace transform,

$$\mathcal{L}\left[\int_0^t f(\tau)\,d\tau\right] = \int_0^\infty \left[\int_0^t f(\tau)\,d\tau\right]e^{-st}dt \qquad (1.1.21)$$

$$= -\frac{e^{-st}}{s}\int_0^t f(\tau)\,d\tau\bigg|_0^\infty + \frac{1}{s}\int_0^\infty f(t)e^{-st}dt \qquad (1.1.22)$$

$$= \frac{F(s)}{s}. \qquad (1.1.23)$$

3

● **Example 1.1.2**

Find the Laplace transform of $f(t) = \int_0^t e^{-4\tau} \sin(3\tau)\, d\tau$.

$$\mathcal{L}\left[\int_0^t e^{-4\tau} \sin(3\tau)\, d\tau\right] = \frac{1}{s}\mathcal{L}\left[e^{-4t} \sin(3t)\right] \qquad (1.1.24)$$

$$= \frac{1}{s}\frac{3}{(s+4)^2 + 9}. \qquad (1.1.25)$$

Finally, there is the mathematical concept of the *convolution product*:

$$f(t) * g(t) = \int_0^t g(t-x)f(x)\, dx = \int_0^t g(x)f(t-x)\, dx. \qquad (1.1.26)$$

Convolution is of fundamental importance because

$$\mathcal{L}[f(t) * g(t)] = \int_0^\infty \left[\int_0^t f(x)g(t-x)\, dx\right] e^{-st} dt \qquad (1.1.27)$$

$$= \int_0^\infty \left[\int_x^\infty f(x)g(t-x)e^{-st} dt\right] dx \qquad (1.1.28)$$

$$= \int_0^\infty f(x)\left[\int_0^\infty g(r)e^{-s(r+x)} dr\right] dx \qquad (1.1.29)$$

$$= \left[\int_0^\infty f(x)e^{-sx} dx\right]\left[\int_0^\infty g(r)e^{-sr} dr\right] \qquad (1.1.30)$$

$$= F(s)G(s) \qquad (1.1.31)$$

where $t = r+x$. In other words, the Laplace transform of the convolution of two functions equals the product of the Laplace transforms of each of the functions. This relationship is often called *Borel's theorem*[2].

● **Example 1.1.3**

Verify Borel's theorem when $f(t) = H(t-1) - H(t-2)$ and $g(t) = e^t$. The convolution of $f(t)$ with $g(t)$ is

$$f(t) * g(t) = \int_0^t e^{t-x}[H(x-1) - H(x-2)]\, dx. \qquad (1.1.32)$$

---

[2] Borel, É., 1901: *Leçons sur les séries divergentes*. Paris: Gauthier-Villars, p. 104.

At this point, we must consider three cases: $t < 1$, $1 < t < 2$ and $t > 2$. If $t < 1$, both step functions remain zero and $f(t) * g(t) = 0$. For $1 < t < 2$,

$$f(t) * g(t) = \int_1^t e^{t-x}dx = e^{t-1} - 1. \tag{1.1.33}$$

Finally, for $t > 2$,

$$f(t) * g(t) = \int_1^2 e^{t-x}dx = e^{t-1} - e^{t-2}. \tag{1.1.34}$$

We can write this convolution as a single expression through the use of step functions:

$$f(t) * g(t) = \left[e^{t-1} - 1\right] H(t-1) + \left[1 - e^{t-2}\right] H(t-2). \tag{1.1.35}$$

The Laplace transform of $f(t)*g(t)$ follows directly from the second shifting theorem:

$$\mathcal{L}[f(t) * g(t)] = \frac{e^{-s}}{s-1} - \frac{e^{-s}}{s} + \frac{e^{-2s}}{s} - \frac{e^{-2s}}{s-1} \tag{1.1.36}$$

$$= \frac{e^{-s} - e^{-2s}}{s}\left(\frac{1}{s-1}\right) = F(s)G(s). \tag{1.1.37}$$

In section 2.1 and 3.1 we will discuss methods for inverting Laplace transforms using complex variables.

Problems

1. Using the definition of the Laplace transform, verify the following transforms:

$$\mathcal{L}(e^{at}) = \frac{1}{s-a} \qquad \mathcal{L}(t^n) = \frac{n!}{s^{n+1}}$$

$$\mathcal{L}[\sin(at)] = \frac{a}{s^2 + a^2} \qquad \mathcal{L}[\cos(at)] = \frac{s}{s^2 + a^2}.$$

Verify the following convolutions and use them to show that the convolution theorem is true.

2.
$$t^2 * \sin(at) = \frac{t^2}{a} - \frac{4}{a^3}\sin^2(at/2)$$

3.
$$H(t-a) * H(t-b) = (t - a - b)H(t - a - b)$$

4.
$$t * [H(t) - H(t-2)] = \frac{t^2}{2} - \frac{(t-2)^2}{2}H(t-2)$$

5. Show that

$$\int_a^b f(t)\delta(t - t_0)\, dt = \begin{cases} f(t_0) & \text{if } a \le t_0 \le b, \\ 0 & \text{otherwise.} \end{cases}$$

1.2 FOURIER TRANSFORMS

When the function $f(t)$ exists over the interval $(-\infty, \infty)$, we replace the Laplace transform with the Fourier transform[3]

$$\mathcal{F}[f(t)] = F(\omega) = \int_{-\infty}^{\infty} f(t)e^{-i\omega t}dt. \tag{1.2.1}$$

In order that $F(\omega)$ exists, $\int_{-\infty}^{\infty} |f(t)|\, dt < \infty$, a rather stringent requirement. For this reason, many functions that possess a Laplace transform do not possess a Fourier transform. A simple example is the step function $H(t)$.

• **Example 1.2.1**

For our first example let us find the transform of the Dirac delta function. From the definition of the Fourier transform,

$$\mathcal{F}[\delta(t)] = \int_{-\infty}^{\infty} \delta(t)e^{-i\omega t}dt = \lim_{\epsilon \to 0} \frac{1}{\epsilon} \int_{-\epsilon/2}^{\epsilon/2} e^{-i\omega t}dt \tag{1.2.2}$$

$$= \lim_{\epsilon \to 0} \frac{-i}{\epsilon \omega} \left(e^{\epsilon \omega i/2} - e^{-\epsilon \omega i/2} \right) \tag{1.2.3}$$

$$= \lim_{\epsilon \to 0} \frac{-i}{\epsilon \omega} \left(1 + \frac{\epsilon \omega i}{2} - \frac{\epsilon^2 \omega^2}{8} - \cdots - 1 + \frac{\epsilon \omega i}{2} + \frac{\epsilon^2 \omega^2}{8} - \cdots \right) \tag{1.2.4}$$

$$= 1. \tag{1.2.5}$$

One possible method for dealing with the convergence problem, especially popular with electrical engineers, is to modify the Laplace transform so that we have a generalized Fourier transform where ω becomes a complex variable:

$$F(\omega) = \int_{0}^{\infty} f(t)e^{-i\omega t}dt, \qquad \text{Im}(\omega) < 0. \tag{1.2.6}$$

In this way the functions $f(t) = H(t)$ and $f(t) = \sin(at)H(t)$ have a Fourier transform.

Consider now the function $f(t)$ such that $e^{-c_1 t}|f(t)| \to 0$ as $t \to \infty$ and $e^{-c_2 t}|f(t)| \to 0$ as $t \to -\infty$ with $c_2 > c_1$. Then, by choosing $c_2 >$

---

[3] The standard reference for Fourier integrals is Titchmarsh, E. C., 1948: *Introduction to the Theory of Fourier Integrals.* Oxford: At the Clarendon Press, 394 pp.

$-\text{Im}(\omega) > c_1$, the generalized Fourier transform or *two-sided Laplace transform*[4] is

$$F(\omega) = \int_{-\infty}^{\infty} f(t)e^{-i\omega t}dt, \qquad c_2 > -\text{Im}(\omega) > c_1. \qquad (1.2.7)$$

● **Example 1.2.2**

Let us find the generalized Fourier transform of the function

$$f(t) = \begin{cases} e^{at} & \text{if } -\infty < t < 0, \\ \cos(bt) & \text{if } 0 \le t < \infty \end{cases} \qquad (1.2.8)$$

where $a > 0$. If $0 > \text{Im}(\omega) > -a$, then

$$F(\omega) = \int_{-\infty}^{0} e^{at}e^{-i\omega t}dt + \int_{0}^{\infty} \cos(bt)e^{-i\omega t}dt \qquad (1.2.9)$$

$$= \frac{1}{a - \omega i} - \frac{\omega i}{\omega^2 - b^2}. \qquad (1.2.10)$$

The most important property of Fourier transforms is convolution:

$$f(t) * g(t) = \int_{-\infty}^{\infty} f(x)g(t - x)\,dx = \int_{-\infty}^{\infty} f(t - x)g(x)\,dx. \qquad (1.2.11)$$

Then,

$$\mathcal{F}[f(t) * g(t)] = \int_{-\infty}^{\infty} f(x)e^{-i\omega x}\left[\int_{-\infty}^{\infty} g(t - x)e^{-i\omega(t - x)}dt\right]dx \qquad (1.2.12)$$

$$= \int_{-\infty}^{\infty} f(x)G(\omega)e^{-i\omega x}dx = F(\omega)G(\omega). \qquad (1.2.13)$$

Thus, the Fourier transform of the convolution of two functions equals the product of the Fourier transforms of each of the functions.

● **Example 1.2.3**

Verify the convolution theorem using the functions $f(t) = H(t + a) - H(t - a)$ and $g(t) = e^{-t}H(t)$ where $a > 0$.

---

[4] See Van der Pol, B., and H. Bremmer, 1964: *Operational Calculus Based on Two-Sided Transform.* Cambridge: At the University Press, 409 pp.

The convolution of $f(t)$ with $g(t)$ is

$$f(t) * g(t) = \int_{-\infty}^{\infty} e^{-(t-x)} H(t-x) \left[H(x+a) - H(x-a) \right] dx$$

$$\text{(1.2.14)}$$

$$= e^{-t} \int_{-a}^{a} e^{x} H(t-x) \, dx. \qquad (1.2.15)$$

If $t < -a$, then the integrand of (1.2.15) is always zero and $f(t)*g(t) = 0$. If $t > a$,

$$f(t) * g(t) = e^{-t} \int_{-a}^{a} e^{x} dx = e^{-(t-a)} - e^{-(t+a)}. \qquad (1.2.16)$$

Finally, for $-a < t < a$,

$$f(t) * g(t) = e^{-t} \int_{-a}^{t} e^{x} dx = 1 - e^{-(t+a)}. \qquad (1.2.17)$$

In summary,

$$f(t) * g(t) = \begin{cases} 0 & \text{if} \quad t < -a, \\ 1 - e^{-(t+a)} & \text{if} \quad -a < t < a, \\ e^{-(t-a)} - e^{-(t+a)} & \text{if} \quad t > a. \end{cases} \qquad (1.2.18)$$

The Fourier transform of $f(t) * g(t)$ is

$$\mathcal{F}[f(t) * g(t)] = \int_{-a}^{a} \left[1 - e^{-(t+a)} \right] e^{-i\omega t} dt$$

$$+ \int_{a}^{\infty} \left[e^{-(t-a)} - e^{-(t+a)} \right] e^{-i\omega t} dt \qquad (1.2.19)$$

$$= \frac{2\sin(\omega a)}{\omega} - \frac{2i\sin(\omega a)}{1+\omega i} \qquad (1.2.20)$$

$$= \frac{2\sin(\omega a)}{\omega} \left(\frac{1}{1+\omega i} \right) = F(\omega)G(\omega) \qquad (1.2.21)$$

and the convolution theorem holds.

● **Example 1.2.4**

For our final example[5], let us find the inverse $f(x)$ of the Fourier transform

$$F(k) = \int_{-\infty}^{\infty} f(x) e^{-ikx} dx = \frac{e^{ikct} - e^{-ikct}}{ki} \exp\left(-\frac{k^2 c^2 \tau t}{2} \right) \qquad (1.2.22)$$

---

[5] Taken from Tanaka, K., and T. Kurokawa, 1973: Viscous property of steel and its effect on strain wave front. *Bull. JSME*, **16**, 188–193.

using the convolution theorem where c, t and τ are positive and real. We begin by noting that

$$\mathcal{F}\left[\frac{1}{c\sqrt{\tau t}}\exp\left(-\frac{x^2}{2c^2\tau t}\right)\right] = \exp\left(-\frac{k^2c^2\tau t}{2}\right) \tag{1.2.23}$$

and

$$\mathcal{F}[H(ct+x) - H(ct-x)] = \frac{e^{ikct} - e^{-ikct}}{ki}. \tag{1.2.24}$$

From the convolution theorem,

$$f(x) = \frac{1}{c\sqrt{\tau t}}\int_{-ct}^{ct}\exp\left[-\frac{(x-\eta)^2}{2c^2\tau t}\right]d\eta \tag{1.2.25}$$

$$= \sqrt{2}\int_{(x-ct)/c\sqrt{2\tau t}}^{(x+ct)/c\sqrt{2\tau t}} e^{-r^2}dr \tag{1.2.26}$$

if $r = (x-\eta)/c\sqrt{2\tau t}$. Let us now introduce the error function defined by

$$\text{erf}(x) = \frac{2}{\pi}\int_0^x e^{-r^2}dr. \tag{1.2.27}$$

Then, if $x > ct$,

$$f(x) = \sqrt{\frac{\pi}{2}}\left[\text{erf}\left(\frac{x+ct}{c\sqrt{2\tau t}}\right) - \text{erf}\left(\frac{x-ct}{c\sqrt{2\tau t}}\right)\right]. \tag{1.2.28}$$

For $|x| \leq ct$,

$$f(x) = \sqrt{\frac{\pi}{2}}\left[\text{erf}\left(\frac{x+ct}{c\sqrt{2\tau t}}\right) + \text{erf}\left(\frac{ct-x}{c\sqrt{2\tau t}}\right)\right]. \tag{1.2.29}$$

Finally, if $x < -ct$,

$$f(x) = \sqrt{\frac{\pi}{2}}\left[\text{erf}\left(\frac{ct-x}{c\sqrt{2\tau t}}\right) - \text{erf}\left(\frac{-ct-x}{c\sqrt{2\tau t}}\right)\right]. \tag{1.2.30}$$

In sections 2.2 and 3.2 we will discuss the inversion of Fourier transforms by complex variables.

Problems

1. Show that the Fourier transform of

$$f(t) = e^{-a|t|}, \qquad a > 0$$

is

$$F(\omega) = \frac{2a}{\omega^2 + a^2}.$$

2. Show that the Fourier transform of

$$f(t) = te^{-a|t|}, \qquad a > 0$$

is

$$F(\omega) = -\frac{4a\omega i}{(\omega^2 + a^2)^2}.$$

3. Show that the Fourier transform of

$$f(t) = \begin{cases} e^{-(1+i)t} & \text{if } t > 0, \\ -e^{(1-i)t} & \text{if } t < 0, \end{cases}$$

is

$$F(\omega) = \frac{-2i(\omega + 1)}{(\omega + 1)^2 + 1}.$$

4. Show that the Fourier transform of

$$f(t) = \begin{cases} \sin(t) & \text{if } 0 \leq t < 1, \\ 0 & \text{otherwise}, \end{cases}$$

is

$$F(\omega) = -\frac{1}{2}\left[\frac{1 - \cos(\omega - 1)}{\omega - 1} + \frac{\cos(\omega + 1) - 1}{\omega + 1}\right]$$
$$- \frac{i}{2}\left[\frac{\sin(\omega - 1)}{\omega - 1} - \frac{\sin(\omega + 1)}{\omega + 1}\right].$$

5. Show that

$$e^t H(-t) * [H(t) - H(t - 2)] = \begin{cases} e^t - e^{t-2} & \text{if } t < 0, \\ 1 - e^{t-2} & \text{if } 0 < t < 2, \\ 0 & \text{if } t > 2. \end{cases}$$

6. Show that

$$[H(t) - H(t - 2)] * [H(t) - H(t - 2)] = \begin{cases} 0 & \text{if } t < 0, \\ t & \text{if } 0 < t < 2, \\ 4 - t & \text{if } 2 < t < 4, \\ 0 & \text{if } t > 4. \end{cases}$$

7. Show that the Fourier transform of a constant K is $2\pi\delta(\omega)K$.

1.3   LINEAR ORDINARY DIFFERENTIAL EQUATIONS

Most analytic techniques for solving a partial differential equation involve reducing it down to ordinary differential equations that are hopefully easier to solve than the original partial differential equation. From the vast number of possible ordinary differential equations we focus on second-order equations. All of the following techniques extend to higher order equations.

Consider the ordinary differential equation

$$a\frac{d^2y}{dx^2} + b\frac{dy}{dx} + cy = f(x) \tag{1.3.1}$$

where a, b and c are real. For the moment let us take $f(x) = 0$. Assuming a solution of the form $y(x) = Ae^{mx}$ and substituting into (1.3.1),

$$am^2 + bm + c = 0. \tag{1.3.2}$$

This purely algebraic equation is the *characteristic* or *auxiliary equation*. Because (1.3.2) is quadratic, there are either two real roots, or else a repeated real root, or else conjugate complex roots. At this point, let us consider each case separately and state the solution. Any undergraduate book on ordinary differential equations will provide the details for obtaining these general solutions.

Case I: *Two distinct real roots m_1 and m_2,*

$$y(x) = c_1 e^{m_1 x} + c_2 e^{m_2 x}. \tag{1.3.3}$$

Case II: *A repeated real root m_1,*

$$y(x) = c_1 e^{m_1 x} + c_2 x e^{m_1 x}. \tag{1.3.4}$$

Case III: *Conjugate complex roots $m_1 = p + qi$ and $m_2 = p - qi$,*

$$y(x) = c_1 e^{px} \cos(qx) + c_2 e^{px} \sin(qx). \tag{1.3.5}$$

• **Example 1.3.1**

One of the most commonly encountered differential equations is

$$\frac{d^2y}{dx^2} - m^2 y = 0 \tag{1.3.6}$$

where m is real and positive. Because there are two distinct roots, $m_{1,2} = \pm m$, the general solution is

$$y(x) = Ae^{mx} + Be^{-mx}. \tag{1.3.7}$$

Although this solution is perfectly correct, it is most useful in semi-infinite domains. For finite domains, such as $0 < x < L$, we introduce the mathematical functions of hyperbolic sine and cosine. A little algebra shows that (1.3.7) also equals

$$y(x) = C \cosh(mx) + D \sinh(mx) \qquad (1.3.8)$$

where

$$\cosh(mx) = \tfrac{1}{2} \left(e^{mx} + e^{-mx} \right) \qquad (1.3.9)$$

and

$$\sinh(mx) = \tfrac{1}{2} \left(e^{mx} - e^{-mx} \right). \qquad (1.3.10)$$

The advantage of using (1.3.8) comes from the facts that $\sinh(0) = 0$ and $\cosh(0) = 1$.

So far we have only found the solution when $f(x) = 0$, the so-called *homogeneous* or *complementary solution* to (1.3.1). When $f(x)$ is nonzero, we must add a *particular solution* to the complementary solution that yields $f(x)$ upon substitution into (1.3.1). The most common technique for determining this particular solution is the *method of undetermined coefficients*. This method goes as follows: (1) assume a particular solution $y_p(x)$ that consists of an arbitrary linear combination of all of the linearly independent functions which arise from repeated differentiations of $f(x)$, (2) substitute $y_p(x)$ into the differential equation and (3) determine the arbitrary constants of $y_p(x)$ so that the equation resulting from the substitution yields $f(x)$.

• **Example 1.3.2**

Let us find the particular solution for the equation

$$\frac{d^2 y}{dx^2} + 2 \frac{dy}{dx} + y = \cos^2(x). \qquad (1.3.11)$$

Our guess for the particular solution is then

$$y_p(x) = A + B \cos(2x) + C \sin(2x) \qquad (1.3.12)$$

because $\cos^2(x) = [1 + \cos(2x)]/2$. Substituting (1.3.12) into (1.3.11) and equating coefficients for the constant, cosine and sine terms, we find that $A = 1/2$, $B = -3/50$ and $C = 2/25$.

Presently we have only dealt with ordinary differential equations that have constant coefficients. In problems involving cylindrical coordinates, we will solve the equation

$$r^2 \frac{d^2 y}{dr^2} + r \frac{dy}{dr} + (\lambda^2 r^2 - n^2)y = 0 \qquad (1.3.13)$$

commonly known as *Bessel's equation of order n with a parameter* λ. The general solution to (1.3.13) is

$$y(r) = c_1 J_n(\lambda r) + c_2 Y_n(\lambda r) \tag{1.3.14}$$

where $J_n(\ )$ and $Y_n(\ )$ are n-th order Bessel functions of the first and second kind, respectively. Bessel functions have been exhaustively studied and a vast literature now exists on them[6]. For our purpose they are tabulated functions, similar to sine and cosine. Here we note two properties of $Y_n(\ )$. First, $Y_n(z) \to \infty$ as $|z| \to 0$. Second, the power series expansion for $Y_n(z)$ contains a logarithm so that Y_n is a multivalued function. See Section 1.5.

Another closely related differential equation is the *modified Bessel equation*

$$r^2 \frac{d^2 y}{dr^2} + r \frac{dy}{dr} - (\lambda^2 r^2 + n^2) y = 0. \tag{1.3.15}$$

The general solution to (1.3.15) is

$$y(r) = c_1 I_n(\lambda r) + c_2 K_n(\lambda r) \tag{1.3.16}$$

where $I_n(\ )$ and $K_n(\ )$ are n-th order, modified Bessel functions of the first and second kind, respectively. Once again, $I_n(z)$ and $K_n(z)$ are tabulated functions with $I_n(z) \to \infty$ as $|z| \to \infty$ and $K_n(z) \to \infty$ as $|z| \to 0$. Furthermore, the power series representation for K_n contains a logarithm and K_n is a multivalued function. Although these results just scratch the known relationships concerning Bessel functions, we will introduce additional results as we need them.

The remaining task is to compute the arbitrary constants in the homogeneous solution. In this book we always have conditions at both ends of a given domain, even if one of these points is at infinity. We now illustrate the procedure used in solving these *boundary value problems*.

• Example 1.3.3

Solve the boundary value problem

$$\frac{d^2 y}{dx^2} - sy = -\frac{1}{s}, \qquad y(0) = y(1) = 0 \tag{1.3.17}$$

where $s > 0$. The general solution to (1.3.17) is

$$y(x) = A \sinh(x\sqrt{s}) + B \cosh(x\sqrt{s}) + \frac{1}{s^2}. \tag{1.3.18}$$

---

[6] *The* standard reference is Watson, G. N., 1966: *A Treatise on the Theory of Bessel Functions.* Cambridge (England): At the University Press, 804 pp.

Table 1.3.1: Some useful relationships involving Bessel functions of integer order.

$$J_{n-1}(z) + J_{n+1}(z) = \frac{2n}{z} J_n(z), \qquad n = 1, 2, 3, \cdots$$

$$J_{n-1}(z) - J_{n+1}(z) = 2J_n'(z), \quad n = 1, 2, 3, \cdots; \quad J_0'(z) = -J_1(z)$$

$$I_{n-1}(z) - I_{n+1}(z) = \frac{2n}{z} I_n(z), \qquad n = 1, 2, 3, \cdots$$

$$I_{n-1}(z) + I_{n+1}(z) = 2I_n'(z), \quad n = 1, 2, 3, \cdots; \quad I_0'(z) = I_1(z)$$

$$K_{n-1}(z) - K_{n+1}(z) = -\frac{2n}{z} K_n(z), \qquad n = 1, 2, 3, \cdots$$

$$K_{n-1}(z) + K_{n+1}(z) = -2K_n'(z), \quad n = 1, 2, 3, \cdots; \quad K_0'(z) = -K_1(z)$$

$$J_n(ze^{m\pi i}) = e^{nm\pi i} J_n(z)$$

$$I_n(ze^{m\pi i}) = e^{nm\pi i} I_n(z)$$

$$K_n(ze^{m\pi i}) = e^{-mn\pi i} K_n(z) - m\pi i \frac{\cos(mn\pi)}{\cos(n\pi)} I_n(z)$$

$$I_n(z) = e^{-n\pi i/2} J_n(ze^{\pi i/2}), \qquad -\pi < \arg(z) \le \pi/2$$

$$I_n(z) = e^{3n\pi i/2} J_n(ze^{-3\pi i/2}), \qquad \pi/2 < \arg(z) \le \pi$$

We have chosen to use hyperbolic functions because the domain lies between $x = 0$ and $x = 1$. Now

$$y(0) = B + \frac{1}{s^2} = 0 \tag{1.3.19}$$

and

$$y(1) = A \sinh(\sqrt{s}) + B \cosh(\sqrt{s}) + \frac{1}{s^2} = 0. \tag{1.3.20}$$

Solving for A and B,

$$A = \frac{\cosh(\sqrt{s}) - 1}{s^2 \sinh(\sqrt{s})} \qquad \text{and} \qquad B = -\frac{1}{s^2}. \tag{1.3.21}$$

Therefore,

$$y(x) = \frac{1 - \cosh(x\sqrt{s})}{s^2} + \frac{\cosh(\sqrt{s}) - 1}{s^2 \sinh(\sqrt{s})} \sinh(x\sqrt{s}). \tag{1.3.22}$$

● **Example 1.3.4**

Solve the boundary value problem

$$\frac{d^2 y}{dr^2} + \frac{1}{r}\frac{dy}{dr} - sy = 0, \qquad y'(a) = -\frac{1}{s}, \quad \lim_{r\to\infty} y(r) \to 0 \qquad (1.3.23)$$

where $s > 0$. The general solution to (1.3.23) is

$$y(r) = AI_0(r\sqrt{s}) + BK_0(r\sqrt{s}) \qquad (1.3.24)$$

from (1.3.15)–(1.3.16). Because $I_0(r\sqrt{s}) \to \infty$ as $r \to \infty$, $A = 0$. At $r = a$,

$$y'(a) = B\sqrt{s}\,K_0'(a\sqrt{s}) = -B\sqrt{s}\,K_1(a\sqrt{s}) = -\frac{1}{s} \qquad (1.3.25)$$

because $K_0'(z) = -K_1(z)$ or

$$B = \frac{1}{s^{3/2}K_1(a\sqrt{s})}. \qquad (1.3.26)$$

Therefore the solution to the boundary value problem is

$$y(r) = \frac{K_0(r\sqrt{s})}{s^{3/2}K_1(a\sqrt{s})}. \qquad (1.3.27)$$

Problems

Find the general solution for the following ordinary differential equations:

1. $y'' + 4y = 8x^3 - 20x^2 + 16x - 18$

2. $y'' + 4y' = -3\cos(3x) + \sin(2x)$

3. $y'' - 6y' + 8y = 3e^x$

4. $y'' + 4y' + 4y = xe^{-x}$

5. $y'' + y = e^x \sin(x)$

6. $y'' + y = \cos(x) + 3\sin(2x)$

7. Solve the boundary value problem

$$\frac{d^2 y}{dx^2} - (a^2 + s)y = 0, \qquad y(-1) = \frac{1}{s}, \qquad y(1) = 0$$

where a and s are real and positive.

8. Solve the boundary value problem

$$\frac{d^2 y}{dr^2} + \frac{1}{r}\frac{dy}{dr} - sy = 0, \qquad |y(0)| < \infty, \qquad y'(a) = -\frac{1}{s}$$

where a and s are real and positive.

1.4   COMPLEX VARIABLES

Complex variables provide analytic tools for the evaluation of integrals with an ease that rarely occurs with real functions. The power of integration on the complex plane has its roots in the basic three C's: the Cauchy-Riemann equations, the Cauchy-Goursat theorem and Cauchy's residue theorem.

The Cauchy-Riemann equations have their origin in the definition of the derivative in the complex plane. Just as we have the concept of the function in real variables, where for a given value of x we can compute a corresponding value of $y = f(x)$, we can define a complex function $w = f(z)$ where for a given value of $z = x + iy$ ($i = \sqrt{-1}$) we may compute $w = f(z) = u(x,y) + iv(x,y)$. In order for $f'(z)$ to exist in some region, $u(x,y)$ and $v(x,y)$ must satisfy the Cauchy-Riemann equations:

$$\frac{\partial u}{\partial x} = \frac{\partial v}{\partial y} \quad and \quad \frac{\partial u}{\partial y} = -\frac{\partial v}{\partial x}. \tag{1.4.1}$$

If $f(z)$ satisfies these conditions around the point z_0, then it is *analytic* there. If a function is analytic everywhere in the complex plane, then it is an *entire* function. Alternatively, if the function is analytic everywhere except at some isolated singularities, then it is *meromorphic*. Note that $f(z)$ must satisfy the Cauchy-Riemann equations in a region and not just at a point. For example, $f(z) = |z|$ satisfies the Cauchy-Riemann equations at $z = 0$ and nowhere else. Consequently, this function is not analytic anywhere on the complex plane.

Integration on the complex plane is more involved than in real, single variables because $dz = dx + idy$. We must specify a path or contour as we integrate from one point to another. Of all of the possible contour integrals, a closed contour is the best. To see why, we introduce the following results:

Cauchy-Goursat theorem[7]: *If $f(z)$ is an analytic function at each point within and on a closed contour C, then $\oint_C f(z)\, dz = 0$.*

This theorem leads immediately to

The principle of deformation of contours: *The value of a line integral of an analytic function around any simple closed contour remains unchanged if we deform the contour in such a manner that we do not pass over a point where $f(z)$ is not analytic.*

---

[7] See Goursat, E., 1900: Sur la définition générale des fonctions analytiques, d'après Cauchy. *Trans. Am. Math. Soc.*, **1**, 14–16.

Consequently we can evaluate difficult integrals by deforming the contour so that the actual evaluation is along a simpler contour or the computations are made easier. See Example 1.4.1.

Most integrations on the complex plane, however, deal with meromorphic functions. Our next theorem involves these functions; it is

Cauchy's residue theorem[8]: *If $f(z)$ is analytic inside a closed contour C (taken in the positive sense) except at points $z_1, z_2, \ldots, z_n$ where $f(z)$ has singularities, then*

$$\oint_C f(z)\,dz = 2\pi i \sum_{j=1}^{n} \text{Residue of } f(z) \text{ at } z_j. \qquad (1.4.2)$$

The question now turns to what is a residue and how do we compute it. The answer involves the nature of the singularity and an extension of the Taylor expansion, called a *Laurent expansion*:

$$f(z) = \sum_{n=0}^{\infty} a_n (z - z_j)^n + \sum_{n=1}^{\infty} a_{-n}(z - z_j)^{-n} \qquad (1.4.3)$$

for $0 < |z - z_j| < a$. The first summation is merely the familiar Taylor expansion; the second summation involves negative powers of $z - z_j$ and gives the behavior at singularity. The *residue* equals the coefficient a_{-1}.

The construction of a Laurent expansion for a given singularity has two practical purposes: (1) it gives the nature of the singularity and (2) we will occasionally use it to give the actual value of the residue. Turning to the nature of the singularity, there are three types:

• *Essential Singularity*: Consider the function $f(z) = \cos(1/z)$. Using the expansion for cosine,

$$\cos\left(\frac{1}{z}\right) = 1 - \frac{1}{2!z^2} + \frac{1}{4!z^4} - \frac{1}{6!z^6} + \cdots \qquad (1.4.4)$$

for $0 < |z| < \infty$. Note that this series never truncates in the inverse powers of z. Essential singularities have Laurent expansions which have an infinite number of inverse powers of $z - z_j$. The value of the residue for this essential singularity at $z = 0$ is zero.

---

[8] See Mitrinovič, D. S., and J. K. Kečkič, 1984: *The Cauchy Method of Residues*. Boston: D. Reidel Publishing Co., 361 pp. Section 10.3 gives the historical development of the residue theorem.

Transform Methods for Solving Partial Differential Equations

- *Removable Singularity:* Consider the function $f(z) = \sin(z)/z$. This function appears, at first blush, to have a singularity at $z = 0$. Upon applying the expansion for sine, we see that

$$\frac{\sin(z)}{z} = 1 - \frac{z^2}{3!} + \frac{z^4}{5!} - \frac{z^6}{7!} + \frac{z^8}{9!} - \cdots \qquad (1.4.5)$$

for all z and we actually have no singularity at all. This is an example of a removable singularity; because a removable singularity is not really a singularity, the residue is zero.

- *Pole of order n:* Consider the function

$$f(z) = \frac{1}{(z-1)^3(z+1)}. \qquad (1.4.6)$$

This function has two singularities: one at $z = 1$ and the other at $z = -1$. We shall only consider the case $z = 1$. After a little algebra,

$$f(z) = \frac{1}{(z-1)^3}\frac{1}{2+(z-1)} \qquad (1.4.7)$$

$$= \frac{1}{2}\frac{1}{(z-1)^3}\frac{1}{1+(z-1)/2} \qquad (1.4.8)$$

$$= \frac{1}{2}\frac{1}{(z-1)^3}\left[1 - \frac{z-1}{2} + \frac{(z-1)^2}{4} - \frac{(z-1)^3}{8} + \cdots\right] \qquad (1.4.9)$$

$$= \frac{1}{2(z-1)^3} - \frac{1}{4(z-1)^2} + \frac{1}{8(z-1)} - \frac{1}{16} + \cdots \qquad (1.4.10)$$

for $0 < |z - 1| < 2$. Because the largest inverse (negative) power is three, the singularity at $z = 1$ is called a third-order pole; the value of the residue is $1/8$. Generally, we refer to a first-order pole as a *simple* pole.

The construction of a Laurent expansion is not the method of choice in computing a residue. (For an essential singularity it is the only method; however, essential singularities are very rare in applications.) The common method for a pole of order n is

$$\text{Residue at } (z = z_j) = \frac{1}{(n-1)!}\lim_{z \to z_j}\frac{d^{n-1}}{dz^{n-1}}\left[(z - z_j)^n f(z)\right].$$

$$(1.4.11)$$

For a simple pole (1.4.11) simplifies to

$$\text{Residue of a simple pole at } (z = z_j) = \lim_{z \to z_j} (z - z_j) f(z).$$

(1.4.12)

Quite often, $f(z) = p(z)/q(z)$. From l'Hospital's rule, it follows that

$$\text{Residue of a simple pole at } (z = z_j) = \frac{p(z_j)}{q'(z_j)}.$$

(1.4.13)

The desirability of dealing with closed contour integrals should be clear by now. This is true to such an extent that mathematicians have devised several theorems which allow us to change a line integral into a closed one by adding an arc at infinity. The one of greatest relevance to us is by C. Jordan[9]:

Jordan's lemma: *Suppose that, on a circular arc C_R with radius R and center at the origin, $f(z) \to 0$ uniformly as $R \to \infty$. Then*

$$(1) \qquad \lim_{R \to \infty} \int_{C_R} f(z) e^{imz} dz = 0, \qquad (m > 0) \qquad (1.4.14)$$

if C_R is in the first and/or second quadrant;

$$(2) \qquad \lim_{R \to \infty} \int_{C_R} f(z) e^{-imz} dz = 0, \qquad (m > 0) \qquad (1.4.15)$$

if C_R is in the third and/or fourth quadrant;

$$(3) \qquad \lim_{R \to \infty} \int_{C_R} f(z) e^{mz} dz = 0, \qquad (m > 0) \qquad (1.4.16)$$

---

[9] Jordan, C., 1894: *Cours D'Analyse de l'École Polytechnique.* Vol. 2, Paris: Gauthier-Villars, pp. 285–286. See also Whittaker, E. T., and G. N. Watson, 1963: *A Course of Modern Analysis.* Cambridge: At the University Press, p. 115.

if C_R is in the second and/or third quadrant; and

$$(4) \qquad \lim_{R \to \infty} \int_{C_R} f(z) e^{-mz} dz = 0, \qquad (m > 0) \qquad (1.4.17)$$

if C_R is in the first and/or fourth quadrant. Technically, only (1) is actually Jordan's lemma while the remaining points are variations.

Proof: We shall prove the first part; the remaining portions follow by analog. We begin by noting that

$$|I_R| = \left| \int_{C_R} f(z) e^{imz} dz \right| \le \int_{C_R} |f(z)| \left| e^{imz} \right| |dz|. \qquad (1.4.18)$$

Now

$$|dz| = R \, d\theta, \qquad |f(z)| \le M_R, \qquad (1.4.19)$$

$$\left| e^{imz} \right| = \left| \exp(imRe^{\theta i}) \right| = \left| \exp\{imR[\cos(\theta) + i\sin(\theta)]\} \right| = e^{-mR\sin(\theta)}. \qquad (1.4.20)$$

Therefore,

$$|I_R| \le RM_R \int_{\theta_0}^{\theta_1} \exp[-mR\sin(\theta)] \, d\theta \qquad (1.4.21)$$

where $0 \le \theta_0 < \theta_1 \le \pi$. Because the integrand is positive, the right side of (1.4.21) is largest if we take $\theta_0 = 0$ and $\theta_1 = \pi$. Then

$$|I_R| \le RM_R \int_0^\pi e^{-mR\sin(\theta)} d\theta = 2RM_R \int_0^{\pi/2} e^{-mR\sin(\theta)} d\theta. \qquad (1.4.22)$$

We cannot evaluate the integrals in (1.4.22) as they stand. However, because $\sin(\theta) \ge 2\theta/\pi$, we can bound the value of the integral by

$$|I_R| \le 2RM_R \int_0^{\pi/2} e^{-2mR\theta/\pi} d\theta = \frac{\pi}{m} M_R \left(1 - e^{-mR}\right). \qquad (1.4.23)$$

If $m > 0$, $|I_R|$ tends to zero with M_R as $R \to \infty$.

● **Example 1.4.1**

To illustrate how useful distorting the original contour may be in evaluating an integral, consider[10]

$$\frac{1}{2\pi i} \oint_C f(z) \, dz = \frac{1}{2\pi i} \oint_C \frac{z^{-n-1}}{1 - ae^z} \, dz, \qquad 0 < a < 1 \qquad (1.4.24)$$

---

[10] Based upon Götze, F., and H. Friedrich, 1980: Berechnungs- und Abschätzungsformeln für verallgemeinerte geometrische Reihen. *Zeit. angew. Math. Mech.*, **60**, 737–739.

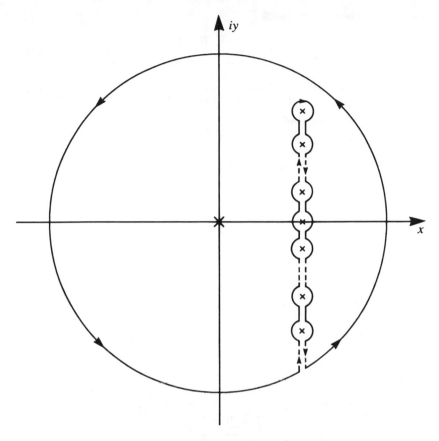

Fig. 1.4.1: Modified contour used to evaluate (1.4.24).

where the contour C is the circle $|z| < -\ln(a)$ and $n > 1$. The integrand has an infinite number of simple poles at $z_m = -\ln(a) + 2m\pi i$ with $m = 0, \pm 1, \pm 2, \ldots$ (which lie outside the original contour) and a $(n+1)$th-order pole at $z = 0$. Because the straightforward evaluation of this integral by the residue theorem would require differentiating the denominator n times, we choose to evaluate (1.4.24) by expanding the contour so that it is a circle of infinite radius with a cut that excludes the simple poles at z_m. See Fig. 1.4.1. Then, by the residue theorem,

$$\frac{1}{2\pi i} \oint_C f(z)\,dz = I - \sum_{m=-\infty}^{\infty} \text{Residue of } f(z) \text{ at } z_m \qquad (1.4.25)$$

where I is the contribution from the circle at infinity. Because the residue of $f(z)$ at z_m is $-z_m^{-1-n}$,

$$\frac{1}{2\pi i} \oint_C f(z)\,dz = I + \sum_{m=-\infty}^{\infty} [2m\pi i - \ln(a)]^{-n-1}. \qquad (1.4.26)$$

21

Turning to the evaluation of I,

$$|I| = \left| \frac{1}{2\pi} \int_0^{2\pi} \frac{(Re^{\theta i})^{-n}}{1 - a \exp\{R[\cos(\theta) + i\sin(\theta)]\}} d\theta \right| \quad (1.4.27)$$

$$\leq \frac{R^{-n}}{2\pi} \int_0^{2\pi} \frac{1}{|1 - a \exp[R\cos(\theta)]|} d\theta. \quad (1.4.28)$$

As $R \to \infty$, we note that

$$|1 - a\exp[R\cos(\theta)]|^{-1}$$

$$\sim \begin{cases} O[e^{-R\cos(\theta)}], & 0 \leq \theta < \pi/2, \ 3\pi/2 < \theta \leq 2\pi, \\ O(1), & \pi/2 < \theta < 3\pi/2 \end{cases} \quad (1.4.29)$$

so that the integral (1.4.28) is finite and equals M, say. Consequently,

$$|I| \leq \lim_{R \to \infty} \frac{MR^{-n}}{2\pi} \to 0. \quad (1.4.30)$$

Therefore,

$$\frac{1}{2\pi i} \oint_C f(z)\, dz = [-\ln(a)]^{-n-1} + \sum_{m=1}^{\infty} [2m\pi i - \ln(a)]^{-n-1}$$

$$+ [-2m\pi i - \ln(a)]^{-n-1}. \quad (1.4.31)$$

- **Example 1.4.2**

Let us evaluate

$$\int_0^{\infty} \frac{\cos(kx)}{x^2 + a^2}\, dx$$

where a and k are real and positive. First,

$$\int_0^{\infty} \frac{\cos(kx)}{x^2 + a^2}\, dx = \frac{1}{2} \int_{-\infty}^{\infty} \frac{\cos(kx)}{x^2 + a^2}\, dx \quad (1.4.32)$$

$$= \frac{1}{2}\text{Re}\left\{ \int_{-\infty}^{\infty} \frac{e^{ikx}}{x^2 + a^2}\, dx \right\}. \quad (1.4.33)$$

We close the line integral along the real axis by introducing an infinite semicircle in the upper half-plane as dictated by Jordan's lemma. Therefore,

$$\int_0^{\infty} \frac{\cos(kx)}{x^2 + a^2}\, dx = \frac{1}{2}\text{Re}\left\{ \oint_C \frac{e^{ikz}}{z^2 + a^2}\, dz \right\} \quad (1.4.34)$$

$$= \text{Re}\left\{ \pi i \left[\text{Residue at } z = ia \text{ of } \frac{e^{ikz}}{z^2 + a^2} \right] \right\} \quad (1.4.35)$$

$$= \text{Re}\left\{ \pi i \lim_{z \to ia} \frac{(z - ia)e^{ikz}}{z^2 + a^2} \right\} = \frac{\pi}{2a} e^{-ka}. \quad (1.4.36)$$

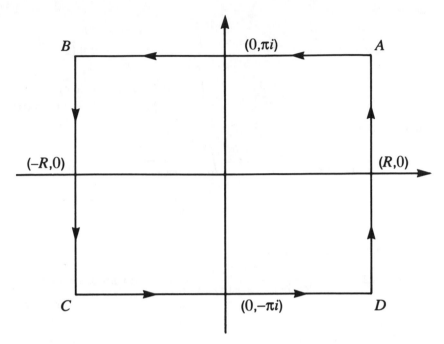

Fig. 1.4.2: Closed contour used in evaluating (1.4.37).

• **Example 1.4.3**

When the definite integral involves hyperbolic functions, a rectangular closed contour is generally the best one to use. For example, consider the contour integral[11]

$$\oint_C \frac{e^{2\lambda z}}{[\cosh(z) + \cos(\pi/2\lambda)]^2} \, dz, \qquad 1/2 < \lambda < 1 \qquad (1.4.37)$$

where C is the closed rectangular contour $ABCD$ shown in Fig. 1.4.2. Along AD as $R \to \infty$,

$$\frac{e^{2\lambda z}}{[\cosh(z) + \cos(\pi/2\lambda)]^2} \to e^{-2(1-\lambda)R} \to 0. \qquad (1.4.38)$$

Similarly, along BC

$$\frac{e^{2\lambda z}}{[\cosh(z) + \cos(\pi/2\lambda)]^2} \to e^{-2(1+\lambda)R} \to 0. \qquad (1.4.39)$$

---

[11] Taken from Hawthorne, W. R., 1954: The secondary flow about struts and airfoils. *J. Aeronaut. Sci.*, **21**, 588–608.

Along AB

$$\int_{\infty}^{-\infty} \frac{e^{2\lambda z}}{[\cosh(z) + \cos(\pi/2\lambda)]^2} \, dz = -\int_0^\infty \frac{e^{2\lambda(-x+\pi i)}}{[\cos(\pi/2\lambda) - \cosh(x)]^2} \, dx$$

$$-\int_0^\infty \frac{e^{2\lambda(x+\pi i)}}{[\cos(\pi/2\lambda) - \cosh(x)]^2} \, dx$$

$$(1.4.40)$$

while along CD

$$\int_{-\infty}^{\infty} \frac{e^{2\lambda z}}{[\cosh(z) + \cos(\pi/2\lambda)]^2} \, dz = \int_0^\infty \frac{e^{2\lambda(x-\pi i)}}{[\cos(\pi/2\lambda) - \cosh(x)]^2} \, dx$$

$$+\int_0^\infty \frac{e^{2\lambda(-x-\pi i)}}{[\cos(\pi/2\lambda) - \cosh(x)]^2} \, dx.$$

$$(1.4.41)$$

Therefore, summing together the four segments,

$$-4i\sin(2\lambda\pi)\int_0^\infty \frac{\cosh(2\lambda x)\, dx}{[\cosh(x) - \cos(\pi/2\lambda)]^2} = 2\pi i \times \text{sum of residues}$$

$$(1.4.42)$$

or

$$\int_0^\infty \frac{\cosh^2(\lambda x)}{[\cosh(x) - \cos(\pi/2\lambda)]^2} \, dx = -\frac{\pi}{4\sin(2\lambda\pi)} \times \text{sum of residues}$$

$$+\frac{1}{2}\int_0^\infty \frac{dx}{[\cosh(x) - \cos(\pi/2\lambda)]^2}.$$

$$(1.4.43)$$

From a table of integrals,

$$\int_0^\infty \frac{dx}{[\cosh(x) - \cos(\pi/2\lambda)]^2} = \int_1^\infty \frac{4y\, dy}{[y^2 - 2y\cos(\pi/2\lambda) + 1]^2}$$

$$(1.4.44)$$

$$= \frac{\pi[1 - (1/2\lambda)]\cot(\pi/2\lambda) + 1}{\sin^2(2\pi/\lambda)}.$$

$$(1.4.45)$$

The only poles located inside the closed contour occur at $z_\pm = \pm\pi i[1 - (1/2\lambda)]$. To compute the residues, we note that

$$\frac{e^{2\lambda z}}{[\cosh(z) - \cosh(z_\pm)]^2} = \frac{e^{2\lambda z_\pm}e^{2\lambda\zeta}}{\{\cosh(z_\pm)[\cosh(\zeta) - 1] + \sinh(z_\pm)\sinh(\zeta)\}^2}$$

$$(1.4.46)$$

$$= \frac{e^{2\lambda z_\pm}}{\cosh^2(z_\pm)}\left\{\frac{1}{\zeta^2} + \frac{[2\lambda - \coth(z_\pm)]}{\zeta} + \cdots\right\}$$

$$(1.4.47)$$

because $\cos(\pi/2\lambda) = -\cosh(z_\pm)$ and $\zeta = z - z_\pm$. Therefore, the poles are second order and the residues equal

$$-e^{\pm\pi i(2\lambda-1)}[2\lambda \mp i\cot(\pi/2\lambda)]\csc^2(\pi/2\lambda).$$

Hence the sum of residues is

$$[4\lambda\cos(2\lambda\pi) + 2\sin(2\lambda\pi)\cot(\pi/2\lambda)]\csc^2(\pi/2\lambda).$$

Substituting this sum and (1.4.45) into (1.4.43), we finally have

$$\int_0^\infty \frac{\cosh^2(\lambda x)}{[\cosh(x) - \cos(\pi/2\lambda)]^2}\, dx = \csc^2(\pi/2\lambda)\{1 - \pi[2\lambda\cot(2\pi\lambda)$$
$$+ \cot(\pi/2\lambda)/2\lambda]\}/2. \qquad (1.4.48)$$

All of the results in this section apply to single-valued functions. In those instances where there are multivalued functions due to the presence of z raised to some rational power, inverse functions or logarithms, we must make them single-valued. The method for doing this is the subject of the next two sections.

Problems

Use the residue theorem and Jordan's lemma to verify the following integrals where a and h are real and positive:

1.

$$\int_{-\infty}^\infty \frac{\sin(x)}{x^2 + 4x + 5}\, dx = -\frac{\pi}{e}\sin(2)$$

2.

$$\int_0^\infty \frac{\cos(x)}{(x^2 + 1)^2}\, dx = \frac{\pi}{2e}$$

3.

$$\int_{-\infty}^\infty \frac{x\sin(ax)}{x^2 + 4}\, dx = \pi e^{-2a}$$

4.

$$\int_{-\infty}^\infty \frac{x\cos(\pi x)}{x^2 + 2x + 5}\, dx = \frac{\pi}{2}e^{-2\pi}$$

5.

$$\int_{-\infty}^\infty \frac{\cos(ax)}{(x^2 + h^2)^2}\, dx = \frac{\pi}{2h^3}(1 + ah)e^{-ah}$$

6.

$$\int_{-\infty}^\infty \frac{\cos(ax)}{(x - a)^2 + h^2}\, dx = \frac{\pi e^{-ah}\cos(a^2)}{h}$$

7.
$$\int_{-\infty}^{\infty} \frac{\sin(ax)}{(x-a)^2 + h^2}\, dx = \frac{\pi e^{-ah}\sin(a^2)}{h}$$

8.
$$\int_0^{\infty} \frac{\cos(ax)}{x^4 + 1}\, dx = \frac{\pi}{4\sqrt{2}} e^{-a/\sqrt{2}}\left[\sin\left(\frac{a}{\sqrt{2}}\right) + \cos\left(\frac{a}{\sqrt{2}}\right)\right]$$

9.
$$\int_0^{\infty} \frac{x\sin(ax)}{x^4 + 1}\, dx = \frac{\pi}{2} e^{-a/\sqrt{2}} \sin\left(\frac{a}{\sqrt{2}}\right)$$

10.
$$\int_0^{\infty} \frac{\cos(ax)}{(x^2 + h^2)(x^2 + 1)}\, dx = \frac{\pi}{2}\left[\frac{e^{-ah}}{h(1 - h^2)} + \frac{e^{-a}}{h^2 - 1}\right], \qquad h \neq 0$$

11.
$$\int_0^{\infty} \frac{x\sin(ax)}{(x^2 + h^2)(x^2 + 1)}\, dx = \frac{\pi}{2}\left[\frac{e^{-a} - e^{-ah}}{h^2 - 1}\right], \qquad h \neq 0$$

12. Use the residue theorem to show that

$$\int_0^{\infty} \frac{8x^2}{(x^2 + 1)^2[(1 + a/h)x^2 + (a/h - 1)]}\, dx = \frac{\pi a}{h}[1 - \sqrt{1 - (h/a)^2}]$$

where $a > h > 0$.

13. By evaluating the integral

$$\oint_C e^{-iz^2}\, dz$$

around the closed rectangular contour that has its sides parallel to the real and imaginary axes and its corners at $(0,0)$, $(\infty, 0)$, $(\infty, -i\sqrt{x})$ and $(0, -i\sqrt{x})$, show[12] that

$$\int_0^{\infty} e^{-2\sqrt{x}\,t - it^2}\, dt = \frac{\sqrt{\pi}(1 - i)}{2\sqrt{2}} e^{-ix}$$

$$+ \frac{i\sqrt{\pi}}{\sqrt{2}} e^{-ix}\left[C\left(\sqrt{\frac{2x}{\pi}}\right) + iS\left(\sqrt{\frac{2x}{\pi}}\right)\right]$$

[12] Taken from Bhattacharyya, A., and R. G. Rastogi, 1987: Faraday polarization fluctuations of VHF geostationary satellite signals near the geomagnetic equator. *J. Geophys. Res.*, **92**, 8821–8826. ©1987 American Geophysical Union.

where $C(\ )$ and $S(\ )$ are the Fresnel cosine and sine functions[13].

14. By evaluating the integral

$$\oint_C \frac{\cosh[(\theta/\pi)z]}{\cosh(z)}\, dz$$

around the closed rectangular contour that consists of (1) the real axis from $(-R,0)$ to $(R,0)$, (2) segments normal to real axis along $(R,0)$ to (R,π) and $(-R,\pi)$ to $(-R,0)$ and (3) a segment parallel to the real axis from (R,π) to $(-R,\pi)$ where $R \to \infty$, show[14] that

$$\frac{2}{\pi} \int_0^\infty \frac{\cosh[(\theta/\pi)x]}{\cosh(x)}\, dx = \sec(\theta/2), \qquad 0 < \theta < \pi.$$

1.5 MULTIVALUED FUNCTIONS, BRANCH POINTS, BRANCH CUTS AND RIEMANN SURFACES

In this section, we introduce functions that yield several different values of w for a given z, i.e., multivalued functions. We must make these functions single-valued so that we can apply the techniques from the previous section. Furthermore, this condition is also necessary for a well-posed physical problem.

Consider the complex function $w = z^{1/2}$. For each value of z there are two possible values of w. For example, if $z = -i$, then w equals either $w_1 = -1/\sqrt{2} + i/\sqrt{2}$ or $w_2 = 1/\sqrt{2} - i/\sqrt{2}$. The points w_1 and w_2 are distinct members from two *branches* of the same function which "branch off" from the same point, $z = 0$. The number of branches depends upon the nature of $f(z)$. For example, $\log(z)$ has an infinite number of branches, namely $\ln(r) + \theta i + n\pi i, n = 0, 1, 2, \ldots$

In the case of real variables, the branches easily separate. For example, the square root of a positive, real number has two distinct branches: a and $-a$ where a is a real, positive number. However, in complex functions the two branches are hardly distinguishable because they do not separate at all. Therefore, if we wish to keep them separate, we must do it artificially.

Consider again the complex function $w = z^{1/2} = r^{1/2}e^{\theta i/2}$. If we move around a closed path that does *not* encircle the origin, the values

---

[13] See Abramowitz, M., and I. A. Stegun, 1968: *Handbook of Mathematical Functions.* New York: Dover Publications, Inc., Subsections 7.3.1 and 7.3.2.

[14] Taken from Tsien, H. S., and M. Finston, 1949: Interaction between parallel streams of subsonic and supersonic velocities. *J. Aeronaut. Sci.*, **16**, 515–528.

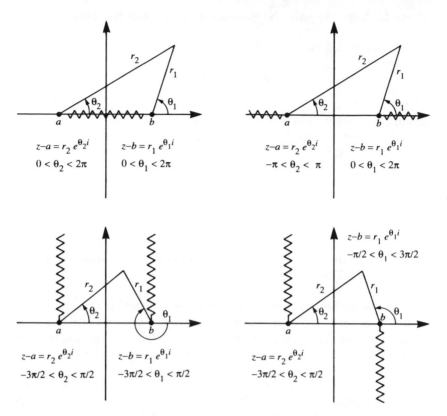

Fig. 1.5.1: Some popular branch cut configurations for the complex square root $\sqrt{(z-a)(z-b)}$. The branch cuts are denoted by wavy lines.

of r and θ vary continuously. At the end, the final value of θ equals our initial value, θ_0. Consequently, we can say that all of the values of w along this contour belong to the same branch, $w_1 = r^{1/2}e^{i\theta_0/2}$.

Let our closed contour now enclose the origin. Because the final value of $\theta = \theta_0 + 2\pi$, the final value of w now equals $w_2 = -r^{1/2}e^{i\theta_0/2} = -w_1$. We have reached the other branch of w in a continuous manner. Consequently, $z = 0$ appears to be a special point; one branch ties to the other there because they have the same value. We reach each branch from the other after a complete turn around the origin. A *branch point* is any point having this property. In this example, infinity is also a branch point. We can show this by the substitution $z = 1/z'$ and an examination of the transformed function about $z' = 0$.

In our example with $w = z^{1/2}$, we reached the other branch by completing a closed contour around the origin. However, we cannot say exactly when, in our journey, we crossed the boundary from w_1 and w_2.

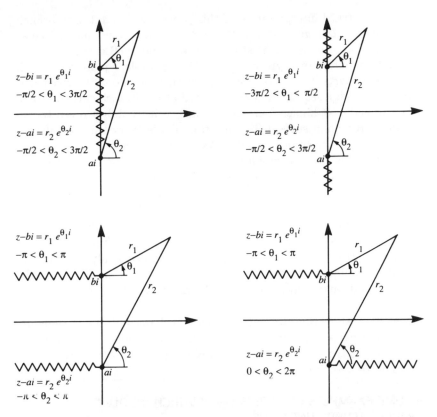

Fig. 1.5.2: Some popular branch cut configurations for the complex square root $\sqrt{(z - ai)(z - bi)}$. The branch cuts are denoted by wavy lines.

For example, it might have been when we crossed the positive real axis or when we crossed the negative real axis. The conclusions are the same either way. This ambiguity leads to the concept of the *branch cut*. The reason why we must define a branch cut between branch points lies in the fact that we cannot make a multivalued function single-valued by excluding the branch point and a small neighborhood around it. Its multi-valuedness does not depend on the mere existence of the branch point itself, but on the possibility of encircling it.

In summary, a branch cut is a line that *we* choose that connects two branch points. Furthermore, it defines the separation between the two branches. For this reason the branch cut is a barrier that we may not cross[15].

---

[15] The Russian mathematical term for the edge of a branch cut is *bereg*, which commonly means "the bank of a river".

A geometrical interpretation of this process is to limit each branch to a particular *Riemann surface*. We can view each Riemann surface as a floor in a large department store. On each floor (Riemann surface) you can only obtain one type of branch (for example, square roots with positive real parts). However, we can reach other floors (surfaces), if desired, through a set of escalators, located at the branch cut, that take you up to the next higher Riemann surface (if it exists) or down to the next lower Riemann surface (if it exists). These (very thin) escalators extend between the branch points. As the architect of your Riemann surface, there is great flexibility in choosing where your branch cuts lie. Figs. 1.5.1 and 1.5.2 show some of the more popular choices for $\sqrt{(z-a)(z-b)}$ and $\sqrt{(z-ai)(z-bi)}$, respectively.

The advantage of introducing a Riemann surface is the fact that every continuous curve on the z-plane maps the multivalued function into a continuous curve on the w-plane. This relationship between the Riemann surfaces allows us to apply to multivalued analytic functions all of the techniques of integration, analytic continuation, and so forth, which depend on a continuous path being drawn from one point to another.

In the next section, we illustrate the mechanics of integrations involving multivalued functions. In certain instances, however, integration across the cut is necessary, even desirable. In section 4.4 we will show how to do this for fun and pleasure.

1.6 SOME EXAMPLES OF INTEGRATION WHICH INVOLVE MULTIVALUED FUNCTIONS

In this section we perform contour integration with multivalued functions. Essentially, the introduction of branch points and cuts requires careful bookkeeping of phases (or arguments). Once we set up our bookkeeping, the evaluation follows directly.

• Example 1.6.1

Let us evaluate the integral[16]

$$\oint_C f(z)\, dz = \oint_C \left(\frac{a+z}{c-z}\right)^{1/2} \frac{dz}{(d-z)(z-b)} \tag{1.6.1}$$

where $-a < b < c < d$. Fig. 1.6.1 shows the contour C [$=C_1 + C_2 + \ldots + C_8$]. The integrand is a multivalued function with branch points at $z = -a$ and $z = c$ and simple poles at $z = b$ and $z = d$.

---

[16] This example is taken from Glagolev, N. I., 1945: Resistance of cylindrical bodies in rolling (in Russian). *Prikl. Mat. Mek.*, **9**, 318–333.

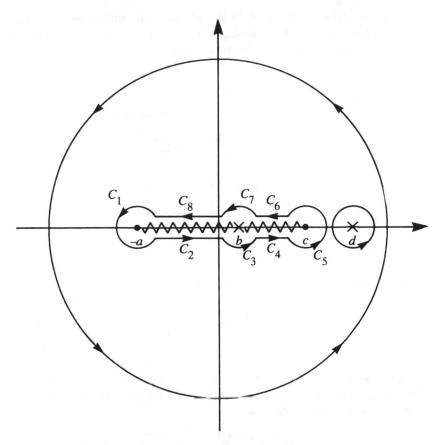

Fig. 1.6.1: The contour used to evaluate (1.6.1).

We begin by expanding our original contour C without bound so that it includes all of the singularities. Fig. 1.6.1 shows this enlarged contour Γ. Next we employ the following theorem:

If $zf(z)$ tends uniformly to a limit k as $|z|$ increases indefinitely, the value of $\oint f(z)\,dz$, taken around a very large circle, center the origin, tends towards $2\pi i k$. [17]

Because $zf(z)$ tends to zero,

$$\oint_{\Gamma} \left(\frac{a+z}{c-z}\right)^{1/2} \frac{dz}{(d-z)(z-b)} = 0. \qquad (1.6.2)$$

We evaluate the integral around the contour C by noting that an integration around the contour Γ is equivalent to integrations along C

---

[17] Forsyth, A. R., 1965: *Theory of Functions of a Complex Variable.* New York: Dover Publications, Inc., p. 14.

and around the pole at $z = d$. If the phase of the functions $z + a$ and $z - c$ lies between 0 and 2π, which is consistent with our choice for the branch cut, then the residue theorem yields

$$\oint_C \left(\frac{a+z}{c-z}\right)^{1/2} \frac{dz}{(d-z)(z-b)} = \oint_\Gamma f(z)\, dz$$

$$- 2\pi i\, [\text{Residue at } f(z) \text{ of } z = d]$$

$$= \frac{2\pi}{d-b} \left(\frac{a+d}{d-c}\right)^{1/2} \qquad (1.6.3)$$

because $(a + z)/(c - z)$ always has the phase of $-\pi$ at the point $z = d$.

Integrals similar to (1.6.3) are important because we can use them to evaluate definite integrals where the integrand contains a square root. In our problem, for example, we can convert this contour integral into a real, definite integral by first noting that

$$\oint_C f(z)\, dz = \int_{C_1} f(z)\, dz + \int_{C_2} f(z)\, dz + \int_{C_3} f(z)\, dz + \int_{C_4} f(z)\, dz$$

$$+ \int_{C_5} f(z)\, dz + \int_{C_6} f(z)\, dz + \int_{C_7} f(z)\, dz + \int_{C_8} f(z)\, dz. \qquad (1.6.4)$$

Next we express $z - c$ and $z + a$ in terms of $re^{\theta i}$ with the phase $0 < \theta < 2\pi$. To evaluate the integral on the contour C_1, for example, $z + a = \epsilon e^{\theta i}$ and $z - c = (a + c)e^{\pi i}$, so that

$$\int_{C_1} f(z)\, dz = \lim_{\epsilon \to 0} \int_0^{2\pi} \frac{\epsilon^{1/2} e^{\theta i/2}}{\sqrt{a+c}} \frac{i\epsilon e^{\theta i}}{(a+d)(-a-b)}\, d\theta = 0 \qquad (1.6.5)$$

whereas for C_5, $z - c = \epsilon e^{\theta i}$ and $z + a = (a+c)e^{0i}$ or $(a+c)e^{2\pi i}$, resulting in

$$\int_{C_5} f(z)\, dz = - \lim_{\epsilon \to 0} \int_{-\pi}^0 \frac{(a+c)^{1/2}}{i\epsilon^{1/2} e^{\theta i/2}} \frac{i\epsilon e^{\theta i}}{(d-c)(c-b)}\, d\theta$$

$$+ \lim_{\epsilon \to 0} \int_0^\pi \frac{(a+c)^{1/2}}{i\epsilon^{1/2} e^{\theta i/2}} \frac{i\epsilon e^{\theta i}}{(d-c)(c-b)}\, d\theta = 0. \qquad (1.6.6)$$

At $z = b$, $z - b = \epsilon e^{\theta i}$, $z - c = (c-b)e^{\pi i}$, $z + a = (a+b)e^{0i}$ on C_7 and $(a+b)e^{2\pi i}$ on C_3 so that

$$\int_{C_3} f(z)\, dz = \lim_{\epsilon \to 0} \left(\frac{a+b}{c-b}\right)^{1/2} \left(\frac{1}{d-b}\right) \int_{-\pi}^0 \frac{i\epsilon e^{\theta i}}{\epsilon e^{\theta i}}\, d\theta \qquad (1.6.7)$$

$$= \left(\frac{a+b}{c-b}\right)^{1/2} \frac{\pi i}{d-b} \qquad (1.6.8)$$

and

$$\int_{C_7} f(z)\, dz = -\lim_{\epsilon \to 0} \left(\frac{a+b}{c-b}\right)^{1/2} \left(\frac{1}{d-b}\right) \int_0^\pi \frac{i\epsilon e^{\theta i}}{\epsilon e^{\theta i}}\, d\theta \qquad (1.6.9)$$

$$= -\left(\frac{a+b}{c-b}\right)^{1/2} \frac{\pi i}{d-b}. \qquad (1.6.10)$$

We evaluate the remaining contour integrals by noting that $z - c = (c-x)e^{\pi i}$ and $z + a = (a+x)e^{2\pi i}$ along C_2 and C_4 and $(a+x)e^{0i}$ along C_6 and C_8 where $-a < x < c$. Upon substituting these definitions into the integrals and simplifying,

$$\int_{C_2 + C_4} f(z)\, dz = \int_{-a}^c \left(\frac{a+x}{c-x}\right)^{1/2} \frac{dx}{(d-x)(x-b)} \qquad (1.6.11)$$

and

$$\int_{C_6 + C_8} f(z)\, dz = -\int_c^{-a} \left(\frac{a+x}{c-x}\right)^{1/2} \frac{dx}{(d-x)(x-b)}. \qquad (1.6.12)$$

Substituting these results together into (1.6.4), our final answer is

$$\int_{-a}^c \left(\frac{a+x}{c-x}\right)^{1/2} \frac{dx}{(d-x)(x-b)} - \int_c^{-a} \left(\frac{a+x}{c-x}\right)^{1/2} \frac{dx}{(d-x)(x-b)}$$
$$= \frac{2\pi}{d-b}\left(\frac{a+d}{d-c}\right)^{1/2} \qquad (1.6.13)$$

or

$$\int_{-a}^c \left(\frac{a+x}{c-x}\right)^{1/2} \frac{dx}{(d-x)(x-b)} = \frac{\pi}{d-b}\left(\frac{a+d}{d-c}\right)^{1/2} \qquad (1.6.14)$$

A similar problem involves the integral

$$\oint_C f(z)\, dz = \oint_C \left(\frac{a+z}{c-z}\right)^\delta \frac{dz}{z-b} \qquad (1.6.15)$$

where $-a < b < c$ and $|\delta| < 1$. This multivalued function has branch points at $z = -a$ and $z = c$. Fig. 1.6.2 shows the contour of integration. To evaluate the contour integral, we again make use of the theorem stated after (1.6.1). Thus,

$$\oint_C f(z)\, dz = 2\pi i (e^{-\pi i})^\delta = 2\pi i e^{-\pi \delta i}. \qquad (1.6.16)$$

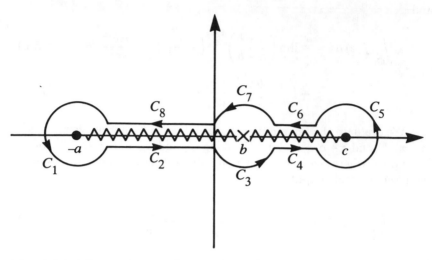

Fig. 1.6.2: The contour used to evaluate (1.6.15).

Once again, we can reduce our contour integral to a real definite integral by breaking the closed contour into eight segments,

$$\oint_C f(z)\,dz = \int_{C_1} f(z)\,dz + \int_{C_2} f(z)\,dz + \int_{C_3} f(z)\,dz + \int_{C_4} f(z)\,dz$$
$$+ \int_{C_5} f(z)\,dz + \int_{C_6} f(z)\,dz + \int_{C_7} f(z)\,dz + \int_{C_8} f(z)\,dz.$$
$$(1.6.17)$$

Proceeding as before, the contribution from the circles at $z = -a$ and $z = c$ vanish. We evaluate the remaining integrals by noting that $z - c = (c - x)e^{\pi i}$ and $z + a = (x + a)e^{2\pi i}$ along C_2, C_3 and C_4 and $z + a = (x + a)e^{0i}$ along C_6, C_7 and C_8 with $-a < x < c$. After substituting into (1.6.17),

$$\int_{-a}^{c} \left(\frac{a+x}{c-x}\right)^{\delta} \frac{dx}{x-b} + e^{-2\pi\delta i} \int_{c}^{-a} \left(\frac{a+x}{c-x}\right)^{\delta} \frac{dx}{x-b}$$
$$+ \pi i \left(\frac{a+b}{c-b}\right)^{\delta} + e^{-2\pi\delta i}\pi i \left(\frac{a+b}{c-b}\right)^{\delta} = 2\pi i e^{-\pi\delta i} \quad (1.6.18)$$

or

$$\int_{-a}^{c} \left(\frac{a+x}{c-x}\right)^{\delta} \frac{dx}{x-b} = \frac{\pi}{\sin(\delta\pi)} - \pi\cot(\delta\pi)\left(\frac{a+b}{c-b}\right)^{\delta}. \quad (1.6.19)$$

• Example 1.6.2

Let us evaluate

$$\oint_C f(z)\,dz = \oint_C \frac{dz}{(z^2 - a^2)^{\mu/2}(z-x)^{1-\mu}} \tag{1.6.20}$$

where $-a < x < a$ and $0 < \mu < 1$. Fig. 1.6.3 shows that the contour runs just below the real axis from $-a$ to a and then back to $-a$ just above the real axis[18]. The integrand is a multivalued function with three branch points at $z = a$, x and $-a$. The branch cut runs along the real axis from $-a$ to a.

The value of (1.6.20) follows from the limit of $zf(z)$ as $|z| \to \infty$. (See the previous example.) If the argument of $z - a$, $z + a$ and $z - x$ lies between 0 and 2π, this limit equals one and

$$\oint_C f(z)\,dz = 2\pi i. \tag{1.6.21}$$

Once again we can use our results to evaluate a real, definite integral by evaluating (1.6.21) along the various legs of the contour shown in Fig. 1.6.3,

$$\int_{C_1} f(z)\,dz + \int_{C_2} f(z)\,dz + \int_{C_3} f(z)\,dz + \int_{C_4} f(z)\,dz$$
$$+ \int_{C_5} f(z)\,dz + \int_{C_6} f(z)\,dz + \int_{C_7} f(z)\,dz + \int_{C_8} f(z)\,dz = 2\pi i. \tag{1.6.22}$$

At $z = -a$, $z + a = \epsilon e^{\theta i}$, $z - x = (x + a)e^{\pi i}$ and $z - a = 2ae^{\pi i}$ so that

$$\int_{C_4} f(z)\,dz = \lim_{\epsilon \to 0} \int_0^{2\pi} \frac{i\epsilon e^{\theta i}\,d\theta}{(2a\epsilon e^{\theta i + \pi i})^{\mu/2}(x + a)^{1-\mu}e^{\pi i - \mu\pi i}} = 0; \tag{1.6.23}$$

while at $z = a$, $z - a = \epsilon e^{\theta i}$, $z + a = 2ae^{0i}$ or $2ae^{2\pi i}$ and $z - x = (a - x)e^{0i}$ or $(a - x)e^{2\pi i}$,

$$\int_{C_8} f(z)\,dz = \lim_{\epsilon \to 0} \int_{-\pi}^0 \frac{i\epsilon e^{\theta i}\,d\theta}{(2a\epsilon e^{\theta i + 2\pi i})^{\mu/2}(a - x)^{1-\mu}e^{2\pi i - 2\mu\pi i}}$$
$$+ \lim_{\epsilon \to 0} \int_0^\pi \frac{i\epsilon e^{\theta i}\,d\theta}{(2a\epsilon e^{\theta i})^{\mu/2}(a - x)^{1-\mu}} = 0. \tag{1.6.24}$$

---

[18] Reprinted from *J. Appl. Math. Mech.*, **23**, V. Kh. Arutiunian, The plane contact problem of the theory of creep, 1283–1313, ©1959, with kind permission from Pergamon Press Ltd., Headington Hill Hall, Oxford OX3 0BX, UK.

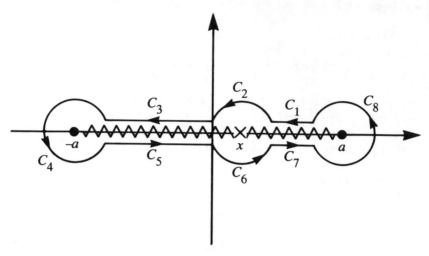

Fig. 1.6.3: The contour used to evaluate (1.6.20).

At $z = x$, $z - x = \epsilon e^{\theta i}$, $z - a = (a - x)e^{\pi i}$ and $z + a = (x + a)e^{0i}$ on C_2 and $(x + a)e^{2\pi i}$ on C_6 so that

$$\int_{C_2} f(z)\, dz = \lim_{\epsilon \to 0} \int_0^\pi \frac{i\epsilon e^{\theta i} d\theta}{(a^2 - x^2)^{\mu/2} e^{\pi \mu i/2} (\epsilon e^{\theta i})^{1-\mu}} = 0 \qquad (1.6.25)$$

and

$$\int_{C_6} f(z)\, dz = \lim_{\epsilon \to 0} \int_\pi^{2\pi} \frac{i\epsilon e^{\theta i} d\theta}{(a^2 - x^2)^{\mu/2} e^{3\pi \mu i/2} (\epsilon e^{\theta i})^{1-\mu}} = 0. \qquad (1.6.26)$$

Finally, for the straight line segments, $z - a = (a - s)e^{\pi i}$, $z + a = (a + s)e^{0i}$ along C_1 and C_3 and $(a + s)e^{2\pi i}$ on C_5 and C_7, and $z - x = |s - x|e^{0i}$ on C_1, $|x - s|e^{\pi i}$ on C_3 and C_5 and $|s - x|e^{2\pi i}$ on C_7. Therefore,

$$\int_{C_1} f(z)\, dz = \int_a^x \frac{e^{-\mu \pi i/2} ds}{(a^2 - s^2)^{\mu/2} |s - x|^{1-\mu}}$$

$$= -\int_x^a \frac{e^{-\mu \pi i/2} ds}{(a^2 - s^2)^{\mu/2} |s - x|^{1-\mu}} \qquad (1.6.27)$$

$$\int_{C_3} f(z)\, dz = \int_x^{-a} \frac{e^{-\pi i + \mu \pi i/2} ds}{(a^2 - s^2)^{\mu/2} |s - x|^{1-\mu}}$$

$$= \int_{-a}^x \frac{e^{\mu \pi i/2} ds}{(a^2 - s^2)^{\mu/2} |s - x|^{1-\mu}} \qquad (1.6.28)$$

$$\int_{C_5} f(z)\, dz = \int_{-a}^{x} \frac{e^{-\pi i - \mu \pi i/2} ds}{(a^2 - s^2)^{\mu/2} |s - x|^{1-\mu}} \tag{1.6.29}$$

and

$$\int_{C_7} f(z)\, dz = \int_{x}^{a} \frac{e^{-2\pi i + \mu \pi i/2} ds}{(a^2 - s^2)^{\mu/2} |s - x|^{1-\mu}}. \tag{1.6.30}$$

Substituting all of these integrals into (1.6.22) and simplifying,

$$\int_{-a}^{a} \sin\left(\frac{\mu\pi}{2}\right) \frac{(a^2 - s^2)^{-\mu/2}}{|s - x|^{1-\mu}}\, ds = \frac{\pi}{2}. \tag{1.6.31}$$

- **Example 1.6.3**

Let us simplify the integral with a singular kernel[19]

$$\int_{1}^{\infty} t^{\alpha-1}(t-1)^{\delta-1} \frac{t^{1/2}}{t - \xi}\, dt \qquad 0 < \xi < 1 \tag{1.6.32}$$

by evaluating the complex integral

$$\oint_C z^{\alpha-1}(z-1)^{\delta-1} \frac{z^{1/2}}{z - \xi}\, dz \tag{1.6.33}$$

where α and δ are nonintegers. The contour C is a circle of infinite radius with appropriate branch cuts as shown in Fig. 1.6.4. From the residue theorem,

$$\oint_C z^{\alpha-1}(z-1)^{\delta-1} \frac{z^{1/2}}{z - \xi}\, dz = 2\pi i \xi^{\alpha-1}(1-\xi)^{\delta-1}\xi^{1/2} e^{(\delta-1)\pi i} \tag{1.6.34}$$

because there is a simple pole at $z = \xi$.

Because the integrals vanish along C_R as $R \to \infty$ and along the arcs C_2 and C_5 as the circles about $z = 0$ and $z = 1$ become infinitesimally small, we need only evaluate the line integrals along C_1, C_3, C_4 and C_6. The branch cuts associated with $z^{\alpha-1}$ and $z^{1/2}$ lie along the negative real axis. Consequently the argument for z runs from $-\pi$ and π. The argument for $z - 1$ used in $(z-1)^{\delta-1}$, however, lies between 0 and 2π. Taking these arguments into account,

$$\int_{C_3} z^{\alpha-1}(z-1)^{\delta-1} \frac{z^{1/2}}{z - \xi}\, dz = \int_{1}^{\infty} x^{\alpha-1}(x-1)^{\delta-1} \frac{x^{1/2}}{x - \xi}\, dx \tag{1.6.35}$$

---

[19] Taken from Liu, P. L.-F., 1986: Hydrodynamic pressures on rigid dams during earthquakes. *J. Fluid Mech.*, **165**, 131–145. Reprinted with the permission of Cambridge University Press.

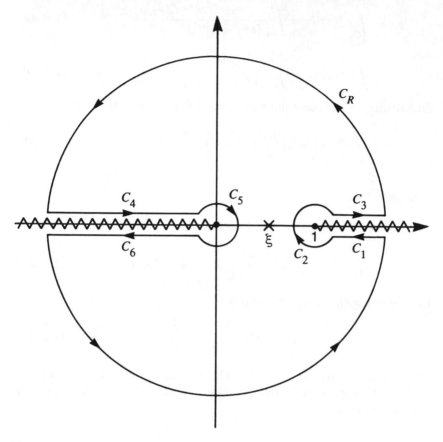

Fig. 1.6.4: The contour used to evaluate (1.6.34).

$$\int_{C_4} z^{\alpha-1}(z-1)^{\delta-1}\frac{z^{1/2}}{z-\xi}\,dz$$

$$= \int_{\infty}^{0} \frac{x^{\alpha-1}e^{(\alpha-1)\pi i}(x+1)^{\delta-1}e^{(\delta-1)\pi i}x^{1/2}e^{\pi i/2}}{x+\xi}\,dx \quad (1.6.36)$$

$$= -ie^{(\alpha+\delta)\pi i}\int_{0}^{\infty}\frac{x^{\alpha-1}(x+1)^{\delta-1}x^{1/2}}{x+\xi}\,dx \qquad (1.6.37)$$

$$\int_{C_6} z^{\alpha-1}(z-1)^{\delta-1}\frac{z^{1/2}}{z-\xi}\,dz$$

$$= \int_{0}^{\infty}\frac{x^{\alpha-1}e^{-(\alpha-1)\pi i}(x+1)^{\delta-1}e^{(\delta-1)\pi i}x^{1/2}e^{-\pi i/2}}{x+\xi}\,dx$$

$$(1.6.38)$$

$$= -ie^{(\delta-\alpha)\pi i} \int_0^\infty \frac{x^{\alpha-1}(x+1)^{\delta-1}x^{1/2}}{x+\xi}\,dx \qquad (1.6.39)$$

$$\int_{C_1} z^{\alpha-1}(z-1)^{\delta-1}\frac{z^{1/2}}{z-\xi}\,dz$$

$$= \int_\infty^1 \frac{x^{\alpha-1}(x-1)^{\delta-1}e^{2\pi(\delta-1)i}x^{1/2}}{x-\xi}\,dx \qquad (1.6.40)$$

$$= -e^{2\delta\pi i}\int_1^\infty \frac{x^{\alpha-1}(x-1)^{\delta-1}x^{1/2}}{x-\xi}\,dx. \qquad (1.6.41)$$

Substituting these results into (1.6.34),

$$\int_1^\infty \frac{x^{\alpha-\frac{1}{2}}(x-1)^{\delta-1}}{x-\xi}\,dx = \frac{\xi^{\alpha-\frac{1}{2}}(1-\xi)^{\delta-1}}{\sin(\pi\delta)}\pi$$
$$- \frac{\cos(\pi\alpha)}{\sin(\pi\delta)}\int_0^\infty \frac{x^{\alpha-1/2}(x+1)^{\delta-1}}{x+\xi}\,dx. \qquad (1.6.42)$$

Although we must still evaluate an integral numerically, the integrand is no longer singular.

Let us now redo this integral with the contour shown in Fig. 1.6.5. The phase of both z and $z-1$ now run from $-\pi$ to π. The contributions from C_R, C_2 and C_{10} are zero. However, for the other contours,

$$\int_{C_1} \frac{z^{\alpha-1}(z-1)^{\delta-1}z^{1/2}}{z-\xi}\,dz$$

$$= \int_\infty^0 \frac{x^{\alpha-1}e^{(\alpha-1)\pi i}(1+x)^{\delta-1}e^{(\delta-1)\pi i}x^{1/2}e^{\pi i/2}}{x+\xi}\,dx \qquad (1.6.43)$$

$$= -ie^{(\alpha+\delta)\pi i}\int_0^\infty \frac{x^{\alpha-1}(1+x)^{\delta-1}x^{1/2}}{x+\xi}\,dx \qquad (1.6.44)$$

$$\int_{C_3+C_5} \frac{z^{\alpha-1}(z-1)^{\delta-1}z^{1/2}}{z-\xi}\,dz$$

$$= \int_0^1 \frac{x^{\alpha-1}(1-x)^{\delta-1}e^{(\delta-1)\pi i}x^{1/2}}{x-\xi}\,dx \qquad (1.6.45)$$

$$= -e^{\delta\pi i}\int_0^1 \frac{x^{\alpha-1}(1-x)^{\delta-1}x^{1/2}}{x-\xi}\,dx \qquad (1.6.46)$$

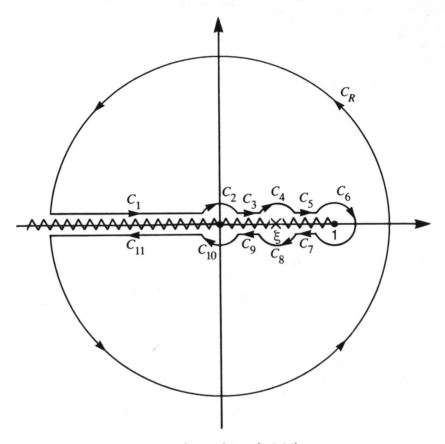

Fig. 1.6.5: The contour used in redoing (1.6.34).

$$\int_{C_7+C_9} \frac{z^{\alpha-1}(z-1)^{\delta-1}z^{1/2}}{z-\xi}\,dz$$

$$= \int_1^0 \frac{x^{\alpha-1}(1-x)^{\delta-1}e^{-(\delta-1)\pi i}x^{1/2}}{x-\xi}\,dx \qquad (1.6.47)$$

$$= e^{-\delta\pi i}\int_0^1 \frac{x^{\alpha-1}(1-x)^{\delta-1}x^{1/2}}{x-\xi}\,dx \qquad (1.6.48)$$

$$\int_{C_{11}} \frac{z^{\alpha-1}(z-1)^{\delta-1}z^{1/2}}{z-\xi}\,dz$$

$$= \int_0^\infty \frac{x^{\alpha-1}e^{-(\alpha-1)\pi i}(1+x)^{\delta-1}e^{-(\delta-1)\pi i}x^{1/2}e^{-\pi i/2}}{x+\xi}\,dx$$

$$(1.6.49)$$

$$= -ie^{-(\alpha+\delta)\pi i} \int_0^\infty \frac{x^{\alpha-1}(1+x)^{\delta-1}x^{1/2}}{x+\xi} dx \qquad (1.6.50)$$

$$\int_{C_4+C_8} \frac{z^{\alpha-1}(z-1)^{\delta-1}z^{1/2}}{z-\xi} dz$$

$$= \lim_{\epsilon\to 0} \int_\pi^0 i\epsilon e^{\theta i} d\theta$$

$$\times \frac{(\xi+\epsilon e^{\theta i})^{\alpha-1}(1-\xi-\epsilon e^{\theta i})^{\delta-1}e^{(\delta-1)\pi i}(\xi+\epsilon e^{\theta i})^{1/2}}{\epsilon e^{\theta i}}$$

$$+ \lim_{\epsilon\to 0} \int_0^{-\pi} i\epsilon e^{\theta i} d\theta$$

$$\times \frac{(\xi+\epsilon e^{\theta i})^{\alpha-1}(1-\xi-\epsilon e^{\theta i})^{\delta-1}e^{-(\delta-1)\pi i}(\xi+\epsilon e^{\theta i})^{1/2}}{\epsilon e^{\theta i}}$$

$$\qquad (1.6.51)$$

$$= -i\pi\xi^{\alpha-1}(1-\xi)^{\delta-1}\xi^{1/2}e^{(\delta-1)\pi i}$$
$$- i\pi\xi^{\alpha-1}(1-\xi)^{\delta-1}\xi^{1/2}e^{-(\delta-1)\pi i} \qquad (1.6.52)$$

$$= 2i\pi\xi^{\alpha-1}(1-\xi)^{\delta-1}\xi^{1/2}\cos(\pi\delta). \qquad (1.6.53)$$

Bringing these results together,

$$\int_0^1 \frac{x^{\alpha-\frac12}(1-x)^{\delta-1}}{x-\xi} dx = \pi\xi^{\alpha-1/2}(1-\xi)^{\delta-1}\frac{\cos(\pi\delta)}{\sin(\pi\delta)}$$

$$- \frac{\cos[\pi(\alpha+\delta)]}{\sin(\pi\delta)} \int_0^\infty \frac{x^{\alpha-1/2}(1+x)^{\delta-1}}{x+\xi} dx.$$

$$\qquad (1.6.54)$$

• **Example 1.6.4**

We now turn to some examples that involve logarithms. For example, consider the contour integral of $2z\log(z)/(z^2+2iaz-1)$ around the contour shown in Fig. 1.6.6 where $a > 1$. The branch cut lies along the negative real axis. The only singularity within the contour is a pole located at $z_1 = i(\sqrt{a^2-1}-a)$. For the contour C_1, $z = xe^{-\pi i}$ and $dz = dx\, e^{-\pi i}$. Therefore,

$$2\int_{C_1} \frac{z\log(z)}{z^2+2iaz-1} dz = 2\int_0^1 \frac{x[\ln(x)-i\pi]}{x^2-2iax-1} dx \qquad (1.6.55)$$

whereas along C_3, $z = xe^{\pi i}$ and $dz = dx\, e^{\pi i}$ so that

$$2\int_{C_3} \frac{z\log(z)}{z^2+2iaz-1} dz = 2\int_1^0 \frac{x[\ln(x)+i\pi]}{x^2-2iax-1} dx. \qquad (1.6.56)$$

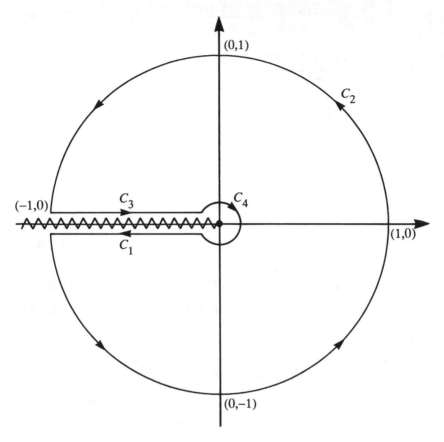

Fig. 1.6.6: The contour used to derive (1.6.65)–(1.6.66).

Along the unit circle, $z = e^{\theta i}$, $dz = ie^{\theta i}d\theta$ and

$$2 \int_{C_2} \frac{z \log(z)}{z^2 + 2iaz - 1} dz = 2 \int_{-\pi}^{\pi} \frac{i\theta e^{\theta i}}{e^{2\theta i} + 2iae^{\theta i} - 1} ie^{\theta i} d\theta \qquad (1.6.57)$$

$$= 2 \int_{-\pi}^{\pi} \frac{i\theta e^{\theta i}}{e^{\theta i} + 2ia - e^{-\theta i}} i\, d\theta \qquad (1.6.58)$$

$$= \int_{-\pi}^{\pi} \frac{\cos(\theta) + i\sin(\theta)}{\sin(\theta) + a} i\theta\, d\theta. \qquad (1.6.59)$$

Finally,

$$2 \int_{C_4} \frac{z \log(z)}{z^2 + 2iaz - 1} dz = \lim_{\epsilon \to 0} \int_{\pi}^{-\pi} \frac{\epsilon e^{\theta i}[\ln(\epsilon) + i\theta]i\epsilon e^{\theta i}}{\epsilon^2 e^{2\theta i} + 2ia\epsilon e^{\theta i} - 1} d\theta = 0.$$
$$(1.6.60)$$

The sum of the integrals (1.6.55), (1.6.56), (1.6.59) and (1.6.60) must equal $2\pi i$ times the residues at the pole z_1:

$$-4\pi i \int_0^1 \frac{x}{x^2 - 2iax - 1} \, dx + \int_{-\pi}^{\pi} \frac{\cos(\theta) + i\sin(\theta)}{a + \sin(\theta)} i\theta \, d\theta$$

$$= 4\pi i \frac{z_1[\ln(|z_1|) + i\arg(z_1)]}{2z_1 + 2ia} \qquad (1.6.61)$$

$$= 2\pi i(\sqrt{a^2 - 1} - a)\frac{\ln[a - \sqrt{a^2 - 1}] - i\pi/2}{\sqrt{a^2 - 1}}. \qquad (1.6.62)$$

Now

$$\int_0^1 \frac{x}{x^2 - 2iax - 1} \, dx = \frac{1}{2} \int_0^1 \frac{(x^2 - 1) \, d(x^2 - 1)}{(x^2 - 1)^2 + 4a^2(x^2 - 1) + 4a^2}$$

$$+ 2ia \int_0^1 \frac{x^2}{(x^2 - 1)^2 + 4a^2 x^2} \, dx \qquad (1.6.63)$$

$$= \frac{a}{2\sqrt{a^2 - 1}} \ln[a - \sqrt{a^2 - 1}] + \frac{1}{2}\ln(2a)$$

$$+ \frac{i}{2\sqrt{a^2 - 1}} \left[\frac{\pi}{2}\sqrt{a^2 - 1} - a\tan^{-1}(\sqrt{a^2 - 1})\right]. \qquad (1.6.64)$$

Substituting (1.6.64) into (1.6.62) and then separating the real and imaginary parts,

$$\int_{-\pi}^{\pi} \frac{\theta \sin(\theta)}{a + \sin(\theta)} \, d\theta = \frac{2\pi a}{\sqrt{a^2 - 1}} \left[\frac{\pi}{2} - \tan^{-1}(\sqrt{a^2 - 1})\right] \qquad (1.6.65)$$

$$\int_{-\pi}^{\pi} \frac{\theta \cos(\theta)}{a + \sin(\theta)} \, d\theta = 2\pi \ln\left[2a^2 - 2a\sqrt{a^2 - 1}\right]. \qquad (1.6.66)$$

• **Example 1.6.6**

Let us evaluate the expression

$$\oint_C f(z) \log(c - z) \, dz \qquad (1.6.67)$$

where Fig. 1.6.7 shows the contour C, $f(z)$ is a single-valued function that vanishes along $|z| = R \to \infty$. Our branch cut has real values for

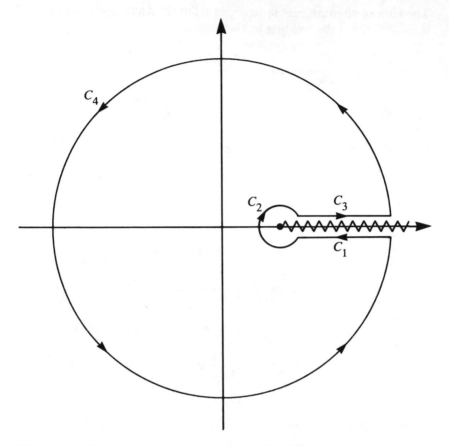

Fig. 1.6.7: The contour used to evaluate (1.6.67).

z lying along the real axis from $-\infty$ and c. Evaluating the various line integrals,

$$\int_{C_1} f(z) \log(c-z)\,dz = \int_\infty^c f(x)[\ln(x-c)+\pi i]\,dx, \qquad \textbf{(1.6.68)}$$

$$\int_{C_2} f(z) \log(c-z)\,dz = 0, \qquad \textbf{(1.6.69)}$$

$$\int_{C_3} f(z) \log(c-z)\,dz = \int_c^\infty f(x)[\ln(x-c)-\pi i]\,dx, \qquad \textbf{(1.6.70)}$$

and

$$\int_{C_4} f(z) \log(c-z)\,dz = 0. \qquad \textbf{(1.6.71)}$$

Therefore,

$$\oint_C f(z) \log(c - z)\, dz = 2\pi i[\text{Residue of } f(z) \log(c - z) \text{ within } C], \tag{1.6.72}$$

$$-2\pi i \int_c^\infty f(x)\, dx = 2\pi i[\text{Residue of } f(z) \log(c - z) \text{ within } C], \tag{1.6.73}$$

or

$$\int_c^\infty f(x)\, dx = -[\text{Residue of } f(z) \log(c - z) \text{ within } C]. \tag{1.6.74}$$

Consequently the value of $\int_c^\infty f(x)\, dx$ equals the negative of the sum of the residues of $f(z) \log(c - z)$ where we only use the poles associated with $f(z)$. This nice result was first published in an obscure paper by Neville[20].

To illustrate this result, consider the case of

$$I = \int_c^\infty \frac{1}{x^2 - a^2}\, dx \tag{1.6.75}$$

with $c > a$. The function $f(z) = 1/(z^2 - a^2)$ has simple poles at $z = \pm a$. Our analysis shows that

$$\int_c^\infty \frac{1}{x^2 - a^2}\, dx = -[\text{Residue of } \frac{\log(c - z)}{z^2 - a^2} \text{ at } z = a \text{ and } z = -a] \tag{1.6.76}$$

$$= -\frac{\log(c - a)}{2a} + \frac{\log(c + a)}{2a} = \frac{1}{2a} \ln \left[\frac{c + a}{c - a} \right] \tag{1.6.77}$$

if $c > a > 0$.

Problems

1. Evaluate $\oint_C dz/[z \log(z)]$ around the illustrated contour and show that

$$\int_0^\infty \frac{dx}{x[\ln(x)^2 + \pi^2]} = 1.$$

Use the principal value of the logarithm, $\log(z) = \ln(r) + \theta i$.

---

[20] Neville, E. H., 1945: Indefinite integration by means of residues. *The Mathematical Student*, **13**, 16–25.

Transform Methods for Solving Partial Differential Equations

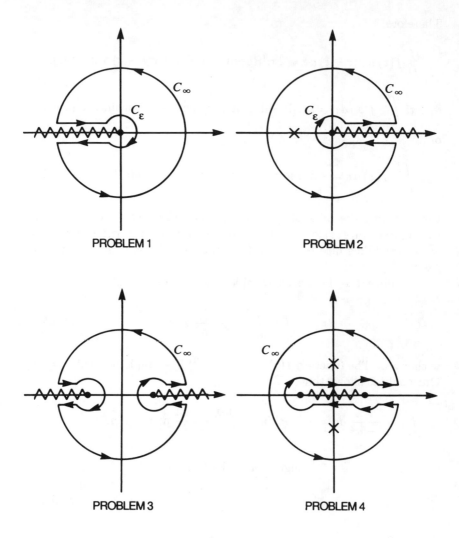

PROBLEM 1

PROBLEM 2

PROBLEM 3

PROBLEM 4

2. Use the illustrated contour to evaluate $\oint_C dz/[(z+1)\sqrt{z}]$ and show that

$$\int_0^\infty \frac{dx}{(x+1)\sqrt{x}} = \pi.$$

3. Evaluate $\oint_C dz/[z\sqrt{z^2-1}]$ around the illustrated contour and verify that

$$\int_1^\infty \frac{dx}{x\sqrt{x^2-1}} = \frac{\pi}{2}.$$

4. Use the illustrated contour and two different complex functions to

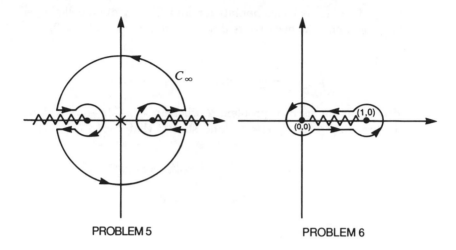

PROBLEM 5   PROBLEM 6

show that

$$\int_{-1}^{1} \frac{dx}{(x^2 + 1)\sqrt{1 - x^2}} = \frac{\pi}{\sqrt{2}}$$

and

$$\int_{-1}^{1} \frac{dx}{(x^2 + 1)^2\sqrt{1 - x^2}} = \frac{3\pi}{4\sqrt{2}}$$

Caution: Remember that the square root will have different signs at the two poles.

5. Use the illustrated contour and show that

$$\int_{1}^{\infty} \frac{1}{x^2} \sqrt{\frac{x - 1}{x + 1}} \, dx = \pi - \int_{1}^{\infty} \frac{1}{x^2} \sqrt{\frac{x + 1}{x - 1}} \, dx.$$

6. Use the illustrated contour and show that

$$\int_{0}^{1} \frac{dx}{(x^2 - x^3)^{1/3}} = \frac{2\pi}{\sqrt{3}}.$$

7. Show[21] that

$$\frac{1}{2\pi i} \oint_{C} \left(\frac{z + 1}{z} \right)^{\alpha} z^{n-1} dz = \frac{\sin[(n - \alpha)\pi]}{\pi} \int_{0+}^{1} x^{n-1-\alpha} (1 - x)^{\alpha} dx$$

---

[21] Taken from Jury, E. I., and C. A. Galtieri, 1961: A note on the inverse z transformation. *IRE Trans. Circuit Theory*, **CT-8**, 371–374. ©1961 IEEE.

where $n \geq 0$, α is real and noninteger and C is a dumbbell-shaped contour lying along the negative real axis from $z = -1$ and $z = 0$.

8. Evaluate

$$\oint_C \frac{z^{p-1}}{z^2 - 2\alpha z + 1} \, dz, \qquad 0 \leq \alpha < 1,$$

around the contour used in problem 2 where $p = \pi/[2\pi - 2\cos^{-1}(\alpha)]$ with the branch $0 \leq \cos^{-1}(\alpha) < 2\pi$ and show[22] that

$$\int_0^\infty \frac{x^{p-1}}{x^2 - 2\alpha x + 1} \, dx = \frac{\pi}{\sqrt{1-\alpha^2}} \{\cos[(p-1)\cos^{-1}(\alpha)]$$
$$- \cot[\pi(p-1)]\sin[(p-1)\cos^{-1}(\alpha)]\}.$$

Hint: $\alpha \pm i\sqrt{1-\alpha^2} = \exp[\pm i\cos^{-1}(\alpha)]$.

9. Show[23] that

$$\int_{-1}^1 \frac{\ln|t - p|}{(t-q)\sqrt{1-t^2}} \, dt = -2\pi \frac{\text{sgn}(p-q)}{\sqrt{1-q^2}} \tan^{-1}\left(\sqrt{\frac{1 \pm q}{1 \mp q}}\right)$$

where $-1 < q < 1$ and the $\pm$ correspond to $p > q$ and $p < q$, respectively.

Step 1. Consider first the case $p = 1$. From the contours shown in Fig. 1.6.8, show that

$$\int_{C_3 + C_4} \frac{\log(z-1)}{(z-q)\sqrt{1-z^2}} \, dz = -\int_{C_1 + C_2} \frac{\log(z-1)}{(z-q)\sqrt{1-z^2}} \, dz.$$

Step 2. By parameterizing each contour as follows:

$$C_1 : z - 1 = \rho e^{\pi i/2}, z + 1 = \sqrt{4 + \rho^2} e^{\theta i}, \log(z-1) = \ln(\rho) + \pi i/2$$

$$C_2 : z - 1 = \rho e^{\pi i/2}, z + 1 = \sqrt{4 + \rho^2} e^{\theta i}, \log(z-1) = \ln(\rho) - 3\pi i/2$$

where $0 < \rho < \infty$ and $\tan(\theta) = \rho/2$ and

$$C_3 : z - 1 = (1 - \rho)e^{\pi i}, z + 1 = (1 + \rho)e^{0i}, \log(z-1) = \ln(1 - \rho) - \pi i$$

---

[22] Taken from Greenwell, R. N., and C. Y. Wang, 1980: Fluid flow through a partially filled cylinder. *Appl. Sci. Res.*, **36**, 61–75. Reprinted by permission of Kluwer Academic Publishers.

[23] Taken from Lewis, P. A., and G. R. Wickham, 1992: The diffraction of SH waves by an arbitrary shaped crack in two dimensions. *Phil. Trans. R. Soc. Lond.*, **A340**, 503–529.

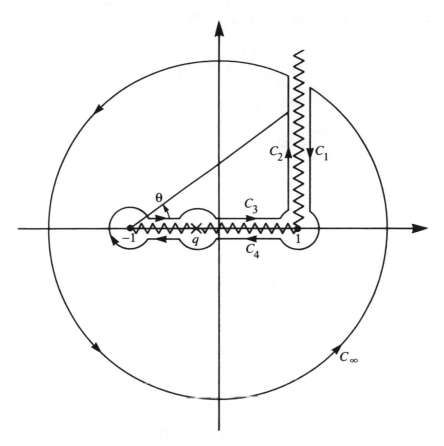

Fig. 1.6.8: Contour used in solving problem 8. Redrawn from Lewis, P. A., and G. R. Wickham, *Phil. Trans. R. Soc. Lond.*, **A340**, 503–529 (1992).

$C_4 : z - 1 - (1 - \rho)e^{\pi i}, z \mid 1 - (1 + \rho)e^{2\pi i}, \log(z - 1) = \ln(1 - \rho) - \pi i$

where $-1 < \rho < 1$, show that

$$\int_{C_1 + C_2} = 2\pi e^{-\pi i/4} \int_0^{\pi/2} \frac{e^{-\theta i/2}}{\sqrt{\sin(\theta)}[(1 - q)\cos(\theta) + 2i\sin(\theta)]} \, d\theta$$

and

$$\int_{C_3 + C_4} = -2i \int_{-1}^1 \frac{\ln(1 - t)}{(t - q)\sqrt{1 - t^2}} \, dt.$$

Note that the simple pole at $z = q$ makes no contribution.

Step 3. Evaluate the integral for $C_1 + C_2$ by noting that $(1 + q) + (1 - q) = 2$, replacing cosine and sine by their exponential equivalents and letting

$1 - e^{2\theta i} = \eta^2$. You should find that

$$\int_{C_1+C_2} = -\frac{4\pi i}{\sqrt{1-q^2}} \tan^{-1}\left(\sqrt{\frac{1+q}{1-q}}\right).$$

Finally, by matching imaginary parts, show that

$$\int_{-1}^{1} \frac{\ln(1-t)}{(t-q)\sqrt{1-t^2}} \, dt = -\frac{2\pi}{\sqrt{1-q^2}} \tan^{-1}\left(\sqrt{\frac{1+q}{1-q}}\right).$$

Step 4. Now redo the first three steps with $p = -1$ and show that

$$\int_{-1}^{1} \frac{\ln(t-1)}{(t-q)\sqrt{1-t^2}} \, dt = \frac{2\pi}{\sqrt{1-q^2}} \tan^{-1}\left(\sqrt{\frac{1-q}{1+q}}\right).$$

Then, use analytic continuity to argue for the more general result.

Chapter 2
Transform Methods
with Single-Valued Functions

For many students of engineering and the sciences, Laplace and Fourier transforms are an important tool in solving ordinary differential equations. In this book we shall show that we may apply these transforms with equal benefit in solving partial differential equations.

For many years scholars attributed the invention of operational mathematics to Oliver Heaviside. Modern research[1,2,3,4,5,6], however,

---

[1] Cooper, J. L. B., 1952: Heaviside and the operational calculus. *Math. Gazette*, **36**, 5–19.

[2] Lützen, J., 1979: Heaviside's operational calculus and the attempts to rigorise it. *Arch. Hist. Exact Sci.*, **21**, 161–200.

[3] Deakin, M. A. B., 1981: The development of the Laplace transform, 1737–1937: I. Euler to Spitzer, 1737–1880. *Arch. Hist. Exact Sci.*, **25**, 343–390.

[4] Deakin, M. A. B., 1982: The development of the Laplace transform, 1737–1937: II. Poincaré to Doetsch, 1880–1937. *Arch. Hist. Exact Sci.*, **26**, 351–381.

[5] Petrova, S. S., 1986: Heaviside and the development of the symbolic calculus. *Arch. Hist. Exact Sci.*, **37**, 1–23.

[6] Deakin, M. A. B., 1992: The ascendancy of the Laplace transform and how it came about. *Arch. Hist. Exact Sci.*, **44**, 265–286.

has shown that during the early nineteenth century both English and French mathematicians developed both symbolic calculus and operational methods. For example, Cauchy used operational methods, applying the Fourier transform to solve the wave equation. Later in the nineteenth century, this knowledge was apparently forgotten until Heaviside rediscovered the Laplace transform[7]. However, because of Heaviside's lack of mathematical rigor, a controversy of legendary proportions[8] developed between him and the mathematical "establishment". It remained for the English mathematician T. J. I'a. Bromwich[9] and the German electrical engineer K. W. Wagner[10] to justify Heaviside's work. Although each was ignorant of the other's work, both used function theory. Wagner concentrated on the expansion formula whereas Bromwich gave a broader explanation of the operational calculus.

Having developed the necessary background material in the previous chapter, we illustrate the salient points of applying operational mathematics to solve partial differential equations by considering the sound waves that arise when a sphere of radius a begins to pulsate at time $t = 0$. The symmetric wave equation in spherical coordinates is

$$\frac{1}{r}\frac{\partial^2(ru)}{\partial r^2} = \frac{1}{c^2}\frac{\partial^2 u}{\partial t^2} \tag{2.0.1}$$

where c is the speed of sound, $u(r, t)$ is the velocity potential and $-\partial u/\partial r$ gives the velocity of the parcel of air. At the surface of the sphere $r = a$, the radial velocity must equal the velocity of the pulsating sphere

$$-\frac{\partial u}{\partial r} = \frac{d\xi}{dt} \tag{2.0.2}$$

where $\xi(t)$, the displacement of the surface of the pulsating sphere, equals $B\sin(\omega t)H(t)$. The air is initially at rest.

As we shall shortly see, transform techniques are quite applicable because there are initial conditions and the partial differential equation is linear. If the Laplace transform of the solution $u(r, t)$ is

$$\overline{u}(r, s) = \int_0^\infty u(r, t)e^{-st}dt, \tag{2.0.3}$$

---

[7] Heaviside, O., 1893: On operators in physical mathematics. *Proc. R. Soc. Lond.*, **A52**, 504–529.

[8] Nahin, P., 1988: *Oliver Heaviside: Sage in Solitude.* New York: IEEE Press, Chapter 10.

[9] Bromwich, T. J. I'a., 1916: Normal coordinates in dynamical systems. *Proc. Lond. Math. Soc.*, Ser. 2, **15**, 401–448.

[10] Wagner, K. W., 1915/16: Über eine Formel von Heaviside zur Berechnung von Einschaltvorgängen. *Arch. Electrotechnik*, **4**, 159–193.

Fig. 2.0.1: Oliver Heaviside (1850–1925). Reproduced by kind permission of the Institution of Electrical Engineers.

then it follows that

$$\mathcal{L}\{u_t(r,t)\} = s\overline{u}(r,s) - u(r,0) \qquad (\textbf{2.0.4})$$

and

$$\mathcal{L}\{u_{tt}(r,t)\} = s^2\overline{u}(r,s) - su(r,0) - u_t(r,0). \qquad (\textbf{2.0.5})$$

Furthermore, derivatives involving r become

$$\mathcal{L}\left\{\frac{\partial[ru(r,t)]}{\partial r}\right\} = \frac{d}{dr}\left\{r\mathcal{L}[u(r,t)]\right\} = \frac{d}{dr}\left[r\overline{u}(r,s)\right] \qquad (\textbf{2.0.6})$$

and

$$\mathcal{L}\left\{\frac{\partial^2[ru(r,t)]}{\partial r^2}\right\} = \frac{d^2}{dr^2}\left\{r\mathcal{L}[u(r,t)]\right\} = \frac{d^2}{dr^2}\left[r\overline{u}(r,s)\right]. \qquad (\textbf{2.0.7})$$

From (2.0.4)–(2.0.7) the Laplace transform of (2.0.1) is

$$\frac{d^2}{dr^2}\left[r\overline{u}(r,s)\right] - \frac{s^2}{c^2}r\overline{u}(r,s) = 0. \qquad (\textbf{2.0.8})$$

The solution of (2.0.8) is

$$r\bar{u}(r, s) = A \exp(-rs/c). \tag{2.0.9}$$

We have discarded the $\exp(rs/c)$ solution because it becomes infinite in the limit of $r \to \infty$. After substituting (2.0.9) into the Laplace transformed (2.0.2),

$$-\frac{d}{dr}\left(A\frac{e^{-sr/c}}{r}\right)\bigg|_{r=a} = \frac{\omega B s}{s^2 + \omega^2} = Ae^{-as/c}\left(\frac{1}{a^2} + \frac{s}{ac}\right). \tag{2.0.10}$$

Therefore,

$$r\bar{u}(r, s) = \frac{\omega B a^2 c s}{(s^2 + \omega^2)(as + c)} e^{-s(r-a)/c} \tag{2.0.11}$$

$$= \frac{\omega B a^2 c}{a^2\omega^2 + c^2} e^{-s(r-a)/c}\left(\frac{cs + \omega^2 a}{s^2 + \omega^2} - \frac{c}{s + c/a}\right). \tag{2.0.12}$$

Applying the second shifting theorem, the inversion of (2.0.12) follows directly

$$ru(r, t) = \frac{\omega B a^2 c^2}{a^2\omega^2 + c^2}\left\{\cos\left[\omega\left(t - \frac{r-a}{c}\right)\right] + \frac{\omega a}{c}\sin\left[\omega\left(t - \frac{r-a}{c}\right)\right]\right.$$
$$\left. - \exp\left[-\frac{c}{a}\left(t - \frac{r-a}{c}\right)\right]\right\}H\left(t - \frac{r-a}{c}\right). \tag{2.0.13}$$

The greatest difficulty with this technique is the inversion. Unlike our example, we cannot do, in general, the inversion by inspection and another method must be sought. Fortunately, we may apply the powerful tools of complex variables. If the transform is single-valued, we show how to do this inversion in section 2.1; for multivalued transforms we illustrate this technique in section 3.1. In a certain sense, we could consider this book a course on applied complex variables[11].

In addition to Laplace transforms, we may also use Fourier transforms to solve of partial differential equations. Again, the most difficult aspect of the analysis is the inversion and, again, we may apply complex variables to find the inverse. We illustrate this in Section 2.2 for single-valued transforms while we treat multivalued transforms in Section 3.2.

---

[11] "Had Heaviside been able to make full use of Cauchy's method of complex integration, then (to quote a well-known saying) 'we should have learned something'." Quote taken from Bromwich, T. J. I'a., 1928: Note on Prof. Carslaw paper. *Math. Gazette*, **14**, p. 227.

Fig. 2.1.1: Thomas John l'Anson Bromwich (1875–1929). By permission of the President and Council of the Royal Society.

2.1 INVERSION OF LAPLACE TRANSFORMS BY CONTOUR INTEGRATION

Laplace transforms are a popular tool for solving initial-value problems. In most undergraduate courses, the use of tables, special theorems, partial fractions and convolution are the methods taught for finding the inverse. In most problems involving partial differential equations, these techniques fail us.

In 1916, Bromwich[9] showed that we can express the inverse of a Laplace transform as the contour integral

$$f(t) = \frac{1}{2\pi i} \int_{c-\infty i}^{c+\infty i} F(z) e^{tz} dz \qquad (2.1.1)$$

where $F(z)$ is the Laplace transform

$$F(s) = \int_{0}^{\infty} f(t) e^{-st} dt \qquad (2.1.2)$$

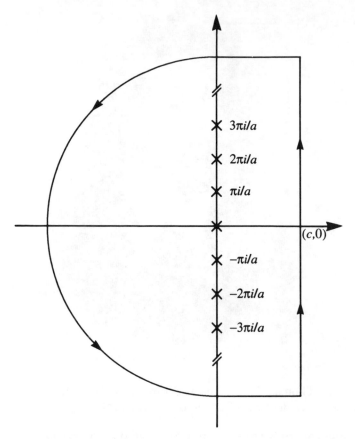

Fig. 2.1.2: Contour used in the inversion of (2.1.7).

and c is the Laplace convergence abscissa. The value of c must be greater than the real part of any of the singularities of $F(s)$.

Proof: Consider the piecewise differentiable function $f(x)$ which vanishes for $x < 0$. We can express the function $e^{-cx}f(x)$ by the complex Fourier representation

$$f(x)e^{-cx} = \frac{1}{2\pi}\int_{-\infty}^{\infty} e^{i\omega x}\left[\int_0^\infty e^{-i\omega t}e^{-ct}f(t)\,dt\right]d\omega \qquad (2.1.3)$$

where we choose c so that $\int_0^\infty e^{-ct}|f(t)|\,dt$ exists. We now multiply (2.1.3) by e^{cx} and then bring this factor inside the first integral. Then

$$f(x) = \frac{1}{2\pi}\int_{-\infty}^{\infty} e^{(c+i\omega)x}\left[\int_0^\infty e^{-(c+i\omega)t}f(t)\,dt\right]d\omega. \qquad (2.1.4)$$

Let $s = c + i\omega$ where s is a new, complex variable of integration and

$$f(x) = \frac{1}{2\pi i} \int_{c-\infty i}^{c+\infty i} e^{sx} \left[\int_0^\infty e^{-st} f(t)\, dt \right] ds. \qquad (2.1.5)$$

The quantity inside the integral is equal to the Laplace transform $F(s)$. Therefore,

$$f(t) = \frac{1}{2\pi i} \int_{c-\infty i}^{c+\infty i} F(z) e^{zt} dz. \qquad (2.1.6)$$

Of course, this brings us no closer to actually finding the inverse unless we can evaluate the integral. Fortunately, complex variables are particularly adept at doing this. In the following examples, we illustrate how these techniques apply when the transform is single-valued.

• **Example 2.1.1**

For our first example of the inversion of Laplace transforms by complex integration, let us find the inverse of

$$F(s) = \frac{1}{s \sinh(as)} \qquad (2.1.7)$$

where a is real. From Bromwich's integral,

$$f(t) = \frac{1}{2\pi i} \int_{c-\infty i}^{c+\infty i} \frac{e^{tz}}{z \sinh(az)}\, dz. \qquad (2.1.8)$$

Here c is greater than the real part of any of the singularities in (2.1.8). Using the Taylor expansion for the hyperbolic sine, we find that $z = 0$ is a second-order pole. Next, we note that because $\sinh(az) = -i\sin(iaz)$, there are an infinite number of simple poles located where $\sin(iaz) = 0$, i.e., $z_n = \pm n\pi i/a$ where $n = 1, 2, 3, \ldots$

We may convert the line integral (2.1.8), with the contour lying parallel and slightly to the right of the imaginary axis, into a closed contour using Jordan's lemma (1.4.16) through the addition of an infinite semicircle joining $i\infty$ to $-i\infty$. See Fig. 2.1.2. We now apply the residue theorem. For the second-order pole at $z = 0$,

$$\text{Res}(z = 0) = \frac{1}{1!} \lim_{z \to 0} \frac{d}{dz} \left\{ \frac{(z-0)^2 e^{tz}}{z \sinh(az)} \right\} \qquad (2.1.9)$$

$$= \lim_{z \to 0} \frac{d}{dz} \left\{ \frac{z e^{tz}}{\sinh(az)} \right\} \qquad (2.1.10)$$

$$= \lim_{z \to 0} \left\{ \frac{e^{tz}}{\sinh(az)} + \frac{zt e^{tz}}{\sinh(az)} - \frac{az \cosh(az) e^{tz}}{\sinh^2(az)} \right\} \qquad (2.1.11)$$

$$= \frac{t}{a} \qquad (2.1.12)$$

after using $\sinh(az) = az + O(z^3)$. For the simple poles $z_n = \pm n\pi i/a$,

$$\text{Res}(z = z_n) = \lim_{z \to z_n} \frac{(z - z_n)e^{tz}}{z \sinh(az)} \tag{2.1.13}$$

$$= \lim_{z \to z_n} \frac{e^{tz}}{\sinh(az) + az \cosh(az)} \tag{2.1.14}$$

$$= \frac{\exp[\pm n\pi it/a]}{(-1)^n(\pm n\pi i)} \tag{2.1.15}$$

because $\cosh(\pm n\pi i) = \cos(n\pi) = (-1)^n$. Thus, summing up all of the residues gives

$$f(t) = \frac{t}{a} + \sum_{n=1}^{\infty} \frac{(-1)^n \exp[n\pi it/a]}{n\pi i} - \sum_{n=1}^{\infty} \frac{(-1)^n \exp[-n\pi it/a]}{n\pi i} \tag{2.1.16}$$

$$= \frac{t}{a} + \frac{2}{\pi} \sum_{n=1}^{\infty} \frac{(-1)^n}{n} \sin(n\pi t/a). \tag{2.1.17}$$

• **Example 2.1.2**

For our second example, we invert

$$F(s) = \frac{\cosh(qx)}{sq \sinh(qL)} \tag{2.1.18}$$

where $q = s^{1/2}/a$ and the constants a, L and x are real. One immediate concern is the presence of $s^{1/2}$ because this is a multivalued function. However, when we replace the hyperbolic cosine and sine functions with their Taylor expansions, $F(s)$ contains only powers of s and is, in fact, a single-valued function.

From Bromwich's integral,

$$f(t) = \frac{1}{2\pi i} \int_{c-\infty i}^{c+\infty i} \frac{\cosh(qx)e^{tz}}{zq \sinh(qL)} \, dz \tag{2.1.19}$$

where $q = z^{1/2}/a$. Using the Taylor expansions for the hyperbolic cosine and sine, we find that $z = 0$ is a second-order pole. The remaining poles are located where $\sinh(qL) = -i \sin(iqL) = 0$. Therefore, $z_n^{1/2}L/a = n\pi i$ or $z_n = -n^2\pi^2 a^2/L^2$ where $n = 1, 2, 3, \ldots$ We have chosen the positive sign because $z^{1/2}$ must be single-valued; if we had chosen the negative sign the answer would be the same. Further analysis reveals that these poles are simple.

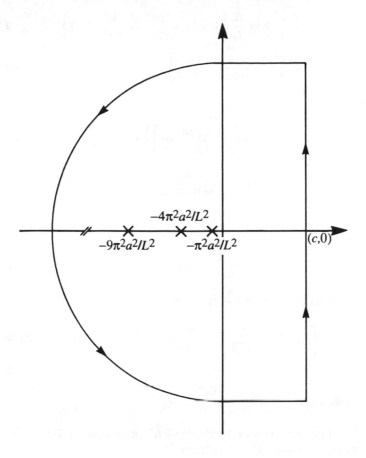

Fig. 2.1.3: Contour used in the inversion of (2.1.18).

Having classified the poles, we now close the line contour which lies slightly to the right of the imaginary axis with an infinite semicircle in the left half-plane and use the residue theorem. See Fig. 2.1.3. The values of the residues are

$$\text{Res}(z = 0) = \frac{1}{1!} \lim_{z \to 0} \frac{d}{dz} \left\{ \frac{(z - 0)^2 \cosh(qx) e^{tz}}{zq \sinh(qL)} \right\} \tag{2.1.20}$$

$$= \lim_{z \to 0} \frac{d}{dz} \left\{ \frac{z \cosh(qx) e^{tz}}{q \sinh(qL)} \right\} \tag{2.1.21}$$

$$= \frac{a^2}{L} \lim_{z \to 0} \frac{d}{dz} \left\{ \frac{z \left[1 + \frac{zx^2}{2!a^2} + \cdots \right] \left[1 + tz + \frac{t^2 z^2}{2!} + \cdots \right]}{z + \frac{L^2 z^2}{3!a^2} + \cdots} \right\}$$
$$\tag{2.1.22}$$

$$= \frac{a^2}{L} \lim_{z \to 0} \frac{d}{dz} \left\{ 1 + tz + \frac{zx^2}{2a^2} - \frac{zL^2}{3!a^2} + \cdots \right\} \tag{2.1.23}$$

$$= \frac{a^2}{L} \left\{ t + \frac{x^2}{2a^2} - \frac{L^2}{6a^2} \right\} \tag{2.1.24}$$

and

$$\text{Res}(z = z_n) = \left[\lim_{z \to z_n} \frac{\cosh(qx)}{zq} e^{tz} \right] \left[\lim_{z \to z_n} \frac{z - z_n}{\sinh(qL)} \right] \tag{2.1.25}$$

$$= \lim_{z \to z_n} \frac{\cosh(qx)e^{tz}}{zq \cosh(qL)L/(2a^2 q)} \tag{2.1.26}$$

$$= \frac{\cosh(n\pi x i/L) \exp[-n^2 \pi^2 a^2 t/L^2]}{(-n^2 \pi^2 a^2/L^2) \cosh(n\pi i)L/(2a^2)} \tag{2.1.27}$$

$$= -\frac{2L(-1)^n}{n^2 \pi^2} \cos(n\pi x/L) e^{-n^2 \pi^2 a^2 t/L^2}. \tag{2.1.28}$$

Summing together all of the residues,

$$f(t) = \frac{a^2}{L} \left\{ t + \frac{x^2}{2a^2} - \frac{L^2}{6a^2} \right\} - \frac{2L}{\pi^2} \sum_{n=1}^{\infty} \frac{(-1)^n}{n^2} \cos(n\pi x/L) e^{-n^2 \pi^2 a^2 t/L^2}. \tag{2.1.29}$$

• Example 2.1.3

To illustrate inversion by Bromwich's integral when Bessel functions are present, consider the transform[12]

$$F(s) = \frac{2I_1(\lambda\sqrt{s})}{\lambda s^{3/2} I_0(\lambda\sqrt{s})} \tag{2.1.30}$$

where $I_0(\ )$ and $I_1(\ )$ are zeroth and first-order, modified Bessel functions of the first kind, respectively. Their power series representation is

$$I_0(z) = 1 + \frac{z^2}{4} + \frac{z^4}{64} + O(z^6) \tag{2.1.31}$$

and

$$I_1(z) = \frac{z}{2} + \frac{z^3}{16} + \frac{z^5}{384} + O(z^7). \tag{2.1.32}$$

---

[12] Taken from Raval, U., 1972: Quasi-static transient response of a covered permeable inhomogeneous cylinder to a line current source. *PA-GEOPH*, **96**, 140–156. Published by Birkhäuser Verlag, Basel, Switzerland.

Consequently,

$$F(s) = \frac{1 + \frac{\lambda^2 s}{8} + \frac{\lambda^4 s^2}{192} + \cdots}{s(1 + \frac{\lambda^2 s}{4} + \frac{\lambda^4 s^2}{64} + \cdots)}. \qquad (2.1.33)$$

From (2.1.33), we see that $F(s)$ is a single-valued function and has a simple pole at $s = 0$. The remaining poles occur where $I_0(\lambda\sqrt{s}) = 0$. If $\lambda\sqrt{s} = -i\alpha$, then these poles occur where $I_0(-i\alpha) = J_0(\alpha) = 0$. Denoting the nth root of $J_0(\alpha) = 0$ by α_n, then $s_n = -\alpha_n^2/\lambda^2$ where $n = 1, 2, 3, \ldots$.

As a result of this analysis, we see that $F(s)$ has simple poles that lie along the negative real axis. Consequently,

$$f(t) = \frac{1}{2\pi i} \oint_C \frac{2I_1(\lambda\sqrt{z})e^{tz}}{\lambda z^{3/2} I_0(\lambda\sqrt{z})} \, dz \qquad (2.1.34)$$

where C consists of a path along, and just to the right of, the imaginary axis and a semicircle of infinite radius in the left side of the complex plane. To evaluate the contour integral, we apply the residue theorem where

$$\text{Res}(z = 0) = \lim_{z \to 0} \frac{(1 + \frac{\lambda^2 z}{8} + \frac{\lambda^4 z^2}{192} + \cdots)(1 + tz + \cdots)}{1 + \frac{\lambda^2 z}{4} + \frac{\lambda^4 z^2}{64} + \cdots} = 1 \qquad (2.1.35)$$

and

$$\text{Res}(z = -\alpha_n^2/\lambda^2) = \lim_{z \to -\alpha_n^2/\lambda^2} \frac{2(z + \alpha_n^2/\lambda^2)I_1(\lambda\sqrt{z})e^{tz}}{\lambda z^{3/2} I_0(\lambda\sqrt{z})} \qquad (2.1.36)$$

$$= \lim_{z \to -\alpha_n^2/\lambda^2} \frac{2I_1(\lambda\sqrt{z})e^{tz}}{\lambda z^{3/2} I_0'(\lambda\sqrt{z})[\lambda/(2\sqrt{z})]} \qquad (2.1.37)$$

$$= -\frac{4\exp(-\alpha_n^2 t/\lambda^2)}{\alpha_n^2} \qquad (2.1.38)$$

because $I_0'(z) = I_1(z)$. Adding the residues together, we find that the inverse of $F(s)$ is

$$f(t) = 1 - 4 \sum_{n=1}^{\infty} \frac{\exp(-\alpha_n^2 t/\lambda^2)}{\alpha_n^2} \qquad (2.1.39)$$

where $J_0(\alpha_n) = 0$ for $n = 1, 2, 3, \ldots$

• **Example 2.1.4**

Consider now the inversion of

$$F(s) = K_0(r\sqrt{s^2 - a^2}) \qquad (2.1.40)$$

where r, a are real, positive and $K_0(\ )$ is a modified Bessel function of the second kind and zeroth order. One of the integral representations of $K_0(\ )$ is

$$K_0(rz) = \int_0^\infty \frac{\cos(r\eta)}{\sqrt{\eta^2 + z^2}}\, d\eta, \quad r > 0, \quad |\arg(z)| < \pi/2. \qquad (2.1.41)$$

Therefore,

$$f(t) = \frac{1}{2\pi i} \int_{c-\infty i}^{c+\infty i} \int_0^\infty \frac{\cos(r\eta)e^{tz}}{\sqrt{\eta^2 - a^2 + z^2}}\, d\eta\, dz \qquad (2.1.42)$$

$$= \int_0^\infty \cos(r\eta) \left[\frac{1}{2\pi i} \int_{c-\infty i}^{c+\infty i} \frac{e^{tz}}{\sqrt{\eta^2 - a^2 + z^2}}\, dz \right] d\eta \qquad (2.1.43)$$

$$= \int_0^\infty \cos(r\eta) J_0(t\sqrt{\eta^2 - a^2})\, d\eta \qquad (2.1.44)$$

$$= \frac{\cosh(a\sqrt{t^2 - r^2})}{\sqrt{t^2 - r^2}} H(t - r) \qquad (2.1.45)$$

because $\mathcal{L}[J_0(bt)] = 1/\sqrt{b^2 + s^2}$.

• **Example 2.1.5**

Let us find the inverse[13] of the Laplace transform

$$F(s) = \frac{1}{s} e^{\gamma s^2 + s^3/3} \qquad (2.1.46)$$

or

$$f(t) = \frac{1}{2\pi i} \int_{c-\infty i}^{c+\infty i} \frac{1}{z} e^{zt + \gamma z^2 + z^3/3}\, dz. \qquad (2.1.47)$$

We begin by noting that

$$\frac{1}{z} e^{zt} = \frac{1}{z} + \int_0^t e^{z\eta}\, d\eta \qquad (2.1.48)$$

---

[13] Reprinted from *Int. J. Solids Structures*, **16**, T. C. T. Ting, The effects of dispersion and dissipation on wave propagation in viscoelastic layered composites, pp. 903–911, ©1980, with kind permission from Pergamon Press Ltd., Headington Hill Hall, Oxford OX3 0BW, UK. See also Cole, J. D., and T. Y. Wu, 1952: Heat conduction in a compressible fluid. *J. Appl. Mech.*, **19**, 209–213.

so that

$$f(t) = \frac{1}{2\pi i} \int_{c-\infty i}^{c+\infty i} \frac{1}{z} e^{\gamma z^2 + z^3/3} dz + \int_0^t \left\{ \frac{1}{2\pi i} \int_{c-\infty i}^{c+\infty i} e^{\eta z + \gamma z^2 + z^3/3} dz \right\} d\eta.$$

(2.1.49)

To evaluate the first integral in (2.1.49) we now deform the original Bromwich integral to the contour shown in Fig. 2.1.4. The contribution from the integrals along the arcs AB and EF vanish as $R \to \infty$. Along BC $z = re^{-\pi i/3}$ and $dz = dr\, e^{-\pi i/3}$ while $z = re^{\pi i/3}$ and $dz = dr\, e^{\pi i/3}$ along DE. Then,

$$\frac{1}{2\pi i} \int_{BC} \frac{1}{z} e^{\gamma z^2 + z^3/3} dz = \frac{1}{2\pi i} \int_\infty^0 \exp\left[\gamma r^2 e^{-2\pi i/3} - r^3/3 \right] \frac{dr}{r},$$

(2.1.50)

$$\frac{1}{2\pi i} \int_{C_\epsilon} \frac{1}{z} e^{\gamma z^2 + z^3/3} dz = \frac{1}{3}$$

(2.1.51)

and

$$\frac{1}{2\pi i} \int_{DE} \frac{1}{z} e^{\gamma z^2 + z^3/3} dz = \frac{1}{2\pi i} \int_0^\infty \exp\left[\gamma r^2 e^{2\pi i/3} - r^3/3 \right] \frac{dr}{r}.$$ (2.1.52)

Combining (2.1.50)–(2.1.52),

$$\frac{1}{2\pi i} \int_{BE} \frac{1}{z} e^{\gamma z^2 + z^3/3} dz$$

$$= \frac{1}{3} + \frac{1}{2\pi i} \int_0^\infty e^{-r^3/3} \frac{dr}{r} \left\{ \exp\left[-\frac{\gamma r^2}{2} + \frac{\sqrt{3} i \gamma r^2}{2} \right] \right.$$

$$\left. - \exp\left[-\frac{\gamma r^2}{2} - \frac{\sqrt{3} i \gamma r^2}{2} \right] \right\}$$

(2.1.53)

$$= \frac{1}{3} + \frac{1}{\pi} \int_0^\infty e^{-\gamma r^2/2 - r^3/3} \sin\left(\frac{\sqrt{3}}{2} \gamma r^2 \right) \frac{dr}{r}.$$ (2.1.54)

Consider now the integral

$$\int_{c-\infty i}^{c+\infty i} e^{\eta z + \gamma z^2 + z^3/3} dz = e^{-\eta \gamma + 2\gamma^3/3} \int_{c'-\infty i}^{c'+\infty i} e^{(\eta - \gamma^2)\chi + \chi^3/3} d\chi$$

(2.1.55)

where $z = \chi - \gamma$. If we set $b = \gamma^2 - \eta$, then

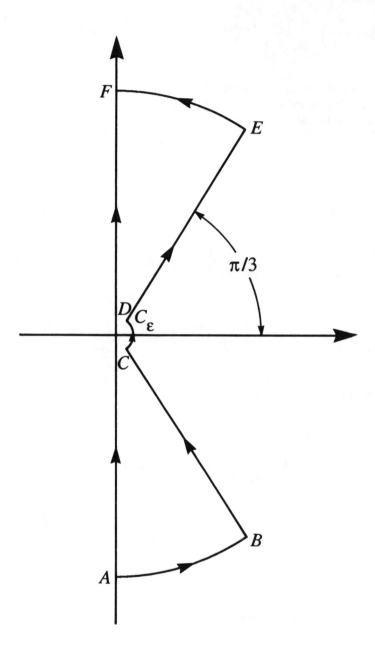

Fig. 2.1.4: Contour used in the inversion of (2.1.46).

$$\frac{1}{2\pi i} \int_{c'-\infty i}^{c'+\infty i} e^{-b\chi+\chi^3/3} d\chi$$

$$= \frac{1}{2\pi i} \int_{AF} e^{-b\chi+\chi^3/3} d\chi \qquad (2.1.56)$$

$$= \frac{1}{2\pi i} \int_0^\infty e^{-ibr-r^3 i/3} i\, dr - \frac{1}{2\pi i} \int_\infty^0 e^{ibr+r^3 i/3} i\, dr$$

$$\qquad\qquad (2.1.57)$$

$$= \frac{1}{\pi} \int_0^\infty \cos(br + r^3/3)\, dr = \mathrm{Ai}(\gamma^2 - \eta) \qquad (2.1.58)$$

where Ai( ) is the Airy function of the first kind. Therefore, the final answer is

$$f(t) = \frac{1}{3} + \frac{1}{\pi} \int_0^\infty e^{-\gamma r^2/2 - r^3/3} \sin\left(\frac{\sqrt{3}}{2}\gamma r^2\right) \frac{dr}{r}$$

$$+ \int_0^t e^{-\eta\gamma + 2\gamma^3/3} \mathrm{Ai}(\gamma^2 - \eta)\, d\eta. \qquad (2.1.59)$$

In this form the inverse is now more amenable to physical interpretation or further asymptotic analysis.

Problems

For the following transforms, use the inversion integral to find the inverse Laplace transform for constant a, M, R and α.

1. $\dfrac{1}{s^3(s+1)^2}$

2. $\dfrac{s+1}{(s+2)^2(s+3)}$

3. $\dfrac{s+2}{s(s-a)(s^2+4)}$

4. $\dfrac{1}{s\sinh(as)}$

5. $\dfrac{\sinh(as/2)}{s\cosh(as/2)}$

6. $\dfrac{\sinh(a\sqrt{s})}{s\sinh(\sqrt{s})}$

7. $\dfrac{\sinh(s)}{s^2\cosh(s)}$

8. $\dfrac{1}{Rs + M^2/\alpha}$

9. $\dfrac{1}{2sa - \sqrt{s}\tanh(\sqrt{s})}$

$$\times \left[1 - \frac{\cosh(a\sqrt{Rs+M^2/\alpha})}{\cosh(\sqrt{Rs+M^2/\alpha})}\right]$$

10. Show that the inverse[14] of the Laplace transform

$$F(s) = \frac{\tanh(as^{1/2})}{s[ms^{1/2} + \tanh(as^{1/2})]}$$

is

$$f(t) = \frac{1}{1+b} + \frac{1}{b} \sum_{n=1}^{\infty} \frac{\sin(2\lambda_n) \exp(-\lambda_n^2 t/\alpha^2)}{\lambda_n[\cos^2(\lambda_n) + 1/b]}$$

where $b = m/\alpha$ and α and m are real.

11. Show that the inverse of the Laplace transform

$$F(s) = \frac{I_0(r\sqrt{s/\kappa})}{I_0(a\sqrt{s/\kappa})}$$

is

$$f(t) = \frac{2\kappa}{a} \sum_{n=1}^{\infty} \frac{\alpha_n J_0(\alpha_n r)}{J_1(\alpha_n a)} e^{-\kappa \alpha_n^2 t}$$

where α_n is the n-th positive root of $J_0(\alpha a) = 0$. The constants a, κ and r are real and positive. The function $I_0(\ )$ is the zeroth-order, modified Bessel function of the first kind while $J_0(\ )$ and $J_1(\ )$ are zeroth and first-order Bessel functions of the first kind, respectively.

12. Show that the inverse[15] of the Laplace transform

$$F(s) = \frac{I_0(r\sqrt{s/\kappa})}{sI_0(a\sqrt{s/\kappa})(k_1^2/\kappa - s)}$$

is

$$f(t) = \frac{\kappa}{k_1^2} - \frac{\kappa}{k_1^2} \frac{I_0(k_1 r/\kappa)}{I_0(k_1 a/\kappa)} e^{k_1^2 t/\kappa} + \frac{2\kappa}{a} \sum_{n=1}^{\infty} \frac{J_0(\alpha_n r)}{\alpha_n J_1(\alpha_n a)(k_1^2 + \kappa^2 \alpha_n^2)} e^{-\kappa \alpha_n^2 t}$$

where α_n is the n-th positive root of $J_0(\alpha a) = 0$. The constants a, κ and r are real and positive. The function $I_0(\ )$ is the zeroth-order, modified Bessel function of the first kind while $J_0(\ )$ and $J_1(\ )$ are zeroth and first-order Bessel functions of the first kind, respectively.

---

[14] Taken from Arutunyan, N. H., 1949: On the research of statically indeterminate systems with vibrating support columns (in Russian). *Prikl. Mat. Mek.*, **13**, 399–500.

[15] Taken from Wadhawan, M. C., 1974: Dynamic thermoelastic response of a cylinder. *PAGEOPH*, **112**, 73–82. Published by Birkhäuser Verlag, Basel, Switzerland.

13. Show that the inverse of the Laplace transform

$$F(s) = \frac{1}{s} - \frac{I_0(r\sqrt{s})}{s^{3/2}I_1(a\sqrt{s})}$$

is

$$f(t) = 1 - a\left[\frac{2t}{a^2} + \frac{r^2}{2a^2} - \frac{1}{4} - 2\sum_{n=1}^{\infty}\frac{J_0(\alpha_n r/a)}{\alpha_n^2 J_0(\alpha_n)}e^{-\alpha_n^2 t/a^2}\right]$$

where α_n is the n-th positive root of $J_1(\alpha) = 0$, $a > 0$, and $I_0(\ )$ and $I_1(\ )$ are the zeroth and first-order, modified Bessel function of the first kind while $J_0(\ )$ and $J_1(\ )$ are zeroth and first-order Bessel functions of the first kind, respectively.

14. Show that the inverse of the Laplace transform

$$F(s) = \frac{I_n(a\sqrt{s})}{\sqrt{s}I_{n+1}(b\sqrt{s})}, \qquad n > 0$$

where $0 < a < b$ and $I_n(z)$ is a modified Bessel function of the first kind and integer order n is

$$f(t) = 2(n + 1)\frac{a^n}{b^{n+1}} + \frac{4}{b}\sum_{m=1}^{\infty}\frac{J_n(a\alpha_m/b)}{J_n(\alpha_m) - J_{n+2}(\alpha_m)}e^{-\alpha_m^2 t/b^2}$$

where α_m is the m-th positive root of $J_{n+1}(\alpha_m) = 0$ and $J_n(\ )$ is the n-order Bessel functions of the first kind, respectively.

15. Find the inverse[16] of the Laplace transform

$$F(s) = \frac{1}{s}\exp[-r\sqrt{s}\tanh(\sqrt{s})]$$

where $r > 0$. This transform has a simple pole at $s = 0$ and an infinite number of essential singularities at $s_n = -(2n - 1)^2\pi^2/4$ where $n = 1, 2, 3, \ldots$ Although we could find an inverse via the residue theorem, a more convenient form follows by deforming Bromwich's contour along the imaginary axis of the s-plane, except for an infinitesimally small semicircle around the simple pole. If we use this contour, show that

$$f(t) = \frac{1}{2} + \frac{2}{\pi}\int_{0+}^{\infty}\exp\left[-\frac{r\eta}{2}\frac{\sinh(\eta) - \sin(\eta)}{\cosh(\eta) + \cos(\eta)}\right]$$
$$\times \sin\left[\frac{\eta^2 t}{2} - \frac{r\eta}{2}\frac{\sinh(\eta) + \sin(\eta)}{\cosh(\eta) + \cos(\eta)}\right]\frac{d\eta}{\eta}.$$

---

[16] Taken from Roshal', A. A., 1969: Mass transfer in a two-layer porous medium. *J. Appl. Mech. Tech. Phys.*, **10**, 551–558.

2.2 INVERSION OF FOURIER TRANSFORMS BY CONTOUR INTEGRATION

In those instances when $f(t)$ exists for all t, is continuous and has a piecewise continuous derivative in any interval, the appropriate transform is the Fourier transform

$$F(\omega) = \int_{-\infty}^{\infty} f(t)e^{-i\omega t}dt. \tag{2.2.1}$$

Given a Fourier transform, we compute its inverse from the inversion integral[17]

$$f(t) = \frac{1}{2\pi} \int_{-\infty}^{\infty} F(\omega)e^{i\omega t}d\omega. \tag{2.2.2}$$

Proof[18]: Consider the integral

$$I = \int_{-\infty}^{\infty} F(k)e^{ikx}dk \tag{2.2.3}$$

for all real values of k. Then

$$I = \int_{-\infty}^{\infty} e^{ikx} \left\{ \int_{-\infty}^{\infty} f(x)e^{-ikx}dx \right\} dk \tag{2.2.4}$$

$$= \int_{-\infty}^{0} e^{ikx} \left\{ \int_{C_1} f(z)e^{-ikz}dz \right\} dk + \int_{0}^{\infty} e^{ikx} \left\{ \int_{C_2} f(z)e^{-ikz}dz \right\} dk \tag{2.2.5}$$

where C_1 is a contour that runs along and just above the real axis and C_2 is a contour that runs along and just below the real axis. In both integrals $\text{Re}(ikz) > 0$. We now reverse the order of integration and obtain

$$I = \int_{C_1} f(z) \left\{ \int_{-\infty}^{0} e^{-ik(z-x)}dx \right\} dz + \int_{C_2} f(z) \left\{ \int_{0}^{\infty} e^{-ik(z-x)}dx \right\} dz \tag{2.2.6}$$

$$= -\int_{C_1} \frac{f(z)}{i(z-x)}dz + \int_{C_2} \frac{f(z)}{i(z-x)}dz = \oint_C \frac{f(z)}{i(z-x)}dz \tag{2.2.7}$$

---

[17] For the generalized Fourier transform or two-sided transform defined by (1.2.6), the inverse is

$$f(t) = \frac{1}{2\pi} \int_{-\infty-\epsilon i}^{\infty-\epsilon i} F(\omega)e^{i\omega t}d\omega$$

where $c_2 > \epsilon > c_1$.

[18] Taken from MacRobert, T. M., 1931: Fourier integrals. *Proc. R. Soc. Edin.*, **51**, 116–126.

where C is a closed, counterclockwise contour between $-\infty + 0i$ and $\infty + 0i$. From Cauchy's integral formula, we finally obtain

$$\int_{-\infty}^{\infty} F(k)e^{ikx} dk = 2\pi f(x). \tag{2.2.8}$$

In section 2.4 we will use Fourier transforms to solve partial differential equations where one or more of the independent variables extend from $-\infty$ to ∞. Although (2.2.2) could be evaluated by direct integration, the most widely encountered technique is through contour integration. In this method we convert the integration along the real axis into a closed contour integration by the addition of an infinite semicircle in the upper or lower half-plane, as dictated by Jordan's lemma. The following examples will illustrate the technique.

• Example 2.2.1

For our first example we find the inverse for

$$F(\omega) = \frac{1}{\omega^4 + 2(b^2 - a^2)\omega^2 + a^4 + 2a^2b^2 + b^4}. \tag{2.2.9}$$

Applying Jordan's lemma, we may replace the integration along the real axis from $-\infty$ to ∞ by an infinite, closed semicircle contour in the upper half-plane for $t > 0$. See Fig. 2.2.1. The inverse Fourier transform is

$$f(t) = \frac{1}{2\pi} \oint_C \frac{e^{itz}}{z^4 + 2(b^2 - a^2)z^2 + a^4 + 2a^2b^2 + b^4} dz. \tag{2.2.10}$$

The value of the contour integral equals $2\pi i$ times the sum of the residues within the contour.

The denominator of the integrand in (2.2.10) has four zeros, located at $\pm a \pm bi$. However, only the residues at $z = \pm a + bi$ are of interest here. For these poles,

$$\text{Res}(a + bi) = \frac{e^{it(a+bi)}}{4(a + bi)^3 + 4(b^2 - a^2)(a + bi)} \tag{2.2.11}$$

$$= e^{-bt}\frac{[a\sin(at) - b\cos(at) - ia\cos(at) - ib\sin(at)]}{8ab(a^2 + b^2)}. \tag{2.2.12}$$

$$\text{Res}(-a + bi) = -e^{-bt}\frac{[a\sin(at) - b\cos(at) + ia\cos(at) + ib\sin(at)]}{8ab(a^2 + b^2)}. \tag{2.2.13}$$

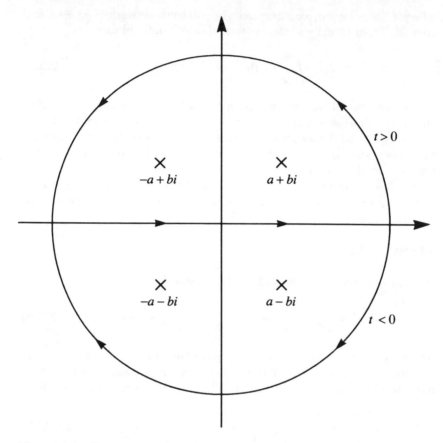

Fig. 2.2.1: Contour used to find the inverse of the Fourier transform (2.2.9)

Consequently the inverse Fourier transform is

$$f(t) = e^{-bt}\frac{a\cos(at) + b\sin(at)}{4ab(a^2 + b^2)} \qquad (2.2.14)$$

for $t > 0$. For $t < 0$ we take the semicircle in the lower half-plane because the contribution from the semicircle vanishes as $R \to \infty$. After evaluating the residues from the two poles in the lower half-plane $-a\pm bi$,

$$f(t) = e^{bt}\frac{a\cos(at) - b\sin(at)}{4ab(a^2 + b^2)} \qquad (2.2.15)$$

for $t < 0$. Therefore, the inverse is

$$f(t) = e^{-b|t|}\frac{a\cos(a|t|) + b\sin(a|t|)}{4ab(a^2 + b^2)}. \qquad (2.2.16)$$

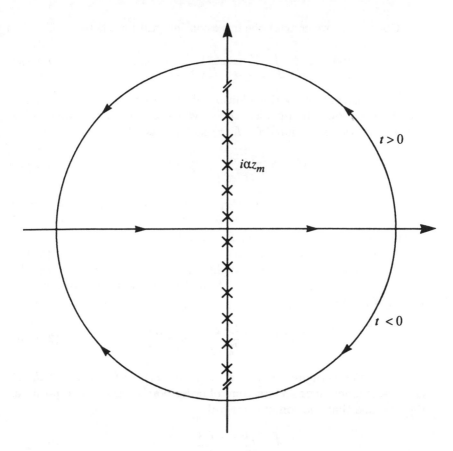

Fig. 2.2.2: Contour used to find the inverse of the Fourier transform (2.2.17)

• **Example 2.2.2**

Let us find the inverse to the transform[19]

$$F(\omega) = -\frac{k + \omega}{\omega} \frac{I_n(\omega/\alpha)}{I_n'(\omega/\alpha)} \qquad (2.2.17)$$

where $I_n(z)$ is a modified Bessel function of the first kind of order n, k and α are constants. We will assume that k and n are nonzero.

---

[19] Taken from Matsuzaki, Y. and Y. C. Fung, 1977: Unsteady fluid dynamic forces on a simply-supported circular cylinder of finite length conveying a flow, with applications to stability analysis. *J. Sound Vib.*, **54**, 317–330. Published by Academic Press Ltd., London, U.K.

Transform Methods for Solving Partial Differential Equations

Once again we convert the inversion integral for $t > 0$

$$f(t) = -\frac{1}{2\pi} \int_{-\infty}^{\infty} \frac{k + \omega}{\omega} \frac{I_n(\omega/\alpha)}{I_n'(\omega/\alpha)} e^{i\omega t} d\omega \qquad (2.2.18)$$

into a closed contour by the addition of an infinite semicircle in either the upper or lower half-plane. To see which one is correct we employ the asymptotic expansion[20] for $I_n(z)$ as $|z| \to \infty$. Because

$$\left| e^{izt} \frac{k + z}{z} \frac{I_n(z/\alpha)}{I_n'(z/\alpha)} \right| \to e^{-Rt \sin(\theta)} \to 0 \qquad (2.2.19)$$

as $R \to \infty$, $t > 0$ and $0 < \theta < \pi$, we close the contour in the upper half-plane. A similar argument shows that the contour should be closed in the lower plane for $t < 0$. See Fig. 2.2.2.

A possible source of difficulty is the presence of z in the denominator of the integral because it might introduce a singularity along the contour of integration. However, from the power series expansion of $I_n(z)$ (Watson's book, p. 77),

$$e^{izt} \frac{k + z}{z} \frac{I_n(z/\alpha)}{I_n'(z/\alpha)} \sim e^{izt} \frac{k}{n} \qquad (2.2.20)$$

as $z \to 0$. Furthermore, because $I_n'(\pm ix) = (\pm i)^{n-1} J_n'(x)$, where $J_n(z)$ is a Bessel function of the first kind of order n and x is real and positive, we conclude that the contour integral

$$\oint e^{izt} \frac{k + z}{z} \frac{I_n(z/\alpha)}{I_n'(z/\alpha)} dz \qquad (2.2.21)$$

has only simple poles. They lie along the imaginary axis at the locations $\omega = \pm i\alpha z_m$ where z_m is the mth root of $J_n'(z_m) = 0$ excluding $z_0 = 0$ and $m = 1, 2, 3, \ldots$ Consequently, the value of the closed contour equals $2\pi i$ times the sum of the residues. This leads to

$$f(t) = \sum_{m=1}^{\infty} \frac{i\alpha z_m + k}{z_m} \frac{J_n(z_m)}{J_n''(z_m)} e^{-\alpha z_m t} \qquad (2.2.22)$$

for $t > 0$. A similar analysis yields the inverse

$$f(t) = \sum_{m=1}^{\infty} \frac{k - i\alpha z_m}{z_m} \frac{J_n(z_m)}{J_n''(z_m)} e^{\alpha z_m t} \qquad (2.2.23)$$

for $t < 0$ when we close the contour in the lower half-plane.

---

[20] See Watson, G. N., 1966: *Theory of Bessel Functions*. Cambridge (England): At the University Press, p. 203.

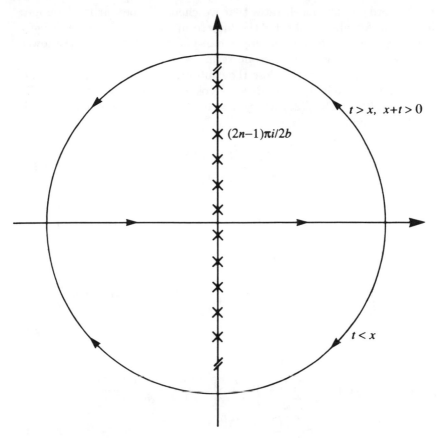

Fig. 2.2.3: Contour used to find the inverse of the Fourier transform (2.2.24)

• **Example 2.2.3**

Next we find the inverse of

$$F(\omega) = \frac{\cos(\omega x)}{\cosh(\omega b)} \tag{2.2.24}$$

when x and t are positive. From the inversion integral,

$$f(t) = \frac{1}{2\pi} \int_{-\infty}^{\infty} \frac{\cos(\omega x)}{\cosh(\omega b)} e^{i\omega t} d\omega \tag{2.2.25}$$

$$= \frac{1}{4\pi} \int_{-\infty}^{\infty} \left[\frac{e^{i\omega(t+x)}}{\cosh(\omega b)} + \frac{e^{i\omega(t-x)}}{\cosh(\omega b)} \right] d\omega \tag{2.2.26}$$

after rewriting $\cos(\omega x)$ in terms of complex exponentials. To convert (2.2.26) into a contour integral, we close the line integral with a semicircle in either the upper or lower half-plane. See Fig. 2.2.3. For x and

$t > 0$, Jordan's lemma dictates that we choose a contour in the upper half-plane for all x and t for the first term in (2.2.26). Alternatively Jordan's lemma requires the upper half-plane for $t > x$ and the lower half-plane for $t < x$ for the second term.

Regardless of how we close the contour, the poles of the integrand lie along the imaginary axis. Because $\cosh(\omega b) = \cos(i\omega b)$, the poles lie at $z_n = \pm(2n + 1)\pi i/(2b)$ where $n = 0, 1, 2, 3, \ldots$ Consequently for all positive x and t,

$$\frac{1}{2\pi i} \oint_C \frac{e^{iz(t+x)}}{\cosh(zb)} \, dz = \sum_{n=0}^{\infty} \frac{\exp\{-\frac{(2n+1)\pi(x+t)}{2b}\}}{b \sinh\{\frac{i(2n+1)\pi}{2}\}} \qquad (2.2.27)$$

$$= \frac{1}{bi} \sum_{n=0}^{\infty} (-1)^n \exp\left\{-\frac{(2n + 1)\pi(t + x)}{2b}\right\}. \qquad (2.2.28)$$

Similarly for the second term in (2.2.26),

$$\frac{1}{2\pi i} \oint_C \frac{e^{iz(t-x)}}{\cosh(zb)} \, dz = \frac{1}{bi} \sum_{n=0}^{\infty} (-1)^n \exp\left\{-\frac{(2n + 1)\pi|t - x|}{2b}\right\}. \qquad (2.2.29)$$

Therefore,

$$f(t) = \frac{1}{2b} \sum_{n=0}^{\infty} (-1)^n \left\{ \exp\left[-\frac{(2n + 1)\pi(t + x)}{2b}\right] \right.$$
$$\left. + \exp\left[-\frac{(2n + 1)\pi|t - x|}{2b}\right] \right\}. \qquad (2.2.30)$$

● **Example 2.2.4**

For our next example, we find the inverse[21] of the generalized Fourier transform

$$F(\omega) = \frac{ia}{\omega(\omega^2 - ia\omega - a)}. \qquad (2.2.31)$$

---

[21] The appearance of singularities along the real axis of the ω-plane indicates that we have a generalized Fourier transform. Usually the contour passes just below the real axis of the ω-plane [see (1.2.7)]. However, the only strictly valid method for determining whether we should pass above or below a singularity on the real axis consists of introducing a small amount of friction in the original formulation of the problem. This results in the singularity moving off the real axis.

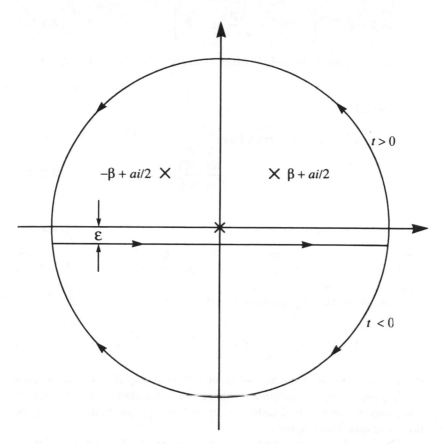

Fig. 2.2.4: Contour used to find the inverse of the Fourier transform (2.2.31)

We can convert the inversion integral

$$f(t) = \frac{1}{2\pi} \int_{-\infty-\epsilon i}^{\infty-\epsilon i} F(\omega) e^{i\omega t} d\omega, \qquad 0 < \epsilon << 1 \qquad (2.2.32)$$

into the closed contour integral

$$f(t) = \frac{1}{2\pi} \oint_C \frac{ia}{z(z^2 - iaz + a)} e^{izt} dz \qquad (2.2.33)$$

by adding an infinite semicircle in the upper half-plane as required by Jordan's lemma for $t > 0$ and then applying the residue theorem. Three simple poles lie within the contour: $z = 0$ and $z = \pm\beta + ia/2$ where $\beta = \sqrt{a - a^2/4}$. Employing the residue theorem,

$$f(t) = 1 - \frac{ae^{-at/2}}{2\beta(\beta^2 + \frac{a^2}{4})}\left[\left(\beta - \frac{ia}{2}\right)e^{i\beta t} + \left(\beta + \frac{ia}{2}\right)e^{-i\beta t}\right] \qquad (2.2.34)$$

$$= 1 - e^{-at/2} \left[\cos(\beta t) + \frac{2\beta}{4 - a} \sin(\beta t) \right] \tag{2.2.35}$$

where we have used the definition of β to simplify the resulting expression.

• **Example 2.2.5**

Let us now invert the transform

$$F(\omega) = \frac{\sinh(\zeta)}{\zeta \cosh(\zeta)} \tag{2.2.36}$$

where

$$\zeta^2 = (\omega - i\delta)^2 - k^2, \quad 0 < \delta \ll 1. \tag{2.2.37}$$

The poles of the transform are located at $\zeta = 0$ and $\zeta_n = \pm(2n + 1)\pi i/2$ where $n = 0, 1, 2, 3, \ldots$ Closer examination revels that $\zeta = 0$ is a removable singularity.

Solving for the ω_n associated with each ζ_n,

$$\omega_n^\pm = \pm\sqrt{k^2 + \zeta_n^2} + i\delta = \pm\sqrt{k^2 - \frac{(2n+1)^2\pi^2}{4}} + i\delta. \tag{2.2.38}$$

From (2.2.38) we see that for sufficiently large n, say $N+1$, ω_n becomes purely imaginary. Consequently there are a number of poles that lie along and above the real axis, i.e., $n = 0$ to N; the vast majority lie along the imaginary axis.

To find the inverse for $t > 0$, we convert the line integral into a closed contour by adding an infinite semicircle in the upper half-plane. Therefore,

$$f(t) = \frac{1}{2\pi} \int_{-\infty}^{\infty} \frac{\sinh(\zeta)}{\zeta \cosh(\zeta)} e^{i\omega t} d\omega = \frac{1}{2\pi} \oint_C \frac{\sinh(\zeta)}{\zeta \cosh(\zeta)} e^{izt} dz. \tag{2.2.39}$$

Upon applying the residue theorem,

$$
\begin{aligned}
f(t) = {} & i \sum_{n=0}^{N} \frac{\sinh(\zeta)e^{-\delta t}}{\frac{d}{d\zeta}[\zeta \cosh(\zeta)]\frac{d\zeta}{dz}|_{z=\omega_n^+}} \exp\left[it\sqrt{k^2 - \frac{(2n+1)^2\pi^2}{4}}\right] \\
& + i \sum_{n=0}^{N} \frac{\sinh(\zeta)e^{-\delta t}}{\frac{d}{d\zeta}[\zeta \cosh(\zeta)]\frac{d\zeta}{dz}|_{z=\omega_n^-}} \exp\left[-it\sqrt{k^2 - \frac{(2n+1)^2\pi^2}{4}}\right] \\
& + i \sum_{n=N+1}^{\infty} \frac{\sinh(\zeta)e^{-\delta t}}{\frac{d}{d\zeta}[\zeta \cosh(\zeta)]\frac{d\zeta}{dz}|_{z=\omega_n^+}} \exp\left[-t\sqrt{\frac{(2n+1)^2\pi^2}{4} - k^2}\right]
\end{aligned}
\tag{2.2.40}
$$

or

$$f(t) = -2 \sum_{n=0}^{N} \frac{e^{-\delta t} \sin\left[t\sqrt{k^2 - \frac{(2n+1)^2\pi^2}{4}}\right]}{\sqrt{k^2 - \frac{(2n+1)^2\pi^2}{4}}}$$

$$+ \sum_{n=N+1}^{\infty} \frac{\exp\left[-\delta t - t\sqrt{\frac{(2n+1)^2\pi^2}{4} - k^2}\right]}{\sqrt{\frac{(2n+1)^2\pi^2}{4} - k^2}}. \qquad (2.2.41)$$

Problems

By taking the appropriate closed contour, find the inverse of the following Fourier transform by contour integration. The parameters a, h and x are real and positive. In problem 8 replace the sine by its complex form and consider the three cases $t > h/2$, $-h/2 < t < h/2$ and $t < -h/2$.

1. $\dfrac{1}{\omega^2 - ia^2\omega}$

2. $\dfrac{1}{\omega^2 + a^2}$

3. $\dfrac{\omega}{\omega^2 + a^2}$

4. $\dfrac{\omega}{(\omega^2 + a^2)^2}$

5. $\dfrac{\omega^2}{(\omega^2 + a^2)^2}$

6. $\dfrac{1}{\omega^2 - 3i\omega - 3}$

7. $\dfrac{1}{(\omega - ia)^{2n+2}}$

8. $\dfrac{2i\sin(\omega h/2)}{\omega^2 + a^2}$

9. $\dfrac{\omega^2}{(\omega^2 - 1)^2 + 4a^2\omega^2}$

10. $\dfrac{1}{I_0(\omega)}$

11. $\dfrac{\cosh(\omega x)}{i\sinh(\omega h)}$

12. $\dfrac{\sin(\omega/a)}{\omega\cos(\omega h/a)}$

13. $\dfrac{m\pi \sinh(\omega a)}{2\omega(m^2\pi^2 - \omega^2)\cosh(\omega a)}$
$\times [1 + (-1)^m \cos(\omega) - i(-1)^m \sin(\omega)]$

14. By replacing $\cos(\omega x)$ with its complex equivalent and taking the appropriate closed contours depending upon whether $x \geq t > 0$ or $0 \leq x < t$, find the inverse of Fourier transform

$$F(\omega) = \frac{2\cos(\omega x)}{\omega i \cosh(\omega a)} - \frac{1}{\omega i}.$$

15. Use the residue theorem to invert the Fourier transform[22]

$$F(\omega) = -\frac{Pa \exp[-\omega^2 b^2 (r-a)/(2c^3)]}{r\rho c(\omega + B - Ai)(\omega - B - Ai)}$$

where a, A, b, B, c, P, r and ρ are constant and positive, and $r > a$.

16. If $P = \sqrt{-\lambda^2 - i\omega}$ and λ is a real constant, show that the inverse[23] of the Fourier transforms

$$F(\omega) = \frac{P \cot(P) + \lambda}{2\lambda P \cot(P) - P^2 + \lambda^2} \quad \text{and} \quad G(\omega) = \frac{1}{P \cot(P) + \lambda}$$

is

$$f(t) = 2H(t) \sum_{n=1}^{\infty} \frac{\overline{\omega}_n - \lambda^2}{\overline{\omega}_n + 2\lambda} e^{-\overline{\omega}_n t}$$

where $\overline{\omega}_n = P_n^2 + \lambda^2$ and $2\lambda \cot(P_n) - P_n^2 + \lambda^2 = 0$ and

$$g(t) = 2H(t) \sum_{n=1}^{\infty} \frac{\overline{\omega}_n - \lambda^2}{\overline{\omega}_n + \lambda} e^{-\overline{\omega}_n t}$$

where $\overline{\omega}_n = P_n^2 + \lambda^2$ and $P_n \cot(P_n) + \lambda = 0$.

17. Find the inverse[24] of Fourier transform

$$F(\omega) = \frac{\cosh(\zeta)}{\zeta \sinh(\zeta)}$$

when $t > 0$ where

$$\zeta^2 = (\omega - i\delta)^2 - k^2, \quad 0 < \delta \ll 1.$$

---

[22] Taken from Chakraborty, S. K., 1961: Disturbances in a viscoelastic medium due to impulsive forces on a spherical cavity. *Geofisica pura e appl.*, **48**, 23–26. Published by Birkhäuser Verlag, Basel, Switzerland.

[23] Taken from Goldman, M. M., and D. V. Fitterman, 1987: Direct time-domain calculation of the transient response for a rectangular loop over a two-layer medium. *Geophysics*, **52**, 997–1006.

[24] A similar problem appeared in Abrahams, I. D., 1982: Scattering of sound by finite elastic surfaces bounding ducts or cavities near resonance. *Quart. J. Mech. Appl. Math.*, **35**, 91–101. Taken by permission of Oxford University Press.

18. Show that the inverse[25] of the Fourier transform

$$F(\omega) = \frac{\sqrt{\omega i}}{\sinh(d\sqrt{\omega i})}$$

is

$$f(t) = -\frac{1}{2d^3}\sum_{n=1}^{\infty}(-1)^n n^2 \exp(-n^2\pi^2 t/d^2)$$

if $t > 0$; $f(t) = 0$ if $t < 0$.

19. Consider the transform[26]

$$F(\omega) = \frac{\cosh(\omega h)}{(U\omega - i\delta)^2 \cosh(\omega h) - g\omega\sinh(\omega h)}$$

with $U^2 < gh$, $0 < \delta << 1$ and U and g are constants.

(a) Using a perturbation expansion, show that the poles are at

$$\omega_I = \frac{i\delta}{U - \sqrt{gh}} + O(\delta^2)$$

$$\omega_{II} = \frac{i\delta}{U + \sqrt{gh}} + O(\delta^2)$$

$$\omega_{III} = \omega_0 - \frac{2i\delta U\omega_0}{g[\omega_0 h \operatorname{sech}^2(\omega_0 h) - \tanh(\omega_0 h)]} + O(\delta^2)$$

$$\omega_{IV} = -\omega_0 - \frac{2i\delta U\omega_0}{g[\omega_0 h \operatorname{sech}^2(\omega_0 h) - \tanh(\omega_0 h)]} + O(\delta^2)$$

where $\tanh(\omega_0 h) = U^2\omega_0/g$.

(b) Because $\omega_0 h \operatorname{sech}^2(\omega_0 h) - \tanh(\omega_0 h)$ is negative, the last three poles lie above the real axis whereas the first pole lies below it. By taking the appropriate contour, find the inverse correct to $O(\delta)$.

20. Find the inverse Fourier transform[27] of

$$F(\omega) = -\frac{iU\omega + \epsilon}{(U\omega - i\epsilon)^2[\omega^2 + (nN/U\kappa)^2] - N^2\omega^2}$$

---

[25] Taken from Stern, R. B., 1988: Time domain calculation of electric field penetration through metallic shields. *IEEE Trans. Electromag. Compat.*, **EC-30**, 307–311. ©1988 IEEE.

[26] Taken from Newman, J. N., 1976: Blockage with a free surface. *J. Ship Res.*, **20**, 199–203.

[27] Taken from Janowitz, G. S., 1981: Stratified flow over a bounded obstacle in a channel. *J. Fluid Mech.*, **110**, 161–170. Reprinted with the permission of Cambridge University Press.

as $\epsilon \to 0$. The constants U, N, κ and n are real and positive. Consider the two cases of $\kappa < n$ and $n > \kappa$.

(a) Using a perturbation expansion, show that the poles are at

$$\omega_I = \frac{n\epsilon i/U\kappa}{n/\kappa - 1} + O(\epsilon^2)$$

$$\omega_{II} = \frac{n\epsilon i/U\kappa}{n/\kappa + 1} + O(\epsilon^2)$$

$$\omega_{III} = \omega_0 - \frac{\epsilon i}{U(n^2/\kappa^2 - 1)} + O(\epsilon^2)$$

$$\omega_{IV} = -\omega_0 - \frac{\epsilon i}{U(n^2/\kappa^2 - 1)} + O(\epsilon^2)$$

where $\omega_0 = N\sqrt{1 - (n/\kappa)^2}/U$.

(b) Show that if $n/\kappa < 1$, the inverse is

$$f(t) = \frac{U}{2N^2} \left\{ \frac{1}{(n/\kappa)(1 + n/\kappa)} + \frac{2\cos[Nt\sqrt{1 - (n/\kappa)^2}/U]}{1 - (n/\kappa)^2} \right\}$$

for $t > 0$ and

$$f(t) = \frac{U}{2N^2(n/\kappa)(1 - n/\kappa)}$$

for $t < 0$, while if $n/\kappa > 1$, the inverse is

$$f(t) = \frac{U}{2N^2} \left\{ \frac{2 - \exp[-Nt\sqrt{(n/\kappa)^2 - 1}/U]}{(n/\kappa)^2 - 1} \right\}$$

for $t > 0$ and

$$f(t) = \frac{U}{2N^2} \frac{\exp[Nt\sqrt{(n/\kappa)^2 - 1}/U]}{(n/\kappa)^2 - 1}$$

for $t < 0$.

2.3 THE SOLUTION OF PARTIAL DIFFERENTIAL EQUATIONS BY LAPLACE TRANSFORMS

Having developed inversion techniques, we now use Laplace transforms in the general solution of partial differential equations. To illustrate this technique, we find the temperature reached in the interface between a brake drum and its lining during a single braking[28]. Ignoring the errors introduced in replacing the cylindrical portion of the drum

---

[28] Newcomb, T. P., 1958: The flow of heat in a parallel-faced infinite solid. *Brit. J. Appl. Phys.*, **9**, 370–372.

by a flat rectangular plate, we model the rate of heat generation during a constant deceleration as $N(1 - Mt)$ and ignore the radiation of heat away from the surface.

Consider the linear flow of heat in a solid initially at temperature zero and bounded by a pair of infinite parallel planes at $x = 0$ and $x = L$. At $x = 0$ there is no flow of heat perpendicular to the plane. At $x = L$ there is a flux $N(1 - Mt)$ into the solid. If $u(x,t)$, κ and a^2 denote the temperature, thermal conductivity and diffusivity, respectively, then the heat equation is

$$\frac{\partial u}{\partial t} = a^2 \frac{\partial^2 u}{\partial x^2}, \qquad 0 < x < L \tag{2.3.1}$$

with the boundary conditions

$$\frac{\partial u(0, t)}{\partial x} = 0, \qquad t > 0 \tag{2.3.2}$$

and

$$\kappa \frac{\partial u(L, t)}{\partial x} = N(1 - Mt), \qquad t > 0. \tag{2.3.3}$$

Introducing the Laplace transform of $u(x,t)$, defined as

$$\bar{u}(x, s) = \int_0^\infty u(x, t) e^{-st} dt, \tag{2.3.4}$$

the equation to be solved becomes

$$\frac{d^2 \bar{u}}{dx^2} - \frac{s}{a^2} \bar{u} = 0, \tag{2.3.5}$$

subject to the boundary conditions that

$$\frac{d\bar{u}(0, s)}{dx} = 0, \tag{2.3.6}$$

and

$$\frac{d\bar{u}(L, s)}{dx} = \frac{N}{\kappa} \left(\frac{1}{s} - \frac{M}{s^2} \right). \tag{2.3.7}$$

The solution of (2.3.5) is

$$\bar{u}(x, s) = A \cosh(qx) + B \sinh(qx) \tag{2.3.8}$$

where $q = s^{1/2}/a$. Using the boundary conditions, the solution becomes

$$\bar{u}(x, s) = \frac{N}{\kappa} \left(\frac{1}{s} - \frac{M}{s^2} \right) \frac{\cosh(qx)}{q \sinh(qL)}. \tag{2.3.9}$$

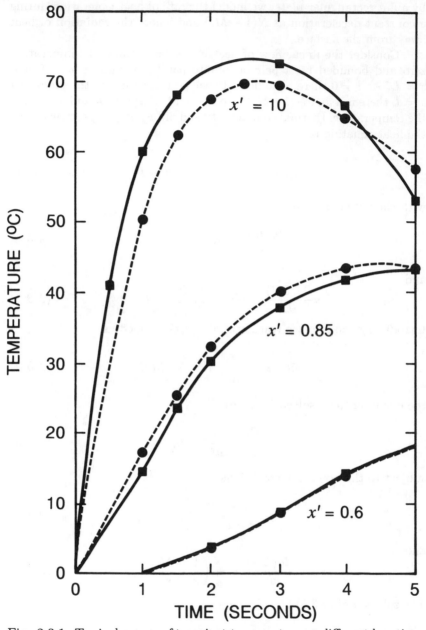

Fig. 2.3.1: Typical curves of transient temperature at different locations in a brake lining. Circles denote computed values while squares are experimental measurements. Taken from Newcomb, T. P., 1958: The flow of heat in a parallel-faced infinite solid. *Brit. J. Appl. Phys.*, **9**, 372.

It now remains to invert the transform (2.3.9). We inverted the first portion $\cosh(qx)/[sq\sinh(qL)]$ in Example 2.1.2. We may invert the second portion in a similar manner. Consequently, the inverse of (2.3.9) is

$$
u(x,t) = \frac{a^2 N}{\kappa L}\left\{ t + \frac{x^2}{2a^2} - \frac{L^2}{6a^2}\right\}
$$
$$
- \frac{2LN}{\kappa\pi^2}\sum_{n=1}^{\infty}\frac{(-1)^n}{n^2}\cos(n\pi x/L)e^{-n^2\pi^2 a^2 t/L^2}
$$
$$
- \frac{a^2 NM}{\kappa L}\left\{ \frac{t^2}{2} + \frac{tx^2}{2a^2} - \frac{tL^2}{6a^2} + \frac{x^4}{24a^4} - \frac{x^2 L^2}{12a^4} + \frac{7L^4}{360a^4}\right\}
$$
$$
- \frac{2L^3 NM}{a^2\kappa\pi^4}\sum_{n=1}^{\infty}\frac{(-1)^n}{n^4}\cos(n\pi x/L)e^{-n^2\pi^2 a^2 t/L^2}. \qquad (2.3.10)
$$

Fig. 2.3.1 shows the temperature in the brake lining at various places within the lining $[x' = x/L]$ if $a^2 = 3.3\times 10^{-3}$ cm^2/sec, $\kappa = 1.8\times 10^{-3}$ cal/(cm sec °C), $L = 0.48$ cm and $N = 1.96$ cal/(cm^2 sec).

For our next example, we use Laplace transforms to solve the wave equation in the semi-infinite domain $0 < x < \infty$. Within this domain, there are two layers in which the physical properties differ. The governing equations are

$$
\frac{\partial^2 u_1}{\partial x^2} = \frac{1}{c_1^2}\frac{\partial^2 u_1}{\partial t^2}, \qquad 0 < x < a \qquad (2.3.11)
$$

and

$$
\frac{\partial^2 u_2}{\partial x^2} = \frac{1}{c_2^2}\frac{\partial^2 u_2}{\partial t^2} + \kappa_0\kappa_2\delta(x-\xi)\delta(t), \qquad a < x < \infty \qquad (2.3.12)
$$

where $\xi > a$. The boundary conditions are

$$
u_1(0,t) = 0 \qquad (2.3.13)
$$

and

$$
u_1(a,t) = u_2(a,t) \quad \text{and} \quad \frac{1}{\kappa_1}\frac{\partial u_1(a,t)}{\partial x} = \frac{1}{\kappa_2}\frac{\partial u_2(a,t)}{\partial x}. \qquad (2.3.14)
$$

This system of equations arises in finding the TM-waves inside and outside of a one-dimensional dielectric slab[29]. The system is initially at rest.

---

[29] Veselov, G. I., A. I. Kirpa and N. I. Platonov, 1986: Transient-field representation in terms of steady-improper solutions. *IEE Proc., Part H*, **133**, 21–25.

Transform Methods for Solving Partial Differential Equations

Taking the Laplace transform of (2.3.11)–(2.3.14),

$$\frac{d^2\bar{u}_1}{dx^2} - \frac{s^2}{c_1^2}\bar{u}_1 = 0 \qquad (2.3.15)$$

$$\frac{d^2\bar{u}_2}{dx^2} - \frac{s^2}{c_2^2}\bar{u}_2 = \kappa_0\kappa_2\delta(x-\xi) \qquad (2.3.16)$$

along with

$$\bar{u}_1(0,s) = 0, \qquad (2.3.17)$$

$$\bar{u}_1(a,s) = \bar{u}_2(a,s) \qquad (2.3.18)$$

and

$$\frac{1}{\kappa_1}\frac{d\bar{u}_1(a,s)}{dx} = \frac{1}{\kappa_2}\frac{d\bar{u}_2(a,s)}{dx}. \qquad (2.3.19)$$

The difficulty in solving this set of differential equations is the presence of the delta function in (2.3.16). We avoid this difficulty by further breaking down the region $a < x < \infty$ into two additional regions $a < x < \xi$ and $\xi < x < \infty$. At $x = \xi$, we integrate (2.3.16) across a very narrow strip from ξ^- to ξ^+. This gives the additional conditions that

$$\bar{u}_2(\xi^-,s) = \bar{u}_2(\xi^+,s) \quad \text{and} \quad \frac{d\bar{u}_2(\xi^+,s)}{dx} - \frac{d\bar{u}_2(\xi^-,s)}{dx} = \kappa_0\kappa_2. \quad (2.3.20)$$

Solutions which satisfy the differential equations (2.3.15)–(2.3.16) plus the boundary conditions (2.3.17)–(2.3.20) are

$$\bar{u}_1(x,s) = -\frac{s\kappa_0\exp[-s(\xi-a)/c_2]\sinh(sx/c_1)}{\Delta(s)}, \quad 0 < x < a \quad (2.3.21)$$

and

$$\begin{aligned}
\bar{u}_2(x,s) = &-\frac{\kappa_0\kappa_2 c_2}{2s}\exp\left(-\frac{s}{c_2}|x-\xi|\right) \\
&-\frac{\kappa_0\kappa_2 c_2}{2s}\exp\left[-\frac{s(\xi-a)}{c_2}\right]\exp\left[-\frac{s(x-a)}{c_2}\right] \\
&\times\left[\frac{2s\sinh(sa/c_1)}{\kappa_2 c_2\Delta(s)} - 1\right], \qquad a < x < \infty
\end{aligned} \qquad (2.3.22)$$

where

$$\Delta(s) = \frac{s}{\kappa_2 c_2}\sinh\left(\frac{sa}{c_1}\right) + \frac{s}{\kappa_1 c_1}\cosh\left(\frac{sa}{c_1}\right). \qquad (2.3.23)$$

We invert the first term of (2.3.22) by inspection and it equals

$$u_2^{(d)} = -\frac{\kappa_0\kappa_2 c_2}{2}H\left(t - \frac{|x-\xi|}{c_2}\right). \qquad (2.3.24)$$

This portion of the solution represents the direct wave emitted from $x = \xi$. The remaining terms contain simple poles which are the zeros of $s^{-1}w(s)$. A little algebra gives these zeros as

$$s_n = \frac{c_1}{2a} \ln \left(\frac{1-\rho}{1+\rho} \right) + i\frac{n\pi c_1}{2a} \qquad (2.3.25)$$

where $\rho = \kappa_1 c_1 / \kappa_2 c_2$, $n = \pm 1, \pm 3, \ldots$ for $0 < \rho < 1$ and $n = 0, \pm 2, \pm 4,$ $\ldots$ for $1 < \rho < \infty$. Furthermore, the second term in $\bar{u}_2(x, s)$ contains a simple pole at $s = 0$. Preforming the inversion using Bromwich's integral, the $s = 0$ pole yields

$$u_2^{(s)} = \frac{\kappa_0 \kappa_2 c_2}{2} H \left(t - \frac{x-a}{c_2} + \frac{\xi-a}{c_2} \right). \qquad (2.3.26)$$

This is a steady-state field set up in region 2 from the reflection of the direct wave from the interface at $x = a$.

The poles at $s = s_n$ give the transient solutions to the problem. Applying the residue theorem in conjunction with Bromwich's integral,

$$u_1^{(t)} = -\frac{\kappa_0 \kappa_1 c_1^2}{a(1-\rho^2)} H \left(t + \frac{x-a}{c_1} + \frac{\xi-a}{c_2} \right)$$
$$\times \left\{ \sum_n \frac{\sinh(s_n x/c_1) \exp[s_n t - s_n(\xi - a)/c_2]}{s_n \sinh(s_n a/c_1)} \right\} \qquad (2.3.27)$$

and

$$u_2^{(t)} = -\frac{\kappa_0 \kappa_1 c_1^2}{a(1-\rho^2)} H \left(t - \frac{x-a}{c_2} - \frac{\xi-a}{c_2} \right)$$
$$\times \left\{ \sum_n \frac{\exp[s_n t - s_n(\xi - a)/c_2 - s_n(x - a)/c_2]}{s_n} \right\}. \qquad (2.3.28)$$

The solution $u_1^{(t)}(x, t)$ gives the total field in region 1 while the total field within region 2 equals the sum of $u_2^{(d)}$, $u_2^{(s)}$ and $u_2^{(t)}$.

In addition to providing a nice example of solving the wave equation by Laplace transforms, our problem also foreshadows some future results concerning open, radiating structures, such as an antenna above an dielectric slab. Since the turn of the twentieth century, when it became evident that radio waves could be used effectively for long distance wireless communications, physicists and electrical engineers have sought to understand the radiation field of antennas near boundaries. Generally, they assumed a temporal behavior of $e^{i\omega t}$ and reduced the wave equation to a spatial one (the Helmholtz equation) where ω appears as another parameter. Theoretically, we could find the exact response by adding up all of the contributions at different frequencies.

Transform Methods for Solving Partial Differential Equations

In solving the Helmholtz equation, two classes of solutions were found; some, the "proper" modes, decayed exponentially as $|x| \to \infty$ while the others, the "improper" modes, grew exponentially. Consequently, these improper modes were always discarded. However, in our problem, the transient solution $u_2^{(t)}$ acts like an improper mode because it behaves as $\exp[-s_n(x-a)/c_2]$ with $\text{Re}(s_n) < 0$. Why is our solution valid? Partly because the Heaviside function only allows nonzero solutions for $x - a < c_2 t - \xi + a$. Thus, the combined temporal and spatial dependence $\exp\{s_n[t - (x-a)/c_2]\}$ always remains finite as $x \to \infty$ and the term "improper" is really a misnomer. Our problem suggests that improper modes might be useful in expressing the radiation field in more complicated problem. We will explore this possibility in Section 4.4

For our final example, we use Laplace transforms to solve a partial differential equation in cylindrical coordinates. It differs from our previous problems in its use of Bessel functions.

Let us solve

$$\frac{\partial u}{\partial t} = \frac{\partial^2 u}{\partial r^2} + \frac{1}{r}\frac{\partial u}{\partial r} - \frac{u}{r^2}, \quad 0 < r < 1, t > 0 \tag{2.3.29}$$

subject to the boundary conditions that $u(0,t) = 0$ and $u(1,t) = H(t)$ and the initial condition that $u(r,0) = 0$. Introducing the Laplace transform of $u(r,t)$,

$$\bar{u}(r,s) = \int_0^\infty u(r,t)e^{-st}dt, \tag{2.3.30}$$

(2.3.29) becomes

$$r^2\frac{d^2\bar{u}}{dr^2} + r\frac{d\bar{u}}{dr} - (r^2 s + 1)\bar{u} = 0 \tag{2.3.31}$$

with the associated boundary conditions $\bar{u}(0,s) = 0$ and $\bar{u}(1,s) = 1/s$. The general solution to (2.3.31) is

$$\bar{u}(r,s) = AI_1(r\sqrt{s}) + BK_1(r\sqrt{s}) \tag{2.3.32}$$

where $I_1(\ )$ and $K_1(\ )$ are first-order, modified Bessel functions of the first and second kind, respectively. Because $K_1(\ )$ becomes unbounded at $r = 0$, we discard it and the boundary condition $\bar{u}(1,s) = 1/s$ leads to

$$\bar{u}(r,s) = \frac{I_1(r\sqrt{s})}{sI_1(\sqrt{s})}. \tag{2.3.33}$$

We now invert $\bar{u}(r,s)$. Because the first few terms of the power series for $I_1(z)$ are

$$I_1(z) = \frac{z}{2} + \frac{z^3}{16} + \frac{z^5}{384} + \cdots, \tag{2.3.34}$$

we find upon substituting this power series into (2.3.33) that

$$\bar{u}(r, s) = \frac{r[\frac{1}{2} + \frac{r^2 s}{16} + \frac{r^4 s^2}{384} + \cdots]}{s[\frac{1}{2} + \frac{s}{16} + \frac{s^2}{384} + \cdots]}. \qquad (2.3.35)$$

Consequently, $\bar{u}(r, s)$ is single-valued function and has a simple pole at $s = 0$. The remaining poles are at $I_1(\sqrt{s}) = 0$. If $\sqrt{s} = -\alpha i$, then $I_1(-\alpha i) = -i J_1(\alpha) = 0$. Denoting the zeros of $J_1(\ )$, the first-order Bessel function of the first kind, by α_n, then additional poles are located along the negative real axis at $s_n = -\alpha_n^2$ with $n = 1, 2, 3, \ldots$

Having located the poles of $\bar{u}(r, s)$, the inverse $u(r, t)$ is

$$u(r, t) = \frac{1}{2\pi i} \oint_C \frac{I_1(r\sqrt{z})e^{tz}}{z I_1(\sqrt{z})} \, dz \qquad (2.3.36)$$

where C consists of a line parallel to the imaginary axis, but slightly to the right of it, and a semicircle of infinite radius in the left side of the complex plane. From (2.3.35) the residue at $z = 0$ is r. The remaining residues are

$$\text{Res}(z = -\alpha_n^2) = \lim_{z \to -\alpha_n^2} \frac{(z + \alpha_n^2)I_1(r\sqrt{z})e^{tz}}{z I_1(\sqrt{z})} \qquad (2.3.37)$$

$$= \frac{2 I_1(-\alpha_n r i)}{(-\alpha_n i) I_1'(\alpha_n i)} e^{-\alpha_n^2 t} \qquad (2.3.38)$$

$$= \frac{2 J_1(\alpha_n r)}{\alpha_n J_0(\alpha_n)} e^{-\alpha_n^2 t} \qquad (2.3.39)$$

because $2 I_1'(z) = I_0(z) + I_2(z)$, $I_0(-i\alpha_n) = J_0(\alpha_n)$ and $I_2(-i\alpha_n) = -J_2(\alpha_n)$. Furthermore, we have $J_2(\alpha_n) = -J_0(\alpha_n)$ because $J_2(\alpha_n) + J_0(\alpha_n) = 2J_1(\alpha_n)/\alpha_n = 0$. Consequently, upon adding up all of the residues, the final solution is

$$u(r, t) = r + 2 \sum_{n=1}^{\infty} \frac{J_1(\alpha_n r)}{\alpha_n J_0(\alpha_n)} e^{-\alpha_n^2 t} \qquad (2.3.40)$$

where $J_1(\alpha_n) = 0$ with $n = 1, 2, 3, \ldots$

Problems

1. Solve the nonhomogeneous heat equation

$$\frac{\partial u}{\partial t} - \frac{\partial^2 u}{\partial x^2} = 1, \qquad 0 < x < 1, t > 0$$

with the boundary conditions $u(0,t) = u(1,t) = 0$ and the initial condition $u(x,0) = 0$.

Step 1. Show that the Laplace transform of the partial differential equation and boundary conditions yields the ordinary differential equation:

$$\frac{d^2\overline{u}(x,s)}{dx^2} - s\overline{u}(x,s) = -\frac{1}{s}, \qquad \overline{u}(0,s) = 0, \qquad \overline{u}(1,s) = 0.$$

Step 2. Show that the solution to step 1 is

$$\overline{u}(x,s) = \frac{1 - \cosh(x\sqrt{s})}{s^2} + \frac{[\cosh(\sqrt{s}) - 1]\sinh(x\sqrt{s})}{s^2 \sinh(\sqrt{s})}.$$

Step 3. Show that the singularities in the Laplace transform are located at $s_n = -n^2\pi^2$ where $n = 0, 1, 2, 3, \ldots$ and are *all* simple poles.

Step 4. Use Bromwich's integral to invert $\overline{u}(x,s)$ and show that

$$u(x,t) = \frac{x(1-x)}{2} - \frac{4}{\pi^3} \sum_{n=1}^{\infty} \frac{\sin[(2n-1)\pi x]}{(2n-1)^3} e^{-(2n-1)^2 t}.$$

2. Solve the heat equation

$$\frac{\partial u}{\partial t} = \frac{\partial^2 u}{\partial x^2}, \qquad 0 < x < 1, t > 0$$

with the boundary conditions $u(0,t) = 0$, $u(1,t) = 1$ and the initial condition $u(x,0) = 0$.

Step 1. Show that the Laplace transform of the partial differential equation and boundary conditions yields the ordinary differential equation:

$$\frac{d^2\overline{u}(x,s)}{dx^2} - s\overline{u}(x,s) = 0, \qquad \overline{u}(0,s) = 0, \qquad \overline{u}(1,s) = \frac{1}{s}.$$

Step 2. Show that the solution to step 1 is

$$\overline{u}(x,s) = \frac{\sinh(x\sqrt{s})}{s \sinh(\sqrt{s})}.$$

Step 3. Show that the singularities in the Laplace transform are located at $s_n = -n^2\pi^2$ where $n = 0, 1, 2, 3, \ldots$ and are *all* simple poles.

Step 4. Use Bromwich's integral to invert $\bar{u}(x, s)$ and show that

$$u(x, t) = x + \frac{2}{\pi} \sum_{n=1}^{\infty} \frac{(-1)^n}{n} \sin(n\pi x) e^{-n^2\pi^2 t}.$$

3. Solve the heat equation

$$\frac{\partial u}{\partial t} = \frac{\partial^2 u}{\partial x^2}, \qquad 0 < x < 1, t > 0$$

with the boundary conditions $u_x(0, t) = 0$, $u(1, t) = t$ and the initial condition $u(x, 0) = 0$.

Step 1. Show that the Laplace transform of the partial differential equation and boundary conditions yields the ordinary differential equation:

$$\frac{d^2\bar{u}(x, s)}{dx^2} - s\bar{u}(x, s) = 0, \qquad \bar{u}'(0, s) = 0, \qquad \bar{u}(1, s) = \frac{1}{s^2}.$$

Step 2. Show that the solution to step 1 is

$$\bar{u}(x, s) = \frac{\cosh(x\sqrt{s})}{s^2 \cosh(\sqrt{s})}.$$

Step 3. Show that the singularities in the Laplace transform consist of a second-order pole located at $s = 0$ and simple poles located at $s_n = -(2n - 1)^2\pi^2/4$ where $n = 1, 2, 3, \ldots$.

Step 4. Use Bromwich's integral to invert $\bar{u}(x, s)$ and show that

$$u(x, t) = t + \frac{x^2 - 1}{2} + \frac{32}{\pi^2} \sum_{n=1}^{\infty} \frac{(-1)^n}{(2n - 1)^3}$$
$$\times \cos\left[\frac{(2n - 1)\pi x}{2}\right] \exp\left[-\frac{(2n - 1)^2\pi^2 t}{4}\right].$$

4. Solve the partial differential equation[30]

$$\frac{\partial u}{\partial t} = \frac{\partial^2 u}{\partial x^2} - a^2 u, \qquad -1 < x < 1, t > 0$$

---

[30] Taken from Jacobs, C., 1971: Transient motions produced by disks oscillating torsionally about a state of rigid rotation. *Quart. J. Mech. Appl. Math.*, **24**, 221–236. By permission of Oxford University Press.

with the boundary conditions $u(-1, t) = 1$, $u(1, t) = 0$ and the initial condition $u(x, 0) = 0$.

Step 1. Show that the Laplace transform of the partial differential equation and boundary conditions yields the ordinary differential equation:

$$\frac{d^2\bar{u}(x, s)}{dx^2} - (a^2 + s)\bar{u}(x, s) = 0, \qquad \bar{u}(-1, s) = \frac{1}{s}, \qquad \bar{u}(1, s) = 0.$$

Step 2. Show that the solution to step 1 is

$$\bar{u}(x, s) = \frac{\sinh[(1 - x)\sqrt{a^2 + s}]}{s \sinh(2\sqrt{a^2 + s})}.$$

Step 3. Show that the singularities in the Laplace transform consist of simple poles located at $s = 0$ and $s_n = -a^2 - n^2\pi^2/4$ where $n = 1, 2, 3, \ldots$.

Step 4. Use Bromwich's integral to invert $\bar{u}(x, s)$ and show that

$$u(x, t) = \frac{\sinh[a(1 - x)]}{\sinh(2a)} + \frac{\pi}{2} \sum_{n=1}^{\infty} \frac{(-1)^n n}{a^2 + n^2\pi^2/4}$$

$$\times \sin\left[\frac{n\pi(1 - x)}{2}\right] e^{-(a^2 + n^2\pi^2/4)t}.$$

5. Solve the nonhomogeneous partial differential equation[31]

$$\frac{\partial u}{\partial t} = b + \frac{\partial^2 u}{\partial x^2} - m^2 u, \qquad -1 < x < 1, t > 0$$

with the boundary conditions $u(-1, t) = u(1, t) = 0$ and the initial condition $u(x, 0) = 0$.

Step 1. Show that the Laplace transform of the partial differential equation and boundary conditions yields the ordinary differential equation:

$$\frac{d^2\bar{u}(x, s)}{dx^2} - (m^2 + s)\bar{u}(x, s) = -\frac{b}{s}$$

[31] Taken from Ram, G., and R. S. Mishra, 1977: Unsteady flow through magnetohydrodynamic porous media. *Indian J. Pure Appl. Math.*, 8, 637–647.

with $\bar{u}(-1, s) = 0$ and $\bar{u}(1, s) = 0$.

Step 2. Show that the solution to step 1 is

$$\bar{u}(x, s) = \frac{b}{s(s + m^2)} \left[1 - \frac{\cosh(x\sqrt{s + m^2})}{\cosh(\sqrt{s + m^2})} \right].$$

Step 3. Show that the singularities in the Laplace transform consist of simple poles located at $s = 0$, $s = -m^2$ and $s_n = -m^2 - (2n - 1)^2\pi^2/4$ where $n = 1, 2, 3, \ldots$.

Step 4. Use Bromwich's integral to invert $\bar{u}(x, s)$ and show that

$$u(x, t) = \frac{b}{m^2} \left[1 - \frac{\cosh(mx)}{\cosh(m)} \right]$$

$$- \frac{16b}{\pi} \sum_{n=1}^{\infty} \frac{(-1)^n}{(2n - 1)[(2n - 1)^2\pi^2 + 4m^2]} \cos[(2n - 1)\pi x/2]$$

$$\times \exp\{-[m^2 + (2n - 1)^2\pi^2/4]t\}.$$

6. Solve the nonhomogeneous partial differential equation

$$\frac{\partial u}{\partial t} = \frac{\partial^2 u}{\partial x^2} - b^2 u + \cos^2\left(\frac{\pi x}{a}\right), \qquad -a/2 < x < a/2, t > 0$$

with the boundary conditions $u(-a/2, t) = u(a/2, t) = 0$ and the initial condition $u(x, 0) = 0$.

Step 1. Show that the Laplace transform of the partial differential equation and boundary conditions yields the ordinary differential equation:

$$\frac{d^2\bar{u}(x, s)}{dx^2} - (b^2 + s)\bar{u}(x, s) = -\frac{1}{2s}\left[1 + \cos\left(\frac{2\pi x}{a}\right)\right]$$

with $\bar{u}(-a/2, s) = 0$ and $\bar{u}(a/2, s) = 0$.

Step 2. Show that the solution to step 1 is

$$\bar{u}(x, s) = \frac{1}{2s(s + b^2)}\left[1 - \frac{\cosh(x\sqrt{s + b^2})}{\cosh(a\sqrt{s + b^2}/2)}\right]$$

$$+ \frac{1}{2s(s + b^2 + 4\pi^2/b^2)}\left[\cos\left(\frac{2\pi x}{a}\right) + \frac{\cosh(x\sqrt{s + b^2})}{\cosh(a\sqrt{s + b^2}/2)}\right].$$

Step 3. Show that the singularities in the Laplace transform consist of simple poles located at $s = 0$, $s = -b^2$, $s = -b^2 - 4\pi^2/a^2$ and $s_n = -b^2 - (2n-1)^2\pi^2/a^2$ where $n = 1, 2, 3, \ldots$.

Step 4. Use Bromwich's integral to invert $\bar{u}(x, s)$ and show that

$$u(x,t) = \frac{b^2 + 2\pi^2/a^2}{b^2(b^2 + 4\pi^2/a^2)} \cos^2\left(\frac{\pi x}{a}\right)$$

$$+ \frac{2\pi^2}{a^2 b^2(b^2 + 4\pi^2/a^2)} \left[\sin^2\left(\frac{\pi x}{a}\right) - \frac{\cosh(bx)}{\cosh(ab/2)}\right]$$

$$- \frac{8a^2}{\pi} \sum_{n=1}^{\infty} \frac{(-1)^n}{(2n-1)[a^2 b^2 + (2n-1)^2\pi^2][(2n-1)^2 - 4]}$$

$$\times \cos[(2n-1)\pi x/a] \exp\{-[b^2 + (2n-1)^2\pi^2/a^2]t\}.$$

7. Solve the wave equation

$$\frac{\partial^2 u}{\partial t^2} = \frac{\partial^2 u}{\partial x^2}, \qquad 0 < x < 1, t > 0$$

with the boundary conditions $u(0,t) = 0$, $u_x(1,t) = \delta(t)$ and the initial conditions $u(x,0) = u_t(x,0) = 0$.

Step 1. Show that the Laplace transform of the partial differential equation and boundary conditions yields the ordinary differential equation:

$$\frac{d^2\bar{u}(x,s)}{dx^2} - s^2\bar{u}(x,s) = 0, \qquad \bar{u}(0,s) = 0, \qquad \bar{u}'(1,s) = 1.$$

Step 2. Show that the solution to step 1 is

$$\bar{u}(x,s) = \frac{\sinh(sx)}{s\,\cosh(s)}.$$

Step 3. Show that the singularities in the Laplace transform are located at $s_n = \pm(2n-1)\pi i/2$ where $n = 1, 2, 3, \ldots$ and are *all* simple poles.

Step 4. Use Bromwich's integral to invert $\bar{u}(x, s)$ and show that

$$u(x,t) = \frac{4}{\pi} \sum_{n=1}^{\infty} \frac{(-1)^{n+1}}{2n-1} \sin\left[\frac{(2n-1)\pi x}{2}\right] \sin\left[\frac{(2n-1)\pi t}{2}\right].$$

8. Solve the wave equation

$$\frac{\partial^2 u}{\partial t^2} = \frac{\partial^2 u}{\partial x^2}, \qquad 0 < x < 1, t > 0$$

with the boundary conditions $u_x(0, t) = H(t)$, $u(1, t) = 0$ and the initial conditions $u(x, 0) = u_t(x, 0) = 0$.

Step 1. Show that the Laplace transform of the partial differential equation and boundary conditions yields the ordinary differential equation:

$$\frac{d^2\bar{u}(x, s)}{dx^2} - s^2\bar{u}(x, s) = 0, \qquad \bar{u}'(0, s) = \frac{1}{s}, \qquad \bar{u}(1, s) = 0.$$

Step 2. Show that the solution to step 1 is

$$\bar{u}(x, s) = \frac{\sinh[s(x - 1)]}{s^2 \cosh(s)}.$$

Step 3. Show that the singularities in the Laplace transform are located at $s = 0$ and $s_n = \pm(2n-1)\pi i/2$ where $n = 1, 2, 3, \ldots$ and are *all* simple poles.

Step 4. Use Bromwich's integral to invert $\bar{u}(x, s)$ and show that

$$u(x, t) = x - 1 + \frac{8}{\pi^2} \sum_{n=1}^{\infty} \frac{(-1)^n}{(2n-1)^2} \sin\left[\frac{(2n-1)\pi(x-1)}{2}\right]$$
$$\times \cos\left[\frac{(2n-1)\pi t}{2}\right].$$

9. Solve the wave equation

$$\frac{\partial^2 u}{\partial t^2} = \frac{\partial^2 u}{\partial x^2}, \qquad 0 < x < 1, t > 0$$

with the boundary conditions $u_x(0, t) = 0$, $u(1, t) = H(t)$ and the initial conditions $u(x, 0) = x$, $u_t(x, 0) = 0$.

Step 1. Show that the Laplace transform of the partial differential equation and boundary conditions yields the ordinary differential equation:

$$\frac{d^2\bar{u}(x, s)}{dx^2} - s^2\bar{u}(x, s) = -sx, \qquad \bar{u}'(0, s) = 0, \qquad \bar{u}(1, s) = \frac{1}{s}.$$

Transform Methods for Solving Partial Differential Equations

Step 2. Show that the solution to step 1 is

$$\bar{u}(x, s) = \frac{x}{s} + \frac{\sinh[s(1 - x)]}{s^2 \cosh(s)}.$$

Step 3. Show that the singularities in the Laplace transform are located at $s = 0$ and $s_n = \pm(2n-1)\pi i/2$ where $n = 1, 2, 3, \ldots$ and are *all* simple poles.

Step 4. Use Bromwich's integral to invert $\bar{u}(x, s)$ and show that

$$u(x, t) = 1 + \frac{8}{\pi^2} \sum_{n=1}^{\infty} \frac{(-1)^n}{(2n-1)^2} \sin\left[\frac{(2n-1)\pi(1-x)}{2}\right]$$
$$\times \cos\left[\frac{(2n-1)\pi t}{2}\right].$$

10. Solve the wave equation

$$\frac{\partial^2 u}{\partial t^2} = \frac{\partial^2 u}{\partial x^2}, \qquad 0 < x < 1, t > 0$$

with the boundary conditions $u(0, t) = 0$, $u_x(1, t) = \cos(\omega t)$ and the initial conditions $u(x, 0) = u_t(x, 0) = 0$. Assume that $\omega \neq (2n-1)\pi/2$ where $n = 1, 2, 3, \ldots$

Step 1. Show that the Laplace transform of the partial differential equation and boundary conditions yields the ordinary differential equation:

$$\frac{d^2\bar{u}(x, s)}{dx^2} - s^2\bar{u}(x, s) = 0, \qquad \bar{u}(0, s) = 0, \qquad \bar{u}'(1, s) = \frac{s}{s^2 + \omega^2}.$$

Step 2. Show that the solution to step 1 is

$$\bar{u}(x, s) = \frac{\sinh(sx)}{(s^2 + \omega^2) \cosh(s)}.$$

Step 3. Show that the singularities in the Laplace transform are located at $s_n = \pm(2n-1)\pi i/2$ where $n = 1, 2, 3, \ldots$ and $s = \pm\omega i$ and that they are *all* simple poles.

Step 4. Use Bromwich's integral to invert $\bar{u}(x,s)$ and show that

$$u(x,t) = \frac{\sin(\omega x)\cos(\omega t)}{\omega \cos(\omega)} - 2\sum_{n=1}^{\infty} \frac{(-1)^n}{\omega^2 - [(2n-1)^2\pi^2/4]}$$

$$\times \sin\left[\frac{(2n-1)\pi x}{2}\right] \cos\left[\frac{(2n-1)\pi t}{2}\right].$$

11. Solve the wave equation

$$\frac{\partial^2 u}{\partial t^2} = \frac{\partial^2 u}{\partial x^2}, \qquad 0 < x < 1, t > 0$$

with the boundary conditions $u_x(0,t) = \sin(\omega t)$, $u(1,t) = 0$ and the initial conditions $u(x,0) = u_t(x,0) = 0$. Assume that $\omega \neq (2n-1)\pi/2$ where $n = 1, 2, 3, \ldots$ Why?

Step 1. Show that the Laplace transform of the partial differential equation and boundary conditions yields the ordinary differential equation:

$$\frac{d^2\bar{u}(x,s)}{dx^2} - s^2\bar{u}(x,s) = 0, \qquad \bar{u}'(0,s) = \frac{\omega}{s^2 + \omega^2}, \qquad \bar{u}(1,s) = 0.$$

Step 2. Show that the solution to step 1 is

$$\bar{u}(x,s) = \frac{\omega \sinh[s(x-1)]}{s(s^2 + \omega^2)\cosh(s)}.$$

Step 3. Show that the singularities in the Laplace transform are located at $s_n = \pm(2n-1)\pi i/2$ where $n = 1, 2, 3, \ldots$ and $s = \pm\omega i$ and that they are *all* simple poles.

Step 4. Use Bromwich's integral to invert $\bar{u}(x,s)$ and show that

$$u(x,t) = \frac{\sin[\omega(x-1)]\sin(\omega t)}{\omega\cos(\omega)}$$

$$+ 2\omega\sum_{n=1}^{\infty} \frac{(-1)^{n+1}}{[(2n-1)\pi/2]\{\omega^2 - [(2n-1)^2\pi^2/4]\}}$$

$$\times \sin\left[\frac{(2n-1)\pi(x-1)}{2}\right] \sin\left[\frac{(2n-1)\pi t}{2}\right]$$

$$= 2\sum_{n=1}^{\infty} \frac{(-1)^{n+1}}{[(2n-1)\pi/2]\{\omega^2 - [(2n-1)^2\pi^2/4]\}}$$

$$\times \sin\left[\frac{(2n-1)\pi(x-1)}{2}\right]$$

$$\times \left\{\omega\sin\left[\frac{(2n-1)\pi t}{2}\right] - \frac{(2n-1)\pi}{2}\sin(\omega t)\right\}.$$

12. Solve the wave equation

$$\frac{\partial^2 u}{\partial t^2} = \frac{\partial^2 u}{\partial x^2}, \qquad 0 < x < 1, t > 0$$

with the boundary conditions $u_x(0, t) = 1 - \cos(\omega t)$, $u(1, t) = 0$ and the initial conditions $u(x, 0) = u_t(x, 0) = 0$. Assume that $\omega \neq (2n - 1)\pi/2$ where $n = 1, 2, 3, \ldots$ Why?

Step 1. Show that the Laplace transform of the partial differential equation and boundary conditions yields the ordinary differential equation:

$$\frac{d^2\overline{u}(x, s)}{dx^2} - s^2\overline{u}(x, s) = 0, \ \ \overline{u}'(0, s) = \frac{1}{s} - \frac{s}{s^2 + \omega^2}, \ \ \overline{u}(1, s) = 0.$$

Step 2. Show that the solution to step 1 is

$$\overline{u}(x, s) = \left(\frac{1}{s} - \frac{s}{s^2 + \omega^2}\right) \frac{\sinh[s(x - 1)]}{s\cosh(s)}.$$

Step 3. Show that the singularities in the Laplace transform are located at $s_n = \pm(2n - 1)\pi i/2$ where $n = 1, 2, 3, \ldots$, $s = 0$ and $s = \pm\omega i$ and that they are *all* simple poles.

Step 4. Use Bromwich's integral to invert $\overline{u}(x, s)$ and show that

$$u(x, t) = x - 1 - \frac{\sin[\omega(x - 1)]\cos(\omega t)}{\omega\cos(\omega)}$$

$$+ \frac{8}{\pi^2}\sum_{n=1}^{\infty}\frac{(-1)^n}{(2n - 1)^2}$$

$$\times \sin\left[\frac{(2n - 1)\pi(x - 1)}{2}\right]\cos\left[\frac{(2n - 1)\pi t}{2}\right]$$

$$- 2\sum_{n=1}^{\infty}\frac{(-1)^{n+1}}{\omega^2 - [(2n - 1)^2\pi^2/4]}$$

$$\times \sin\left[\frac{(2n - 1)\pi(x - 1)}{2}\right]\cos\left[\frac{(2n - 1)\pi t}{2}\right]$$

or

$$u(x, t) = x - 1 - \frac{\sin[\omega(x - 1)] \cos(\omega t)}{\omega \cos(\omega)}$$

$$+ \frac{8}{\pi^2} \sum_{n=1}^{\infty} \frac{(-1)^n}{(2n - 1)^2}$$

$$\times \sin\left[\frac{(2n - 1)\pi(x - 1)}{2}\right] \cos\left[\frac{(2n - 1)\pi t}{2}\right]$$

$$+ 2 \sum_{n=1}^{\infty} \frac{(-1)^{n+1}}{\omega^2 - [(2n - 1)^2\pi^2/4]} \sin\left[\frac{(2n - 1)\pi(x - 1)}{2}\right]$$

$$\times \left\{\cos(\omega t) - \cos\left[\frac{(2n - 1)\pi t}{2}\right]\right\}.$$

13. Solve the wave equation[32]

$$\frac{\partial^2 u}{\partial t^2} = c^2 \frac{\partial^2 u}{\partial x^2}, \qquad 0 < x < L, t > 0$$

with the boundary conditions $u(0, t) = 0$ and

$$M\frac{\partial^2 u(L, t)}{\partial t^2} - kL\frac{\partial u(L, t)}{\partial x} = -k[1 - \cos(\omega t)]$$

and the initial conditions $u(x, 0) = u_t(x, 0) = 0$.

Step 1. Show that the Laplace transform of the partial differential equation yields the ordinary differential equation:

$$\frac{d^2\bar{u}(x, s)}{dx^2} - \frac{s^2}{c^2}\bar{u}(x, s) = 0,$$

$$\bar{u}(0, s) = 0, \quad s^2 M\bar{u}(L, s) - kL\frac{d\bar{u}(L, s)}{dx} = -\frac{k\omega^2}{s(s^2 + \omega^2)}.$$

Step 2. Show that the solution to step 1 is

$$\bar{u}(x, s) = \frac{\beta c\omega^2}{L} \frac{\sinh(sx/c)}{s^2(s^2 + \omega^2)[\beta \cosh(sL/c) - (sL/c)\sinh(sL/c)]}$$

---

[32] Taken from Durant, N. J., 1960: Stress in a dynamically loaded helical spring. *Quart. J. Mech. Appl. Math.*, **13**, 251–256. By permission of Oxford University Press.

where $\beta = kL^2/Mc^2$.

Step 3. Show that the singularities in the Laplace transform are located at $s = 0$, $s = \pm wi$ and $\alpha_n \tan(\alpha_n) = -\beta$ with $s_n = \pm ic\alpha_n/L$ where $n = 0, 1, 2, 3, \ldots$ and are *all* simple poles.

Step 4. Use Bromwich's integral to invert $\bar{u}(x, s)$ and show that

$$u(x,t) = \frac{x}{L} - \frac{\beta c}{Lw} \frac{\sin(wx/c)\cos(wt)}{\beta \cos(wL/c) + (wL/c)\sin(wL/c)}$$
$$- \frac{2\beta w^2 L^2}{c^2} \sum_{n=1}^{\infty} \frac{\sin(\alpha_n x/L)\cos(\alpha_n ct/L)}{\alpha_n[\alpha_n^2 - (wL/c)^2](\beta^2 + \alpha_n^2 - \beta)\cos(\alpha_n)}.$$

Assume that $\alpha_n \neq wL/c$. Why?

14. Solve the wave equation[33]

$$\frac{\partial^2 (ru)}{\partial t^2} = c^2 \frac{\partial^2 (ru)}{\partial r^2}, \qquad a < r < \infty, t > 0$$

with the boundary condition that

$$-\rho c^2 \left(\frac{\partial^2 u}{\partial r^2} + \frac{2}{3r} \frac{\partial u}{\partial r} \right)\bigg|_{r=a} = p_0 e^{-\alpha t} H(t)$$

where $\alpha > 0$ and $\lim_{r \to \infty} u(r, t) \to 0$. The initial conditions are $u(r, 0) = u_t(r, 0) = 0$.

Step 1. Show that the Laplace transform of the partial differential equation yields the ordinary differential equation:

$$\frac{d^2[r\bar{u}(r, s)]}{dr^2} - \frac{s^2}{c^2}[r\bar{u}(r, s)] = 0, \qquad a < r < \infty$$

with

$$-\rho c^2 \left[\frac{d^2\bar{u}(a, s)}{dr^2} + \frac{2}{3a} \frac{d\bar{u}(a, s)}{dr} \right] = \frac{p_0}{s + \alpha}$$

as well as $\lim_{r \to \infty} \bar{u}(r, s) \to 0$.

Step 2. Show that the solution to step 1 is

$$\bar{u}(r, s) = -\frac{a p_0 \exp[-s(r - a)/c]}{\rho r(s + \alpha)[s^2 + 4sc/(3a) + 4c^2/(3a^2)]}.$$

---

[33] See Sharpe, J. A., 1942: The production of elastic waves by explosion pressures. I. Theory and empirical field observations. *Geophysics*, **7**, 144–154.

Step 3. Show that the singularities in the Laplace transform are simple poles located at $s = -\alpha$ and $s = -2c/(3a) \pm 2\sqrt{2}ci/(3a)$.

Step 4. Use Bromwich's integral to invert $\bar{u}(r, s)$ and show that

$$
u(r, t) = \frac{ap_0}{\rho r[(\beta/\sqrt{2} - \alpha)^2 + \beta^2]} \left\{ e^{-\beta\tau/\sqrt{2}} \left[\left(\frac{1}{\sqrt{2}} - \frac{\alpha}{\beta} \right) \sin(\beta\tau) \right. \right.
$$
$$
\left. \left. + \cos(\beta\tau) \right] - e^{-\alpha\tau} \right\} H(\tau)
$$

where $\tau = t - (r - a)/c$ and $\beta = 2\sqrt{2}c/(3a)$.

15. Solve the wave equation[34]

$$
\frac{\partial^2(ru)}{\partial t^2} = c^2 \frac{\partial^2(ru)}{\partial r^2}, \qquad a < r < \infty, t > 0
$$

with the boundary condition that

$$
-\rho c^2 \left(\frac{\partial^2 u}{\partial r^2} + \frac{2}{3r} \frac{\partial u}{\partial r} \right) \bigg|_{r=a} = p_0 \sin(\omega t)[H(t) - H(t - \pi/\omega)]
$$

and $\lim_{r \to \infty} u(r, t) \to 0$. The initial conditions are $u(r, 0) = u_t(r, 0) = 0$.

Step 1. Show that the Laplace transform of the partial differential equation yields the ordinary differential equation:

$$
\frac{d^2[r\bar{u}(r, s)]}{dr^2} - \frac{s^2}{c^2}[r\bar{u}(r, s)] = 0, \qquad a < r < \infty
$$

with

$$
\rho c^2 \left[\frac{d^2\bar{u}(a, s)}{dr^2} + \frac{2}{3a} \frac{d\bar{u}(a, s)}{dr} \right] = -p_0 \frac{\omega}{s^2 + \omega^2} \left[1 + e^{-s\pi/\omega} \right]
$$

as well as $\lim_{r \to \infty} \bar{u}(r, s) \to 0$.

Step 2. Show that the solution to step 1 is

$$
\bar{u}(r, s) = -\frac{awp_0 \exp[-s(r - a)/c]}{\rho r(s^2 + \omega^2)[s^2 + 4sc/(3a) + 4c^2/(3a^2)]} \left[1 + e^{-s\pi/\omega} \right].
$$

---

[34] See Ghosh, S. S., 1969: On the disturbances in a viscoelastic medium due to blast inside a spherical cavity. *PAGEOPH*, **75**, 93–97.

Step 3. Show that the singularities in the Laplace transform are simple poles located at $s = \pm \omega i$ and $s = -\delta \pm \beta i$ where $\delta = 2c/(3a)$ and $\beta = 2\sqrt{2}c/(3a)$.

Step 4. Show that the inverse of

$$F(s) = -\frac{a\omega p_0}{\rho r(s^2 + \omega^2)[s^2 + 4sc/(3a) + 4c^2/(3a^2)]}$$

is

$$f(t) = \frac{ap_0\omega}{\rho r[\omega^2 - 4c^2/(3a^2)]^2 + 16c^2\omega^2/(9a^2)}$$
$$\times \left[\left(\omega^2 - \frac{4c^2}{3a^2}\right)\frac{\sin(\omega t)}{\omega} + \frac{4c}{3a}\cos(\omega t)\right.$$
$$\left. - \left(\omega^2 - \frac{4c^2}{9a^2}\right)e^{-\delta t}\frac{\sin(\beta t)}{\beta} - \frac{4c}{3a}e^{-\delta t}\cos(\beta t)\right].$$

Step 5. Use the second shifting theorem to show that

$$u(r,t) = f(\tau)H(\tau) + f(\tau - \pi/\omega)H(\tau - \pi/\omega)$$

where $\tau = t - (r-a)/c$.

16. Solve the nonhomogeneous heat equation[35] in cylindrical coordinates

$$\frac{\partial u}{\partial t} - \frac{1}{r}\frac{\partial}{\partial r}\left(r\frac{\partial u}{\partial r}\right) = H(t), \qquad 0 < r < 1, t > 0$$

with the boundary conditions $\partial u(0,t)/\partial r = 0$ and $u(1,t) = 0$ and the initial condition $u(r,0) = 0$.

Step 1. Show that the Laplace transform of the partial differential equation yields the ordinary differential equation:

$$\frac{1}{r}\frac{d}{dr}\left(r\frac{d\bar{u}}{dr}\right) - s\bar{u} = -\frac{1}{s}$$

with $d\bar{u}(0,s)/dr = \bar{u}(1,s) = 0$.

---

[35] Taken from Achard, J. L., and G. M. Lespinard, 1981: Structure of the transient wall-friction law in one-dimensional models of laminar pipe flows. *J. Fluid Mech.*, **113**, 283–298. Reprinted with the permission of Cambridge University Press.

Step 2. Show that the solution to step 1 is

$$\bar{u}(r,s) = \frac{1}{s^2}\left[1 - \frac{I_0(r\sqrt{s})}{I_0(\sqrt{s})}\right]$$

where $I_0(\ )$ is the modified Bessel function of the first kind and order zero.

Step 3. Show that the singularities in the Laplace transform are located at $s = 0$ and $s_n = -\alpha_n^2$ with $\sqrt{s_n} = -\alpha_n i$ where α_n is the nth positive zero of the Bessel function $J_0(\ )$ and $n = 0, 1, 2, 3, \ldots$ and that they are *all* simple poles.

Step 4. Use Bromwich's integral to invert $\bar{u}(r,s)$ and show that

$$u(r,t) = \tfrac{1}{4}(1 - r^2) - 2\sum_{n=1}^{\infty} \frac{J_0(\alpha_n r)}{\alpha_n^3 J_1(\alpha_n)}e^{-\alpha_n^2 t}.$$

17. Solve the heat equation[36]

$$\frac{\partial u}{\partial t} = \frac{\partial^2 u}{\partial r^2} + \frac{1}{r}\frac{\partial u}{\partial r}, \qquad 0 \le r < a, t > 0$$

with the boundary conditions $|u(0,t)| < \infty$ and

$$\frac{\partial u(a,t)}{\partial r} = -H(t)$$

and the initial condition $u(r,0) = 1$.

Step 1. Show that the Laplace transform of the partial differential equation yields the ordinary differential equation:

$$\frac{d^2\bar{u}}{dr^2} + \frac{1}{r}\frac{d\bar{u}}{dr} - s\bar{u} = -1, \qquad 0 \le r < a$$

with $|\bar{u}(0,s)| < \infty$ and $\bar{u}'(a,s) = -1/s$.

Step 2. Show that the solution to step 1 is

$$\bar{u}(r,s) = \frac{1}{s} - \frac{I_0(r\sqrt{s})}{s^{3/2}I_1(a\sqrt{s})}$$

---

[36] Taken from Jaeger, J. C., 1944: Note on a problem in radial flow. *Proc. Phys. Soc.*, **56**, 197–203.

where $I_0(\ )$ and $I_1(\ )$ are the modified Bessel functions of the first kind and order zero and one, respectively.

Step 3. Show that the singularities in the Laplace transform are located at $s = 0$ and $s_n = -\alpha_n^2/a^2$ with $a\sqrt{s_n} = -\alpha_n i$ where α_n is the nth positive zero of the Bessel function $J_1(\ )$ and $n = 0, 1, 2, 3, \ldots$. The pole at $s = 0$ is second order while the remaining poles are simple.

Step 4. Use Bromwich's integral to invert $\overline{u}(r, s)$ and show that

$$u(r,t) = 1 - a\left[\frac{2t}{a^2} + \frac{r^2}{2a^2} - \frac{1}{4} - 2\sum_{n=1}^{\infty}\frac{J_0(\alpha_n r/a)}{\alpha_n^2 J_0(\alpha_n)}e^{-\alpha_n^2 t/a^2}\right].$$

18. Solve the heat equation[37]

$$\frac{\partial u}{\partial t} = \frac{\partial^2 u}{\partial r^2} + \frac{1}{r}\frac{\partial u}{\partial r}, \qquad 0 \leq r < a, t > 0$$

with the boundary conditions $|u(0,t)| < \infty$ and $u(a,t) = \cos(\omega t)\, H(t)$ and the initial condition $u(r,0) = 0$.

Step 1. By taking the Laplace transform of the partial differential equation and boundary conditions, show that they reduce to the ordinary differential equation

$$\frac{d^2\overline{u}}{dr^2} + \frac{1}{r}\frac{d\overline{u}}{dr} - s\overline{u} = 0, \qquad 0 \leq r < a$$

with the boundary conditions $|\overline{u}(0, s)| < \infty$ and

$$\overline{u}(a, s) = \frac{s}{s^2 + \omega^2}.$$

Step 2. Show that the solution to step 1 is

$$\overline{u}(r, s) = \frac{s}{s^2 + \omega^2}\frac{I_0(r\sqrt{s})}{I_0(a\sqrt{s})}$$

where $I_0(\ )$ is a zeroth-order, modified Bessel function of the first kind.

---

[37] Reprinted from *Int. J. Mech. Sci.*, **8**, P. G. Bhuta and L. R. Koval, A viscous ring damper for a freely precessing satellite, pp. 383–395, ©1966, with kind permission from Pergamon Press Ltd., Headington Hill Hall, Oxford OX3 0BW, UK.

Step 3. Show that the singularities of $\bar{u}(r, s)$ are the simple poles at $s = \pm \omega i$ and $s = -\alpha_n^2$ with $n = 1, 2, 3, \ldots$ where $J_0(\alpha_n a) = 0$ and $J_0( )$ is a Bessel function of the first kind and zeroth order.

Step 4. Use Bromwich's integral to invert $\bar{u}(r, s)$ and show that

$$u(r, t) = \frac{I_0(r\sqrt{\omega i})}{2I_0(a\sqrt{\omega i})} e^{i\omega t} + \frac{I_0(r\sqrt{-\omega i})}{2I_0(a\sqrt{-\omega i})} e^{-i\omega t}$$

$$- \frac{2}{a} \sum_{n=1}^{\infty} \frac{\alpha_n^3}{\alpha_n^4 + \omega^2} \frac{J_0(\alpha_n r)}{J_1(\alpha_n a)} e^{-\alpha_n^2 t}$$

where $J_1( )$ is a Bessel function of the first kind and first order.

19. Solve the heat equation[38]

$$\frac{\partial u}{\partial t} = \frac{\partial^2 u}{\partial r^2} + \frac{1}{r} \frac{\partial u}{\partial r} - \frac{u}{r^2}, \qquad 0 \leq r < 1, t > 0$$

with the boundary conditions $|u(0, t)| < \infty$ and $u(1, t) = \sin(\omega t)$ and the initial condition $u(r, 0) = 0$.

Step 1. By taking the Laplace transform of the partial differential equation and boundary conditions, show that they reduce to the ordinary differential equation

$$\frac{d^2 \bar{u}}{dr^2} + \frac{1}{r} \frac{d\bar{u}}{dr} - \left(s + \frac{1}{r^2} \right) \bar{u} = 0, \qquad 0 \leq r < 1$$

with the boundary conditions $|\bar{u}(0, s)| < \infty$ and

$$\bar{u}(1, s) = \frac{\omega}{s^2 + \omega^2}.$$

Step 2. Show that the solution to step 1 is

$$\bar{u}(r, s) = \frac{\omega}{s^2 + \omega^2} \frac{I_1(r\sqrt{s})}{I_1(\sqrt{s})}$$

where $I_1( )$ is a modified Bessel function of the first order and kind.

---

[38] Taken from Schwarz, W. H., 1963: The unsteady motion of an infinite oscillating cylinder in an incompressible Newtonian fluid at rest. *Appl. Sci. Res.*, **A11**, 115–124. Reprinted by permission of Kluwer Academic Publishers.

Step 3. Show that the singularities of $\bar{u}(r, s)$ are simple poles at $s = \pm\omega i$ and $s = -\alpha_n^2$ with $n = 1, 2, 3, \ldots$ where $J_1(\alpha_n) = 0$ and $J_1(\ )$ is a Bessel function of the first order and kind.

Step 4. Use Bromwich's integral to invert $\bar{u}(r, s)$ and show that

$$u(r, t) = \frac{M_1(r\sqrt{\omega})}{M_1(\sqrt{\omega})} \sin[\omega t + \theta_1(r\sqrt{\omega}) - \theta_1(\sqrt{\omega})]$$

$$+ 2\omega \sum_{n=1}^{\infty} \frac{\alpha_n}{\alpha_n^4 + \omega^2} \frac{J_1(\alpha_n r)}{J_0(\alpha_n)} e^{-\alpha_n^2 t}$$

where

$$M_1(z) = \sqrt{\mathrm{ber}_1^2(z) + \mathrm{bei}_1^2(z)}, \quad \theta_1(z) = \tan^{-1}[\mathrm{bei}_1(z)/\mathrm{ber}_1(z)],$$

$\mathrm{ber}_1(z)$, $\mathrm{bei}_1(z)$ are Kelvin's ber and bei functions, respectively, and $J_0(\ )$ and $J_1(\ )$ are Bessel functions of the first kind and zeroth and first order, respectively.

20. Solve the partial differential equation[39]

$$\epsilon \frac{\partial^2 u}{\partial t^2} + \frac{\partial u}{\partial t} = \frac{\partial^2 u}{\partial x^2}, \qquad 0 < x < \infty, t > 0$$

with $\epsilon > 0$, the boundary conditions $\lim_{x \to \infty} u(x, t) \to 0$ and

$$\frac{\partial u(0, t)}{\partial x} = -H(t) - \epsilon\delta(t)$$

and the initial conditions $u(x, 0) = u_t(x, 0) = 0$.

Step 1. By taking the Laplace transform of the partial differential equation and boundary conditions, show that they reduce to

$$\frac{d^2\bar{u}}{dx^2} - s(1 + \epsilon s)\bar{u} = 0$$

with the boundary conditions $\lim_{x \to \infty} \bar{u}(x, s) \to 0$ and

$$\bar{u}'(0, s) = -\epsilon - 1/s.$$

---

[39] Taken from Maurer, M. J., and H. A. Thompson, 1973: Non-Fourier effects at high heat flux. *J. Heat Transfer*, **95**, 284–286.

Step 2. Show that the solution to step 1 is

$$\bar{u}(x, s) = \sqrt{\epsilon}\left(1 + \frac{1}{\epsilon s}\right)\frac{\exp[-\sqrt{\epsilon x}\sqrt{s(s + 1/\epsilon)}]}{\sqrt{s(s + 1/\epsilon)}}.$$

Step 3. Use tables to invert the Laplace transform in step 2 and show that

$$u(x, t) = \sqrt{\epsilon}H(t - \sqrt{\epsilon x})\left\{e^{-t/2\epsilon}I_0\sqrt{\left(\frac{t}{2\epsilon}\right)^2 - \frac{x^2}{4\epsilon}}\right.$$

$$\left. + \frac{1}{\epsilon}\int_0^t e^{-\eta/2\epsilon}I_0\sqrt{\left(\frac{\eta}{2\epsilon}\right)^2 - \frac{x^2}{4\epsilon}}\,d\eta\right\}$$

where $I_0(\ )$ is a modified Bessel function of the first kind and zeroth order.

21. Solve the partial differential equation[40]

$$\frac{\partial^2 u}{\partial t^2} + 2\frac{\partial u}{\partial t} = \frac{\partial^2 u}{\partial x^2}, \qquad 0 < x < \infty, t > 0$$

with the boundary conditions

$$u(0, t) = 1 \qquad \text{and} \qquad \lim_{x \to \infty} u(x, t) \to 0$$

and the initial conditions $u(x, 0) = u_t(x, 0) = 0$.

Step 1. By taking the Laplace transform of the partial differential equation and boundary conditions, show that they reduce to

$$\frac{d^2\bar{u}}{dx^2} - s(s + 2)\bar{u} = 0$$

with the boundary conditions

$$\bar{u}(0, s) = \frac{1}{s} \qquad \text{and} \qquad \lim_{x \to \infty} \bar{u}(x, s) \to 0.$$

Step 2. Show that the solution to step 1 is

$$\bar{u}(x, s) = \frac{1}{s}\exp[-x\sqrt{s(s + 2)}].$$

---

[40] Taken from Baumeister, K. J., and T. D. Hamill, 1969: Hyperbolic heat-conduction equation – a solution for the semi-infinite body problem. *J. Heat Transfer*, **91**, 543–548.

Step 3. Show that

$$e^{-x\sqrt{s(s+2)}} = -\frac{d}{dx}\left\{\mathcal{L}\left[e^{-t}I_0(\sqrt{t^2-x^2})H(t-x)\right]\right\}$$

$$= \mathcal{L}\left[xe^{-t}\frac{I_1(\sqrt{t^2-x^2})}{\sqrt{t^2-x^2}}H(t-x)\right.$$

$$\left. + e^{-t}I_0(\sqrt{t^2-x^2})\delta(t-x)\right]$$

where $I_0(\ )$ and $I_1(\ )$ are modified Bessel functions of the first kind and zeroth and first order, respectively.

Step 4. Finish the problem by noting that

$$u(x,t) = \int_0^t \mathcal{L}^{-1}\left\{e^{-x\sqrt{s(s+2)}}\right\}d\tau$$

$$= \left\{e^{-x} + x\int_x^t e^{-\eta}\frac{I_1(\sqrt{\eta^2-x^2})}{\sqrt{\eta^2-x^2}}\,d\eta\right\}H(t-x).$$

22. Solve the nonhomogeneous heat equation[41]

$$\frac{\partial u}{\partial t} = \frac{1}{R}\frac{\partial^2 u}{\partial x^2} + P(1-e^{-\alpha t}), \qquad -1 < x < 1, t > 0$$

with the boundary conditions $u(-1,t) = 0$, $u(1,t) = 0$ and the initial condition $u(x,0) = 0$.

Step 1. Show that the Laplace transform of the partial differential equation yields the ordinary differential equation:

$$\frac{d^2\bar{u}(x,s)}{dx^2} - Rs\bar{u}(x,s) = -\frac{\alpha RP}{s(s+\alpha)}, \qquad \bar{u}(-1,s) = \bar{u}(1,s) = 0.$$

Step 2. Show that the solution to step 1 is

$$\bar{u}(x,s) = \frac{\alpha P}{s^2(s+\alpha)}\left[1 - \frac{\cosh(x\sqrt{Rs})}{\cosh(\sqrt{Rs})}\right].$$

---

[41] Taken from Prakash, S., 1967: Nonsteady parallel viscous flow through a straight channel. *Bull. Cal. Math. Soc.*, **59**, 55–59.

Step 3. Show that the singularities in the Laplace transform are located at $s = 0$, $s = -\alpha$ and $s_n = -(2n-1)^2\pi^2/4R$ where $n = 1, 2, 3, \ldots$ and are *all* simple poles.

Step 4. Use Bromwich's integral to invert $\overline{u}(x, s)$ and show that

$$u(x, t) = \frac{PR}{2}(1 - x^2) + \frac{P}{\alpha}\left[1 - \frac{\cos(x\sqrt{R\alpha})}{\cos(\sqrt{R\alpha})}\right]e^{-\alpha t} + \frac{64\alpha RP}{\pi^3}$$
$$\times \sum_{n=1}^{\infty} \frac{(-1)^n \cos[(2n-1)\pi x/2] \exp[-(2n-1)^2\pi^2 t/4R]}{(2n-1)^3[4\alpha - (2n-1)^2\pi^2/R]}.$$

23. Solve the partial differential equation[42]

$$\frac{\partial^2 u}{\partial x^2} - \frac{1}{c^2}\frac{\partial^2 u}{\partial t^2} - \frac{\omega_0^2}{c^2}u = 0, \qquad -\infty < x < \infty, t > 0$$

subject to the boundary conditions $\lim_{|x|\to\infty} u(x, t) \to 0$ and the initial conditions $u(x, 0) = u_0(x)$ and $u_t(x, 0) = u_1(x)$.

Step 1. By taking the Laplace transform of the partial differential equation and boundary conditions, show that they reduce to the ordinary differential equation

$$\frac{d^2\overline{u}}{dx^2} - \left(\frac{s^2 + \omega_0^2}{c^2}\right)\overline{u} = -\frac{s}{c^2}u_0(x) - \frac{1}{c^2}u_1(x)$$

with $\lim_{|x|\to\infty} \overline{u}(x, s) \to 0$.

Step 2. Show that the solution to step 1 is

$$\overline{u}(x, s) = \frac{1}{2c^2 q}\int_{-\infty}^{x} e^{-q(x-\eta)}[su_0(\eta) + u_1(\eta)]\,d\eta$$
$$+ \frac{1}{2c^2 q}\int_{x}^{\infty} e^{-q(\eta-x)}[su_0(\eta) + u_1(\eta)]\,d\eta$$

where $q^2 = (s^2 + \omega_0^2)/c^2$.

Step 3. Using the transforms that

$$\mathcal{L}\left[cJ_0(\omega_0\sqrt{t^2 - x^2/c^2})H(t - x/c)\right] = e^{-qx}/q, \qquad x > 0$$

---

[42] Taken from Jaeger, J. C., and K. C. Westfold, 1949: Transients in an ionized medium with applications to bursts of solar noise. *Australian J. Sci. Res.*, **A2**, 322–334.

and

$$\mathcal{L}\left\{\left[c\delta(t-x/c) - \frac{c\omega_0 t J_1(\omega_0\sqrt{t^2-x^2/c^2})}{\sqrt{t^2-x^2/c^2}}\right]H(t-x/c)\right\} = se^{-qx}/q$$

if $x > 0$, then show that

$$u(x,t) = \tfrac{1}{2}\left[u_0(x-ct) + u_0(x+ct)\right]$$

$$+ \frac{1}{2c}\int_{x-ct}^{x+ct}\left\{u_1(\eta)J_0[\omega_0\sqrt{t^2-(x-\eta)^2/c^2}]\right.$$

$$\left.- u_0(\eta)\frac{\omega_0 t J_1[\omega_0\sqrt{t^2-(x-\eta)^2/c^2}]}{\sqrt{t^2-(x-\eta)^2/c^2}}\right\}d\eta.$$

This corresponds to disturbances, each of half the original waveform, moving in opposite directions, and each followed by a tail.

24. Solve the partial differential equation[43]

$$\frac{\partial u}{\partial t} - \nu\frac{\partial^2 u}{\partial x^2} - l^2\frac{\partial^3 u}{\partial x^2 \partial t} + \nu l^2\frac{\partial^4 u}{\partial x^4} = 0, \qquad 0 < x < \infty, t > 0$$

with the boundary conditions $u(0,t) = V_0$, $\mu l^2\partial^2 u(0,t)/\partial x^2 = M_0$, $\lim_{x\to\infty} u(x,t) \to 0$ and $\lim_{x\to\infty}\mu l^2\partial^2 u(x,t)/\partial x^2 \to 0$ and the initial condition $u(x,0) = 0$.

Step 1. Show that the Laplace transform of the partial differential equation yields the ordinary differential equation:

$$\nu l^2\frac{d^4\overline{u}}{dx^4} - (\nu + l^2 s)\frac{d^2\overline{u}}{dx^2} + s\overline{u} = 0$$

with $\lim_{x\to\infty}\overline{u}(x,s) \to 0$, $\lim_{x\to\infty}\mu l^2\overline{u}''(x,s) \to 0$, $\overline{u}(0,s) = V_0/s$ and $\mu l^2\overline{u}''(0,s) = M_0/s$.

Step 2. Show that the solution to step 1 is

$$\overline{u}(x,s) = \left(\frac{V_0}{s} + \frac{a}{s} - \frac{a}{s-\nu/l^2}\right)\exp(-x/l)$$

$$- \left(\frac{a}{s} - \frac{a}{s-\nu/l^2}\right)\exp(x\sqrt{s/\nu})$$

where $a = M_0/\mu - V_0$.

---

[43] Taken from Guram, G. S., 1983: Flow of a dipolar fluid due to suddenly accelerated flat plate. *Acta Mech.*, **49**, 133–138.

Step 3. Using a table of Laplace transforms and the convolution theorem, show that $u(x, t)$ is

$$u(x, t) = \left(V_0 + a - ae^{\nu t/l^2}\right) \exp(-x/l) - a \ \mathrm{erfc}[x/(2\sqrt{\nu t}\,)]$$

$$+ \frac{ax}{2\sqrt{\nu\pi}} \exp(\nu t/l^2) \int_0^1 \frac{1}{\beta^{3/2}} \exp\left(-\frac{\nu\beta}{l^2} - \frac{x^2}{4\nu\beta}\right) d\beta$$

where erfc( ) is the complementary error function.

25. Solve the partial differential equations

$$\frac{\partial^2 u_1}{\partial t^2} = a_1^2 \frac{\partial^2 u_1}{\partial x^2} \qquad 0 < x < \infty, t > 0$$

and

$$\frac{\partial^2 u_2}{\partial t^2} = a_2^2 \frac{\partial^2 u_2}{\partial x^2} \qquad -\infty < x < 0, t > 0$$

with the boundary conditions $\lim_{x \to \infty} u_1(x, t) \to 0$, $\lim_{x \to -\infty} u_2(x, t) \to 0$, $\partial u_1(0, t)/\partial x = \partial u_2(0, t)/\partial x$ and $\partial u_1(0, t)/\partial x = P_0 e^{-\alpha t}$ with $\alpha > 0$ and the initial conditions $u_1(x, 0) = u_2(x, 0) = 0$.

Step 1. Show that the Laplace transform of the partial differential equation yields the ordinary differential equation:

$$\frac{d^2 \bar{u}_1}{dx^2} - \frac{s^2}{a_1^2} \bar{u}_1 = 0, \qquad 0 < x < \infty$$

and

$$\frac{d^2 \bar{u}_2}{dx^2} - \frac{s^2}{a_2^2} \bar{u}_2 = 0, \qquad -\infty < x < 0$$

with $\lim_{x \to \infty} \bar{u}_1(x, s) \to 0$, $\lim_{x \to -\infty} \bar{u}_2(x, s) \to 0$, $\bar{u}_1'(0, s) = \bar{u}_2'(0, s)$ and $\bar{u}_1(0, s) = P_0/(s + \alpha)$.

Step 2. Show that the solution to step 1 is

$$\bar{u}_1(x, s) = -\frac{a_1 P_0}{s(s + \alpha)} \exp(-sx/a_1)$$

and

$$\bar{u}_2(x, s) = \frac{a_2 P_0}{s(s + \alpha)} \exp(sx/a_2).$$

Step 3. By taking the inverse Laplace transform of step 2, show that $u_1(x, t)$ and $u_2(x, t)$ are

$$u_1(x, t) = -\frac{a_1 P_0}{\alpha} \left[1 - e^{-(t - x/a_1)}\right] H(l - x/a_1)$$

and

$$u_2(x,t) = \frac{a_2 P_0}{\alpha} \left[1 - e^{-(t+x/a_2)}\right] H(t + x/a_2).$$

26. Solve the partial differential equation[44]

$$\frac{\partial u}{\partial t} - fv = \nu \frac{\partial^2 u}{\partial z^2}, \qquad 0 < z < \infty, t > 0$$

$$\frac{\partial v}{\partial t} + fu = \nu \frac{\partial^2 v}{\partial z^2}, \qquad 0 < z < \infty, t > 0$$

with the boundary conditions $\lim_{z \to \infty} u(z,t) \to 0$, $\lim_{z \to \infty} v(z,t) \to 0$,

$$\frac{\partial u(0,t)}{\partial z} = -g(t) \qquad \text{and} \qquad \frac{\partial v(0,t)}{\partial z} = 0$$

and the initial conditions $u(z,0) = v(z,0) = 0$ where f and ν are real and positive.

Step 1. If you introduce $w = u + iv$, show that we can rewrite the system of partial differential equations and boundary and initial conditions as

$$\frac{\partial w}{\partial t} + ifw = \nu \frac{\partial^2 w}{\partial z^2}, \qquad 0 < z < \infty, t > 0$$

with the boundary conditions $\lim_{z \to \infty} w(z,t) \to 0$ and

$$\frac{\partial w(0,t)}{\partial z} = -g(t)$$

and the initial condition $w(z,0) = 0$.

Step 2. Take the Laplace transform of the partial differential equation and boundary conditions in step 1 and show that they reduce to

$$\nu \frac{d^2 \overline{w}}{dz^2} - (s + if)\overline{w} = 0, \qquad 0 < z < \infty$$

with $\lim_{z \to \infty} \overline{w}(z,s) \to 0$ and

$$\frac{d\overline{w}(0,s)}{dz} = -\overline{g}(s).$$

---

[44] First posed and solved by Ekman, V. W., 1905: On the influence of the earth's rotation on ocean-currents. *Ark. Math. Astr. och Fys.*, **2**, No. 11, 52 pp.

Step 3. Show that the solution to step 2 is

$$\overline{w}(z, s) = \frac{\overline{g}(s)}{\sqrt{(s+if)/\nu}} \exp\left[-z\sqrt{\frac{s+if}{\nu}}\right].$$

Step 4. Use the convolution theorem and the first shifting theorem to invert the Laplace transform and show that

$$w(z, t) = \sqrt{\frac{\nu}{\pi}} e^{-ift} \int_0^t \frac{g(\eta)}{\sqrt{t-\eta}} \exp\left[-\frac{z^2}{4\nu(t-\eta)}\right] d\eta$$

or

$$u(z, t) = \sqrt{\frac{\nu}{\pi}} \cos(ft) \int_0^t \frac{g(\eta)}{\sqrt{t-\eta}} \exp\left[-\frac{z^2}{4\nu(t-\eta)}\right] d\eta$$

and

$$v(z, t) = \sqrt{\frac{\nu}{\pi}} \sin(ft) \int_0^t \frac{g(\eta)}{\sqrt{t-\eta}} \exp\left[-\frac{z^2}{4\nu(t-\eta)}\right] d\eta.$$

27. Solve the partial differential equations

$$\frac{\partial u}{\partial t} = \frac{\partial^2 u}{\partial x^2} - v, \qquad 0 < x < \infty, t > 0$$

and

$$\frac{\partial v}{\partial t} = \frac{\partial^2 v}{\partial x^2} + u, \qquad 0 < x < \infty, t > 0$$

with the boundary conditions

$$u(0, t) = H(t), \qquad v(0, t) = 0,$$

$$\lim_{x \to \infty} u(x, t) \to 0, \qquad \lim_{x \to \infty} v(x, t) \to 0$$

and the initial conditions $u(x, 0) = v(x, 0) = 0$.

Step 1. By defining the new variable $w(x, t) = u(x, t) + iv(x, t)$, show that you can write the system of partial differential equations as

$$\frac{\partial w}{\partial t} = \frac{\partial^2 w}{\partial x^2} + iw, \qquad 0 < x < \infty, t > 0$$

with the boundary conditions

$$w(0, t) = H(t), \qquad \lim_{x \to \infty} w(x, t) \to 0$$

and the initial condition $w(x, 0) = 0$.

Step 2. By taking the Laplace transform of the partial differential equation in step 1, show that you can reduce it to

$$\frac{d^2\overline{w}}{dx^2} - (s-i)\overline{w} = 0, \quad \overline{w}(0,s) = \frac{1}{s}, \quad \lim_{x\to\infty} \overline{w}(x,s) \to 0.$$

Step 3. Show that the solution to step 2 is

$$\overline{w}(x,s) = \frac{1}{s}e^{-x\sqrt{s-i}}.$$

Step 4. From a table of Laplace transforms, show that

$$u(x,t) = \frac{1}{2}\text{Re}\left\{e^{(x/\sqrt{2}-ix/\sqrt{2})}\text{erfc}\left[\frac{x}{2\sqrt{t}} + (1-i)\sqrt{\frac{t}{2}}\right]\right.$$
$$\left. + e^{-(x/\sqrt{2}-ix/\sqrt{2})}\text{erfc}\left[\frac{x}{2\sqrt{t}} - (1-i)\sqrt{\frac{t}{2}}\right]\right\}$$

and

$$v(x,t) = \frac{1}{2}\text{Im}\left\{e^{(x/\sqrt{2}-ix/\sqrt{2})}\text{erfc}\left[\frac{x}{2\sqrt{t}} + (1-i)\sqrt{\frac{t}{2}}\right]\right.$$
$$\left. + e^{-(x/\sqrt{2}-ix/\sqrt{2})}\text{erfc}\left[\frac{x}{2\sqrt{t}} - (1-i)\sqrt{\frac{t}{2}}\right]\right\}$$

where $\text{erfc}(z)$ is the complex complementary error function.

28. Solve the partial differential equation[45]

$$\left(1 + \frac{1}{\omega_0}\frac{\partial}{\partial t}\right)\frac{\partial^2(ru)}{\partial r^2} = \frac{1}{c^2}\frac{\partial^2(ru)}{\partial t^2}, \qquad a < r < \infty, t > 0$$

with the boundary conditions

$$\mu\left(1 + \frac{1}{\omega_0}\frac{\partial}{\partial t}\right)\left[3\frac{\partial^2 u(a,t)}{\partial r^2} + \frac{2}{a}\frac{\partial u(a,t)}{\partial r}\right] = -\delta(t)$$

and

$$\lim_{r\to\infty} ru(r,t) \to 0$$

---

[45] Taken from Clark, G. B., and G. B. Rupert, 1966: Plane and spherical waves in a Voigt medium. *J. Geophys. Res.*, **71**, 2047–2053. ©1966 American Geophysical Union.

and the initial conditions $u(r, 0) = u_t(r, 0) = 0$.

Step 1. Show that the partial differential equation and boundary conditions reduce to the ordinary differential equation

$$\frac{d^2(r\bar{u})}{dr^2} - \frac{s^2}{c^2(1 + s/\omega_0)} r\bar{u} = 0$$

with the boundary conditions

$$\mu\left(1 + \frac{s}{\omega_0}\right)\left[3\frac{d^2\bar{u}(a, s)}{dr^2} + \frac{2}{a}\frac{d\bar{u}(a, s)}{dr}\right] = -1$$

and

$$\lim_{r\to\infty} r\bar{u}(r, s) \to 0.$$

Step 2. Show that the solution to step 1 is

$$r\bar{u}(r, s) = -\frac{a\exp[-s(r - a)/(c\sqrt{1 + s/\omega_0}\,)]}{\mu(1 + s/\omega_0)}$$

$$\times \frac{1}{\{4/a^2 + 4s/(ac\sqrt{1 + s/\omega_0}\,) \mid 3s^2\omega_0^2/[c^2(1 + s/\omega_0)]\}}.$$

Step 3. Show that

$$\mathcal{L}\left[\frac{\mu r}{a\omega_0}e^{-t}u(r, t/\omega_0)\right] = -\frac{\exp[-\alpha(s - 1)/\sqrt{s}\,]}{s[A(s - 1)^2/s + B(s - 1)/\sqrt{s} + C]}$$

where $A = 3\omega_0^2/c^2$, $B = 4\omega_0/(ac)$, $C = 4/a^2$ and $\alpha = \omega_0(r - a)/c$.

Step 4. Use[46]

$$F(\sqrt{s}\,) = \mathcal{L}\left[\int_0^\infty \frac{\tau}{2\sqrt{\pi}\,t^{3/2}}e^{-\tau^2/4t}f(\tau)\,d\tau\right]$$

to show that

$$\frac{\mu r}{a\omega_0}e^{-t}u(r, t/\omega_0) = -\frac{1}{2\sqrt{\pi}\,t^{3/2}}\int_0^\infty \tau e^{-\tau^2/4t}\mathcal{L}^{-1}[G(s)]\,d\tau$$

where

$$G(s) = \frac{\exp[-\alpha(s - 1/s)]}{s^2[A(s - 1/s)^2 + B(s - 1/s) + C]}.$$

---

[46] Doetsch, G., 1961: *Guide to the Application of Laplace Transforms.* New York: D. van Nostrand Co., Ltd., p. 229.

Step 5. Use[47]

$$\mathcal{L}\left\{\int_0^t \left(\frac{t-\tau}{a\tau}\right)^\nu J_{2\nu}[2\sqrt{a\tau t - a\tau^2}]f(\tau)\,d\tau\right\} = s^{-2\nu-1}F\left(s+\frac{a}{s}\right)$$

to show that

$$g(t) = \frac{1}{\sqrt{4AC-B^2}}\int_0^t \sqrt{\frac{\eta-\tau}{\eta}}J_1[2\sqrt{\eta^2-\eta\tau}]\exp\left[-\frac{B(\eta-\alpha)}{2A}\right]$$

$$\times \sin\left[\frac{\sqrt{4AC-B^2}\,(\eta-\alpha)}{2A}\right]H(\eta-\alpha)\,d\eta.$$

Therefore, the complete solution is

$$\frac{\mu r}{a\omega_0}e^{-t}u(r,t/\omega_0) = -\frac{1}{2\sqrt{\pi(4AC-B^2)t^3}}\int_0^\infty \tau e^{-\tau^2/4t}$$

$$\left\{\int_0^t \sqrt{\frac{\eta-\tau}{\eta}}J_1[2\sqrt{\eta^2-\eta\tau}]\exp\left[-\frac{B(\eta-\alpha)}{2A}\right]\right.$$

$$\left.\times \sin\left[\frac{\sqrt{4AC-B^2}\,(\eta-\alpha)}{2A}\right]H(\eta-\alpha)\,d\eta\right\}d\tau.$$

2.4 THE SOLUTION OF PARTIAL DIFFERENTIAL EQUATIONS BY FOURIER TRANSFORMS

In the previous section we used Laplace transforms to reduce a partial differential equation to an ordinary differential equation. One reason why we chose Laplace transforms was the presence of initial conditions. Another popular method of solving partial differential equations is the Fourier transform. Of course, if this technique is to work, the Fourier transform must exist. This means that the dependent variables must die away to zero as $|x|$ or $|t| \to \infty$.

To illustrate this technique, let us calculate the sound waves[48] radiated by a sphere of radius a whose surface expands radially with an impulsive acceleration $v_0\delta(t)$ where $\delta(\ )$ is the Dirac impulse function.

---

[47] Erdélyi, A., W. Magnus, F. Oberhettinger and F. G. Tricomi, 1954: *Table of Integral Transforms. Volume 1.* New York: McGraw-Hill Book Co., Inc., p. 133.

[48] Taken from Hodgson, D. C., and J. E. Bowcock, 1975: Billet expansion as a mechanism for noise production in impact forming machines. *J. Sound Vib.,* **42**, 325–335. Published by Academic Press Ltd., London, U.K.

This problem has a number of applications; one of them would be the sound waves generated by an exploding depth charge[49].

If we assume radial symmetry, the corresponding wave equation is

$$\frac{1}{r^2}\frac{\partial}{\partial r}\left(r^2\frac{\partial u}{\partial r}\right) = \frac{1}{c^2}\frac{\partial^2 u}{\partial t^2} \tag{2.4.1}$$

subject to the boundary condition

$$\frac{\partial u}{\partial r} = -\rho v_0 \delta(t) \tag{2.4.2}$$

at $r = a$ where $u(r,t)$ is the pressure field, c is the speed of sound and ρ is the average density of the fluid. Assuming that the pressure field possesses a Fourier transform, we may reexpress it by the Fourier integral

$$u(r,t) = \frac{1}{2\pi}\int_{-\infty}^{\infty} U(r,\omega)e^{i\omega t}d\omega. \tag{2.4.3}$$

Substituting into (2.4.1), it becomes

$$\frac{1}{r^2}\frac{d}{dr}\left(r^2\frac{dU}{dr}\right) + k^2 U = 0 \tag{2.4.4}$$

and

$$\frac{dU(a,\omega)}{dr} = -\rho v_0 \tag{2.4.5}$$

with $k = \omega/c$. The most general solution of (2.4.4) is

$$U(r,\omega) = A(\omega)\frac{e^{ikr}}{r} + B(\omega)\frac{e^{-ikr}}{r} \tag{2.4.6}$$

and

$$u(r,t) = \frac{1}{2\pi}\int_{-\infty}^{\infty}\left[A(\omega)\frac{e^{i\omega t+ikr}}{r} + B(\omega)\frac{e^{i\omega t-ikr}}{r}\right]d\omega. \tag{2.4.7}$$

At this point, we note that the first term on the right side of (2.4.7) represents an inward propagating wave while the second term is an outward propagating wave. Because there is no source of energy at infinity, the inward propagating wave is aphysical and we discard it. This boundary

---

[49] Probst (Probst, W., 1972: Die Schallerzeugung durch eine expandierende Kugel. *Acustica*, **27**, 299–306.) used a complicated form of this problem to explain the high sound levels generated by an inflating air-bag system.

condition is often referred to as the *Sommerfeld radiation condition* because Sommerfeld[50] used it first in his study of electromagnetic waves on the surface of a conducting earth.

Upon substituting (2.4.6) into (2.4.5) with $A(\omega) = 0$,

$$U(r,\omega) = \frac{\rho a^2 v_0}{2\pi r} \frac{e^{-i\omega(r-a)/c}}{1 + i\omega a/c} \qquad (2.4.8)$$

or

$$u(r,t) = \frac{1}{2\pi} \int_{-\infty}^{\infty} U(r,\omega)e^{i\omega t}d\omega = \frac{\rho a^2 v_0}{2\pi r} \int_{-\infty}^{\infty} \frac{e^{i\omega[t-(r-a)/c]}}{1 + i\omega a/c} d\omega. \qquad (2.4.9)$$

To evaluate (2.4.9), we employ the residue theorem. For $t < (r-a)/c$ Jordan's lemma dictates that we close the line integral by an infinite semicircle in the lower half-plane. For $t > (r-a)/c$, we close the contour with an semicircle in the upper half-plane. The final result is

$$u(r,t) = \frac{\rho a c v_0}{r} \exp\left[-\frac{c}{a}\left(t - \frac{r-a}{c}\right)\right] H\left(t - \frac{r-a}{c}\right) \qquad (2.4.10)$$

where $H(\ )$ is the Heaviside step function.

For our second example[51], we find the solution to the scalar, reduced (Helmholtz), inhomogeneous wave equation in free space:

$$\nabla^2 u(x,y,z) + \kappa_0^2 u(x,y,z) = -4\pi\delta(x)\delta(y)\delta(z) \qquad (2.4.11)$$

where $\kappa_0 = \omega/c$, ω is the frequency of the wave and c is the phase speed. By direct substitution, we find[52] that

$$u(x,y,z) = \frac{e^{i\omega R/c}}{R} \qquad (2.4.12)$$

is a solution of (2.4.11) where $R = \sqrt{x^2 + y^2 + z^2}$. However, in this section we derive solutions to (2.4.11) using Fourier transforms.

---

[50] Sommerfeld, A., 1909: Über die Ausbreitung der Wellen in der draftlosen Telegraphie. *Ann. Phys., Vierte Folge*, **28**, 665–736.

[51] Patterned after Biggs, A. W., 1977: Fourier transforms in propagation and scattering problems. *IEEE Trans. Antennas Propagat.*, **AP-25**, 585–586. ©1977 IEEE.

[52] For a derivation, see Aki, K., and P. G. Richards, 1980: *Quantitative Seismology: Theory and Methods. Vol. I.* San Francisco: W. H. Freeman and Co., Section 4.1.

We begin by assuming that a Fourier transform exists for $u(x, y, z)$. Therefore, we can write $u(x, y, z)$ as

$$u(x, y, z) = \frac{1}{(2\pi)^3} \int_{-\infty}^{\infty} \int_{-\infty}^{\infty} \int_{-\infty}^{\infty} U(k, l, m) e^{i(kx+ly+mz)} dk\, dl\, dm.$$
(2.4.13)

Direct substitution of (2.4.13) into (2.4.11) yields

$$U(k, l, m) = \frac{4\pi}{\kappa^2 - \kappa_0^2}$$
(2.4.14)

where $\kappa^2 = k^2 + l^2 + m^2$ so that

$$u(x, y, z) = \frac{1}{2\pi^2} \int_{-\infty}^{\infty} \int_{-\infty}^{\infty} \int_{-\infty}^{\infty} \frac{e^{i(kx+ly+mz)}}{\kappa^2 - \kappa_0^2} dk\, dl\, dm.$$
(2.4.15)

To evaluate (2.4.15) we first perform the integration in m by closing the line integral along the real axis as dictated by Jordan's lemma. For $z > 0$, this requires an infinite semicircle in the upper half-plane; for $z < 0$, an infinite semicircle in the lower half-plane. One source of difficulty is the presence of singularities along the real axis at $m = \pm\sqrt{\kappa_0^2 - k^2 - l^2}$. We avoid this difficulty by introducing some friction (making it a slightly lossy medium) so that $\kappa_0 = \kappa_0' + i\kappa_0''$, where $\kappa_0' \gg \kappa_0''$. This lifts the singularities off the real axis and a straightforward application of the residue theorem yields

$$u(x, y, z) = \frac{1}{2\pi} \int_{-\infty}^{\infty} \int_{-\infty}^{\infty} \frac{e^{ikx+ily-|z|\sqrt{k^2+l^2-\omega^2/c^2}}}{\sqrt{k^2 + l^2 - \omega^2/c^2}} dk\, dl$$
(2.4.16)

with the condition that the square root has a real part ≥ 0. Because the $u(x, y, z)$ representations given by (2.4.12) and (2.4.16) must be equivalent, we immediately obtain the *Weyl integral*[53]:

$$\frac{e^{i\omega R/c}}{R} = \frac{1}{2\pi} \int_{-\infty}^{\infty} \int_{-\infty}^{\infty} \frac{e^{ikx+ily-|z|\sqrt{k^2+l^2-\omega^2/c^2}}}{\sqrt{k^2 + l^2 - \omega^2/c^2}} dk\, dl.$$
(2.4.17)

We can further simplify (2.4.17) by introducing $k = \rho\cos(\phi)$, $l = \rho\sin(\phi)$, $x = r\cos(\theta)$ and $y = r\sin(\theta)$. Then,

$$\frac{e^{i\omega R/c}}{R} = \frac{1}{2\pi} \int_0^{\infty} \int_0^{2\pi} \frac{e^{i\rho r\cos(\theta-\phi)-i|z|\sqrt{\omega^2/c^2-\rho^2}}}{-i\sqrt{\omega^2/c^2 - \rho^2}} d\phi\, \rho\, d\rho.$$
(2.4.18)

---

[53] Weyl, H., 1919: Ausbreitung elektromagnetischer Wellen über einer ebenen Leiter. *Ann. Phys., Vierte Folge*, **60**, 481 500.

Transform Methods for Solving Partial Differential Equations

If we now carry out the integration in ϕ, we obtain the *Sommerfeld integral*[50]:

$$i\frac{e^{i\omega R/c}}{R} = \int_0^\infty \frac{J_0(\rho r)e^{-i|z|\sqrt{\omega^2/c^2-\rho^2}}}{\sqrt{\omega^2/c^2-\rho^2}}\,\rho\,d\rho \qquad (2.4.19)$$

where the imaginary part of the square root must be positive. Both the Weyl and Sommerfeld integrals are used extensively in electromagnetism and elasticity as an integral representation of spherical waves propagating from a point source.

Problems

1. The nondimensional shallow water equations in one spatial dimension:

$$\frac{\partial u}{\partial t} - v = -\frac{\partial h}{\partial x}, \qquad \frac{\partial v}{\partial t} + u = 0 \quad \text{and} \quad \frac{\partial h}{\partial t} + c^2\frac{\partial u}{\partial x} = 0$$

describe wave motions within a homogeneous ocean whose depth is small compared to the radius of the rotating earth. The zonal and meridional velocities are $u(x,t)$ and $v(x,t)$; the height of the free surface is $h(x,t)$. The phase speed c equals the square root of the depth of the ocean times gravity divided by the Coriolis parameter.

In this problem let us find the wave motion in a quiescent ocean after we raise its interface by h_0 in the region $-a < x < a$. Mathematically, this is equivalent to solving the shallow water equations subject to the initial condition that

$$u(x,0) = v(x,0) = 0 \quad \text{and} \quad h(x,0) = \begin{cases} h_0 & \text{if } |x| < a, \\ 0 & \text{if } |x| > a \end{cases}$$

and assuming that $u(x,t)$, $v(x,t)$ and $h(x,t)$ vanish as $|x| \to \infty$.

Step 1. By defining the Fourier transform of $u(x,t)$, $v(x,t)$ and $h(x,t)$ as

$$U(k,t) = \int_{-\infty}^\infty u(x,t)e^{-ikx}dx, \quad V(k,t) = \int_{-\infty}^\infty v(x,t)e^{-ikx}dx$$

and

$$H(k,t) = \int_{-\infty}^\infty h(x,t)e^{-ikx}dx,$$

show that we can transform the shallow water equations into the ordinary differential equations

$$\frac{dU}{dt} - kV = -ikH, \qquad \frac{dV}{dt} + U = 0$$

and

$$\frac{dH}{dt} + ikc^2 U = 0$$

with

$$U(k,0) = V(k,0) = 0 \quad \text{and} \quad H(k,0) = \frac{2h_0 \sin(ka)}{k}.$$

Step 2. Show that we can combine the three equations in step 1 into the single equation

$$\frac{d^3 H}{dt^3} + (1 + k^2 c^2)\frac{dH}{dt} = 0.$$

Step 3. Show that the solution to the ordinary differential equation and initial conditions in steps 1 and 2 is

$$H(k,t) = \frac{2h_0 \sin(ka)}{k}\left[\frac{1 + k^2 c^2 \cos(t\sqrt{1 + k^2 c^2})}{1 + k^2 c^2}\right].$$

Step 4. Show that $h(x,t)$ is

$$h(x,t) = \frac{h_0}{\pi}\int_{-\infty}^{\infty}\frac{\sin(ka)}{k}\left[\frac{1 + k^2 c^2 \cos(t\sqrt{1 + k^2 c^2})}{1 + k^2 c^2}\right]e^{ikx}dk$$

or

$$h(x,t) = \frac{2h_0}{\pi}\int_{0}^{\infty}\frac{\sin(ka)}{k}\left[\frac{1 + k^2 c^2 \cos(t\sqrt{1 + k^2 c^2})}{1 + k^2 c^2}\right]\cos(kx)\,dk.$$

2. Solve the nonhomogeneous equation of telegraphy[54]

$$\frac{\partial^2 u}{\partial x^2} - \frac{1}{c^2}\frac{\partial^2 u}{\partial t^2} - \frac{4\pi\sigma}{c^2}\frac{\partial u}{\partial t} = \frac{4\pi}{c}f(x), \quad -\infty < x < \infty, t > 0$$

where the initial conditions are $u(x,0) = u_t(x,0) = 0$.

Step 1. Assuming that $f(x)$ has a Fourier transform and

$$U(k,t) = \int_{-\infty}^{\infty} u(x,t)e^{-ikx}dx,$$

---

[54] Karzas, W. J., and R. Latter, 1962: The electromagnetic signal due to the interaction of nuclear explosions with the earth's magnetic field. *J. Geophys. Res.*, **67**, 4635–4640. ©1962 American Geophysical Union.

show that we can reduce the partial differential equation to the ordinary differential equation

$$\frac{d^2 U}{dt^2} + 4\pi\sigma \frac{dU}{dt} + k^2 c^2 U = -4\pi c F(k)$$

where $F(k)$ is the Fourier transform of $f(x)$.

Step 2. Show that we can write the solution to step 1 as

$$U(k,t) = -4\pi c F(k) \int_0^t e^{-2\pi\sigma(t-\tau)} \frac{\sin[(t-\tau)\sqrt{k^2 c^2 - 4\pi^2\sigma^2}]}{\sqrt{k^2 c^2 - 4\pi^2\sigma^2}} \, d\tau.$$

[Hint: Use Laplace transforms.]

Step 3. Show that we can write the solution $u(x,t)$ as

$$u(x,t) = -2 \int_0^t e^{-2\pi\sigma(t-\tau)}$$

$$\times \left\{ \int_{-\infty}^{\infty} F(k) \frac{c \sin[(t-\tau)\sqrt{k^2 c^2 - 4\pi^2\sigma^2}]}{\sqrt{k^2 c^2 - 4\pi^2\sigma^2}} e^{ikx} dk \right\} d\tau$$

$$= -4 \int_0^t \left\{ \int_{-\infty}^{\infty} f(\chi) I(|x-\chi|, t-\tau) \, d\chi \right\} e^{-2\pi\sigma(t-\tau)} d\tau$$

where

$$I(x,t) = \int_0^{\infty} \frac{\sin[ct\sqrt{k^2 - (2\pi\sigma/c)^2}]}{\sqrt{k^2 - (2\pi\sigma/c)^2}} \cos(kx) \, dk$$

$$= \frac{\pi}{2} I_0[2\pi\sigma\sqrt{t^2 - (x/c)^2}] H(t - x/c).$$

3. Solve the partial differential equation[55]

$$\frac{\partial^2 u}{\partial x^2} + \frac{\partial^2 u}{\partial z^2} - u = 0, \qquad -\infty < x < \infty, 0 < z < 1$$

with the boundary conditions

$$\frac{\partial u(x,0)}{\partial z} = 0 \text{ and } u(x,1) = e^{-x} H(x), \qquad -\infty < x < \infty.$$

Step 1. By taking the Fourier transform

$$U(k,z) = \int_{-\infty}^{\infty} u(x,z) e^{-ikx} dx$$

---

[55] Taken from Horvay, G., 1961: Temperature distribution in a slab moving from a chamber at one temperature to a chamber at another temperature. *J. Heat Transfer*, **83**, 391–402.

of the partial differential equation and boundary conditions, show that they reduce to the ordinary differential equation

$$\frac{d^2U}{dz^2} - m^2U = 0, \qquad 0 < z < 1$$

where $m^2 = k^2 + 1$ with the boundary conditions $U'(k,0) = 0$ and

$$U(k,1) = \frac{1}{1 + ki}.$$

Step 2. Show that the solution to step 1 is

$$U(k,z) = \frac{\cosh(mz)}{i(k - i)\cosh(m)}.$$

Step 3. Show that the singularities of $U(k,z)$ are simple poles at $k = i$ and $k_n = \pm i\sqrt{4 + (2n + 1)^2\pi^2}/2$ where $n = 0, 1, 2, \ldots$

Step 4. Use the residue theorem to invert $U(k,z)$ and show that

$$u(x, z) = e^{-x}H(x)$$
$$+ 2\pi \sum_{n=0}^{\infty} \frac{(-1)^n(2n + 1)}{\sigma_n[2 - \operatorname{sgn}(x)\sigma_n]}e^{-\sigma_n|x|/2}\cos\left[\frac{(2n + 1)\pi z}{2}\right]$$

where $\sigma_n = \sqrt{4 + (2n + 1)^2\pi^2}$.

4. Find the particular solution to the partial differential equation[56]

$$\frac{\partial^2 u_p}{\partial x^2} + \frac{\partial^2 u_p}{\partial z^2} - i\omega\mu\sigma u_p = \sigma B_0[\delta(z + h/2) - \delta(z - h/2)]\sin(kx)$$

where $-\infty < x < \infty, -\infty < z < \infty$.

Step 1. By taking the Fourier transform in the z-direction

$$U_p(x, m) = \int_{-\infty}^{\infty} u_p(x, z)e^{-imz}dz,$$

show that the partial differential equation reduces to the ordinary differential equation

$$\frac{d^2U_p}{dx^2} - (m^2 + i\omega\mu\sigma)U_p = 2i\sigma B_0 \sin(mh/2)\sin(kx).$$

---

[56] Taken from Robey, D. H., 1953: Magnetic dispersion of sound in electrically conducting plates. *J. Acoust. Soc. Am.*, **25**, 603–609.

Step 2. Show that the particular solution $U_p(x, m)$ is

$$U_p(x, m) = -\frac{2i\sigma B_0 \sin(mh/2)\sin(kx)}{m^2 + \alpha^2}$$

where $\alpha^2 = k^2 + i\omega\mu\sigma$.

Step 3. Use the residue theorem to find the inverse of $U_p(x, m)$ and show that

$$u_p(x, z) = \frac{\sigma B_0}{\alpha} \sinh(\alpha z)\sin(kx)e^{-\alpha h/2}.$$

5. Find the forced solution to the partial differential equation[57]

$$\frac{\partial^2 u}{\partial r^2} + \frac{1}{r}\frac{\partial u}{\partial r} + \frac{\partial^2 u}{\partial z^2} = \frac{\partial u}{\partial t}, \quad 0 < r < 1, -\infty < z < \infty, t > 0$$

subject to the boundary conditions $\lim_{|z|\to 0} u(r, z, t) \to 0$, $|u(0, z, t)| < \infty$ and

$$u(1, z, t) = \begin{cases} 0 & \text{if } z < 0, \\ 1 & \text{if } 0 < z < t, \\ 0 & \text{if } z > t. \end{cases}$$

Step 1. If

$$u(r, z, t) = \frac{1}{2\pi}\int_{-\infty}^{\infty} U(r, k, t)e^{ikz}dk,$$

show that the original partial differential equation reduces to the two-dimensional partial differential equation

$$\frac{\partial^2 U}{\partial r^2} + \frac{1}{r}\frac{\partial U}{\partial r} - k^2 U = \frac{\partial U}{\partial t}$$

with the boundary condition

$$U(1, k, t) = \frac{1 - e^{-ikt}}{ik}.$$

Step 2. Show that the forced solution to step 1 is

$$U(r, k, t) = \frac{I_0(kr)}{ikI_0(k)} - \frac{I_0(r\sqrt{k^2 - ik})}{ikI_0(\sqrt{k^2 - ik})}e^{-ikt}$$

---

[57] Taken from Singh, H., 1981: Thermal stresses in an infinite cylinder. *Indian J. Pure Appl. Math.*, **12**, 405–418.

where $I_0(\ )$ is the modified Bessel function of the first kind and zeroth order.

Step 3. Using the inversion formula

$$f(t) = \frac{1}{4\pi}\left[\int_{-\infty-\epsilon i}^{\infty-\epsilon i} F(\omega)e^{i\omega t}d\omega + \int_{-\infty+\epsilon i}^{\infty+\epsilon i} F(\omega)e^{i\omega t}d\omega\right]$$

where $0 < \epsilon << 1$ and the residue theorem, show that

$$\mathcal{F}^{-1}\left[\frac{I_0(kr)}{ikI_0(k)}\right] = \tfrac{1}{2}\text{sgn}(z) - \text{sgn}(z)\sum_{n=1}^{\infty} \frac{J_0(\alpha_n r)}{\alpha_n J_1(\alpha_n)}e^{-\alpha_n|z|}$$

where $J_0(\ )$ and $J_1(\ )$ are Bessel functions of the first kind and zeroth and first order, respectively, and α_n is the n-th root of $J_0(\alpha) = 0$. Finally, show that

$$\mathcal{F}^{-1}\left[\frac{I_0(r\sqrt{k^2 - ik}\,)}{ikI_0(\sqrt{k^2 - ik}\,)}\right] = -\tfrac{1}{2}\text{sgn}(t-z) + \text{sgn}(t-z)$$

$$\times \sum_{n=1}^{\infty} \frac{2\alpha_n J_0(\alpha_n r)}{\lambda_n(2\lambda_n - 1)J_1(\alpha_n)}e^{\lambda_n(t-z)}$$

where

$$\lambda_n = \begin{cases} (1 - \sqrt{1 + 4\alpha_n^2}\,)/2 & \text{if} \quad t - z > 0, \\ (1 + \sqrt{1 + 4\alpha_n^2}\,)/2 & \text{if} \quad t - z < 0. \end{cases}$$

2.5 THE SOLUTION OF PARTIAL DIFFERENTIAL EQUATIONS BY HANKEL TRANSFORMS

Although Fourier and Laplace transforms are the most commonly known ones, there are others. When there are cylindrical coordinates, the application of Bessel functions within an integral transform appears a logical choice. This leads to the Hankel transform defined by

$$F(k) = \int_0^\infty f(r)J_\nu(kr)r\,dr, \qquad \nu > -1/2 \qquad (2.5.1)$$

and its inverse[58]

$$f(r) = \int_0^\infty F(k)J_\nu(kr)k\,dk. \qquad (2.5.2)$$

---

[58] Watson, G. N., 1966: *A Treatise on the Theory of Bessel Functions.* Cambridge: At the University Press, Section 14.4. See also MacRobert, T. M., 1931: Fourier integrals. *Proc. R. Soc. Edin.*, **51**, 116–126.

Transform Methods for Solving Partial Differential Equations

Unlike Fourier and Laplace transforms, we do not generally use complex variables to invert the Hankel transform.

Consider the surface waves[59] generated within an incompressible ocean of infinite depth due to an explosion above it which generates the pressure field $f(r, t)$. Assuming potential flow, the continuity equation is

$$\frac{1}{r}\frac{\partial}{\partial r}\left(r\frac{\partial u}{\partial r}\right) + \frac{\partial^2 u}{\partial z^2} = 0, \quad 0 < r < \infty, z < 0. \tag{2.5.3}$$

This velocity potential $u(r, z, t)$ must satisfy the boundary condition that

$$\frac{\partial^2 u}{\partial t^2} + g\frac{\partial u}{\partial z} = \frac{1}{\rho}\frac{\partial f}{\partial t}\{H[r] - H[r - r_0(t)]\} \tag{2.5.4}$$

at $z = 0$ where ρ is the density of the liquid and $r_0(t)$ is the extent of the blast. The fluid is initially at rest. A general solution to (2.5.3) is clearly

$$u(r, z, t) = \int_0^\infty A(k, t)e^{kz}J_0(kr)k\,dk. \tag{2.5.5}$$

Because the Hankel transform representation of $f(r, t)$ is

$$f(r, t) = \int_0^\infty \left\{\int_0^{r_0(t)} f(\alpha, t)J_0(k\alpha)\alpha\,d\alpha\right\}J_0(kr)k\,dk, \tag{2.5.6}$$

(2.5.4) becomes

$$\frac{\partial^2 A}{\partial t^2} + gkA = \frac{1}{\rho}\frac{\partial}{\partial t}\int_0^{r_0(t)} f(\alpha, t)J_0(k\alpha)\alpha\,d\alpha. \tag{2.5.7}$$

The general solution to (2.5.7) is

$$A(k, t) = A_0(k)\cos(\sigma t) + B_0(k)\sin(\sigma t)$$
$$+ \frac{1}{\rho\sigma}\int_0^t \frac{\partial}{\partial s}\left\{\int_0^{r_0(s)} f(\alpha, s)J_0(k\alpha)\alpha\,d\alpha\right\}\sin[\sigma(t-s)]\,ds \tag{2.5.8}$$

where $\sigma^2 = gk$. From the initial conditions, $A_0(k) = B_0(k) = 0$. After an integration by parts, (2.5.8) takes on the final form

$$u(r, z, t) = \frac{1}{\rho}\int_0^\infty e^{kz}J_0(kr)\left\{\int_0^t\left[\int_0^{r_0(s)} f(\alpha, s)J_0(k\alpha)\alpha\,d\alpha\right]\right.$$
$$\left. \times \cos[\sigma(t-s)]\,ds\right\}k\,dk. \tag{2.5.9}$$

---

[59] Sen, A. R., 1963: Surface waves due to blasts on and above liquids. *J. Fluid Mech.*, **16**, 65–81. Reprinted with the permission of Cambridge University Press.

For our second example[60], we find the free, small transverse oscillations of a thin elastic plate of infinite radius and of uniform density. Assuming symmetry about the z-axis, the differential equation in polar coordinates is

$$b^2 \left(\frac{\partial^2}{\partial r^2} + \frac{1}{r}\frac{\partial}{\partial r} \right)^2 u + \frac{\partial^2 u}{\partial t^2} = 0. \qquad (2.5.10)$$

If $U(k,t)$ denotes the Hankel transform of $u(r,t)$ so that

$$U(k,t) = \int_0^\infty ru(r,t)J_0(kr)\,dr, \qquad (2.5.11)$$

then, as the result of a series of integrations by parts,

$$\int_0^\infty r\left(\frac{\partial^2 u}{\partial r^2} + \frac{1}{r}\frac{\partial u}{\partial r} \right) J_0(kr)\,dr$$

$$= r\frac{\partial u}{\partial r}J_0(kr)\Big|_0^\infty - k\int_0^\infty r\frac{\partial u}{\partial r}J_0'(kr)\,dr$$

$$+ \int_0^\infty \frac{\partial u}{\partial r}J_0(kr)\,dr \qquad (2.5.12)$$

$$= k\int_0^\infty u(r,t)\left[J_0'(kr) + krJ_0''(kr) \right] dr. \qquad (2.5.13)$$

However, because $J_0(kr)$ satisfies the differential equation:

$$krJ_0''(kr) + J_0'(kr) + krJ_0(kr) = 0, \qquad (2.5.14)$$

$$\int_0^\infty r\left(\frac{\partial^2 u}{\partial r^2} + \frac{1}{r}\frac{\partial u}{\partial r} \right) J_0(kr)\,dr = -k^2U(k,t). \qquad (2.5.15)$$

Applying this result twice,

$$\int_0^\infty r\left(\frac{\partial^2}{\partial r^2} + \frac{1}{r}\frac{\partial}{\partial r} \right)^2 u(r,t)J_0(kr)\,dr = k^4 U(k,t). \qquad (2.5.16)$$

Consequently, if we multiply (2.5.10) through by $rJ_0(kr)$ and integrate with respect to r over the range $(0,\infty)$, the solution of the ordinary differential equation

$$\frac{d^2 U}{dt^2} + b^2 k^4 U = 0 \qquad (2.5.17)$$

---

[60] Sneddon, I. N., 1945: The Fourier transform solution of an elastic wave equation. *Proc. Camb. Phil. Soc.*, **41**, 239–243. Reprinted with the permission of Cambridge University Press.

determines $U(k,t)$. We then find the displacement $u(r,t)$ by the Hankel inversion formula

$$u(r,t) = \int_0^\infty kU(k,t)J_0(kr)\,dk. \tag{2.5.18}$$

If the initial conditions are $u(r,0) = f(r)$ and $u_t(r,0) = 0$, the solution of (2.5.17) is

$$U(k,t) = A(k)\cos(bk^2 t) + B(k)\sin(bk^2 t). \tag{2.5.19}$$

From the definition of $U(k,t)$,

$$A(k) = U(k,0) = \int_0^\infty \eta f(\eta)J_0(k\eta)\,d\eta \tag{2.5.20}$$

and

$$B(k) = \frac{dU(k,0)}{dt} = 0. \tag{2.5.21}$$

Therefore,

$$u(r,t) = \int_0^\infty k\cos(bk^2 t)\left\{ \int_0^\infty \eta f(\eta)J_0(k\eta)\,d\eta \right\} J_0(kr)\,dk \tag{2.5.22}$$

or, by changing the order of integration,

$$u(r,t) = \int_0^\infty \eta f(\eta)\left\{ \int_0^\infty kJ_0(k\eta)J_0(kr)\cos(bk^2 t)\,dk \right\} d\eta \tag{2.5.23}$$

$$= \frac{1}{2bt}\int_0^\infty \eta f(\eta)J_0\left(\frac{r\eta}{2bt}\right)\sin\left(\frac{r^2 + \eta^2}{4bt}\right) d\eta \tag{2.5.24}$$

because

$$\int_0^\infty kJ_0(k\eta)J_0(kr)\cos(btk^2)\,dk = \frac{1}{2bt}J_0\left(\frac{r\eta}{2bt}\right)\sin\left(\frac{r^2 + \eta^2}{4bt}\right). \tag{2.5.25}$$

Problems

1. Show that the Hankel transform of $\delta(r)$ is $1/2\pi$.

2. Solve the partial differential equation

$$\frac{1}{r}\frac{\partial}{\partial r}\left(r\frac{\partial u}{\partial r}\right) + \frac{\partial^2 u}{\partial z^2} = 0, \qquad 0 < r < \infty, 0 < z < \infty$$

by Hankel transforms with the boundary conditions $|u(0, z)| < \infty$, $\lim_{z \to \infty} u(r, z) \to 0$, $u(r, 0) = u_0$ for $0 \le r < a$ and $u_z(r, 0) = 0$ for $r > a$.

Step 1. By defining the Hankel transform

$$U(k, t) = \int_0^\infty u(r, t) J_0(kr) r \, dr,$$

show that we may convert the partial differential equation into the ordinary differential equation

$$\frac{dU^2}{dz^2} - k^2 U = 0$$

with $\lim_{z \to \infty} U(k, z) \to 0$.

Step 2. Show that the solution to step 1 is $U(k, z) = Ae^{-kz}$.

Step 3. Using the remaining boundary condition and tables, show that

$$U(k, z) = \frac{2u_0}{\pi k^2} \sin(ka).$$

Step 4. Show that

$$u(r, z) = \frac{2u_0 a}{\pi} \int_0^\infty \frac{\sin(ka)}{ka} e^{-kz} J_0(kr) \, dk.$$

3. Solve the partial differential equation[61]

$$\left(\frac{\partial}{\partial t} + 1 \right) \left[\frac{1}{r} \frac{\partial}{\partial r} \left(r \frac{\partial u}{\partial r} \right) \right] - \left(\frac{1}{S} \frac{\partial u}{\partial t} + \gamma \right) u = \frac{1}{r} \frac{\partial m_0}{\partial r},$$

where $0 < r < \infty$ and $t > 0$ by Hankel transforms with the boundary conditions $|u(0, t)| < \infty$ and $\lim_{r \to \infty} u(r, t) \to 0$ and the initial condition $u(r, 0) = 0$.

Step 1. By defining the Hankel transforms

$$U(k, t) = \int_0^\infty u(r, t) J_0(kr) r \, dr$$

---

[61] Taken from Ou, H. W., and A. L. Gordon, 1986: Spin-down of baroclinic eddies under sea ice. *J. Geophys. Res.*, **91**, 7623–7630. ©1986 American Geophysical Union.

and

$$M_0(k,t) = \int_0^\infty m_0(r,t) J_0(kr) r \, dr,$$

show that we may convert the partial differential equation into the ordinary differential equation

$$\left(k^2 + \frac{1}{S}\right) \frac{dU(k,t)}{dt} + \left(k^2 + \gamma\right) U(k,t) = -kM_0(k).$$

Step 2. Show that the solution to step 1 is

$$U(k,t) = -\frac{kM_0(k)}{k^2 + \gamma} \left[1 - \exp\left(-\frac{k+\gamma}{k^2 + 1/S} t\right)\right].$$

Step 3. Show that the inverse Hankel transform is

$$u(r,t) = -\int_0^\infty \frac{k^2 M_0(k)}{k^2 + \gamma} \left[1 - \exp\left(-\frac{k+\gamma}{k^2 + 1/S} t\right)\right] J_0(kr) \, dk.$$

4. Solve the partial differential equation[62]

$$\frac{\partial^2 u}{\partial z^2} + \frac{\partial^2 u}{\partial r^2} + \frac{1}{r} \frac{\partial u}{\partial r} - \frac{u}{r^2} = -\frac{f(z)\delta(r)}{r^2}, \qquad 0 < r < \infty, z > 0$$

by Hankel transforms with the boundary conditions $\lim_{r\to\infty} u(r,z) \to 0$, $\lim_{z\to\infty} u(r,z) \to 0$, $|u(0,z)| < \infty$ and $u_z(r,0) = 0$.

Step 1. By defining the Hankel transform

$$U(k,z) = \int_0^\infty u(r,z) J_1(kr) r \, dr,$$

show that we may convert the partial differential equation into the ordinary differential equation

$$\frac{d^2 U(k,z)}{dz^2} - k^2 U(k,z) = -\frac{kf(z)}{4\pi}$$

with the boundary conditions

$$\lim_{z\to\infty} U(k,z) \to \infty \qquad \text{and} \qquad U'(k,0) = 0.$$

---

[62] Taken from Agrawal, G. L., and W. G. Gottenberg, 1971: Response of a semi-infinite elastic solid to an arbitrary torque along the axis. *PAGEOPH*, **91**, 34–39. Published by Birkhäuser Verlag, Basel, Switzerland.

Transform Methods with Single-Valued Functions

Step 2. Show that the solution to step 1 is

$$U(k, z) = \frac{1}{8\pi} \left[\int_0^\infty f(\tau) e^{-k(z+\tau)} d\tau + \int_z^\infty f(\tau) e^{-k(\tau-z)} d\tau \right.$$
$$\left. + \int_0^z f(\tau) e^{-k(z-\tau)} d\tau \right].$$

Step 3. Invert $U(k, z)$ by noting that

$$\int_0^\infty \left[\int_0^L e^{-k(z+\tau)} f(\tau) d\tau \right] J_1(kr) k \, dk$$

$$= \int_0^L f(\tau) \left[\int_0^\infty e^{-k(z+\tau)} J_1(kr) k \, dk \right] d\tau$$

$$= \int_0^L f(\tau) \frac{r}{[(z+\tau)^2 + r^2]^{3/2}} d\tau$$

and show that

$$u(r, z) = \frac{1}{8\pi} \left[\int_0^\infty \frac{r f(\tau)}{[(z+\tau)^2 + r^2]^{3/2}} d\tau + \int_0^\infty \frac{r f(\tau)}{[(z-\tau)^2 + r^2]^{3/2}} d\tau \right].$$

Papers Using Laplace Transforms
to Solve Partial Differential Equations

Abarband, S. S., 1960: Time dependent temperature distribution in radiating solids. *J. Math. and Phys.*, **39**, 246–257.

Abu-Abdou, K., 1982: Decay of laminar flow inside a retarding rotating cylindrical vessel. *Acta Mech.*, **45**, 197–204.

Achard, J. L., and G. M. Lespinard, 1981: Structure of the transient wall-friction law in one-dimensional models of laminar pipe flows. *J. Fluid Mech.*, **113**, 283–298.

Acheson, D. J., 1975: On hydromagnetic oscillations within the earth and core-mantle coupling. *Geophys. J. R. Astr. Soc.*, **43**, 253–268.

Afanas'ev, E. F., 1962: Diffraction of a nonstationary pressure wave on a moving plate. *J. Appl. Math. Mech.*, **26**, 268–276.

Agarwal, J. P., and S. K. Roy, 1977: On the effect of magnetic field on viscous lifting and drainage of conducting fluid. *Appl. Sci. Res.*, **33**, 141–149.

Agarwal, M., 1978: Stresses and displacements in an infinite elastic medium due to some types of temperature distribution inside a spherical cavity. *Bull. Cal. Math. Soc.*, **70**, 1–10.

Alterman, Z., and P. Kornfeld, 1966: Effect of a fluid core on propagation of an SH-torque pulse from a point-source in a sphere. *Geophysics*, **31**, 741–763.

Alterman, Z., and F. Abramovici, 1967: The motion of a sphere caused by an impulsive force and by an explosive point-source. *Geophys. J. R. Astr. Soc.*, **13**, 117–148.

Alterman, Z. S., and J. Aboudi, 1969: Seismic pulse in a layered sphere: Normal modes and surface waves. *J. Geophys. Res.*, **74**, 2618–2636.

Alzheimer, W. E., M. J. Forrestal and W. B. Murfin, 1968: Transient response of cylindrical, shell-core systems. *AIAA J.*, **6**, 1861–1866.

Amada, S., 1985: Dynamic shear stress analysis of discs subjected to variable rotations. *Bull. JSME*, **28**, 1029–1035.

Amada, S., 1986: Dynamic stress analysis of hollow rotating discs. *Bull. JSME*, **29**, 1383–1389.

Ansari, J. S., and R. Oldenburger, 1967: Propagation of disturbance in fluid lines. *J. Basic Engng.*, **89**, 415–422.

Ansari, M. A. A., 1978: Longitudinal dispersion in saturated porous media. *Indian J. Pure Appl. Math.*, **9**, 436–446.

Ansari, M. A. A., 1978: Longitudinal dispersion in an isotropic porous media flow. *Indian J. Pure Appl. Math.*, **9**, 588–599.

Arenz, R. J., 1965: Two-dimensional wave propagation in realistic viscoelastic materials. *J. Appl. Mech.*, **32**, 303–314.

Atabek, H. B., 1964: Start-up flow of a Bingham plastic in a circular tube. *Zeit. angew. Math. Mech.*, **44**, 332–333.

Baines, P. G., 1967: Forced oscillations of an enclosed rotating fluid. *J. Fluid Mech.*, **30**, 533–546.

Bakshi, S. K., 1969: Magnetoelastic disturbances in a perfectly conducting cylindrical shell in a constant magnetic field. *PAGEOPH*, **76**, 56–64.

Banerjee, A., 1980: Torsional oscillation of a semi-infinite non-homogeneous transversely isotropic circular cylinder one of whose ends is acted on by an impulsive twist. *Bull. Cal. Math. Soc.*, **72**, 309–314.

Barakat, R. G., 1960: Transient diffraction of scalar waves by a fixed sphere. *J. Acoust. Soc. Am.*, **32**, 61–66.

Barakat, R. G., 1961: Propagation of acoustic pulses from a circular cylinder. *J. Acoust. Soc. Am.*, **33**, 1759–1764.

Barber, A. D., J. H. Weiner and B. A. Boley, 1957: An analysis of the effect of thermal contact resistance in a sheet-stringer structure. *J. Aeronaut. Sci.*, **24**, 232–234.

Bathaiah, D., 1978: Forced oscillations of an enclosed rotating fluid under a uniform magnetic field. *Indian J. Pure Appl. Math.*, **9**, 996–1003.

Bathaiah, D., 1980: MHD flow through a porous straight channel. *Acta Mech.*, **35**, 223–229.

Batu, V., 1982: Time-dependent, linearized two-dimensional infiltration and evaporation from nonuniform and nonperiodic strip sources. *Water Resour. Res.*, **18**, 1725–1733.

Batu, V., 1983: Time-dependent linearized two-dimensional analytical infiltration and evaporation from nonuniform and periodic strip sources. *Water Resour. Res.*, **19**, 1523–1529.

Batu, V., 1989: A generalized two-dimensional analytical solution for hydrodynamic dispersion in bounded media with the first-type boundary condition at the source. *Water Resour. Res.*, **25**, 1125–1132.

Batu, V., and M. T. van Genuchten, 1990: First- and third-type boundary conditions in two-dimensional solute transport modeling. *Water Resour. Res.*, **26**, 339–350.

Baumeister, K. J., and T. D. Hamill, 1969: Hyperbolic heat-conduction equation – A solution for the semi-infinite body problem. *J. Heat Transfer*, **91**, 543–548.

Bell, R. P., 1945: A problem of heat conduction with spherical symmetry. *Proc. Phys. Soc.*, **57**, 45–48.

Bellman, R., R. E. Marshak and G. W. Wing, 1949: Laplace transform solution of two-medium neutron ageing problem. *Phil. Mag.*, *Ser. 7*, **40**, 297–308.

Belluigi, A., 1958: La non stazionarietà dei fenomeni elettrogeosmotici. *Geofisica pura e appl.*, **40**, 97–119.

Benfield, A. E., 1951: The temperature in an accumulating snow field. *Mon. Not. R. Astron. Soc., Geophys. Suppl.*, **6**, 139–147.

Berger, H., and J. W. E. Griemsmann, 1968: Transient electromagnetic guided wave propagation in moving media. *IEEE Trans. Microwave Theory Tech.*, **MTT-16**, 842–849.

Bhaduri, S., and M. Kanoria, 1981: Forced vibration of an isotropic circular cylinder having a rigid cylindrical inclusion. *Rev. Roum. Sci. Tech. - Méc. Appl.*, **26**, 475–480.

Bhaduri, S., and M. Kanoria, 1982: Forced vibration of an isotropic sphere having a rigid spherical inclusion. *Indian J. Theoret. Phys.*, **30**, 93–98.

Bhanja, N., 1971: Torsional vibration of a solid finite cylinder having transverse isotropy. *PAGEOPH*, **85**, 182–188.

Bhat, A. M., R. Prakash and J. S. Saini, 1983: Heat transfer in nucleate pool boiling at high heat flux. *Int. J. Heat Mass Transfer*, **26**, 833–840.

Bhattacharya, J., and S. C. Das Gupta, 1963: On leaking modes coupled with shear waves. *Zeit. Geophys.*, **29**, 101–114.

Bhattacharyya, B. K., 1957: Propagation of an electric pulse through a homogeneous and isotropic medium. *Geophysics*, **22**, 905–921.

Bhattacharyya, P., 1962: Note on the flow of a viscous fluid standing on a rigid base due to a single periodic pulse of tangential force on the surface. *Indian J. Theoret. Phys.*, **10**, 73–76.

Bhattacharyya, P., 1966: Note on the flow of two incompressible immiscible viscous fluids due to the pulses of tangential force on the upper surface. *Indian J. Theoret. Phys.*, **14**, 45–58.

Bhattacharya, R., 1975: On forced vibration of anisotropic spherical shell of variable density. *Bull. Cal. Math. Soc.*, **67**, 87–95.

Bhuta, P. G., and L. R. Koval, 1966: A viscous ring damper for a freely precessing satellite. *Int. J. Mech. Sci.*, **8**, 383–395.

Binnie, A. M., 1951: The effect of friction on surges in long pipe-lines. *Quart. J. Mech. Appl. Math.*, **4**, 330–343.

Birchfield, G. E., 1969: Response of a circular model Great Lake to a suddenly imposed wind stress. *J. Geophys. Res.*, **74**, 5547–5554.

Biswas, P., 1976: Large deflection in heated equilateral triangular plate. *Indian J. Pure Appl. Math.*, **7**, 257–264.

Blackwell, B. F., 1990: Temperature profile in semi-infinite body with exponential source and convective boundary condition. *J. Heat Transfer*, **112**, 567–571.

Bochever, F. M., A. E. Oradovskaya and V. I. Pagurova, 1966: Convective diffusion of salts in a radial flow of groundwater. *J. Appl. Mech. Tech. Phys.*, **7(2)**, 87–88.

Bödvarsson, G. S., and C. F. Tsang, 1982: Injection and thermal breakthrough in fractured geothermal reservoirs. *J. Geophys. Res.*, **87**, 1031–1048.

Bodvarsson, G. S., S. M. Benson and P. A. Witherspoon, 1982: Theory of the development of geothermal systems charged by vertical faults. *J. Geophys. Res.*, **87**, 9317–9328.

Boehringer, J. C., and J. Spindler, 1963: Radiant heating of semitransparent materials. *AIAA J.*, **1**, 84–88.

Borisov, V. V., 1970: Transition in the limit to the velocity of light in the problem of the incidence of a plane wave on a moving ionization front. *Radiophys. Quantum Electron.*, **13**, 1059–1061.

Bouthillon, L., 1947: Oscillations et phénomènes transitoires. Leur étude par les transformations de Laplace et de Cauchy. *Ann. Radioélec.*, **2**, 283–328.

Braude, S. Ya., and I. L. Verbitskii, 1977: Derivation of the function describing the energy distribution of particles in a turbulent synchrotron pile. *Radiophys. Quantum Electr.*, **20**, 663–668.

Bromwich, T. J. I'a., 1916: Normal coordinates in dynamical systems. *Proc. Lond. Math. Soc., Ser. 2*, **15**, 401–448.

Bromwich, T. J. I'a., 1919: Examples of operational methods in mathematical physics. *Phil. Mag., Ser. 6*, **37**, 407–419.

Bromwich, T. J. I'a., 1921: Symbolical methods in the theory of conduction of heat. *Proc. Camb. Phil. Soc.*, **20**, 411–427.

Bromwich, T. J. I'a., 1928: Some solutions of the electromagnetic equations, and of the elastic equations, with applications to the problem of secondary waves. *Proc. Lond. Math. Soc., Ser. 2*, **28**, 438–475.

Brown, A., 1965: Diffusion of heat from a sphere to a surrounding medium. *Aust. J. Phys.*, **18**, 483–489.

Dujurke, N. M., 1982: Effect of frequency of residual stresses of oscillating plate problem in MHD. *Indian J. Pure Appl. Math.*, **13**, 1492–1496.

Burka, A. L., 1966: Asymmetric radiative-convective heating of an infinite plate. *J. Appl. Math. Tech. Phys.*, **7(2)**, 85–86.

Carney, J. F., 1968: Dynamic response of a spherical structural system in fluid media. *Int. J. Mech. Sci.*, **10**, 583–591.

Carslaw, H. S., 1920: Bromwich's method of solving problems in the conduction of heat. *Phil. Mag., Ser. 6*, **39**, 603–611.

Carslaw, H. S., 1928: Operational methods in mathematical physics. *Math. Gazette*, **14**, 216–228.

Carslaw, H. S., and J. C. Jaeger, 1938: Some problems in the mathematical theory of the conduction of heat. *Phil. Mag., Ser. 7*, **26**, 473–495.

Carslaw, H. S., 1938: Operational methods in mathematical physics. *Math. Gazette*, **22**, 264–280.

Carslaw, H. S., 1940: A simple application of the Laplace transform. *Phil. Mag., Ser. 7*, **30**, 414–417.

Carslaw, H. S., and J. C. Jaeger, 1940: Some two-dimensional problems in conduction of heat with circular symmetry. *Proc. Lond. Math. Soc.*,

Ser. 2, **46**, 361–388.

Carslaw, H. S., and J. C. Jaeger, 1940: The determination of Green's function for the equation of conduction of heat in cylindrical coordinates by the Laplace transformation. *J. Lond. Math. Soc., Ser. 1*, **15**, 273–281.

Carslaw, H. S., and J. C. Jaeger, 1941: The determination of Green's function for line sources for the equation of conduction of heat in cylindrical coordinates by the Laplace transformation. *Phil. Mag., Ser. 7*, **31**, 204–208.

Carslaw, H. S., and J. C. Jaeger, 1957: *Conduction of Heat in Solids.* Second Ed. Oxford: At the Clarendon Press, Chapters 12–15.

Carslaw, H. S., and J. C. Jaeger, 1963: *Operational Methods in Applied Mathematics.* New York: Dover Publications, Inc., Chapters 5–10.

Carson, J. R., 1922: The Heaviside operational calculus. *Bell Syst. Tech. J.*, **1(2)**, 43–55.

Carson, J. R., 1925: Electric circuit theory and the operational calculus. *Bell Syst. Tech. J.*, **4**, 685–761.

Carson, J. R., 1926: The Heaviside operational calculus. *Bull. Am. Math. Soc., Ser. 2*, **32**, 43–68.

Carson, J. R., 1926: Electric circuit theory and the operational calculus. *Bell Syst. Tech. J.*, **5**, 50–95, 336–384.

Chakrabarti, R., 1974: Forced vibrations of a non-homogeneous isotropic elastic spherical shell. *PAGEOPH*, **112**, 52–57.

Chakrabarty, A., 1971: Mechanical response in a piezoelectric transducer subjected to a current flowing in a semi-conducting boundary layer characterized by a polarization gradient. *Indian J. Theoret. Phys.*, **19**, 65–70.

Chakraborty, S. K., 1958: On disturbances generated by a pulse of pressure on the surface of a spherical cavity in an elastic medium. *Indian J. Theoret. Phys.*, **6**, 85–89.

Chakraborty, S. K., and A. B. Roy, 1965: Note on the propagation of waves from a spherical cavity in an infinite solid. *PAGEOPH*, **61**, 23–28.

Chakraborty, S. K., and A. B. Roy, 1967: Propagation of waves in anisotropic visco-elastic solid from a spherical cavity. *Bull. Cal. Math. Soc.*, **59**, 81–84.

Chakravorty, J. G., and P. K. Chaudhuri, 1983: On the propagation of waves due to a sudden impulse on the boundary of the spherical cavity. *Indian J. Pure Appl. Math.*, **14**, 965–973.

Chakravorty, M., and S. Bhaduri, 1981: Dynamic response of isotropic non-homogeneous spherical and cylindrical thick-walled shells to applied pressure on the inner boundary. *Bull. Cal. Math. Soc.*, **73**, 232–236.

Chang, C. C., and J. T. Yen, 1959: Rayleigh's problem in magnetohydrodynamics. *Phys. Fluids*, **2**, 393–403.

Chang, C. C., and H. B. Atabek, 1962: Flow between two co-axial tubes near the entry. *Zeit. angew. Math. Mech.*, **42**, 425–430.

Charles, M., and Ph. R. Smith, 1970: A general solution to unsteady coupled magnetohydrodynamic Couette flow. *Appl. Sci. Res.*, **22**, 44–59.

Chase, C. A., D. Gidaspow and R. E. Peck, 1969: A regenerator-prediction of Nusselt numbers. *Int. J. Heat Mass Transfer*, **12**, 727–736.

Chattopadhyay, N. C., 1967: Note on thermo-elastic stresses in a thin semi-infinite rod due to some time-dependent temperature applied to its free end. *Indian J. Theoret. Phys.*, **15**, 49–55.

Chaudhuri, B., 1969: On magnetoelastic disturbances in an aeolotropic elastic cylinder placed in a magnetic field. *PAGEOPH*, **73**, 60–68.

Chaudhuri, B. R., 1971: Dynamical problems of thermo-magnetoelasticity for cylindrical regions. *Indian J. Pure Appl. Math.*, **2**, 631–656.

Chawla, S. S., 1967: Magnetohydrodynamic unsteady free convection. *Zeit. angew. Math. Mech.*, **47**, 499–508.

Chen, C.-S., 1985: Analytical and approximate solutions to radial dispersion from an injection well to a geological unit with simultaneous diffusion into adjacent strata. *Water Resour. Res.*, **21**, 1069–1076.

Chen, C.-S., 1986: Solutions for radionuclide transport from an injection well into a single fracture in a porous formation. *Water Resour. Res.*, **22**, 508–518.

Chen, S.-Y., 1958: Transient temperature and thermal stresses in skin of hypersonic vehicle with variable boundary conditions. *Trans. ASME*, **80**, 1389–1394.

Chen, S.-Y., 1961: One-dimensional heat conduction with arbitrary heating rate. *J. Aerospace Sci.*, **28**, 336–337.

Chester, W., 1952: The reflection of a transient pulse by a parabolic cylinder and a paraboloid of revolution. *Quart. J. Mech. Appl. Math.*, **5**, 190–205.

Chin, J. H., 1962: Effect of uncertainties in thermocouple location on computing surface heat fluxes. *ARS J.*, **32**, 273–274.

Chonan, S., 1977: Response of an elastically supported finite beam to a moving load with consideration of the mass of the foundation. *Bull. JSME*, **20**, 1566–1571.

Chou, P. C., and P. F. Gordon, 1967: Radial propagation of axial shear waves in nonhomogeneous elastic media. *J. Acoust. Soc. Am.*, **42**, 36–41.

Choudhuri, S. K. R., 1970: Note on the thermoelastic stress in a thin rod of finite length due to some constant temperature applied to its free end, the other end being fixed and insulated. *Indian J. Theoret. Phys.*, **18**, 99–106.

Choudhuri, S. K. R., 1972: Thermoelastic stress in a rod due to distributed time-dependent heat sources. *AIAA J.*, **10**, 531–533.

Choudhury, N. K. D., and Z. U. A. Warsi, 1964: Weighting function and transient thermal response of buildings. Part I – Homogeneous structure. *Int. J. Heat Mass Transfer*, **7**, 1309–1321.

Chouhury, A. R., 1971: On disturbances in a piezoelectric layer permeated by a magnetic field. *PAGEOPH*, **87**, 111–116.

Chouhury, A. R., 1971: On the electrical response of a non-homogeneous piezo-electric transducer, due to shock-loaded stress. *PAGEOPH*, **87**, 117–123.

Chu, H. S., C. K. Chen and C. I. Weng, 1983: Transient response of circular pins. *J. Heat Transfer*, **105**, 205–208.

Chu, J.-Y., 1963: Elasto-dynamic stresses produced by a source of heat located around a spherical cavity. *Arch. Mech. Stos.*, **15**, 565–582.

Chu, S. C., and S. G. Bankoff, 1964/65: Heat transfer to slug flows with finite wall thickness. *Appl. Sci. Res.*, **A14**, 379–395.

Chun, D. H., 1970: Distribution of concentration in flow through a circular pipe. *Int. J. Heat Mass Transfer*, **13**, 717–723.

Chun, K. R., 1972: Evaporation from a semi-infinite region with a non-volatile solute. *J. Heat Transfer*, **94**, 238–240.

Churchill, R. V., 1936/37: The inversion of the Laplace transformation by a direct expansion in series and its application to boundary-value problems. *Math. Zeit.*, **42**, 567–579.

Churchill, R. V., 1941: A heat conduction problem introduced by C. J. Tranter. *Phil. Mag.*, *Ser. 7*, **31**, 81–87.

Ciałkowski, M. J., and K. W. Grysa, 1980: On a certain inverse problem of temperature and thermal stress fields. *Acta Mech.*, **36**, 169–185.

Clark, G. B., and G. B. Rupert, 1966: Plane and spherical waves in a Voigt medium. *J. Geophys. Res.*, **71**, 2047–2053.

Cohen, L., 1923: Electrical oscillations on lines. *J. Franklin Inst.*, **195**, 45–58.

Cohen, L., 1923: Alternating current cable telegraphy. *J. Franklin Inst.*, **195**, 165–182.

Cole, J. D., and T. Y. Wu, 1952: Heat conduction in a compressible fluid. *J. Appl. Mech.*, **19**, 209–213.

Compagnone, N. F., 1991: A new equation for the limiting capacity of the lead/acid cell. *J. Power Sources*, **35**, 97–111.

Cooper, F., 1977: Heat transfer from a sphere to an infinite medium. *Int. J. Heat Mass Transfer*, **20**, 991–993.

Copley, J. A., and W. C. Thomas, 1974: Two-dimensional transient temperature distribution in cylindrical bodies with pulsating time and space-dependent boundary conditions. *J. Heat Transfer*, **96**, 300–306.

Craggs, J. W., 1945: Heat conduction in semi-infinite cylinders. *Phil. Mag., Ser. 7*, **36**, 220–222.

Crossley, A. F., 1928: Operational solution of some problems in viscous fluid motion. *Proc. Camb. Phil. Soc.*, **24**, 231–235.

Csanady, G. T., 1968: Motions in a model Great Lake due to a suddenly imposed wind. *J. Geophys. Res.*, **73**, 6435–6447.

Daimaruya, M., M. Naitoh and K. Hamada, 1984: Propagation of elastic waves in a finite length bar with a variable cross section. *Bull. JSME*, **27**, 872–878.

Daimaruya, M., M. Naitoh and S. Tanimura, 1988: Impact end stress and elastic response of a finite length bar with a variable cross-section colliding with a rigid wall. *J. Sound Vib.*, **121**, 105–115.

Danilovskaya, V. I., and V. N. Zubchaninova, 1968: Temperature fields and stresses created in a plate by radiant energy. *Sov. Applied Mech.*, **4(1)**, 63–66.

Das, A., and A. Ray, 1981: Stress response in a piezoelectric circular plate. *J. Math. Phys. Sci.*, **15**, 517–524.

Das, B. R., 1961: Note on thermoelastic stresses in a thin semi-infinite rod due to some time dependent temperature applied to its free end. *Indian J. Theoret. Phys.*, **9**, 49–55.

Das, N. C., 1967: A note on disturbances in an inhomogeneous elastic bar acted upon by a magnetic field. *Arch. Mech. Stos.*, **19**, 765–769.

Das Gupta, S. C., 1954: Waves and stresses produced in an elastic medium due to impulse radial forces and twist on the surface of spherical cavity. *Geofisica pura e appl.*, **27**, 1–6.

Das Gupta, S. P., 1968: Effect of low velocity layer in earthquakes. *Zeit. Geophys.*, **34**, 1–8.

Datta, B. K., and P. R. Sengupta, 1984: Note on wave propagation in an infinite elastic space due to explosion at the cavity centre. *Rev. Roum. Sci. Tech. - Méc. Appl.*, **29**, 487–492.

Datta, N., and S. K. Mishra, 1988: Couette flow of a dusty fluid in a rotating frame of reference. *J. Math. Phys. Sci.*, **22**, 421–430.

Datta, N., and D. C. Dalal, 1992: Unsteady flow of a dusty fluid through a circular pipe with impulsive pressure gradient. *Acta Mech.*, **95**, 51–57.

Davies, B., 1978: *Integral Transforms and Their Applications.* New York: Springer-Verlag, Chapter 4.

Davies, R. M., 1948: A critical study of the Hopkinson pressure bar. *Phil. Trans. R. Soc. Lond.*, **A240**, 376–457.

Davies, W., 1959: Thermal transients in graphite-copper contacts. *Brit. J. Appl. Phys.*, **10**, 516–522.

De, S., 1970: Forced vibration of a thin non-homogeneous circular plate having a central hole. *PAGEOPH*, **80**, 84–91.

De, S., 1971: Disturbances produced in an infinite cylindrically-aeolotropic medium. *PAGEOPH*, **90**, 23–29.

De, T. K., 1982: On the phase boundary motion in the earth due to pressure and temperature excitations at the earth's surface. *Zeit. angew. Math. Mech.*, **62**, 249–255.

Debnath, L., 1970: Unsteady hydromagnetic flow induced by an oscillating disk. *Bull. Cal. Math. Soc.*, **62**, 173–182.

Debnath, L., 1972: On unsteady magnetohydrodynamic boundary layers in a rotating flow. *Zeit. angew. Math. Mech.*, **52**, 623–626.

Debnath, L., 1972: On an unsteady hydromagnetic channel flow. *Bull. Cal. Math. Soc.*, **64**, 143–149.

Debnath, L., 1973: On an unsteady hydromagnetic boundary layer flow. *Bull. Cal. Math. Soc.*, **65**, 109–114.

Debnath, L., 1973: On Ekman and Hartmann boundary layers in a rotating fluid. *Acta Mech.*, **18**, 333–341.

Derski, W., 1958: The state of stress in a thin circular ring, due to a non-steady temperature field. *Arch. Mech. Stos.*, **10**, 255–269.

Derski, W., 1958: On transient thermal stresses in a thin circular plate. *Arch. Mech. Stos.*, **10**, 551–558.

Derski, W., 1959: Non-steady-state of thermal stresses in a layered elastic space with a spherical cavity. *Arch. Mech. Stos.*, **11**, 303–316.

Derski, W., 1961: A dynamic problem of thermoelasticity concerning a thin circular plate. *Arch. Mech. Stos.*, **13**, 177–184.

Devanathan, R., and A. R. Rao, 1973: Forced oscillations of a contained rotating stratified fluid. *Zeit. angew. Math. Mech.*, **53**, 617–623.

Dhaliwal, R. S., and K. L. Chowdhury, 1968: Dynamic problems of thermoelasticity for cylindrical regions. *Arch. Mech. Stos.*, **20**, 46–66.

Dhar, A. K., 1982: Heat-source problem of thermoelasticity in a non-simple elastic medium. *Indian J. Pure Appl. Math.*, **13**, 1384–1392.

Dhar, A. K., 1983: Non-simple quasi-static thermoelastic deflection of a thin clamped circular plate. *Rev. Roum. Sci. Tech. - Méc. Appl.*, **28**, 431–437.

Domingos, H., and D. Voelker, 1976: Transient temperature rise in layered media. *J. Heat Transfer*, **98**, 329–330.

Doorly, J. E., and M. L. G. Oldfield, 1987: The theory of advanced multilayer thin film heat transfer gauges. *Int. J. Heat Mass Transfer*, **30**, 1159–1168.

Dörr, J., 1943: Der unendliche, federnd gebettete Balken unter dem Einfluss einer gleichförmig bewegten Last. *Ing.-Arch.*, **14**, 167–192.

Drake, D. G., 1960: Rayleigh's problem in magnetohydrodynamics for a non-perfect conductor. *Appl. Sci. Res.*, **B8**, 467–477.

Dube, S. K., and M. A. A. Khan, 1968: On unsteady hydromagnetic flow of an electrically conducting incompressible fluid in contact with a harmonically oscillating plate. *Bull. de la Cl. de Scienc. l'Acad. roy.*

de Belgique, Ser. 5, **54**, 732–744.

Dube, S. K., 1970: Hydromagnetic flow near an accelerated flat plate. *Indian J. Pure Appl. Math.*, **1**, 170–177.

Dube, S. K., 1970: Temperature distribution in Poiseuille flow between two parallel flat plates. *Indian J. Pure Appl. Math.*, **1**, 277–283.

Dube, S. N., 1972: On the flow of Rivlin-Ericksen fluids in a channel bounded by two parallel flat plates. *Indian J. Pure Appl. Math.*, **3**, 396–401.

Dube, S. N., and J. Singh, 1972: Unsteady flow of a dusty fluid between two parallel flat plates. *Indian J. Pure Appl. Math.*, **3**, 1175–1182.

Dube, S. N., and C. L. Sharma, 1976: Exact solution of the transient forced convection energy equation for timewise variation of inlet temperature in a circular pipe. *Indian J. Pure Appl. Math.*, **7**, 610–615.

Dubinkin, M. V., 1959: The propagation of waves in infinite plates. *J. Appl. Math. Mech.*, **23**, 1409–1413.

Duffey, T. A., and J. N. Johnson, 1981: Transient response of a pulsed spherical shell surrounded by an infinite elastic medium. *Int. J. Mech. Sci.*, **23**, 589–593.

Duffy, D., 1985: The temperature distribution within a sphere placed in a directed uniform heat flux and allowed to radiatively cool. *J. Heat Transfer*, **107**, 28–32.

Dunn, H. S., and W. G. Dove, 1967: Focusing of aperiodic waves in a linear viscous medium. *J. Acoust. Soc. Am.*, **42**, 613–615.

Durant, N. J., 1960: Stress in a dynamically loaded helical spring. *Quart. J. Mech. Appl. Math.*, **13**, 251–256.

Dutta, K. L., and S. K. Chakraborty, 1990: Dynamic response of an elastic medium to random cylindrical sources. *Indian J. Pure Appl. Math.*, **21**, 867–877.

Dutta, S., 1972: The effects of Hall current on unsteady slip flow over a flat plate under transverse magnetic field. *Arch. Mech.*, **24**, 269–276.

Dźygadło, Z., 1968: Pressure on a cylindrical shell performing unsteady oscillation in external or internal linearized supersonic flow. *Proc. Vibr. Prob.*, **9**, 129–146.

Eason, G., 1963: Propagation of waves from spherical and cylindrical cavities. *Zeit. angew. Math. Phys.*, **14**, 12–23.

Eason, G., 1964: Transient thermal stresses in anisotropic bodies with spherical symmetry. *Appl. Sci. Res.*, **A13**, 1–15.

Ehlers, F. E., 1955: The lift and moment on a ring concentric to a cylindrical body in supersonic flow. *J. Aeronaut. Sci.*, **22**, 239–248.

Elrick, D. E., 1961: Transient two-phase capillary flow in porous media. *Phys. Fluids*, **4**, 572–575.

Esmeijer, W. L., 1949: On the dynamic behaviour of an elastically supported beam of infinite length, loaded by a concentrated force. *Appl. Sci. Res.*, **A1**, 151–168.

Everdingen, A. F. van, and W. Hurst, 1949: The application of the Laplace transform to flow problems in reservoirs. *Trans. Am. Inst. Min. Metall. Pet. Eng.*, **186**, 305–324.

Filippov, G. V., and V. G. Shakov, 1967: The unsteady spatial laminar boundary layer on magnetohydrodynamics. *J. Appl. Mech. Tech. Phys.*, **8(6)**, 42–45.

Frank, I., 1958: Transient temperature distribution in aircraft structures. *J. Aeronaut. Sci.*, **25**, 265–267.

Fukumoto, Y., 1990: General unsteady circulatory flow outside a porous circular cylinder with suction or injection. *J. Phys. Soc. Japan*, **59**, 918–926.

Ganguly, A., 1978: A note on the commencement of yielding due to suddenly applied pressure in a transversely isotropic medium. *Bull. Cal. Math. Soc.*, **70**, 215–219.

Garg, P. C., 1978: Flow of a conducting dusty gas past an accelerated plate. *Bull. Cal. Math. Soc.*, **70**, 265–270.

Gaunaurd, G., 1977: One-dimensional model for acoustic absorption in a viscoelastic medium containing short cylindrical cavities. *J. Acoust. Soc. Am.*, **62**, 298–307.

Gembarovič, J., and V. Majerník, 1987: Determination of thermal parameters of relaxation materials. *Int. J. Heat Mass Transfer*, **30**, 199–201.

Gerbes, W., 1951: Zur instationären, laminaren Strömung einer inkompressiblen, zähen Flüssigkeit in kreiszylindrischen Rohren. *Zeit. angew. Phys.*, **3**, 267–271.

Gershunov, E. M., 1972: Hydrodynamic pressure of a liquid on a shell with hydraulic shock. *Sov. Appl. Mech.*, **8**, 715–721.

Gershunov, E. M., 1975: Hydrodynamic pressure of a liquid against a shell in the case of hydraulic shock. *Sov. Appl. Mech.*, **11**, 405–409.

Ghosh, A. K., 1967: Flow of a viscous liquid between two coaxial circular porous cylinders due to longitudinal motion of the inner cylinder. *Indian J. Pure Appl. Phys.*, **5**, 179–181.

Ghosh, A. K., 1968: Note on the temperature distribution in a viscous fluid flowing through a channel bounded by two coaxial pipes. *Rev. Roum. Sci. Tech. - Méc. Appl.*, **13**, 1073–1084.

Ghosh, S. K., S. K. Banerjee and T. K. Banerjee, 1984: Transient response in a piezo-electric transducer. *Indian J. Theoret. Phys.*, **31(1)**, 27–31; **31**, 199–211; **32**, 21–30, 113–128.

Ghosh, S. L., 1966: Note on the longitudinal propagation of elastic disturbance in a thin inhomogeneous elastic rod. *Indian J. Theoret. Phys.*, **14**, 21–26.

Ghosh, S. L., 1966: Note on the longitudinal vibration of a paraboloidal bar. *Indian J. Theoret. Phys.*, **14**, 81–85.

Ghosh, S. L., 1966: Stress waves in a rod of viscoelastic linear Maxwell material. *Indian J. Pure Appl. Phys.*, **4**, 441–442.

Ghosh, S. S., 1969: On the disturbances in a viscoelastic medium due to blast inside a spherical cavity. *PAGEOPH*, **75**, 93–97.

Gibson, R. E., R. L. Schiffman and S. L. Pu, 1970: Plane strain and axially symmetric consolidation of a clay layer on a smooth impervious bar. *Quart. J. Mech. Appl. Math.*, **23**, 505–520.

Giedt, W. H., and D. R. Hornbaker, 1962: Transient temperature variation in a thermally orthotropic plate. *ARS J.*, **32**, 1902–1909.

Giere, A. C., 1964/65: Transient heat flow in a composite slab – constant flux, zero flux boundary conditions. *Appl. Sci. Res.*, **A14**, 191–198.

Glauz, R. D., and E. H. Lee, 1954: Transient wave analysis in a linear time-dependent material. *J. Appl. Phys.*, **25**, 947–953.

Goldenberg, H., 1951: A problem in radial heat flow. *Brit. J. Appl. Phys.*, **2**, 233–237.

Goldman, S., 1949: *Laplace Transform Theory and Electrical Transients.* New York: Dover Publications, Inc., Chapter 10.

Goldstein, S., 1932: Some two-dimensional diffusion problems with circular symmetry. *Proc. Lond. Math. Soc., Ser. 2*, **34**, 51–88.

Gopinath, M., and L. Debnath, 1973: On the growth of unsteady boundary layers on porous flat plates. *PAGEOPH*, **109**, 1810–1818.

Gopinath, M., and L. Debnath, 1973: On unsteady motion of a rotating fluid bounded by flat porous plates. *PAGEOPH*, **110**, 1996–2004.

Gournay, L. S., 1966: Conversion of electromagnetic to acoustic energy by surface heating. *J. Acoust. Soc. Am.*, **40**, 1322–1330.

Griffith, M. V., and G. K. Horton, 1946: The transient flow of heat through a two-layer wall. *Proc. Phys. Soc.*, **58**, 481–487.

Groves, R. N., 1966: On the unsteady motion of a viscous hydromagnetic fluid contained in a cylindrical vessel. *J. Appl. Mech.*, **33**, 748–752.

Guha, D. K., 1982: Flow of a dusty viscous gas through a circular tube with pressure gradient of any function of time. *Bull. Cal. Math. Soc.*, **74**, 333–338.

Gupta, M., R. Kant and H. S. Sharma, 1979: Unsteady Hele-Shaw flow of non-Newtonian fluid. *J. Math. Phys. Sci.*, **13**, 189–197.

Gupta, M., 1980: Unsteady flow of dusty viscous fluid through equilateral triangular ducts with pressure gradient as any function of time. *Indian J. Theoret. Phys.*, **28**, 151–159.

Gupta, M., and S. L. Khandpur, 1980: Unsteady flow of non-Newtonian fluid through elliptical long ducts with pressure gradient as any function of time. *Indian J. Theoret. Phys.*, **28**, 319–327.

Gupta, M., and N. K. Varshneya, 1981: Unsteady flow of non-Newtonian fluid through equilateral triangular ducts with pressure gradient as any function of time. *Indian J. Theoret. Phys.*, **29**, 57–64.

Gupta, M., and H. S. Sharma, 1981: Unsteady flow of a dusty viscous fluid through confocal elliptical ducts. *Indian J. Theoret. Phys.*, **29**,

227–237.

Hall, M., and L. Debnath, 1973: Some exact solutions of unsteady boundary layer equations - I. *PAGEOPH*, **102**, 167–174.

Hanin, M., 1957: Propagation of an aperiodic wave in a compressible viscous medium. *J. Math. and Phys.*, **36**, 234–249.

Hantush, M. S., 1960: Modification of the theory of leaky aquifers. *J. Geophys. Res.*, **65**, 3713–3725.

Hayasi, N., and K. Inouye, 1965: Transient heat transfer through a thin circular pipe due to unsteady flow in the pipe. *J. Heat Transfer*, **87**, 513–520.

Hearn, C. J., 1993: Response of a uniform unbounded ocean to a moving tropical cyclone. *Appl. Math. Modelling*, **17**, 205–212.

Heasley, J. H., 1965: Transient heat flow between contacting solids. *Int. J. Heat Mass Transfer*, **8**, 146–154.

Hemmings, J. W., 1979: The non-adiabatic calorimeter problem and its application to transfer processes in suspensions of solid. *Int. J. Heat Mass Transfer*, **22**, 99–109.

Hill, J. L., 1966: Torsional-wave propagation from a rigid sphere semiembedded in an elastic half-space. *J. Acoust. Soc. Am.*, **40**, 376–379.

Horvay, G., 1954: Transient thermal stresses in circular disks and cylinders. *Trans. ASME*, **76**, 127–135.

Howell, W. H., 1939: A note on the solution of some partial differential equations in the finite domain. *Phil. Mag.*, *Ser. 7*, **28**, 396–402.

Hsieh, D.-Y., and M. S. Plesset, 1961: Theory of rectified diffusion of mass into gas bubbles. *J. Acoust. Soc. Am.*, **33**, 206–215.

Hsu, T. R., 1969: Thermal shock on a finite disk due to an instantaneous point heat source. *J. Appl. Mech.*, **36**, 113–120.

Hu, C.-L., 1969: Spherical model for an acoustical wave generated by rapid laser heating in a liquid. *J. Acoust. Soc. Am.*, **46**, 728–736.

Huang, C.-L., 1969: On forced vibration of isotropic cylinders. *Appl. Sci. Res.*, **20**, 1–15.

Hume, J. R., and V. J. Skoglund, 1962: Theoretical solution of a transient, fluid-cooled heat generator. *J. Aerospace Sci.*, **29**, 1156–1163.

Jacobs, C., 1971: Transient motions produced by disks oscillating torsionally about a state of rigid rotation. *Quart. J. Mech. Appl. Math.*, **24**, 221–236.

Jacobs, C., 1972: Transient and steady state vorticity generated by horizontal temperature gradients. *Quart. J. Mech. Appl. Math.*, **25**, 303–318.

Jaeger, J. C., 1940: Magnetic screening by hollow circular cylinders. *Phil. Mag.*, *Ser. 7*, **29**, 18–31.

Jaeger, J. C., 1940: The solution of boundary value problems by a double Laplace transformation. *Bull. Am. Math. Soc.*, *Ser. 2*, **46**, 687–693.

Jaeger, J. C., 1941: Heat conduction in composite circular cylinders. *Phil. Mag., Ser. 7*, **32**, 324–335.

Jaeger, J. C., 1941: Conduction of heat in regions bounded by planes and cylinders. *Bull. Am. Math. Soc., Ser. 2*, **47**, 734–741.

Jaeger, J. C., 1942: Heat conduction in a wedge, or an infinite cylinder whose cross-section is a circle or a sector of a circle. *Phil. Mag., Ser. 7*, **33**, 527–536.

Jaeger, J. C., 1944: Note on a problem in radial flow. *Proc. Phys. Soc.*, **56**, 197–203.

Jaeger, J. C., 1944: Some problems involving line sources in conduction of heat. *Phil. Mag., Ser. 7*, **35**, 169–179.

Jaeger, J. C., 1945: Conduction of heat in a slab in contact with well-stirred fluid. *Proc. Camb. Phil. Soc.*, **41**, 43–49.

Jaeger, J. C., 1945: On thermal stresses in circular cylinders. *Phil. Mag., Ser. 7*, **36**, 418–428.

Jaeger, J. C., and K. C. Westfold, 1949: Transients in an ionized medium with applications to bursts of solar noise. *Australian J. Sci. Res.*, **A2**, 322–334.

Jaeger, J. C., 1949: *An Introduction to the Laplace Transform with Engineering Applications.* New York: John Wiley & Sons, Inc., Chapter 4.

Jaeger, J. C., 1953: Pulsed surface heating of a semi-infinite solid. *Quart. Appl. Math.*, **11**, 132–137.

Jahagirdar, M. D., and R. M. Lahurikar, 1989: Transient forced and free convection flow past an infinite vertical plate. *Indian J. Pure Appl. Math.*, **20**, 711–715.

Jayasinghe, D. A. P., and H. J. Leutheusser, 1977: Pulsatile waterhammer subject to laminar friction. *J. Basic Engng.*, **94**, 467–472.

Jeffreys, H., 1931: Damping in bodily seismic waves. *Mon. Not. R. Astron. Soc., Geophys. Suppl.*, **2**, 318–323.

Jeffreys, H., 1931: On the cause of oscillatory movement in seismograms. *Mon. Not. R. Astron. Soc., Geophys. Suppl.*, **2**, 407–416.

Jeffreys, H., and E. R. Lapwood, 1957: The reflexion of a pulse within a sphere. *Proc. R. Soc. Lond.*, **A241**, 455–479.

Jeffreys, H., 1964: *Operational Methods in Mathematical Physics.* New York: Stechert-Hafner Service Agency, Chapters 4–6.

Jeffreys, H., and B. S. Jeffreys, 1972: *Methods of Mathematical Physics.* Cambridge: At the University Press, Chapters 18 and 19.

Jelesnianski, C. P., 1970: "Bottom stress time-history" in linearized equations of motion for storm surges. *Mon. Wea. Rev.*, **98**, 462–478.

Jha, P. K., 1980: Unsteady flow of a dusty viscous fluid through a long duct whose cross-section is a cardioid. *Indian J. Theoret. Phys.*, **28**, 89–94.

Jingu, T., K. Hisada, I. Nakahara and S. Machida, 1985: Transient stress in a circular disk under diametrical impact loads. *Bull. JSME*, **28**, 13–19.

Jingu, T., and K. Nezu, 1985: Transient stress in an elastic sphere under diametrical concentrated impact loads. *Bull. JSME*, **28**, 2553–2561.

Jingu, T., and K. Nezu, 1985: Stress waves in an infinite medium under the diametrical concentrated impact loads on the spherical cavity. *Bull. JSME*, **28**, 2592–2598.

Johri, A. K., 1978: Unsteady channel flow of an elastico-viscous liquid. *Indian J. Pure Appl. Math.*, **9**, 481–489.

Kabadi, S. A., and B. Siddappa, 1984: On the flow of an electrically conducting Maxwell fluid past an infinite flat plate in the presence of a transverse magnetic field. *Rev. Roum. Sci. Tech. - Méc. Appl.*, **29**, 593–605.

Kahn, M. A., 1973: Cooling of an initially hot semi-infinite piezoelectric rod by radiation. *Rev. Roum. Sci. Tech. - Méc. Appl.*, **18**, 683–697.

Khare, Km. S., 1980: Flow of perfect gas over porous flat plate. *Indian J. Theoret. Phys.*, **28**, 191–195.

Kaliski, S., and E. Włodarczyk, 1967: Resonance of a longitudinal elastic-visco-plastic wave in a finite bar. *Proc. Vibr. Prob.*, **8**, 113–128.

Kaliski, S., and E. Włodarczyk, 1967: On certain closed-form solutions of the propagation and reflection problem of an elastic-visco-plastic wave in a bar. *Arch. Mech. Stos.*, **19**, 434–452.

Kant, R., 1980: Hydromagnetic flow of an electrically conducting viscous fluid near a time-varying accelerated porous plate with Hall effects. *Indian J. Theoret. Phys.*, **28**, 31–39.

Kantola, R., 1971: Transient response of fluid lines including frequency modulated inputs. *J. Basic Engng.*, **93**, 274–281.

Kao, T., 1977: Non-Fourier heat conduction in thin surface layers. *J. Heat Transfer*, **99**, 343–345 and 501.

Kardas, A., 1966: On a problem in the theory of the unidirectional regenerator. *Int. J. Heat Mass Transfer*, **9**, 567–579.

Kasiviswanathan, S. R., and A. Ramachandra Rao, 1982: On exact solutions of unsteady MHD flow between eccentrically rotating disks. *Arch. Mech.*, **39**, 411–418.

Kaye, J., and V. C. M. Yeh, 1955: Design charts for transient temperature distribution resulting from aerodynamic heating at supersonic speeds. *J. Aeronaut. Sci.*, **22**, 755–762.

Kecs, W. W., 1986: On the longitudinal vibration of the homogeneous, elastic rods. *Rev. Roum. Sci. Tech. - Méc. Appl.*, **31**, 63–70.

Khan, M. A., 1972: Heating of a piezoelectric rod of semi-infinite length due to a prescribed heat flux into it through radiation condition. *PAGEOPH*, **94**, 66–73.

Kishore, N., S. Tejpal and H. K. Katiyar, 1981: On unsteady MHD flow through two parallel porous flat plates. *Indian J. Pure Appl. Math.*,

12, 1372–1379.

Knudsen, H. L., 1947: Pressure and oil flow in oil-filled cables at load variations. *J. Appl. Phys.*, **18**, 545–562.

Kumar, I. J., and H. N. Narang, 1966: Drying of a moist capillary-porous body in moving air. *Int. J. Heat Mass Transfer*, **9**, 95–102.

Kumar, P., and N. P. Singh, 1989: MHD Hele-Shaw flow of an elasticoviscous fluid through porous media. *Bull. Cal. Math. Soc.*, **81**, 32–41.

Kuznetszov, P. E., 1947: The propagation of electromagnetic waves along a line (in Russian). *Prikl. Mat. Mek.*, **11**, 615–620.

Kuznetszov, P. E., 1948: The propagation of electromagnetic waves along two parallel, single conductor lines (in Russian). *Prikl. Mat. Mek.*, **12**, 141–148.

Lahiri, S., and S. K. Dhar, 1984: On the motion of conducting liquid down an inclined plane. *Indian J. Theoret. Phys.*, **32**, 255–260.

Lal, G., P. C. Gupta and R. G. Sharma, 1982: Laminar flow of a viscous fluid between two parallel plates upper plate being laid with a charge density and the lower plate moving with a constant velocity. *Indian J. Theoret. Phys.*, **30**, 271–278.

Lee, E. H., and I. Kanter, 1953: Wave propagation in finite rods of viscoelastic material. *J. Appl. Phys.*, **24**, 1115–1122.

Leibowitz, M. A., and R. C. Ackerberg, 1963: The vibration of a conducting wire in a magnetic field. *Quart. J. Mech. Appl. Math.*, **16**, 507 519.

Levinson, N., 1935: The Fourier transform solution of ordinary and partial differential equations. *J. Math. and Phys.*, **14**, 195–227.

Li, C.-H., 1986: Exact transient solutions of parallel-current transfer processes. *J. Heat Transfer*, **108**, 365–369.

Lindstrom, F. T., and L. Boersma, 1989: Analytical solutions for convective-dispersive transport in confined aquifers with different initial and boundary conditions. *Water Resour. Res.*, **25**, 241–256.

Lowan, A. N., 1934: On the problem of the heat recuperator. *Phil. Mag.*, *Ser. 7*, **17**, 914–933.

Lowan, A. N., 1934: On the operational treatment of certain mechanical and electrical problems. *Phil. Mag.*, *Ser. 7*, **17**, 1134–1144.

Lowan, A. N., 1935: On transverse oscillations of beams under the action of moving variable loads. *Phil. Mag.*, *Ser. 7*, **19**, 708–715.

Lowan, A. N., 1937: On the operational determination of Green's functions in the theory of heat conduction. *Phil. Mag.*, *Ser. 7*, **24**, 62–70.

Lowan, A. N., 1937: On some two-dimensional problems in heat conduction. *Phil. Mag.*, *Ser. 7*, **24**, 410–424.

Lowan, A. N., 1938: On wave-motion for infinite domains. *Phil. Mag.*, *Ser. 7*, **26**, 340–360.

Lowan, A. N., 1939: On the problem of wave-motion for sub-infinite domains. *Phil. Mag.*, *Ser. 7*, **27**, 182–194.

Lowan, A. N., 1940: On some problems in the diffraction of heat. *Phil. Mag.*, *Ser. 7*, **29**, 93–99.

Lowan, A. N., 1941: On the problem of wave-motion for the wedge of an angle. *Phil. Mag.*, *Ser. 7*, **31**, 373–381.

Lowan, A. N., 1945: On the problem of heat conduction in thin plates. *J. Math. and Phys.*, **24**, 22–29.

Ludford, G. S. S., 1959: Rayleigh's problem in hydromagnetics: The impulsive motion of a pole-piece. *Arch. Rational Mech. Anal.*, **3**, 14–27.

Luikov, A., 1936: The application of the Heaviside-Bromwich operational method to the solution of a problem in heat conduction. *Phil. Mag.*, *Ser. 7*, **22**, 239–248.

Luthin, J. N., and J. W. Holmer, 1960: An analysis of the flow of water in a shallow, linear aquifer, and of the approach to a new equilibrium after intake. *J. Geophys. Res.*, **65**, 1573–1576.

Macey, H. H., 1940: Clay-water relationships. *Proc. Phys. Soc.*, **52**, 625–656.

Madejski, J., 1967: Simultaneous mass and heat transfer on an absorbing porous sphere. *Arch. Mech. Stos.*, **19**, 183–196.

Mahalanabis, M., 1971: On disturbances in an inhomogeneous viscoelastic bar of Reiss type in a magnetic field. *PAGEOPH*, **92**, 52–61.

Mahalanabis, R. K., 1968: On temperature changes due to application of impulsive pressure on the surface of a spherical cavity. *Rev. Roum. Sci. Tech. - Méc. Appl.*, **13**, 121–125.

Mahapatra, J. R., 1973: A note on the unsteady motion of a viscous conducting liquid between two porous concentric circular cylinders acted on by a radial magnetic field. *Appl. Sci. Res.*, **27**, 274–282.

Maiti, P. C., 1978: Quasi-static thermal deflection of a thin clamped circular plate subjected to random temperature distribution on its upper face. *Indian J. Pure Appl. Math.*, **9**, 541–547.

Majumder, S. R., 1962: Impulsive rotatory motion of a sphere in a viscous fluid. *Rev. Roum. Sci. Tech. - Méc. Appl.*, **7**, 893–897.

Mareschal, J.-C., 1983: Uplift and heat flow following the injection of magnas into the lithosphere. *Geophys. J. R. Astr. Soc.*, **73**, 109–127.

Marino, M. A., 1967: Hele-Shaw model study of the growth and decay of groundwater ridges. *J. Geophys. Res.*, **72**, 1195–1205.

Mason, D. P., A. Solomon and L. O. Nicolaysen, 1991: Evolution of stress and strain during the consolidation of a fluid-saturated porous elastic sphere. *J. Appl. Phys.*, **70**, 4724–4740.

Masket, A. V., 1946: Forced vibrations of a whirling wire. *Phil. Mag.*, *Ser. 7*, **37**, 426–432.

Matsumoto, H., and S. Ujihashi, 1972: Deformations and stresses in a hollow sphere with spherical transversal isotropy under impulsive pressure. *Bull. JSME*, **15**, 1324–1332.

Matsumoto, H., E. Tsuchida, S. Miyao and N. Tsunadu, 1976: Torsional stress wave propagation in a semi-infinite conical bar. *Bull. JSME*, **19**, 8–14.

Matsumoto, H., I. Nakahara and H. Sekino, 1980: Identification of the dynamic visco-elastic properties under longitudinal impact. *Bull. JSME*, **23**, 1086–1091.

McKay, A. T., 1930: Diffusion into an infinite plane sheet subject to a surface condition, with a method of application to experimental data. *Proc. Phys. Soc.*, **42**, 547–555.

McKay, A. T., 1932: Diffusion for the infinite plane sheet. *Proc. Phys. Soc.*, **44**, 17–24.

McLachlan, N. W., 1939: *Complex Variables & Operational Calculus with Technical Applications*. Cambridge: At the University Press, Chapters 13–15.

McLachlan, N. W., 1948: *Modern Operational Calculus with Applications in Technical Mathematics*. London: MacMillan and Co., Ltd., Chapter 4.

Mersman, W. A., 1942: Heat conduction in a semi-infinite slab. *Phil. Mag., Ser. 7*, **33**, 303–309.

Miles, J. W., 1971: *Integral Transforms in Applied Mathematics*. Cambridge: At the University Press, Chapter 2.

Mintzer, D., and B. S. Tanenbaum, 1960: Spatial and temporal absorption in a viscous medium. *J. Acoust. Soc. Am.*, **32**, 67–71.

Mishra, S. P., and P. Mohapatra, 1973: Flow near an accelerated plate in the presence of a magnetic field. *J. Appl. Phys.*, **44**, 1194–1199.

Mishra, S. P., and D. G. Sahoo, 1978: Magnetohydrodynamic unsteady free convection past a hot vertical plate. *Appl. Sci. Res.*, **34**, 1–16.

Mithal, K. G., 1960: Unsteady flow of a viscous homogeneous incompressible fluid in a circular pipe of uniform cross-section. *Bull. Cal. Math. Soc.*, **52**, 147–154.

Mitra, P., and P. Bhattacharyya, 1981: Unsteady hydromagnetic laminar flow of a conducting dusty fluid between two parallel plates started impulsively from rest. *Acta Mech.*, **39**, 171–182.

Mitra, P., 1981: Unsteady flow of a conducting dusty gas through a circular tube under time dependent pressure gradient in presence of a transverse magnetic field. *Rev. Roum. Sci. Tech. - Méc. Appl.*, **26**, 795–803.

Mitra, P., and P. Bhattacharyya, 1982: On the unsteady flow of a dusty gas between two parallel plates, one being at rest and the other oscillating. *Rev. Roum. Sci. Tech. - Méc. Appl.*, **27**, 57–68.

Mitra, P., 1984: Note on the problem of unsteady dusty viscous fluid flow past a flat plate. *Bull. Cal. Math. Soc.*, **76**, 162–166.

Mohammad, H. K., T. Oroveanu and A. Stan, 1981: Temperature distribution in an oil layer due to hot water injection. *Rev. Roum. Sci. Tech. - Méc. Appl.*, **26**, 673–685.

Mohapatra, P., 1971: Magnetohydrodynamic flow near a time-varying accelerated plate. *Indian J. Phys.*, **45**, 421–431.

Morrison, J. A., 1956: Wave propagation in rods of Voigt material and visco-elastic materials with three-parameter models. *Quart. Appl. Math.*, **14**, 153–169.

Mudford, B. S., 1988: Modeling the occurrence of overpressures on the Scotian Shelf, offshore eastern Canada. *J. Geophys. Res.*, **93**, 7845–7855.

Mukherjee, J., 1970: Waves produced in an infinite elastic medium due to normal forces and twists on the surface of a buried spherical source. *PAGEOPH*, **82**, 11–18.

Mukherjee, S., 1968: Note on the propagation of waves in infinite spherically-aeolotropic medium produced by a blast in a spherical cavity. *PAGEOPH*, **70**, 39–46.

Mukherjee, S., and S. Mukherjee, 1983: Unsteady axisymmetric rotational flow of elastico-viscous liquid. *Indian J. Pure Appl. Math.*, **14**, 1534–1541.

Murthy, M. G., 1966: A dynamical problem of thermoelasticity concerning a circular disc. *Indian J. Pure Appl. Phys.*, **4**, 367–370.

Muthuswamy, V. P., and T. Raghavan, 1984: Forced vibrations of an isotropic, perfectly plastic circular cylinder having a rigid cylindrical inclusion. *Rev. Roum. Sci. Tech. - Méc. Appl.*, **29**, 651–655.

Nag, S. K., R. N. Jana and N. Datta, 1979: Couette flow of a dusty gas. *Acta Mech.*, **33**, 179–187.

Naidu, K. B., 1974: Stratified viscous flow between two oscillating cylinders. *Indian J. Pure Appl. Math.*, **5**, 1127–1136.

Naitoh, M., and M. Daimaruya, 1985: The influence of a rise time of longitudinal impact on the propagation of elastic waves in a bar. *Bull. JSME*, **28**, 20–25.

Nakagawa, N., R. Kawai and M. Akao, 1977: Behavior of impact waves at the interface of an elastic-viscoelastic bar. *Bull. JSME*, **20**, 1402–1408.

Nakamura, K., 1961: Motion of water due to long waves in a rectangular bay of uniform depth. *Sci. Rep. Tohoku Univ., Ser. 5 Geophys.*, **12**, 191–213.

Nalesso, G. F., and A. R. Jacobson, 1988: Shaping of an ion cloud's velocity field by differential braking due to Alfvén wave dissipation in the ionosphere. 1. Coupling with an infinite ionosphere. *J. Geophys. Res.*, **93**, 5794–5802.

Nalesso, G. F., and A. R. Jacobson, 1988: Shaping of an ion cloud's velocity field by differential braking due to Alfvén wave dissipation in the ionosphere. 2. Reflections from the E layer. *J. Geophys. Res.*, **93**, 5803–5809.

Nandy, S. C., 1970: On unsteady flow of a viscous fluid between two parallel plates under variable surface charges. *PAGEOPH*, **83**, 56–61.

Nandy, S. C., 1972: On the unsteady flow of fluid between two parallel plates acted upon by an electric field. *Indian J. Pure Appl. Math.*, **3**, 890–895.

Negi, J. G., and T. Lal, 1969: Deformation of the shape of seismic pulses by a layer of non-uniform velocity distributions. *Zeit. Geophys.*, **35**, 589–610.

Newcomb, T. P., 1958: The flow of heat in a parallel-faced infinite solid. *Brit. J. Appl. Phys.*, **9**, 370–372.

Newcomb, T. P., 1958: The radial flow of heat in an infinite cylinder. *Brit. J. Appl. Phys.*, **9**, 456–458.

Newcomb, T. P., 1959: Flow of heat in a composite solid. *Brit. J. Appl. Phys.*, **10**, 204–206.

Newcomb, T. P., 1959: Transient temperatures attained in disk brakes. *Brit. J. Appl. Phys.*, **10**, 339–340.

Newcomb, T. P., 1960: Temperatures reached in disc brakes. *J. Mech. Engng. Sci.*, **2**, 167–177.

Newman, M. K., 1955: Effect of rotating inertia and shear on maximum strain in cantilever impact excitation. *J. Aeronaut. Sci.*, **22**, 313–320.

Nowacki, W., 1959: Some three-dimensional problems of thermoelasticity. *J. Appl. Math. Mech.*, **23**, 651–665.

Nowacki, W., 1959: Two one-dimensional problems of thermoelasticity. *Arch. Mech. Stos.*, **11**, 333–345.

Odaka, T., and I. Nakahara, 1967: Stresses in an infinite beam impacted by an elastic bar. *Bull. JSME*, **10**, 863–872.

Ogibalov, P. M., 1941: On the spread of plastico-viscous flow about a rotating cylinder (in Russian). *Prikl. Mat. Mek.*, **5**, 13–29.

Olunloya, V. O. S., and K. Hutter, 1979: Forced vibration of a pre-stressed rectangular membrane: Near resonance response. *Acta Mech.*, **32**, 63–77.

Omer, G. C., 1950: Volcanic tremor. Part two: The theory of volcanic tremor. *Bull. Seism. Soc. Am.*, **40**, 175–194.

O'Neill, K., 1982: One-dimensional transport from a highly concentrated, transfer type source. *Int. J. Heat Mass Transfer*, **25**, 27–36.

Pack, D. C., 1956: The oscillations of a supersonic gas jet embedded in a supersonic stream. *J. Aeronaut. Sci.*, **23**, 747–753.

Pal, B. R., 1987: Rotatory vibration of a thin non-homogeneous circular plate under shearing forces. *Rev. Roum. Sci. Tech. - Méc. Appl.*, **32**, 389–396.

Pal, R., 1972: Transient stresses in inhomogeneous elastic hollow circular cylinder due a thermal shock. *Bull. Cal. Math. Soc.*, **64**, 107–119.

Pal, S. K., and P. R. Sengupta, 1986: On the motion of a visco-elastic Maxwell fluid subjected to uniform or periodic body force acting for a finite times. *Indian J. Theoret. Phys.*, **34**, 349–358.

Pan, M., 1974: Forced vibrations of an non-homogeneous anisotropic elastic spherical shell. *Bull. Cal. Math. Soc.*, **66**, 223–232.

Pan, M., 1975: Forced vibrations of an inhomogeneous anisotropic cylindrical shell. *Bull. Cal. Math. Soc.*, **67**, 165–174.

Paria, G., 1963: Study of the transient effect on a superconducting sphere under the influence of a varying magnetic field. *Indian J. Pure Appl. Phys.*, **1**, 87–90.

Parkus, H., 1962: Wärmespannungen bei zufallsabhängiger Oberflächentemperatur. *Zeit. angew. Math. Mech.*, **42**, 499–507.

Pascal, H., 1973: Non-steady multiphase flow through a porous medium. *Rev. Roum. Sci. Tech. - Méc. Appl.*, **18**, 329–343.

Pascal, H., 1973: Nonsteady gas flow in a pipeline network. *Rev. Roum. Sci. Tech. - Méc. Appl.*, **18**, 491–510.

Pascal, H., 1973: Nonsteady gas flow in interconnected pipelines. *Rev. Roum. Sci. Tech. - Méc. Appl.*, **18**, 851–871.

Pascal, H., 1982: Nonsteady gas flow through pipeline systems. *Acta Mech.*, **42**, 49–69.

Paterson, S., 1947: The heating or cooling of a solid sphere in a well-stirred fluid. *Proc. Phys. Soc.*, **59**, 50–58.

Payne, W. T., and L. N. Zadoff, 1965: Electromagnetic diffusion into a moving conductor. *AIAA J.*, **3**, 1294–1297.

Peskin, E., and E. Weber, 1948: The d. c. thermal characteristics of microwave bolometers. *Rev. Sci. Inst.*, **19**, 188–195.

Petrenko, V. G., 1967: On heat conduction of an inhomogeneous plate. *Sov. Appl. Mech.*, **3(11)**, 38–40.

Phythian, J. E., 1963: Cylindrical heat flow with arbitrary heat rates. *AIAA J.*, **1**, 925–927.

Pipes, L. A., 1938: Operational solution of the wave equation. *Phil. Mag.*, *Ser. 7*, **26**, 333–340.

Pipes, L. A., 1939: The operational calculus. III. *J. Appl. Phys.*, **10**, 301–312.

Pohle, F. V., and H. Oliver, 1954: Temperature distribution and thermal stresses in a model of a supersonic wing. *J. Aeronaut. Sci.*, **21**, 8–16.

Polubarinova-Kochina, P. Ia., 1959: Ground water movements at water level fluctuations in a reservoir with a vertical boundary. *J. Appl. Math. Mech.*, **23**, 762–769.

Poppendiek, H. F., 1953: Transient and steady-state heat transfer in irradiated citrus fruit. *ASME Trans.*, **75**, 421–425.

Poppendiek, H. F., 1968: Two-dimensional transport models for the lower layers of the atmosphere. *Int. J. Heat Mass Transfer*, **11**, 67–79.

Prager, W., 1933: Über die Verwendung symbolischer Methoden in der Mechanik. *Ing.-Arch.*, **4**, 16–34.

Prakash, S., 1967: Non-steady parallel viscous flow through a straight channel. *Bull. Cal. Math. Soc.*, **59**, 55–59.

Prakash, S., 1971: Note on the problem of unsteady viscous flow past a flat plate. *Indian J. Pure Appl. Math.*, **2**, 283–289.

Prasada Rao, D. R. V., D. V. Krishna and L. Debnath, 1981: A theory of convective heat transfer in a rotating hydromagnetic viscous flow. *Acta Mech.*, **39**, 225–240.

Pratt, A. W., and E. F. Ball, 1963: Transient cooling of a heated enclosure. *Int. J. Heat Mass Transfer*, **6**, 703–718.

Pratt, A. W., 1965: Fundamentals of heat transmission through the external walls of buildings. *J. Mech. Engng. Sci.*, **7**, 357–366.

Pratt, A. W., 1965: Variable heat flow through walls of cavity construction, naturally exposed. *Int. J. Heat Mass Transfer*, **8**, 861–872.

Pratt, A. W., and R. E. Lacy, 1966: Measurement of the thermal diffusivities of some single-layer walls in buildings. *Int. J. Heat Mass Transfer*, **9**, 345–353.

Pugh, H. Ll. D., and A. J. Harris, 1942: The temperature distribution around a spherical hole in an infinite conducting medium. *Phil. Mag.*, *Ser. 7*, **33**, 661–666.

Puri, P., and P. K. Kulshrestha, 1976: Unsteady hydromagnetic boundary layer in a rotating medium. *J. Appl. Mech.*, **43**, 205–208.

Puri, P., and P. K. Kulshrestha, 1983: Structure of waves in a time dependent hydromagnetic plane Couette flow. *Zeit. angew. Math. Mech.*, **63**, 489–495.

Purohit, G. N., and M. C. Goyal, 1976: Unsteady heat transfer for flow over a flat plate with suction. *Indian J. Pure Appl. Math.*, **7**, 977–982.

Raghavan, T., and V. P. Muthuswamy, 1985: Radial vibrations in cylindrical soil massive in plastic yield state. *Rev. Roum. Sci. Tech. - Méc. Appl.*, **30**, 31–36.

Ram, G., and R. S. Mishra, 1977: Unsteady flow through magnetohydrodynamic porous media. *Indian J. Pure Appl. Math.*, **8**, 637–647.

Ramamurthy, V., 1990: Free convection effects on the Stokes problem for an infinite vertical plate in a dusty fluid. *J. Math. Phys. Sci.*, **24**, 297–312.

Rao, P. B., 1967: Motion of a viscous fluid through a tube subjected to a series of longitudinal pulses. *Indian J. Pure Appl. Phys.*, **5**, 1–5.

Raval, U., and K. N. N. Rao, 1973: Quasi-stationary electromagnetic response of covered permeable conductors to some pulses of spatially uniform magnetic field. *PAGEOPH*, **104**, 553 565.

Reinheimer, J., 1962: Ablation with volume distribution of heat sources. *ARS J.*, **32**, 1106–1107.

Reismann, H., 1962: Temperature distribution in a spinning sphere during atmospheric entry. *J. Aerospace Sci.*, **29**, 151–159.

Reismann, H., and J. Gideon, 1971: Forced wave motion in an unbounded space surrounding a lined, spherical cavity. *PAGEOPH*, **85**, 189–213.

Rich, G. R., 1945: Water-hammer analysis by the Laplace-Mellin transformation. *Trans. ASME*, **67**, 361–376.

Roshal', A. A., 1969: Mass transfer in a two-layer porous medium. *J. Appl. Mech. Tech. Phys.*, **10**, 551–558.

Rotem, Z., J. Gildor and A. Solan, 1963: Transient heat dissipation from storage reservoirs. *Int. J. Heat Mass Transfer*, **6**, 129–141.

Roy, P., 1970: On longitudinal disturbances in a viscoelastic solid of Reiss type placed in a magnetic field. *PAGEOPH*, **82**, 29–33.

Ruisseau, N. R. des, and R. D. Zerkle, 1970: Temperature in semi-infinite and cylindrical bodies subjected to moving heat sources and surface cooling. *J. Heat Transfer*, **92**, 456–464.

Sacheti, N. C., and B. S. Bhatt, 1975: Unsteady motion of a second order fluid between parallel plates. *Indian J. Pure Appl. Math.*, **6**, 996–1006.

Salm, B., 1964: Anlage zur Untersuchung dynamischer Wirkungen von bewegtem Schnee. *Zeit. angew. Math. Phys.*, **15**, 357–374.

Sarker, A. K., 1965: Disturbance due to a shearing-stress applied to the boundary of a spherical cavity in an infinite viscoelastic medium. *Bull. Cal. Math. Soc.*, **57**, 5–15.

Sastry, D. V. S., and R. Seetharamaswamy, 1982: MHD dusty viscous flow through a circular pipe. *Indian J. Pure Appl. Math.*, **13**, 811–817.

Sattarov, R. M., 1977: Some cases of unsteady motion of viscoplastic media in an infinitely long viscoelastic tube. *J. Appl. Mech. Tech. Phys.*, **18**, 355–359.

Saxena, S., and G. C. Sharma, 1987: Unsteady flow of an electrically conducting dusty viscous liquid between two parallel plates. *Indian J. Pure Appl. Math.*, **18**, 1131–1138.

Schatz, E. R., and E. M. Williams, 1950: Pulse transients in exponential transmission lines. *Proc. IRE*, **38**, 1208–1212.

Schetz, J. A., and R. Eichhorn, 1962: Unsteady natural convection in the vicinity of a doubly infinite vertical plate. *J. Heat Transfer*, **84**, 334–338.

Schuder, C. B., and R. C. Binder, 1959: The response of pneumatic transmission lines to step inputs. *J. Basic Engng.*, **81**, 578–583.

Schwarz, W. H., 1963: The unsteady motion of an infinite oscillating cylinder in an incompressible Newtonian fluid at rest. *Appl. Sci. Res.*, **A11**, 115–124.

Sen, B., 1962: Note on the transient response of a linear visco-elastic plate in the form of an equilateral triangle. *Indian J. Theoret. Phys.*, **10**, 77–81.

Sen, S. K., 1983: On the unsteady flow of conducting liquid between two co-axial circular cylinders. *Indian J. Theoret. Phys.*, **31**, 225–232.

Seth, G. S., R. N. Jana and M. K. Maiti, 1981: Unsteady hydromagnetic flow past a porous plate in a rotating medium with time-dependent free stream. *Rev. Roum. Sci. Tech. - Méc. Appl.*, **26**, 383–400.

Sharma, H. S., 1972: Unsteady flow of viscoelastic fluids through circular and coaxial circular ducts with pressure gradients as any function of time. *Indian J. Pure Appl. Math.*, **3**, 535–542.

Sharma, R. S., 1975: Flow over an oscillating porous plate. *Arch. Mech.*, **27**, 115–123.

Sharma, R. S., 1979: Unsteady flow of an elastic-viscous fluid past an infinite porous plate. *Arch. Mech.*, **31**, 199–208.

Sheehan, J. P., and L. Debnath, 1972: Transient vibrations of an isotropic elastic sphere. *PAGEOPH*, **99**, 37–48.

Sheehan, J. P., and L. Debnath, 1972: Forced vibrations of an anisotropic elastic sphere. *Arch. Mech.*, **24**, 117–125.

Shibata, T., and M. Kugo, 1983: Generalization and application of Laplace transformation formulas for diffusion. *Int. J. Heat Mass Transfer*, **26**, 1017–1027.

Shibuya, T., 1975: On the torsional impact of a thick elastic plate. *Int. J. Solids Structures*, **11**, 803–811.

Shipitsina, E. M., 1972: Investigation of the strain of a hollow ball by a pulse loading according to three dimensional elasticity theory and the theory of shells. *Sov. Appl. Mech.*, **8**, 1194–1197.

Siegel, R., and M. Perlmutter, 1962: Heat transfer for pulsating laminar duct flow. *J. Heat Transfer*, **84**, 111–122.

Sigrist, M. W., and F. K. Kneubühl, 1978: Laser-generated stress waves in liquids. *J. Acoust. Soc. Am.*, **64**, 1652–1663.

Singh, C. B., and P. C. Ram, 1977: Unsteady flow of an electrically conducting dusty viscous liquid through a channel. *Indian J. Pure Appl. Math.*, **8**, 1022–1028.

Singh, D., 1963/64: Unsteady motion of a viscous conducting liquid contained between two infinite coaxial cylinders in the presence of an axial magnetic field. *Appl. Sci. Res.*, **B10**, 412–416.

Singh, J., and S. N. Dube, 1975: Unsteady flow of a dusty fluid through a circular pipe. *Indian J. Pure Appl. Math.*, **6**, 69–79.

Singh, K. D., and S. N. Dube, 1986: Exact solution of the mass transfer equation for time variation of inlet concentration in a circular pipe. *J. Math. Phys. Sci.*, **20**, 335–342.

Sinha, D. K., 1962: Note on responses in a piezoelectric plate transducer with a periodic step input. *Indian J. Theoret. Phys.*, **10**, 21–28.

Sinha, D. K., 1963: Note on electrical and mechanical responses with ramp-type input signal in piezoelectric plate transducer. *Indian J. Theoret. Phys.*, **11**, 93–99.

Sinha, D. K., 1965: A note on torsional disturbances in an elastic cylinder in a magnetic field. *Proc. Vibr. Prob.*, **6**, 91–97.

Sinha, K. S., and V. Gupta, 1979: Slow motion of a spheroid on a rotating fluid. *Indian J. Pure Appl. Math.*, **10**, 1183–1195.

Sloan, D. M., 1971: An unsteady MHD duct flow. *Appl. Sci. Res.*, **25**, 126–136.

Smith, E. G., 1941: A simple and rigorous method for the determination of the heat requirements of simple intermittently heated exterior walls. *J. Appl. Phys.*, **12**, 638–642.

Smith, E. G., 1941: The heat requirements of simple intermittently heated interior walls and furniture. *J. Appl. Phys.*, **12**, 642–644.

Smith, J. J., 1923: The solution of differential equations by a method similar to Heaviside's. *J. Franklin Inst.*, **195**, 815–850.

Solarz, L., 1966: Aero-magneto-flutter of a plane duct of finite length. *Proc. Vibr. Prob.*, **7**, 347–362.

Soundalgekar, V. M., S. Ravi and S. B. Miremath, 1980: Hall effects on MHD flow past an accelerated plate. *J. Plasma Phys.*, **23**, 495–500.

Soundalgekar, V. M., 1965: Hydrodynamic flow near an accelerated plate in the presence of a magnetic field. *Appl. Sci. Res.*, **B12**, 151–156.

Sozou, C., 1972: The development of magnetohydrodynamic flow due to the passage of an electric current past a sphere immersed in a fluid. *J. Fluid Mech.*, **56**, 497–503.

Spiga, G., and M. Spiga, 1986: Two-dimensional transient solutions for crossflow heat exchangers with neither gas mixed. *J. Heat Transfer*, **109**, 281–286.

Srinivasan, V., and D. Bathoiath, 1978: The flow of a conducting viscous incompressible fluid between two parallel plates under a uniform transverse magnetic field. *Indian J. Pure Appl. Math.*, **9**, 511–517.

Srivastava, R., and T.-C. J. Yeh, 1991: Analytical solutions for one-dimensional, transient infiltration toward the water table in homogeneous and layered soils. *Water Resour. Res.*, **27**, 753–762.

Starr, A. T., 1930: Lag in a thermometer when the temperature of the external medium is varying. *Phil. Mag.*, Ser. 7, **9**, 901–912.

Steketee, J. A., 1964: An application of the operational calculus to the equations of the Rayleigh problem in MHD. *Appl. Sci. Res.*, **B11**, 255–272.

Steketee, J. A., 1973: EM induction in a semi-infinite solid, impulsively moving in a uniform magnetic field. *Appl. Sci. Res.*, **27**, 307–320.

Stewartson, K., 1953: A weak spherical source in a rotating fluid. *Quart. J. Mech. Appl. Math.*, **6**, 45–49.

Sucec, J., 1986: Transient heat transfer in the laminar thermal entry region of a pipe: An analytical solution. *Appl. Sci. Res.*, **43**, 115–125.

Sur, S. P., 1961: A note on the longitudinal propagation of elastic disturbance in a thin inhomogeneous elastic rod. *Indian J. Theoret. Phys.*, **9**, 61–67.

Suryanarayana, N. V., 1975: Transient response of straight fins. *J. Heat Transfer*, **97**, 417–423.

Suryanarayana, N. V., 1976: Transient response of straight fins. Part II. *J. Heat Transfer*, **98**, 324–327.

Switick, D. M., and L. A. Kennedy, 1968: A solution for unsteady magnetohydrodynamic flow including ion-slip effects. *Zeit. angew. Math. Phys.*, **19**, 145–148.

Szekely, J., 1963: Notes on the transfer at the interface of two independently stirred liquids. *Int. J. Heat Mass Transfer*, **6**, 833–840.

Tandon, P. N., 1970: Flow of a conducting viscoelastic fluid between two parallel plates under a transverse magnetic field. *Indian J. Theoret. Phys.*, **18**, 45–53.

Thomson, W. T., 1950: *Laplace Transforms*. Englewood Cliffs, NJ: Prentice-Hall, Inc., Chapter 8.

Ting, T. C. T., and P. S. Symonds, 1964: Longitudinal impact on viscoplastic rods – Linear stress-strain rate laws. *J. Appl. Mech.*, **31**, 199–207.

Tomski, L., 1981: Longitudinal mass impact of hydraulic servo. *Zeit. angew. Math. Mech.*, **61**, T191–T193.

Touryan, K. J., 1964: Transient temperature variation in a thermally orthotropic cylindrical shell. *AIAA J.*, **2**, 124–126.

Tranter, C. J., 1942: The application of the Laplace transformation to a problem on elastic vibrations. *Phil. Mag., Ser. 7*, **33**, 614–622.

Tranter, C. J., 1944: On a problem in heat conduction. *Phil. Mag., Ser. 7*, **35**, 102–105.

Tranter, C. J., 1947: Heat flow in an infinite medium heated by a cylinder. *Phil. Mag., Ser. 7*, **38**, 131–134.

Tranter, C. J., 1947: Note on a problem in heat conduction. *Phil. Mag., Ser. 7*, **38**, 530–531.

Travelho, J. S., and W. F. N. Santos, 1991: Solution for transient conjugated forced convection in the thermal entrance region of a duct with periodically varying inlet temperature. *J. Heat Transfer*, **113**, 558–562.

Tsui, T., and H. Kraus, 1965: Thermal stress-wave propagation in hollow elastic spheres. *J. Acoust. Soc. Am.*, **37**, 730–737.

Venkatasiva Murthy, K. N., 1979: MHD viscous flow between torsionally oscillating disks. *Appl. Sci. Res.*, **35**, 111–125.

Verma, S. K., and R. N. Singh, 1970: Transient electromagnetic response of an inhomogeneous conducting sphere. *Geophysics*, **35**, 331–336.

Verma, S. K., 1972: Quasi-static time-domain electromagnetic response of a homogeneous conducting infinite cylinder. *Geophysics*, **37**, 92–97.

Verma, S. K., 1973: Time-dependent electromagnetic fields of an infinite, conducting cylinder excited by a long current-carrying cable. *Geophysics*, **38**, 369–379.

Veronis, G., 1958: On the transient response of a β-plane ocean. *J. Oceanogr. Soc. Japan*, **14**, 1–5.

Wadhawan, M. C., 1974: Dynamic thermoelastic response of a cylinder. *PAGEOPH*, **112**, 73–82.

Wagner, K. W., 1940: *Operatorenrechnung nebst Anwendungen in Physik und Technik.* Leipzig: Johann Ambrosius Berth Verlag, Chapter 7.

Wah, T., 1971: Analysis of heat conduction in bridge slab. *Acta Mech.*, **11**, 9–26.

Wait, J. R., 1953: A transient magnetic dipole source in a dissipative medium. *J. Appl. Phys.*, **24**, 341–343.

Warsi, Z. U. A., and N. K. D. Choudhury, 1964: Weighting function and transient response of buildings. Part II – Composite structure. *Int. J. Heat Mass Transfer*, **7**, 1323–1334.

Washio, S., and T. Konishi, 1984: Theoretical investigation on end correction problems (Part IV, Transient flow analysis by multi-mode wave equation). *Bull. JSME*, **27**, 196–203.

Wheeler, L., 1973: Focusing of stress waves in an elastic sphere. *J. Acoust. Soc. Am.*, **53**, 521–524.

Whitehead, S., 1944: An approximate method for calculating heat flow in an infinite medium heated by a cylinder. *Proc. Phys. Soc.*, **56**, 357–366.

Wood, F. M., 1937: The application of Heaviside's operational calculus to the solution of problems in water hammer. *Trans. ASME*, **59**, 707–713.

Yang, W. J., J. A. Clark and V. S. Arpaci, 1961: Dynamic response of heat exchangers having internal heat sources. Part IV. *J. Heat Transfer*, **83**, 321–338.

Yang, W.-J., 1964: Transient heat transfer in a vapor-heated heat exchanger with arbitrary timewise-variant flow perturbation. *J. Heat Transfer*, **86**, 133–142.

Yang, W.-J., and J. A. Clark, 1964: On the application of the source theory to the solution of problems involving phase change. Part 2. Transient interface heat and mass transfer in multi-component liquid-vapor system. *J. Heat Transfer*, **86**, 443–448.

Yuen, W. W., and S. C. Lee, 1989: Non-Fourier heat conduction in a semi-infinite solid subjected to oscillatory surface thermal disturbances. *J. Heat Transfer*, **111**, 178–181.

Zaidel, R. M., 1968: Electrical phenomena on exposing a cable to neutrons and gamma rays. *J. Appl. Mech. Tech. Phys.*, **9**, 278–284.

Zarubin, V. S., 1959: One problem in nonstationary heat conduction. *ARS J.*, **29**, 773–776.

Zatzkis, H., 1953: A certain problem in heat conduction. *J. Appl. Phys.*, **24**, 895–896.

Zinsmeister, G. E., and J. R. Dixon, 1965: An extension of linear moving heat source solutions to a transient case in a composite system. *Int. J. Heat Mass Transfer*, **8**, 1–6.

Zverev, I. N., 1950: Propagation of perturbations in a visco-elastic and visco-plastic rod (in Russian). *Prik. Mat. Mek.*, **14**, 295–302.

Papers Using Fourier Transforms
to Solve Partial Differential Equations

Beaudet, P. R., 1970: Elastic wave propagation in heterogeneous media. *Bull. Seism. Soc. Am.*, **60**, 769–784.

Blake, F. G., 1952: Spherical wave propagation in solid media. *J. Acoust. Soc. Am.*, **24**, 211–215.

Bose, S. K., 1975: Transmission of SH waves across a rectangular step. *Bull. Seism. Soc. Am.*, **65**, 1779–1786.

Bremmer, H., 1964: Long waves associated with disturbances produced in plasmas. *J. Res. NBS*, **68D**, 47–58.

Chabravorty, S. K., 1961: Disturbances in a viscoelastic medium due to impulsive forces on a spherical cavity. *Geofisica pura e appl.*, **48**, 23–26.

Chaudhuri, K. S., 1976: Unsteady wave motions on a sloping beach with application to under-water explosions. *Quart. J. Mech. Appl. Math.*, **29**, 89–100.

Cheng, D. H., and J. E. Benveniste, 1966: Transient response of structural elements to traveling pressure waves of arbitrary shape. *Int. J. Mech. Sci.*, **8**, 607–618.

Chimonas, G., and C. O. Hines, 1970: Atmospheric gravity waves launched by auroral currents. *Planet. Space Sci.*, **18**, 565–582.

Davies, B., 1978: *Integral Transforms and Their Applications.* New York: Springer-Verlag, Chapter 8.

Delsante, A. E., A. N. Stokes and P. J. Walsh, 1983: Application of Fourier transforms to periodic heat flow into the ground under a building. *Int. J. Heat Mass Transfer*, **26**, 121–132.

Ehlers, F. E., 1961: Linearized magnetogasdynamic channel flow with axial symmetry. *ARS J.*, **31**, 334–342.

Grigor'yan, F. E., 1974/75: Analysis of a plane waveguide and derivation of inseparable solutions of the Helmholtz equation. *Sov. Phys. Acoust.*, **20**, 132–136.

Herman, H., and J. M. Klosner, 1965: Transient response of a periodically supported cylindrical shell immersed in a fluid medium. *J. Appl. Mech.*, **32**, 562–568.

Horvay, G., 1961: Temperature distribution in a slab moving from a chamber at one temperature to a chamber at another temperature. *J. Heat Transfer*, **83**, 391–402.

Ignetik, R., Y.-C. Thio and K. C. Westfold, 1985: Transient electromagnetic field above a permeable and conducting half-space. *Geophys. J. R. Astr. Soc.*, **81**, 623–639.

Jahanshahi, A., 1964: Thin plates and shallow cylindrical shells subjected to hot spots. *J. Appl. Mech.*, **31**, 79–82.

Janowitz, G. S., 1975: The effect of bottom topography on a stratified flow in the beta plane. *J. Geophys. Res.*, **80**, 4163–4168.

Jung, H., 1950: Über eine Anwendung der Fouriertransformation in der Elastizitätstheorie. *Ing.-Arch.*, **18**, 263–271.

Kahalas, S. L., 1965: Excitation of extremely low frequency electromagnetic waves in the earth-ionosphere cavity by high-altitude nuclear detonations. *J. Geophys. Res.*, **70**, 3587–3594.

Kajiura, K., 1953: On the influence of bottom topography on ocean currents. *J. Oceanogr. Soc. Japan*, **8**, 1–14.

Karzas, W. J., and R. Latter, 1962: The electromagnetic signal due to the interaction of nuclear explosions with the earth's magnetic field. *J. Geophys. Res.*, **67**, 4635–4640.

Lamb, H., 1905: On deep-water waves. *Proc. Lond. Math. Soc.*, *Ser. 2*, **2**, 371–400.

Lamb, H., 1909: On the theory of waves propagated vertically in the atmosphere. *Proc. Lond. Math. Soc.*, *Ser. 2*, **7**, 122–141.

Leehey, P., and H. G. Davies, 1975: The direct and reverberant response of strings and membranes to convecting, random pressure fields. *J. Sound Vib.*, **38**, 163–184.

Lyra, G., 1943: Theorie der stationären Leewellen Strömung in freier Atmosphäre. *Zeit. angew. Math. Mech.*, **23**, 1–28.

Manaker, A. M., and G. Horvay, 1975: Thermal response in laminated composites. *Zeit. angew. Math. Mech.*, **55**, 503–513.

Matsuzaki, Y., and Y. C. Fung, 1977: Unsteady fluid dynamic forces on a simply-supported circular cylinder of finite length conveying a flow, with applications to stability analysis. *J. Sound Vib.*, **54**, 317–330.

Matthews, P. M., 1958: Vibrations of a beam on elastic foundation. *Zeit. angew. Math. Mech.*, **38**, 105–115.

Matthews, P. M., 1959: Vibrations of a beam on elastic foundation. II. *Zeit. angew. Math. Mech.*, **39**, 13–19.

Mattice, H. C., and P. Lieber, 1954: On attenuation of waves produced in viscoelastic materials. *Trans. Am. Geophys. Union*, **35**, 613–624.

Miles, J. W., 1956: The compressible flow past an oscillating airfoil in a wind tunnel. *J. Aeronaut. Sci.*, **23**, 671–678.

Miles, J. W., 1971: *Integral Transforms in Applied Mathematics*. Cambridge: At the University Press, Chapter 3.

Momoi, T., 1987: Scattering of Rayleigh waves by a semi-circular rough surface in layered media. *Bull. Earthq. Res. Inst.*, **62**, 163–200.

Mow, C. C., 1965: Transient response of a rigid spherical inclusion in an elastic medium. *J. Appl. Mech.*, **32**, 637–642.

Officer, C. B., W. S. Newman, J. M. Sullivan and D. R. Lynch, 1988: Glacial isostatic adjustment and mantle viscosity. *J. Geophys. Res.*, **93**, 6397–6409.

Patil, S. P., 1987: Natural frequencies of a railroad track. *J. Appl. Mech.*, **54**, 299–304.

Prasada Rao, D. R. V., and D. V. Krishna, 1977: Point source in a MHD rotating fluid. *Appl. Sci. Res.*, **33**, 177–185.

Robey, D. H., 1953: Magnetic dispersion of sound in electrically conducting plates. *J. Acoust. Soc. Am.*, **25**, 603–609.

Sawyers, K. N., 1968: Underwater sound pressure from sonic booms. *J. Acoust. Soc. Am.*, **44**, 523–524.

Schoenstadt, A. L., 1977: The effect of spatial discretization on the steady-state and transient solutions of a dispersive wave equation. *J. Comp. Phys.*, **23**, 364–379.

Seeger, G., and H.-G. Stäblein, 1964: Über den Strom-Anlaufvorgang im zylindrischen Leiter. *Zeit. angew. Phys.*, **16**, 419–424.

Sen, A. R., 1963: Surface waves due to blasts on and above liquids. *J. Fluid Mech.*, **16**, 65–81.

Seshadri, S. R., 1963: TEM mode in a parallel-plate waveguide filled with a gyrotropic dielectric. *IEEE Trans. Microwave Theory Tech.*, **MTT-11**, 436–437.

Sezawa, K., 1929: Formation of shallow-water waves due to subaqueous shocks. *Bull. Earthq. Res. Inst.*, **7**, 15–40.

Sezawa, K., and K. Kanai, 1932: Possibility of free oscillations of strata excited by seismic waves. Part IV. *Bull. Earthq. Res. Inst.*, **10**, 273–299.

Sezginer, A., and W. C. Chew, 1984: Closed form expression of the Green's function for the time-domain wave equation for a lossy two-dimensional medium. *IEEE Trans. Antennas Propagat.*, **AP-32**, 527–528.

Sharpe, J. A., 1942: The production of elastic waves by explosion pressures. I. Theory and empirical field observations. *Geophysics*, **7**, 144–154.

Singh, H., 1981: Thermal stresses in an infinite cylinder. *Indian J. Pure Appl. Math.*, **12**, 405–418.

Sneddon, I. N., 1945: The symmetrical vibrations of a thin elastic plate. *Proc. Camb. Phil. Soc.*, **41**, 27–43.

Tanaka, K., and T. Kurokawa, 1973: Viscous property of steel and its effect on strain wave front. *Bull. JSME*, **16**, 188–193.

Tranter, C. J., and J. W. Craggs, 1945: The stress distribution in a long circular cylinder when a discontinuous pressure is applied to the curved surface. *Phil. Mag., Ser. 7*, **36**, 241–250.

Tsien, H. S., and M. Finston, 1949: Interaction between parallel streams of subsonic and supersonic velocities. *J. Aeronaut. Sci.*, **16**, 515–528.

Papers Using Hankel Transforms
to Solve Partial Differential Equations

Agrawal, G. L., and W. G. Gottenberg, 1971: Response of a semi-infinite elastic solid to an arbitrary torque along the axis. *PAGEOPH*, **91**, 34–39.

Ben-Menahem, A., 1961: Radiation of seismic surface-waves from finite moving sources. *Bull. Seism. Soc. Am.*, **51**, 401–435.

Das Gupta, S., 1966: Generation of transverse waves in an elastic medium due to distribution of surface forces. *Proc. Vibr. Prob.*, **7**, 221–235.

Hron, M., 1970: Radiation of seismic surface waves from extending circular sources. *Bull. Seism. Soc. Am.*, **60**, 517–537.

Kajiura, K., 1963: The leading wave of a tsunami. *Bull. Earthq. Res. Inst.*, **41**, 535–571.

Krimigis, S. M., 1965: Interplanetary diffusion model for the time behavior of intensity in a solar cosmic ray event. *J. Geophys. Res.*, **70**, 2943–2960.

Lamb, H., 1905: On deep-water waves. *Proc. Lond. Math. Soc., Ser. 2*, **2**, 371–400.

Lamb, H., 1909: On the theory of waves propagated vertically in the atmosphere. *Proc. Lond. Math. Soc., Ser. 2*, **7**, 122–141.

Lamb, H., 1922: On water waves due to disturbance beneath the surface. *Proc. Lond. Math. Soc., Ser. 2*, **21**, 359–372.

Miles, J. W., 1971: *Integral Transforms in Applied Mathematics*. Cambridge: At the University Press, Chapter 4.

Newlands, M., 1954: Lamb's problem with internal dissipation. I. *J. Acoust. Soc. Am.*, **26**, 434–448.

Nishida, Y., Y. Shindo and A. Atsumi, 1984: Diffraction of horizontal shear waves by a moving interface crack. *Acta Mech.*, **54**, 23–34.

Officer, C. B., W. S. Newman, J. M. Sullivan and D. R. Lynch, 1988: Glacial isostatic adjustment and mantle viscosity. *J. Geophys. Res.*, **93**, 6397–6409.

Olesiak, Z., and I. N. Sneddon, 1960: The distribution of thermal stress in an infinite elastic solid containing a penny-shaped crack. *Arch. Rational Mech. Anal.*, **4**, 237–254.

Ou, H. W., and A. L. Gordon, 1986: Spin-down of baroclinic eddies under sea ice. *J. Geophys. Res.*, **91**, 7623–7630.

Paul, M. K., and B. Banerjee, 1970: Electrical potentials due to a point source upon models of continuously varying conductivity. *PAGEOPH*, **80**, 218–237.

Press, F., and M. Ewing, 1950: Propagation of explosive sound in a liquid layer overlying a semi-infinite elastic solid. *Geophysics*, **15**, 426–446.

Reißner, E., 1937: Freie und erzwungene Torsionsschwingungen des elastischen Halbraumes. *Ing.-Arch.*, **8**, 229–245.

Sen, A. R., 1960: Problems of deep-water waves. Part I. The exact and asymptotic solutions. *Bull. Cal. Math. Soc.*, **52**, 127–146.

Sen, A. R., 1963: Surface waves due to blasts on and above liquids. *J. Fluid Mech.*, **16**, 65–81.

Shindo, Y., 1981: Normal compression waves scattering at a flat annular crack in an infinite elastic solid. *Quart. Appl. Math.*, **39**, 305–315.

Sidhu, R. S., 1971: SH waves from torsional sources in semi-infinite heterogeneous media. *PAGEOPH*, **87**, 55–65.

Singh, B. M., J. Rokne and R. S. Dhaliwal, 1983: Diffraction of torsional wave by a circular rigid disc at the interface of two bonded dissimilar elastic solids. *Acta Mech.*, **49**, 139–146.

Singh, S. J., 1966: Point source in a layer overlying a semi-infinite solid with special reference to diffracted waves. *Geophys. J. R. Astr. Soc.*, **11**, 433–452.

Sinha, A. K., and P. K. Bhattacharya, 1967: Electric dipole over an isotropic and inhomogeneous earth. *Geophysics*, **32**, 652–667.

Terazawa, K., 1915: On deep-sea water waves caused by a local disturbance on or beneath the surface. *Proc. R. Soc. Lond.*, **A92**, 57–81.

Wadhwa, S. K., 1971: Reflected and refracted waves from a linear transition layer. *PAGEOPH*, **89**, 45–66.

Watts, R. G., 1969: Temperature distributions in solid and hollow cylinders due to a moving circumferential ring heat source. *J. Heat Transfer*, **91**, 465–470.

Chapter 3
Transform Methods
with Multivalued Functions

In the previous chapter, we showed how we can use the techniques from an undergraduate complex variable course to solve partial differential equations. When transients are present, multivalued functions such as $z^{1/2}$ appear in the transform and certain special efforts are necessary. The inversion of Laplace and Fourier transforms that contain a multivalued function is shown in sections 3.1 and 3.2, respectively. Then with these new techniques we solve partial differential equations where these multivalued functions arise in sections 3.3 and 3.4.

3.1 INVERSION OF LAPLACE TRANSFORMS BY CONTOUR INTEGRATION

In section 2.1 we showed that we can always use Bromwich's inversion integral to invert Laplace transforms. In the previous chapter we restricted ourselves to transforms that contained only single-valued functions. In many applications the transform will have a multivalued function in it. In this section we show how we may apply Bromwich's integral in the inversion of these transforms.

• **Example 3.1.1**

Let us find the inverse[1] of

$$F(s) = \frac{\exp(-\beta y)}{\beta s}, \tag{3.1.1}$$

where $\beta = \sqrt{\omega^2 + s/x}$, $\mathrm{Re}(\beta) > 0$, $x > 0$ and $y > 0$, by evaluating the Bromwich integral

$$f(t) = \frac{1}{2\pi i} \int_{c-\infty i}^{c+\infty i} \frac{e^{-\beta y + zt}}{\beta z} \, dz. \tag{3.1.2}$$

The branch cut runs from the branch point $z = -x\omega^2$ out to $-\infty$. Consequently, we can close the contour as shown in Fig. 3.1.1 where the point $z = 0$ is a simple pole.

Because the arcs AB and IJ vanish as the semicircle expands to infinity and the integral along FG cancels the integral along DE,

$$f(t) = -\frac{1}{2\pi i} \left(\int_{BC} + \int_{CD} + \int_{EF} + \int_{GH} + \int_{HI} \right) \tag{3.1.3}$$

$$= \frac{1}{\pi} \int_{\infty}^{0} \frac{d\lambda}{\omega^2 + \lambda^2} \exp[-i\lambda y - x(\omega^2 + \lambda^2)t]$$

$$- \frac{\sqrt{x}}{2\pi} \lim_{r \to 0} \int_{\pi}^{0} \frac{\sqrt{r}e^{\theta i/2}}{re^{\theta i} - x\omega^2} \, d\theta \exp\left[-y\sqrt{\frac{r}{x}} e^{\theta i/2} (re^{\theta i} - x\omega^2)t \right]$$

$$+ \left. \frac{e^{-\beta y}}{\beta} \right|_{z=0}$$

$$- \frac{\sqrt{x}}{2\pi} \lim_{r \to 0} \int_{0}^{-\pi} \frac{\sqrt{r}e^{\theta i/2}}{re^{\theta i} - x\omega^2} \, d\theta \exp\left[-y\sqrt{\frac{r}{x}} e^{\theta i/2} + (re^{\theta i} - x\omega^2)t \right]$$

$$- \frac{1}{\pi} \int_{0}^{\infty} \frac{d\lambda}{\omega^2 + \lambda^2} \exp[i\lambda y - x(\omega^2 + \lambda^2)t] \tag{3.1.4}$$

$$= \frac{e^{-\omega y}}{\omega} - \frac{2}{\pi} \int_{0}^{\infty} \frac{\cos(\lambda y)}{\omega^2 + \lambda^2} \exp[-x(\omega^2 + \lambda^2)t] \, d\lambda \tag{3.1.5}$$

$$= \frac{e^{-\omega y}}{\omega} - \frac{1}{2\omega} \left\{ e^{-\omega y} \mathrm{erf}\left[\omega\sqrt{xt} - \frac{y}{2\sqrt{xt}} \right] + e^{\omega y} \mathrm{erf}\left[\omega\sqrt{xt} + \frac{y}{2\sqrt{xt}} \right] \right\}. \tag{3.1.6}$$

---

[1] Adapted from Koizumi, T., 1970: Transient thermal stresses in a semi-infinite body heated axi-symmetrically. *Zeit. angew. Math. Mech.*, **50**, 747–757. See also Koizumi, T., 1970: Thermal stresses in a semi-infinite body with an instantaneous heat source on its surface. *Bull. JSME*, **13**, 26–33.

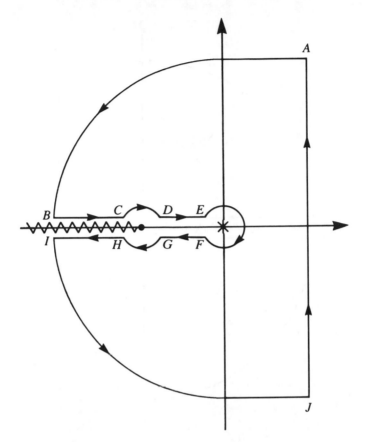

Fig. 3.1.1: The contour used to invert (3.1.1).

- **Example 3.1.2**

In this example, let us find the inverse of the Laplace transform

$$F(s) = \frac{K_1(sr)}{\alpha s K_0(s) + K_1(s)} \qquad (3.1.7)$$

by contour integration where α and r are constants. The functions $K_0(s)$ and $K_1(s)$ are modified Bessel functions of the second kind of the zeroth and first order, respectively. Because we can express them in terms of infinite power series[2], they are tabulated functions, like sine and cosine. For the present, however, we need only note that the power series for $K_n(z)$ contains a logarithm. Consequently, $K_0(s)$ and $K_1(s)$

---

[2] Watson, G. N., 1966: *A Treatise on the Theory of Bessel Functions.* Cambridge (England): At the University Press, p. 80.

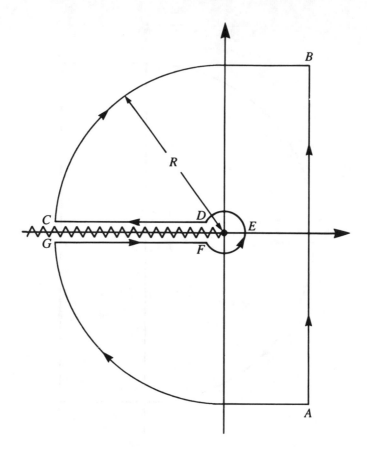

Fig. 3.1.2: The contour used to invert (3.1.7).

are multivalued functions with branch points at $s = 0$ and infinity. Fig. 3.1.2 shows the contour used in the inversion integral with the branch cut along the negative real axis.

We begin by noting that for large z, $K_n(z)$ behaves as

$$K_n(z) \sim \left(\frac{\pi}{2z}\right)^{1/2} e^{-z} \tag{3.1.8}$$

(Watson's book, p. 202) so that

$$e^{tz} F(z) \sim \frac{r^{-1/2}}{1 + \alpha z} \exp[z(t - r + 1)] \tag{3.1.9}$$

in the limit $|z| \to \infty$. Consequently, the contribution along the arcs BC and GA vanishes as $R \to \infty$ if $t > r - 1$. For $t < r - 1$, $f(t) = 0$.

Turning to the nature of $F(z)$, there is a conjugate pair $z_{1,2} = -x \pm iy$ of simple poles if $\alpha > 0$. Fig. 3.1.3 gives the values of x and

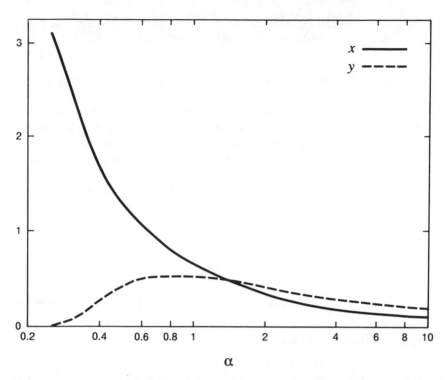

Fig. 3.1.3: The values of x and y that give the position of the simple poles $z_{1,2} = -x \pm iy$ of the transform (3.1.7).

y. For the branch cut integrals, we have along FG $z = \eta e^{-\pi i}$ and $dz = d\eta\, e^{-\pi i}$ so that

$$\int_{FG} F(z)e^{tz}dz = \int_{\infty}^{0} \frac{K_1(\eta) - i\pi I_1(\eta)}{-\alpha\eta[K_0(\eta) + i\pi I_0(\eta)] - K_1(\eta) + i\pi I_1(\eta)}e^{-\eta t}d\eta \tag{3.1.10}$$

and along CD, $z = \eta e^{\pi i}$ and $dz = d\eta\, e^{\pi i}$,

$$\int_{CD} F(z)e^{tz}dz = \int_{0}^{\infty} \frac{K_1(\eta) + i\pi I_1(\eta)}{-\alpha\eta[K_0(\eta) - i\pi I_0(\eta)] - K_1(\eta) - i\pi I_1(\eta)}e^{-\eta t}d\eta \tag{3.1.11}$$

because

$$K_\nu(ze^{m\pi i}) = e^{-m\nu\pi i}K_\nu(z) - i\pi\frac{\sin(m\nu\pi)}{\sin(\nu\pi)}I_\nu(z) \tag{3.1.12}$$

(Watson's book, p. 80). The integral for the small circle DEF vanishes. After a little algebra, we combine (3.1.10)–(3.1.11) together and find

$$f(t) = 2\alpha e^{-xt}\frac{[\alpha^2(x^2 + y^2) + 2\alpha - 1][x\cos(yt) - y\sin(yt)]}{[\alpha^2(x^2 - y^2) + 2\alpha - 1]^2 + 4\alpha^4 x^2 y^2}H(\alpha)$$

167

$$-\int_0^\infty \frac{I_1(r\eta)[\alpha\eta K_0(\eta) + K_1(\eta)] + K_1(r\eta)[\alpha\eta I_0(\eta) - I_1(\eta)]}{[\alpha\eta K_0(\eta) + K_1(\eta)]^2 + \pi^2[\alpha\eta I_0(\eta) - I_1(\eta)]^2} e^{-\eta t} d\eta$$

$$(3.1.13)$$

if $t > r-1$. The first term in (3.1.13) arises from the sum of the residues.

In the previous examples we found the inverse of the Laplace transform when the inversion integral had branch points and cuts that lay along the negative real axis. In those cases when the multivalued functions have branch points and cuts that are more complicated in their structure, a more formal bookkeeping system is necessary to keep track of the phase of these functions. In the following examples we handle these more difficult situations.

● **Example 3.1.3**

Let us find the inverse of the Laplace transform

$$F(s) = \frac{1}{s} \exp\left(-\frac{s}{\sqrt{s^2 + b^2}} x\right). \qquad (3.1.14)$$

From the inversion integral

$$f(t) = \frac{1}{2\pi i} \int_{c-\infty i}^{c+\infty i} \frac{1}{z} \exp\left(-\frac{z}{\sqrt{z^2 + b^2}} x + zt\right) dz \qquad (3.1.15)$$

$$= \frac{1}{2\pi i} \int_C \frac{1}{z} \exp\left(-\frac{z}{\sqrt{z^2 + b^2}} x + zt\right) dz \qquad (3.1.16)$$

where the line integral C comprises the line integrals $C_1, C_2, \ldots, C_8$ plus the integrations around the infinitesimally small circles at $z = 0$, $z = ib$ and $z = -ib$. See Fig. 3.1.4.

As in our previous examples the integrals along C_1 and C_8 vanish as the semicircle expands without bound. Furthermore, the integral along C_2 cancels that along C_7.

Turning to the integrals along C_3, C_4, C_5 and C_6, we introduce the following notation:

$$z = \rho e^{\theta i} \qquad\qquad 0 < \theta < 2\pi \qquad\qquad (3.1.17)$$

$$z - ib = r_1 e^{\theta_1 i} \qquad\qquad -\pi/2 < \theta_1 < 3\pi/2 \qquad\qquad (3.1.18)$$

$$z + ib = r_2 e^{\theta_2 i} \qquad\qquad -\pi/2 < \theta_2 < 3\pi/2. \qquad\qquad (3.1.19)$$

Therefore $\sqrt{z^2 + b^2} = \sqrt{r_1 r_2} e^{\psi i}$ where $\psi = (\theta_1 + \theta_2)/2$ and $-\pi/2 < \psi < 3\pi/2$. With these definitions of the phase, we have from purely geometric considerations:

Along C_3, $z = \rho e^{\pi i/2}$, $z - ib = (b-\rho)e^{3\pi i/2}$, $z + ib = (b+\rho)e^{\pi i/2}$. **(3.1.20)**

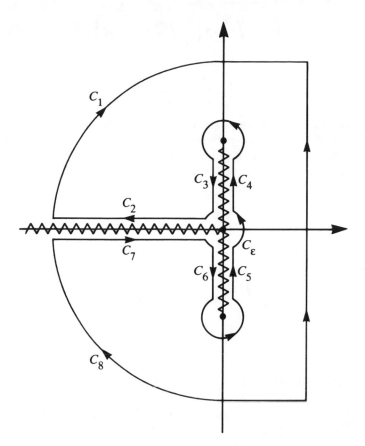

Fig. 3.1.4: The contour used to invert (3.1.14).

Along C_4, $z = \rho e^{\pi i/2}$, $z - ib = (b - \rho)e^{-\pi i/2}$, $z + ib = (b + \rho)e^{\pi i/2}$.

$$(3.1.21)$$

Along C_5, $z = \rho e^{-\pi i/2}$, $z - ib = (b + \rho)e^{-\pi i/2}$, $z + ib = (b - \rho)e^{\pi i/2}$.

$$(3.1.22)$$

Along C_6, $z = \rho e^{-\pi i/2}$, $z - ib = (b + \rho)e^{3\pi i/2}$, $z + ib = (b - \rho)e^{\pi i/2}$

$$(3.1.23)$$

with $0 \leq \rho \leq b$. Upon substituting these relationships into the integrals and simplifying,

$$\int_{C_3} = \int_b^0 \exp\left(\frac{i\rho x}{\sqrt{b^2 - \rho^2}} + i\rho t\right) \frac{d\rho}{\rho} \qquad (3.1.24)$$

$$\int_{C_4} = \int_0^b \exp\left(-\frac{i\rho x}{\sqrt{b^2 - \rho^2}} + i\rho t\right) \frac{d\rho}{\rho} \qquad (3.1.25)$$

$$\int_{C_5} = \int_b^0 \exp\left(\frac{i\rho x}{\sqrt{b^2 - \rho^2}} - i\rho t\right)\frac{d\rho}{\rho} \qquad (3.1.26)$$

and

$$\int_{C_6} = \int_0^b \exp\left(-\frac{i\rho x}{\sqrt{b^2 - \rho^2}} - i\rho t\right)\frac{d\rho}{\rho}. \qquad (3.1.27)$$

Turning to the integration around the three infinitesimal circles, the integrations around $z = ib$ and $z = -ib$ are zero. However, because of the singularity at the origin, the integration around $z = 0$ is nonzero. When we combine the contributions from the four quarter circles, the value of the integral equals

$$\int_{C_\epsilon} \frac{1}{z} \exp\left(-\frac{z}{\sqrt{z^2 + b^2}}x + zt\right) dz = 2\pi i. \qquad (3.1.28)$$

Combining our results,

$$f(t) = 1 - \frac{2}{\pi}\int_0^b \cos(\rho t)\sin\left(\frac{\rho x}{\sqrt{b^2 - \rho^2}}\right)\frac{d\rho}{\rho}. \qquad (3.1.29)$$

- **Example 3.1.4**

In this example we find the inverse[3] of

$$F(s) = \frac{\sqrt{s^2 + a^2}}{s^2} \qquad (3.1.30)$$

by contour integration. The inverse is

$$f(t) = \frac{1}{2\pi i}\int_{c-\infty i}^{c+\infty i} \frac{\sqrt{z^2 + a^2}}{z^2}e^{tz}\,dz = \frac{1}{2\pi i}\oint_\Gamma \frac{\sqrt{z^2 + a^2}}{z^2}e^{tz}\,dz \qquad (3.1.31)$$

where Γ is the closed contour shown in Fig. 3.1.5. As in the previous problem there are branch points associated with the square root at $z = \pm ai$. However, we have chosen the branch cuts to run from each of the branch points out to infinity along paths that are parallel to the real axis rather than along a segment of the imaginary axis.

Within the contour Γ there is a second-order pole at $z = 0$. Consequently, the inverse equals

$$f(t) = \frac{d}{dz}\left(e^{tz}\sqrt{z^2 + a^2}\right)\Bigg|_{z=0} - \frac{1}{2\pi i}\int_C \frac{\sqrt{z^2 + a^2}}{z^2}e^{tz}\,dz \qquad (3.1.32)$$

---

[3] A more complicated version appeared in Dimaggio, F. L., 1956: Effect of an acoustic medium on the dynamic buckling of plates. *J. Appl. Mech.*, **23**, 201–206.

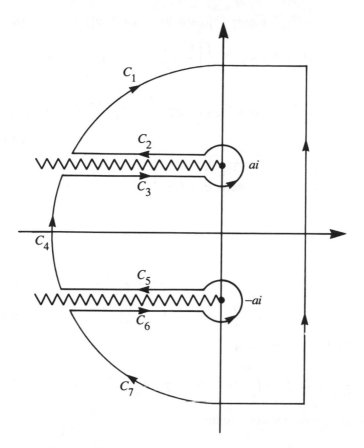

Fig. 3.1.5: The contour used to invert (3.1.30).

where C is the contour comprising the line integrals over $C_1, \ldots, C_7$ plus the integrations around the infinitesimal circles at $z = \pm ai$.

We only need to evaluate C_2, C_3, C_5 and C_6. The integrals over C_1, C_4 and C_7 vanish as the semicircle expands without bound by Jordan's lemma. Similarly, the integrals around the circles at $z = \pm ai$ vanish as they shrink to zero.

Let us consider the integral along C_2. From purely geometric considerations,

$$z = -\rho + ia, \quad z - ia = \rho e^{\pi i} \text{ and } z + ia = \sqrt{\rho^2 + 4a^2}\, e^{\theta i} \quad \textbf{(3.1.33)}$$

where $\theta = \tan^{-1}(-2a/\rho)$ and $0 < \rho < \infty$. Now

$$\sqrt{z^2 + a^2} = \sqrt{(z - ia)(z + ia)} = \rho^{1/2}(\rho^2 + 4a^2)^{1/4} e^{\pi i/2} e^{\theta i/2}$$

$$\textbf{(3.1.34)}$$

171

$$= i\rho^{1/2}(\rho^2 + 4a^2)^{1/4} \left[\cos(\theta/2) + i\sin(\theta/2)\right] \tag{3.1.35}$$

$$= i\rho^{1/2}(\rho^2 + 4a^2)^{1/4} \left\{ \left[\frac{1 + \cos(\theta)}{2}\right]^{1/2} + i \left[\frac{1 - \cos(\theta)}{2}\right]^{1/2} \right\} \tag{3.1.36}$$

$$= i\rho^{1/2} \left[\left(\sqrt{a^2 + \rho^2/4} - \rho/2\right)^{1/2} + i \left(\sqrt{a^2 + \rho^2/4} + \rho/2\right)^{1/2} \right] \tag{3.1.37}$$

because $\cos(\theta) = -\rho/\sqrt{4a^2 + \rho^2}$.

Substituting these results into the line integral and simplifying,

$$\int_{C_3} = \int_{C_2} = -\int_0^\infty i\rho^{1/2} \frac{f_1 + if_2}{\rho^2 - a^2 - 2ia} e^{-\rho t} e^{iat} d\rho \tag{3.1.38}$$

where

$$f_1 = \left(\sqrt{a^2 + \rho^2/4} - \rho/2\right)^{1/2} \tag{3.1.39}$$

$$f_2 = \left(\sqrt{a^2 + \rho^2/4} + \rho/2\right)^{1/2}. \tag{3.1.40}$$

From similar considerations,

$$\int_{C_6} = \int_{C_5} = -\int_0^\infty i\rho^{1/2} \frac{f_1 - if_2}{\rho^2 - a^2 + 2ia} e^{-\rho t} e^{-iat} d\rho. \tag{3.1.41}$$

Combining these results, the inverse is

$$f(t) = at - \frac{2}{\pi} \int_0^\infty \rho^{1/2} e^{-\rho t} d\rho$$
$$\times \frac{(\rho^2 - a^2)[f_1 \cos(at) - f_2 \sin(at)] - 2a\rho[f_1 \sin(at) + f_2 \cos(at)]}{(\rho^2 + a^2)^2}. \tag{3.1.42}$$

• **Example 3.1.5**

Our next example arises in the analysis of wave solutions in beams. We shall find the inverse of the Laplace transform[4]

$$F(s) = \frac{\left[s - N(s^2 - a^2)^{1/2}\right]^{1/2}}{(s^2 - a^2)^{1/2} s^{1/2}}, \quad N < 1, \tag{3.1.43}$$

---

[4] A similar problem occurred in Boley, B. A., and C. C. Chao, 1955: Some solutions to the Timoshenko beam equation. *J. Appl. Mech.*, **22**, 579–586.

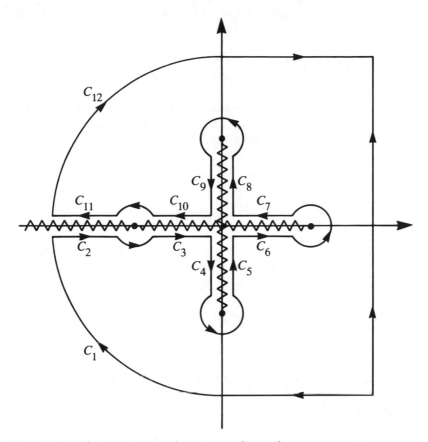

Fig. 3.1.6: The contour used to invert (3.1.43).

by the inversion integral

$$f(t) = \frac{1}{2\pi i} \int_{c-\infty i}^{c+\infty i} \frac{\left[z - N(z^2 - a^2)^{1/2}\right]^{1/2}}{(z^2 - a^2)^{1/2} z^{1/2}} e^{zt} dz \qquad \textbf{(3.1.44)}$$

which we transform into the contour integral

$$f(t) = \frac{1}{2\pi i} \int_C \frac{\left[z - N(z^2 - a^2)^{1/2}\right]^{1/2}}{(z^2 - a^2)^{1/2} z^{1/2}} e^{zt} dz \qquad \textbf{(3.1.45)}$$

where Fig. 3.1.6 shows C. The integrand of (3.1.45) has several multivalued functions, so we must introduce several branch cuts. The function $z^{1/2}$ results in a branch cut along the negative real axis from the branch point $z = 0$ out to $-\infty$. Similarly the function $(z^2 - a^2)^{1/2}$ results in a branch cut along the real axis between the branch points $z = \pm a$.

Let us now turn our attention to $[z - N(z^2 - a^2)^{1/2}]^{1/2}$. We note that

$$[z - N(z^2 - a^2)^{1/2}]^{1/2} = \frac{(1 - N^2)(z^2 + b^2)^{1/2}}{[z + N(z^2 - a^2)^{1/2}]^{1/2}} \qquad (3.1.46)$$

where $b = aN/(1 - N^2)^{1/2}$. The numerator dictates a branch cut that runs from $z = -bi$ to $z = bi$. It only remains to show that the denominator of (3.1.46) does not introduce any further branch points. This would occur if $z + N(z^2 + a^2)^{1/2}$ vanished or $z^2 = N^2(z^2 - a^2)$ or $z = \pm bi$. Consider the case $z = ib$. From our definition of the branch cuts so far, $(z^2 - a^2)^{1/2} = (a^2 + b^2)i$ and both z and $(z^2 - a^2)^{1/2}$ have imaginary parts greater than zero. Therefore their sum cannot be zero and we need no further branch cuts. The phase of $z + ib$ and $z - ib$ lies between $-\pi/2$ and $3\pi/2$.

With our branch cuts now clearly defined, we begin the evaluation of the various line integrals. From Jordan's lemma the contribution from paths C_1 and C_{12} are zero. Now, along C_2 we have for $a < u < \infty$,

$$z = ue^{-\pi i}, z - a = (u + a)e^{-\pi i}, z + a = (u - a)e^{-\pi i} \qquad (3.1.47)$$

$$(z^2 - a^2)^{1/2} = (u^2 - a^2)^{1/2}e^{-\pi i} \qquad (3.1.48)$$

$$\left[z - N(z^2 - a^2)^{1/2}\right]^{1/2} = \left[u - N(u^2 - a^2)^{1/2}\right]^{1/2} e^{-\pi i/2} \qquad (3.1.49)$$

so that

$$\int_{C_2} \frac{\left[z - N(z^2 - a^2)^{1/2}\right]^{1/2}}{(z^2 - a^2)^{1/2}z^{1/2}} e^{zt} dz = \int_{\infty}^{a} \frac{\left[u - N(u^2 - a^2)^{1/2}\right]^{1/2}}{(u^2 - a^2)^{1/2}u^{1/2}} e^{-ut} du \qquad (3.1.50)$$

while along C_{11} for $a < u < \infty$,

$$z = ue^{\pi i}, z - a = (u + a)e^{\pi i}, z + a = (u - a)e^{\pi i} \qquad (3.1.51)$$

$$(z^2 - a^2)^{1/2} = (u^2 - a^2)^{1/2}e^{\pi i} \qquad (3.1.52)$$

$$\left[z - N(z^2 - a^2)^{1/2}\right]^{1/2} = \left[u - N(u^2 - a^2)^{1/2}\right]^{1/2} e^{\pi i/2} \qquad (3.1.53)$$

so that

$$\int_{C_6} \frac{\left[z - N(z^2 - a^2)^{1/2}\right]^{1/2}}{(z^2 - a^2)^{1/2}z^{1/2}} e^{zt} dz = \int_{a}^{\infty} \frac{\left[u - N(u^2 - a^2)^{1/2}\right]^{1/2}}{(u^2 - a^2)^{1/2}u^{1/2}} e^{-ut} du. \qquad (3.1.54)$$

Consequently the contribution from C_{11} cancels that from C_2 and we will discard them at this point.

Along C_3 we have for $0 < u < a$,

$$z = ue^{-\pi i}, z - a = (u + a)e^{-\pi i}, z + a = (a - u) \tag{3.1.55}$$

$$(z^2 - a^2)^{1/2} = (a^2 - u^2)^{1/2}e^{-\pi i/2} \tag{3.1.56}$$

$$z - N(z^2 - a^2)^{1/2} = -u + iN(a^2 - u^2)^{1/2} \tag{3.1.57}$$
$$= -u + i(r^2 - u^2)^{1/2} \tag{3.1.58}$$

where $r = (1 - N^2)^{1/2}(u^2 + b^2)^{1/2}$. Consequently,

$$\left[z - N(z^2 - a^2)^{1/2} \right]^{1/2} = \left(\frac{r - u}{2} \right)^{1/2} + i\left(\frac{r + u}{2} \right)^{1/2}. \tag{3.1.59}$$

Similarly, along C_{10} for $0 < u < a$,

$$z = ue^{\pi i}, z - a = (a + u)e^{\pi i}, z + a = (a - u) \tag{3.1.60}$$

$$(z^2 - a^2)^{1/2} = (a^2 - u^2)^{1/2}e^{\pi i/2} \tag{3.1.61}$$

$$z - N(z^2 - a^2)^{1/2} = -u + iN(a^2 - u^2)^{1/2} \tag{3.1.62}$$

$$\left[z - N(z^2 - a^2)^{1/2} \right]^{1/2} = \left(\frac{r - u}{2} \right)^{1/2} - i\left(\frac{r + u}{2} \right)^{1/2}. \tag{3.1.63}$$

Along C_6 with $0 < u < a$,

$$z = u, z - a = (a - u)e^{-\pi i}, z + a = (a + u) \tag{3.1.64}$$

$$(z^2 - a^2)^{1/2} = (a^2 - u^2)^{1/2}e^{-\pi i/2} \tag{3.1.65}$$

$$z - N(z^2 - a^2)^{1/2} = u + iN(a^2 - u^2)^{1/2} \tag{3.1.66}$$

$$\left[z - N(z^2 - a^2)^{1/2} \right]^{1/2} = \left(\frac{r + u}{2} \right)^{1/2} + i\left(\frac{r - u}{2} \right)^{1/2}. \tag{3.1.67}$$

Along C_7 with $0 < u < a$,

$$z = u, z - a = (a - u)e^{\pi i}, z + a = (a + u) \tag{3.1.68}$$

$$(z^2 - a^2)^{1/2} = (a^2 - u^2)^{1/2}e^{\pi i/2} \tag{3.1.69}$$

$$z - N(z^2 - a^2)^{1/2} = u - iN(a^2 - u^2)^{1/2} \tag{3.1.70}$$

$$\left[z - N(z^2 - a^2)^{1/2} \right]^{1/2} = \left(\frac{r + u}{2} \right)^{1/2} - i\left(\frac{r - u}{2} \right)^{1/2}. \tag{3.1.71}$$

Consequently, direct substitution leads to

$$\frac{1}{2\pi i}\int_{C_3+C_6+C_7+C_{10}}\frac{\left[z-N(z^2-a^2)^{1/2}\right]^{1/2}}{(z^2-a^2)^{1/2}z^{1/2}}e^{zt}dz$$
$$=\frac{2}{\pi}\int_0^a\left(\frac{r+u}{2}\right)^{1/2}\frac{\cosh(ut)}{(a^2-u^2)^{1/2}u^{1/2}}\,du.$$
$$(3.1.72)$$

Finally, turning to the contours along the imaginary axis, we have for C_4 with $0<u<b$,

$$z=ue^{-\pi i/2}, z+a=(a^2+u^2)^{1/2}e^{-\theta i}, z-a=(a^2+u^2)^{1/2}e^{-\pi i+\theta i},$$
$$(3.1.73)$$

where $\theta=\tan^{-1}(u/a)$. Therefore,

$$(z^2-a^2)^{1/2}=(a^2+u^2)^{1/2}e^{-\pi i/2}. \qquad (3.1.74)$$

For $z-N(z^2-a^2)^{1/2}$,

$$z+N(z^2-a^2)^{1/2}=\left[u+N(u^2+a^2)^{1/2}\right]e^{-\pi i/2}. \qquad (3.1.75)$$

Now
$$z-ib=(b+u)e^{3\pi i/2} \qquad (3.1.76)$$
$$z+ib=(b-u)e^{\pi i/2} \qquad (3.1.77)$$

so that

$$z-N(z^2-a^2)^{1/2}=\frac{b^2-u^2}{u+N(a^2+u^2)^{1/2}}e^{5\pi i/2}. \qquad (3.1.78)$$

Along C_5 with $0<u<b$,

$$z=ue^{-\pi i/2},(z^2-a^2)^{1/2}=(a^2+u^2)^{1/2}e^{-\pi i/2}, \qquad (3.1.79)$$

$$z+ib=(b-u)e^{\pi i/2}, z-ib=(b+u)e^{-\pi i/2} \qquad (3.1.80)$$

and
$$z-N(z^2-a^2)=\frac{b^2-u^2}{u+N(a^2+u^2)^{1/2}}e^{\pi i/2}. \qquad (3.1.81)$$

Along C_8 with $0<u<b$,

$$z=ue^{\pi i/2}, z+a=(u^2+a^2)^{1/2}e^{\theta i}, z-a=(u^2+a^2)^{1/2}e^{\pi i-\theta i}, \qquad (3.1.82)$$

$$z+ib=(b+u)e^{\pi i/2}, z-ib=(b-u)e^{-\pi i/2} \qquad (3.1.83)$$

and

$$z - N(z^2 - a^2)^{1/2} = \frac{b^2 - u^2}{u + N(a^2 + u^2)^{1/2}} e^{-\pi i/2}. \tag{3.1.84}$$

Along C_9 with $0 < u < b$,

$$z = u e^{\pi i/2}, (z^2 - a^2)^{1/2} = (u^2 + a^2)^{1/2} e^{\pi i/2}, \tag{3.1.85}$$

$$z + ib = (b + u)e^{\pi i/2}, z - ib = (b - u)e^{3\pi i/2} \tag{3.1.86}$$

and

$$z - N(z^2 - a^2)^{1/2} = \frac{b^2 - u^2}{u + N(a^2 + u^2)^{1/2}} e^{3\pi i/2}. \tag{3.1.87}$$

Substitution of these results into the line integrals and simplification results in

$$\frac{1}{2\pi i} \int_{C_4 + C_5 + C_8 + C_9} \frac{\left[z - N(z^2 - a^2)^{1/2}\right]^{1/2}}{(z^2 - a^2)^{1/2} z^{1/2}} e^{zt} dz$$
$$= \frac{2}{\pi} \int_0^b \left(\frac{b^2 - u^2}{u + N(a^2 + u^2)^{1/2}}\right)^{1/2} \frac{\cos(ut)}{(a^2 + u^2)^{1/2} u^{1/2}} du. \tag{3.1.88}$$

We omit the demonstration that the small circles around the branch points do not contribute to the inverse. The final result is the sum of (3.1.72) and (3.1.88).

• **Example 3.1.6**

We now employ the inversion formula on the transform[5]

$$F(s) = \frac{I_0(\zeta)}{\zeta I_0'(\zeta)} = \frac{I_0(\zeta)}{\zeta I_1(\zeta)} \tag{3.1.89}$$

where

$$\zeta = \sqrt{M^2(s + ik)^2 - s^2} \tag{3.1.90}$$

or

$$s = \frac{-ikM^2 \pm \sqrt{(M^2 - 1)\zeta^2 - M^2 k^2}}{M^2 - 1} \tag{3.1.91}$$

---

[5] Adapted from Widnall, S. E., and E. H. Dowell, 1967: Aerodynamic forces on an oscillating cylindrical duct with an internal flow. *J. Sound Vib.*, **6**, 71–85. Published by Academic Press Ltd., London, U.K.

with $M > 1$ and $k \neq 0$. Upon converting the transform into a contour integration,

$$f(t) = \frac{1}{2\pi i} \oint_C \frac{I_0(\zeta)}{\zeta I_1(\zeta)} e^{zt} dz \qquad (3.1.92)$$

where Fig. 3.1.7 shows the contour C and ζ is now a complex variable. From the definition of ζ branch points are located at $z = s_1 i$ and $z = s_2 i$, where $s_1 = -kM/(M+1)$ and $s_2 = -kM/(M-1)$, with the branch cut running along the imaginary axis between these two points. Because $I_0'(z) = I_1(z) = -iJ_1(iz)$, an infinite number of simple poles are located at

$$z_{n1} = \frac{-ikM^2 + i\sqrt{(M^2 - 1)\alpha_n^2 + M^2 k^2}}{M^2 - 1}, \qquad n = 1, 2, 3, \ldots \qquad (3.1.93)$$

above the branch point $z = s_1 i$ and

$$z_{n2} = \frac{-ikM^2 - i\sqrt{(M^2 - 1)\alpha_n^2 + M^2 k^2}}{M^2 - 1}, \qquad n = 1, 2, 3, \ldots \qquad (3.1.94)$$

below the branch point $z = s_2 i$ with $J_1(\alpha_n) = 0$ except for $\alpha_0 = 0$.

From the asymptotic expansion for $I_0(z)$, the contribution from the line integrals on C_1 and C_7 vanish as we allow the semicircle to approach infinity. Furthermore, the integration along C_2 cancels the integration along C_6. Now along C_4,

$$z = (s_1 - s_0 + s_0 u)e^{\pi i/2} \qquad (3.1.95)$$

$$z - is_1 = s_0(1 - u)e^{-\pi i/2} \qquad (3.1.96)$$

$$z - is_2 = s_0(1 + u)e^{\pi i/2} \qquad (3.1.97)$$

and

$$\zeta = s_0 \sqrt{(M^2 - 1)(1 - u^2)} = x \qquad (3.1.98)$$

where $s_0 = kM/(M^2 - 1)$ and $-1 < u < 1$. The argument of $z - is_1$ and $z - is_2$ lies between $-\pi/2$ and $3\pi/2$. Therefore,

$$\int_{C_4} F(z)e^{tz} dz = s_0 i \int_{-1}^{1} \frac{I_0(x) \exp\left[i(s_1 - s_0 + s_0 u)\right]}{x I_1(x)} du. \qquad (3.1.99)$$

Along both C_3 and C_5,

$$z = (s_1 - s_0 + s_0 u)e^{\pi i/2} \qquad (3.1.100)$$

$$z - is_1 = s_0(1 - u)e^{3\pi i/2} \qquad (3.1.101)$$

$$z - is_2 = s_0(1 + u)e^{\pi i/2} \qquad (3.1.102)$$

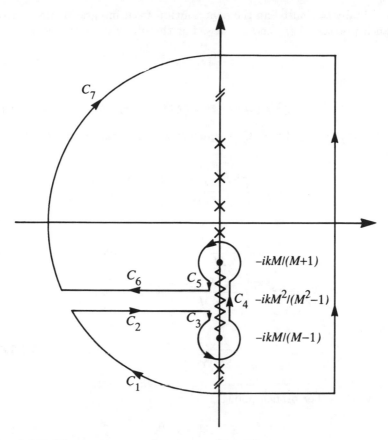

Fig. 3.1.7: The contour used to invert (3.1.89).

and

$$\zeta = -s_0\sqrt{(M^2 - 1)(1 - u^2)} = -x \qquad (3.1.103)$$

so that

$$\int_{C_3+C_5} F(z)e^{tz}dz = s_0i \int_1^{-1} \frac{I_0(-x)\exp\left[i(s_1 - s_0 + s_0u)\right]}{(-x)I_1(-x)}\,du \qquad (3.1.104)$$

$$= -s_0i \int_{-1}^1 \frac{I_0(x)\exp\left[i(s_1 - s_0 + s_0u)\right]}{xI_1(x)}\,du. \qquad (3.1.105)$$

Consequently,

$$\int_{C_3+C_4+C_5} F(z)e^{tz}dz = 0. \qquad (3.1.106)$$

Transform Methods for Solving Partial Differential Equations

Finally we must find the contribution from integrating around the branch points $z = is_1$ and $z = is_2$. For the integration around s_1,

$$z = is_1 + \epsilon e^{\theta i} \tag{3.1.107}$$

and

$$\zeta^2 = 2ikM\epsilon e^{\theta i} + (M^2 - 1)\epsilon^2 e^{2\theta i} \tag{3.1.108}$$

where $-\pi/2 \leq \theta \leq 3\pi/2$. Consequently the integral around the branch cut is

$$\int_{C_{\epsilon_1}} F(z)e^{tz}dz = \lim_{\epsilon \to 0} 2 \int_{-\pi/2}^{3\pi/2} \frac{\exp\left[-ikMt/(M+1) + \epsilon e^{\theta i}t\right]}{2ikM\epsilon e^{\theta i} + (M^2 - 1)\epsilon^2 e^{2\theta i}} i\epsilon e^{\theta i}d\theta \tag{3.1.109}$$

$$= \frac{\exp[-ikMt/(M+1)]}{kM} \int_{-\pi/2}^{3\pi/2} d\theta \tag{3.1.110}$$

$$= 2\pi \frac{\exp[-ikMt/(M+1)]}{kM}, \tag{3.1.111}$$

because

$$\frac{I_0(\zeta)}{\zeta I_1(\zeta)} \simeq \frac{2}{\zeta^2} \tag{3.1.112}$$

as $\zeta \to 0$.

At the other branch point,

$$z = is_2 + \epsilon e^{\theta i}, \tag{3.1.113}$$

$$\zeta^2 = -2ikM\epsilon e^{\theta i} + (M^2 - 1)\epsilon^2 e^{2\theta i} \tag{3.1.114}$$

with $-3\pi/2 < \theta < \pi/2$. Therefore the integral around this point is

$$\int_{C_{\epsilon_2}} F(z)e^{tz}dz = \lim_{\epsilon \to 0} \int_{-3\pi/2}^{\pi/2} \frac{2\exp\left[-ikMt/(M-1) + \epsilon e^{\theta i}t\right]}{-2ikM\epsilon e^{\theta i} + (M^2 - 1)\epsilon^2 e^{2\theta i}} i\epsilon e^{\theta i}d\theta \tag{3.1.115}$$

$$= -2\pi \frac{\exp[-ikMt/(M-1)]}{kM}. \tag{3.1.116}$$

Applying the residue theorem, the final result is

$$f(t) = \sum_{n=1}^{\infty} \frac{I_0(\zeta_{n1})\exp(z_{n1}t)}{\frac{d}{d\zeta}[\zeta I_1(\zeta)]\frac{d\zeta}{dz}|_{\zeta=\zeta_{n1}, z=z_{n1}}} + \sum_{n=1}^{\infty} \frac{I_0(\zeta_{n2})\exp(z_{n2}t)}{\frac{d}{d\zeta}[\zeta I_1(\zeta)]\frac{d\zeta}{dz}|_{\zeta=\zeta_{n2}, z=z_{n2}}}$$

$$- \frac{i}{kM}\left[\exp\left(\frac{-ikMt}{M+1}\right) - \exp\left(\frac{-ikMt}{M-1}\right)\right]. \tag{3.1.117}$$

or

$$f(t) = 2\exp\left[-ikM^2 t/(M^2-1)\right]$$
$$\times \sum_{n=1}^{\infty} \frac{\sin[t\sqrt{(M^2-1)\alpha_n^2 + M^2 k^2/(M^2-1)}]}{\sqrt{(M^2-1)\alpha_n^2 + M^2 k^2}}$$
$$- \frac{i}{kM}\left[\exp\left(\frac{-ikMt}{M+1}\right) - \exp\left(\frac{-ikMt}{M-1}\right)\right] \qquad (3.1.118)$$

because $I_1'(\zeta_{ni}) = [I_0(\zeta_{ni}) + I_2(\zeta_{ni})]/2 = I_0(\zeta_{ni})$ since $I_0(\zeta_{ni}) - I_2(\zeta_{ni}) = 2I_1(\zeta_{ni})/\zeta_{ni} = 0$.

• **Example 3.1.7**

In this example, we invert the transform[6]

$$F(s) = e^{-\mu(s,k)z}\frac{s}{s^2+k^2} \qquad (3.1.119)$$

where

$$\mu(s,k) = \frac{1}{s}\left[s^4 + \left(\frac{1}{4}+k^2\right)s^2 + \beta^2 k^2\right]^{1/2} \qquad (3.1.120)$$

and k and β are real, $0 < \beta < 1/2$, $z > 0$ and $\mathrm{Re}(\mu) > 0$. For convenience, we can rewrite $\mu(s,k)$ as

$$\mu(s,k) = \frac{1}{s}\left\{\left[s^2 + \omega_1^2(k)\right]\left[s^2 + \omega_2^2(k)\right]\right\}^{1/2} \qquad (3.1.121)$$

where

$$\omega_{1,2}(k) = \frac{1}{2}\left[\left(\tfrac{1}{4}+k^2+2\beta k\right)^{1/2} \mp \left(\tfrac{1}{4}+k^2-2\beta k\right)^{1/2}\right]. \qquad (3.1.122)$$

The point of interest in this example is the presence of an essential singularity at $s = 0$.

Fig. 3.1.8 illustrates the contour integration used to invert (3.1. 119). Note that this contour excludes the essential singularity and therefore avoids the necessity of computing its residue. The arguments for the various functions are

$$s + i\omega_2 = re^{\theta i}, \quad -\pi/2 < \theta < 3\pi/2 \qquad (3.1.123)$$

---

[6] Taken from Cole, J. D., and C. Greifinger, 1969: Acoustic-gravity waves from an energy source at the ground in an isothermal atmosphere. *J. Geophys. Res.*, **74**, 3693–3703. ©1969 American Geophysical Union.

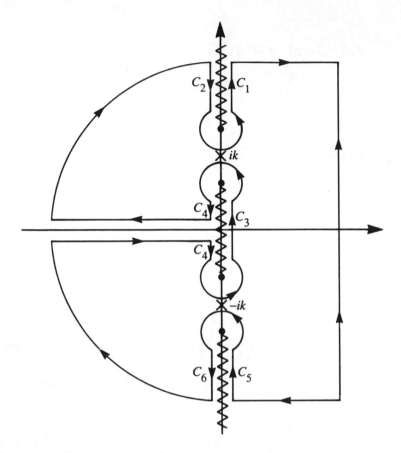

Fig. 3.1.8: The contour used to invert (3.1.119).

$$s - i\omega_2 = re^{\theta i}, -3\pi/2 < \theta < \pi/2 \tag{3.1.124}$$

$$s + i\omega_1 = re^{\theta i}, -3\pi/2 < \theta < \pi/2 \tag{3.1.125}$$

$$s - i\omega_1 = re^{\theta i}, -\pi/2 < \theta < 3\pi/2. \tag{3.1.126}$$

The contribution from the poles at $z = \pm ik$ is

$$\text{Res}(z = \pm ik) = \tfrac{1}{2} e^{-\left(\frac{1}{4} - \beta^2\right)^{1/2} z} e^{\pm ikt} \tag{3.1.127}$$

or

$$\text{Res}(z = ik) + \text{Res}(z = -ik) = e^{-\left(\frac{1}{4} - \beta^2\right)^{1/2} z} \cos(kt). \tag{3.1.128}$$

From the branch cut integrals,

$$\int_{C_1} = \frac{1}{2\pi i} \int_{\omega_2}^{\infty} e^{i(\omega t - \mu_1 z)} \frac{\omega \, d\omega}{\omega^2 - k^2} \tag{3.1.129}$$

and

$$\int_{C_2} = \frac{1}{2\pi i} \int_{\infty}^{\omega_2} e^{i(\omega t + \mu_1 z)} \frac{\omega \, d\omega}{\omega^2 - k^2} \qquad (3.1.130)$$

where

$$\mu_1 = \frac{1}{\omega} \sqrt{\omega^4 - \left(\frac{1}{4} + k^2\right)\omega^2 + \beta^2 k^2}. \qquad (3.1.131)$$

Next,

$$\int_{C_5} = \frac{1}{2\pi i} \int_{-\infty}^{-\omega_2} e^{i(\omega t + \mu_5 z)} \frac{\omega \, d\omega}{\omega^2 - k^2} \qquad (3.1.132)$$

and

$$\int_{C_6} = \frac{1}{2\pi i} \int_{-\omega_2}^{-\infty} e^{i(\omega t - \mu_5 z)} \frac{\omega \, d\omega}{\omega^2 - k^2} \qquad (3.1.133)$$

where $\mu_5 = -\mu_1$. Therefore,

$$\int_{C_1+C_2+C_5+C_6} = \frac{1}{2\pi i} \int_{-\infty}^{\infty} \frac{\omega \, d\omega}{\omega^2 - k^2} \left[e^{i(\omega t - \mu_1 z)} - e^{i(\omega t + \mu_1 z)} \right]$$

$$(3.1.134)$$

where we include only the frequencies $|\omega| > \omega_2$. In a similar manner,

$$\int_{C_3+C_4} = \frac{1}{2\pi i} \int_{-\omega_1}^{\omega_1} \frac{\omega \, d\omega}{k^2 - \omega^2} \left[e^{i(\omega t - \mu_1 z)} - e^{i(\omega t + \mu_1 z)} \right]. \qquad (3.1.135)$$

Therefore,

$$\int_{C_1+\ldots+C_6} = \frac{1}{2\pi i} \int_{-\infty}^{\infty} \frac{\omega \, d\omega}{|\omega^2 - k^2|} \left[e^{i(\omega t - \mu_1 z)} - e^{i(\omega t + \mu_1 z)} \right] \qquad (3.1.136)$$

with $|\omega| > \omega_2$ and $|\omega| < \omega_1$. The final answer equals the sum of (3.1.128) and (3.1.136).

● **Example 3.1.8**

Let us now invert the Laplace transform[7]

$$F(s) = \frac{\gamma_3 e^{-\gamma_3 x} - \gamma_4 e^{-\gamma_4 x}}{(\gamma_3 - \gamma_4)[c_3^2(\gamma_3^2 + \gamma_4^2 + \gamma_3 \gamma_4) - s^2]} \qquad (3.1.137)$$

where

$$\gamma_{3,4} = \frac{1}{c_3 c_4 \sqrt{2}} \sqrt{s^2(c_3^2 + c_4^2) + c_5^2 \nu^2 \pm A}, \qquad (3.1.138)$$

---

[7] Taken from Eason, G., 1970: The displacement produced in a semi-infinite linear Cosserat continuum by an impulsive force. *Proc. Vibr. Prob.*, **11**, 199–220.

$$A^2 = [s^2(c_3^2 + c_4^2) + c_5^2\nu^2]^2 - 4c_3^2c_4^2s^2(s^2 + \nu^2) \tag{3.1.139}$$

$$= \left[s^2(c_3^2 - c_4^2) + \frac{\nu^2}{c_3^2 - c_4^2}(c_5^2c_3^2 + c_5^2c_4^2 - 2c_3^2c_4^2)\right]^2$$

$$+ \frac{4\nu^4 c_3^2 c_4^2 (c_3^2 - c_5^2)(c_5^2 - c_4^2)}{(c_3^2 - c_4^2)^2}, \tag{3.1.140}$$

ν is a real, $x > 0$, γ_3, γ_4 correspond to the upper and lower signs, respectively, $c_3 > c_5 > c_4$ and we use the positive square root of A^2 for A. Under these conditions, $\gamma_{3,4}$ have branch points at $s = \pm i\nu$ and $s = R[\pm\cos(\phi) \pm i\sin(\phi)]$ where

$$R = \frac{\nu c_5}{\sqrt{c_3^2 - c_4^2}} \tag{3.1.141}$$

and

$$\phi = \tan^{-1}\left[\frac{c_3\sqrt{c_5^2 - c_4^2}}{c_4\sqrt{c_3^2 - c_5^2}}\right]. \tag{3.1.142}$$

Fig. 3.1.9 shows our choices for the branch cuts.

In addition to the branch points, there may be poles if the denominator of (3.1.137) vanishes. One case would be $\gamma_3 = \gamma_4$; however, further analysis shows that this does not occur. The other possibility would be

$$G(s) = c_3^2(\gamma_3^2 + \gamma_4^2 + \gamma_3\gamma_4) - s^2 \tag{3.1.143}$$

$$= [s^2c_3^2 + \nu^2c_5^2 + c_3c_4s\sqrt{s^2 + \nu^2}]/c_4^2 = 0. \tag{3.1.144}$$

With the branch cut for $\sqrt{s^2 + \nu^2}$ lying along the imaginary axis, $G(s)$ has no zeros on either the real or imaginary axes. Finally, because $\text{Im}[G(s)] > 0$ in the first and third quadrants and < 0 in the second and fourth quadrants, $G(s)$ has no zeros at all.

To find $f(x,t)$, we write the inversion as the contour integral

$$f(x,t) = \frac{1}{2\pi i}\int_{c-\infty i}^{c+\infty i}\frac{\gamma_3 e^{st - \gamma_3 x} - \gamma_4 e^{st - \gamma_4 x}}{(\gamma_3 - \gamma_4)[c_3^2(\gamma_3^2 + \gamma_4^2 + \gamma_3\gamma_4) - s^2]}\,ds. \tag{3.1.145}$$

To close the contour in the left half-plane, we must examine γ_3 and γ_4 for large s. Because $c_3 > c_4$, γ_4 behaves like s/c_3 for large s. We must close the $e^{st-\gamma_4 x}$ portion of the line integral (3.1.145) in the right half-plane when $c_3 t < x$ and in the left half-plane when $c_3 t > x$ according to Jordan's lemma. Because there are no singularities in the right half-plane, the contribution from this term is zero until $c_3 t > x$. Similar considerations lead to the conclusion that the contribution from $e^{st-\gamma_3 x}$ term will be zero until $t > x/c_4$. When we close the contour in the left half-plane, the original Bromwich contour equals the negative of integral along the contour $KJIHGFEDCBLMNPQRSTU$.

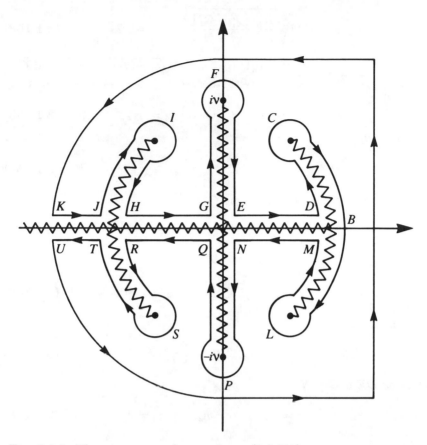

Fig. 3.1.9: The contour used to evaluate (3.1.137).

Let us define the following quantities:

$$\eta_1 = \frac{1}{c_3 c_4 \sqrt{2}} \sqrt{J + (c_3^2 - c_4^2)\sqrt{r^4 + R^4 - 2R^2 r^2 \cos(2\phi)}} \qquad \textbf{(3.1.146)}$$

$$\eta_2 = \frac{1}{c_3 c_4 \sqrt{2}} \sqrt{J - (c_3^2 - c_4^2)\sqrt{r^4 + R^4 - 2R^2 r^2 \cos(2\phi)}} \qquad \textbf{(3.1.147)}$$

$$\chi_1 = \frac{1}{c_3 c_4 \sqrt{2}} \left\{ \Omega_1^2 + [(c_3^2 + c_4^2)R^2 \sin(2\theta) + 2R^2(c_3^2 - c_4^2)B]^2 \right\}^{1/4} \qquad \textbf{(3.1.148)}$$

$$\chi_2 = \frac{1}{c_3 c_4 \sqrt{2}} \left\{ \Omega_2^2 + [(c_3^2 + c_4^2)R^2 \sin(2\theta) - 2R^2(c_3^2 - c_4^2)B]^2 \right\}^{1/4} \qquad \textbf{(3.1.149)}$$

$$\Omega_1 = \nu^2 c_5^2 + (c_3^2 + c_4^2)R^2 \cos(2\theta) + 2R^2(c_3^2 - c_4^2)C \qquad \textbf{(3.1.150)}$$

$$\Omega_2 = \nu^2 c_5^2 + (c_3^2 + c_4^2)R^2 \cos(2\theta) - 2R^2(c_3^2 - c_4^2)C \qquad \textbf{(3.1.151)}$$

$$\Theta_i = \tan^{-1}\left[\sqrt{\frac{2c_3^2 c_4^2 \chi_i^2 - \Omega_i}{2c_3^2 c_4^2 \chi_i^2 + \Omega_i}}\right], \qquad i = 1, 2 \qquad (3.1.152)$$

where $J = \nu^2 c_5^2 + r^2(c_3^2 + c_4^2)$, $B = \sin(\theta)\sqrt{\cos^2(\theta) - \cos^2(\phi)}$ and $C = \cos(\theta)\sqrt{\cos^2(\theta) - \cos^2(\phi)}$. With these definitions, we find that along AB that

$$s = r, \quad \gamma_3 = \eta_1, \quad \gamma_4 = \eta_2 \qquad (3.1.153)$$

Along BC,

$$s = Re^{\theta i}, \qquad \gamma_3 = \chi_1 e^{\Theta_1 i}, \qquad \gamma_4 = \chi_2 e^{\Theta_2 i} \qquad (3.1.154)$$

where $0 < \theta < \phi$. Similarly, along CD

$$s = Re^{\theta i}, \qquad \gamma_3 = \chi_2 e^{\Theta_2 i}, \qquad \gamma_4 = \chi_1 e^{\Theta_1 i}. \qquad (3.1.155)$$

Along BL,

$$s = Re^{-\theta i}, \qquad \gamma_3 = \chi_1 e^{-\Theta_1 i}, \qquad \gamma_4 = \chi_2 e^{-\Theta_2 i}. \qquad (3.1.156)$$

Along LM,

$$s = Re^{-\theta i}, \qquad \gamma_3 = \chi_2 e^{-\Theta_2 i}, \qquad \gamma_4 = \chi_1 e^{-\Theta_1 i}. \qquad (3.1.157)$$

Along DE and MN,

$$s = r, \qquad \gamma_3 = \eta_2, \qquad \gamma_4 = \eta_1. \qquad (3.1.158)$$

Along EF,

$$s = ri, \qquad \gamma_3 = \delta_1 i, \qquad \gamma_4 = \delta_2. \qquad (3.1.159)$$

Along FG,

$$s = ri, \qquad \gamma_3 = \delta_1 i, \qquad \gamma_4 = -\delta_2. \qquad (3.1.160)$$

Along NP,

$$s = -ri, \qquad \gamma_3 = -\delta_1 i, \qquad \gamma_4 = \delta_2. \qquad (3.1.161)$$

Along PQ,

$$s = -ri, \qquad \gamma_3 = -\delta_1 i, \qquad \gamma_4 = -\delta_2. \qquad (3.1.162)$$

Along GH and QR,

$$s = -r, \qquad \gamma_3 = -\eta_2, \qquad \gamma_4 = -\eta_1. \qquad (3.1.163)$$

Along HI,

$$s = -Re^{-\theta i}, \qquad \gamma_3 = -\chi_2 e^{-\Theta_2 i}, \qquad \gamma_4 = -\chi_1 e^{-\Theta_1 i}. \qquad (3.1.164)$$

Along IJ,

$$s = -Re^{-\theta i}, \qquad \gamma_3 = -\chi_1 e^{-\Theta_1 i}, \qquad \gamma_4 = -\chi_2 e^{-\Theta_2 i}. \qquad (3.1.165)$$

Along RS,

$$s = -Re^{\theta i}, \qquad \gamma_3 = -\chi_2 e^{\Theta_1 i}, \qquad \gamma_4 = -\chi_1 e^{\Theta_2 i}. \qquad (3.1.166)$$

Along ST,

$$s = -Re^{\theta i}, \qquad \gamma_3 = -\chi_1 e^{\Theta_1 i}, \qquad \gamma_4 = -\chi_2 e^{\Theta_2 i}. \qquad (3.1.167)$$

Along JK and TU,

$$s = -r, \qquad \gamma_3 = -\eta_1, \qquad \gamma_4 = -\eta_2, \qquad (3.1.168)$$

where

$$\delta_1 = \frac{1}{c_3 c_4 \sqrt{2}} \sqrt{r^2(c_3^2 + c_4^2) - \nu^2 c_5^2 + |c_3^2 - c_4^2| D}, \qquad (3.1.169)$$

$$\delta_2 = \frac{1}{c_3 c_4 \sqrt{2}} \sqrt{|c_3^2 - c_4^2| D - r^2(c_3^2 + c_4^2) + \nu^2 c_5^2}, \qquad (3.1.170)$$

$$k_1 = \frac{\nu c_4 \sqrt{c_3^2 - c_5^2}}{|c_3^2 - c_4^2|}, \qquad (3.1.171)$$

$$k_2 = \frac{\nu c_3 \sqrt{c_4^2 - c_5^2}}{|c_3^2 - c_4^2|} \qquad (3.1.172)$$

and $D = \sqrt{4r^2 k_1^2 + (k_1^2 - k_2^2 - r^2)^2}$.

As stated earlier, prior to the arrival of the first wave at time $t = x/c_3$, $f(x,t) = 0$. For $x/c_3 < t < x/c_4$, there are contributions from cuts CL, FP and IS. Performing the integrations along each of the contours gives

$$f(x,t) =$$

$$I_2 + \int_0^\phi \frac{R\chi_2}{E^2 + F^2} \Big\{ \cos[Rt\sin(\theta) - x\chi_2 \sin(\Theta_2)]$$

$$\times \sinh[Rt\cos(\theta) - x\chi_2 \cos(\Theta_2)][E\cos(\Theta_2 + \theta) + F\sin(\Theta_2 + \theta)]$$

$$- \sin[Rt\sin(\theta) - x\chi_2 \sin(\Theta_2)]$$

$$\times \cosh[Rt\cos(\theta) - x\chi_2\cos(\Theta_2)][E\sin(\Theta_2 + \theta) - F\cos(\Theta_2 + \theta)]\Big\}d\theta$$

$$+ \int_0^\phi \frac{R\chi_1}{E^2 + F^2}\Big\{\cos[Rt\sin(\theta) - x\chi_1\sin(\Theta_1)]$$

$$\times \sinh[Rt\cos(\theta) - x\chi_1\cos(\Theta_1)][E\cos(\Theta_1 + \theta) + F\sin(\Theta_1 + \theta)]$$

$$- \sin[Rt\sin(\theta) - x\chi_1\sin(\Theta_1)]$$

$$\times \cosh[Rt\cos(\theta) - x\chi_1\cos(\Theta_1)][E\sin(\Theta_1 + \theta) - F\cos(\Theta_1 + \theta)]\Big\}d\theta$$

$$(3.1.173)$$

where

$$I_2 = -\int_0^\nu \frac{\delta_2}{\delta_2^2(r^2 + c_3^2\delta_2^2)^2 + \delta_1^2(r^2 - c_3^2\delta_1^2)^2}$$

$$\times \Big[\delta_1(r^2 - c_3^2\delta_1^2)\sin(rt)\cosh(\delta_2 x)$$

$$+ \delta_2(r^2 - c_3^2\delta_2^2)\cos(rt)\sinh(\delta_2 x)\Big]\,dr, \qquad (3.1.174)$$

$$E = -c_3^2\chi_1^3\cos(3\Theta_1) + c_3^2\chi_2^3\cos(3\Theta_2)$$
$$+ R^2\chi_1\cos(2\theta + \Theta_1) - R^2\chi_2\cos(2\theta + \Theta_2) \qquad (3.1.175)$$

and

$$F = -c_3^2\chi_1^3\sin(3\Theta_1) + c_3^2\chi_2^3\sin(3\Theta_2)$$
$$+ R^2\chi_1\sin(2\theta + \Theta_1) - R^2\chi_2\sin(2\theta + \Theta_2). \qquad (3.1.176)$$

For $t > x/c_4$, there are contributions from both the $e^{\gamma_3 x}$ and $e^{\gamma_4 x}$ terms. Because of symmetries present in the values of γ_3 and γ_4, the contribution from cuts CL and IL cancel out and there is only the contribution from cut FP. This yields

$$f(x,t) = I_2 + \int_0^\nu \frac{\delta_1\delta_2(r^2 + c_3^2\delta_2^2)\sin(rt - \delta_1 x)}{\delta_2^2(r^2 + c_3^2\delta_2^2)^2 + \delta_1^2(r^2 - c_3^2\delta_1^2)^2}\,dr. \qquad (3.1.177)$$

• **Example 3.1.9**

During a study of solute transport from an injection well into an aquifer, C.-S. Chen[8] employed a contour integration to invert the Laplace transform

$$F(s) = \frac{\mathrm{Ai}(\lambda)}{s[\mathrm{Ai}(\lambda_0)/2 - s^{1/3}\mathrm{Ai}'(\lambda_0)]} \qquad (3.1.178)$$

---

[8] Taken from Chen, C.-S., 1987: Analytical solutions for radial dispersion with Cauchy boundary at injection well. *Water Resour. Res.*, **23**, 1217–1224. ©1987 American Geophysical Union.

where $\lambda = (1 + 4rs)/(4s^{2/3})$, $\lambda_0 = (1 + 4as)/(4s^{2/3})$, Ai$(z)$ is the Airy function of the first kind and $r > a > 0$.

We begin by closing the Bromwich integral as shown in Fig. 3.1.2 where the branch cut for $s^{1/3}$ lies along the negative real axis of the s-plane. By Jordan's lemma the contribution from the arcs at infinity are zero. Consequently the inversion consists of two possible parts. First, we have terms arising from the poles of $F(s)$. If we denote each pole by s_n, then s_n is a solution of

$$\text{Ai}(\lambda_n)/2 - s_n^{1/3}\text{Ai}'(\lambda_n) = 0 \qquad (3.1.179)$$

where $\lambda_n = (1 + 4as_n)/(4s_n^{2/3})$. Note that in the present case $s = 0$ is outside the contour because it is a branch point of $s^{1/3}$. Upon differentiating (3.1.179) with respect to a and using the differential equation that defines the Airy function, we can rewrite (3.1.179)

$$(1/4 - s_n^{2/3}\lambda_n)\text{Ai}(\lambda_n) = 0. \qquad (3.1.180)$$

Because Ai(λ_n) does not contain any zeros within the closed contour,

$$\frac{1}{4} = s_n^{2/3}\lambda_n = \frac{1 + 4as_n}{4} \text{ or } s_n = 0 \qquad (3.1.181)$$

which is outside of our closed contour. Consequently the only contribution to the inversion will come from the branch cut integral.

Turning to these integrals, we have along CD that $s = \eta^2 e^{\pi i}$. Thus,

$$\frac{1}{2\pi i}\int_{CD} = \frac{1}{\pi i}\int_{\infty}^{0} \frac{\exp(-\eta^2 t)}{\eta} d\eta$$

$$\times \frac{\text{Ai}[\chi(\eta)e^{-2\pi i/3}]}{\text{Ai}[\chi_0(\eta)e^{-2\pi i/3}] - \eta^{2/3}e^{\pi i/3}\text{Ai}'[\chi_0(\eta)e^{-2\pi i/3}]}$$

$$(3.1.182)$$

where $\chi_0(\eta) = (1 - 4a\eta^2)/(4\eta^{4/3})$ and $\chi(\eta) = (1 - 4r\eta^2)/(4\eta^{4/3})$. Substituting the relationships that

$$\text{Ai}(\eta e^{\pm 2\pi i/3}) = e^{\pm \pi i/3}[\text{Ai}(\eta) \mp i\text{Bi}(\eta)]/2 \qquad (3.1.183)$$

and

$$\text{Ai}'(\eta e^{\pm 2\pi i/3}) = e^{\mp \pi i/3}[\text{Ai}'(\eta) \mp i\text{Bi}'(\eta)]/2 \qquad (3.1.184)$$

into (3.1.182) and simplifying,

$$\frac{1}{2\pi i}\int_{CD} = -\frac{1}{\pi i}\int_0^{\infty} e^{-\eta^2 t}\frac{d\eta}{\eta}$$

$$\times \left\{ \frac{\text{Ai}(\chi) + i\text{Bi}(\chi)}{[\text{Ai}(\chi_0) + i\text{Bi}(\chi_0)]/2 + \eta^{2/3}[\text{Ai}'(\chi_0) + i\text{Bi}'(\chi_0)]} \right\}$$

$$(3.1.185)$$

where $\mathrm{Bi}(z)$ denotes the Airy function of the second kind. Similarly, along FG, $s = \eta^2 e^{-\pi i}$ and

$$\frac{1}{2\pi i}\int_{FG} = \frac{1}{\pi i}\int_0^\infty e^{-\eta^2 t}\frac{d\eta}{\eta}$$
$$\times \left\{\frac{\mathrm{Ai}(\chi) - i\mathrm{Bi}(\chi)}{[\mathrm{Ai}(\chi_0) - i\mathrm{Bi}(\chi_0)]/2 + \eta^{2/3}[\mathrm{Ai}'(\chi_0) - i\mathrm{Bi}'(\chi_0)]}\right\}.$$

$$(3.1.186)$$

Thus, the integrals along the branch cuts yields

$$\frac{1}{2\pi i}\int_{CD+FG} = \frac{4}{\pi}\int_0^\infty e^{-\eta^2 t}\left[\frac{\mathrm{Ai}(\chi)f_1 - \mathrm{Bi}(\chi)f_2}{f_1^2 + f_2^2}\right]\frac{d\eta}{\eta} \qquad (3.1.187)$$

where

$$f_1 = \mathrm{Bi}(\chi_0) + 2\eta^{2/3}\mathrm{Bi}'(\chi_0) \qquad (3.1.188)$$

and

$$f_2 = \mathrm{Ai}(\chi_0) + 2\eta^{2/3}\mathrm{Ai}'(\chi_0). \qquad (3.1.189)$$

Finally we must deal with the integration around the branch point $z = 0$:

$$\frac{1}{2\pi i}\int_{DEF} = \frac{1}{2\pi i}\int_\pi^{-\pi}\left\{\frac{\mathrm{Ai}(\xi)}{\mathrm{Ai}(\xi_0)/2 - [\epsilon\exp(\theta i)]^{1/3}\mathrm{Ai}'(\xi_0)}\right\}i\,d\theta$$

$$(3.1.190)$$

$$= \frac{1}{2\pi i}\int_\pi^{-\pi}e^{(a-r)/2}i\,d\theta = -e^{(a-r)/2} \qquad (3.1.191)$$

where $\xi = (1 + 4r\epsilon e^{\theta i})/[4(\epsilon e^{\theta i})^{2/3}]$ and $\xi_0 = (1 + 4a\epsilon e^{\theta i})/[4(\epsilon e^{\theta i})^{2/3}]$. Therefore, the inverse equals

$$f(t) = e^{(a-r)/2} - \frac{4}{\pi}\int_0^\infty e^{-\eta^2 t}\left[\frac{\mathrm{Ai}(\chi)f_1 - \mathrm{Bi}(\chi)f_2}{f_1^2 + f_2^2}\right]\frac{d\eta}{\eta}. \qquad (3.1.192)$$

- **Example 3.1.10**

Let us invert the Laplace transform[9]

$$F(s) = \frac{\exp[-x\sqrt{k^2 + s^2/c^2}\,]}{k^2 + s^2/c^2} \qquad (3.1.193)$$

---

[9] Taken from Watanabe, K., 1981: Transient response of an inhomogeneous elastic half space to a torsional load. *Bull. JSME*, **24**, 1537–1542.

by contour integration where c, k and x are real and positive. If $t > x/c$, Jordan's lemma allows us to close Bromwich's contour as shown in Fig. 3.1.10. Thus, if C denotes the contour $ABCDEF$ shown in Fig. 3.1.10, then

$$f(t) = \frac{c^2}{2\pi i} \int_C \frac{\exp[tz - (x/c)\sqrt{z^2 + k^2c^2}]}{z^2 + k^2c^2} dz \qquad (3.1.194)$$

if $t > x/c$; if $t < x/c$, $f(t) = 0$.

Our first task is to compute the phases of the various functions along the four legs that comprise the branch cut integration; the contribution from the integrations along the arcs at infinity vanish. Along C_1,

$$z - kci = kc(u + 1)e^{-\pi i/2}, z + kci = kc(u - 1)e^{-\pi i/2}, 1 < u < \infty \qquad (3.1.195)$$

so that

$$\sqrt{z^2 + k^2c^2} = kc\sqrt{u^2 - 1}\, e^{-\pi i/2}. \qquad (3.1.196)$$

Along C_2,

$$z - kci = kc(u - 1)e^{-\pi i/2}, z + kci = kc(u + 1)e^{3\pi i/2}, 1 < u < \infty \qquad (3.1.197)$$

and

$$\sqrt{z^2 + k^2c^2} = kc\sqrt{u^2 - 1}\, e^{\pi i/2}. \qquad (3.1.198)$$

Next, along C_3,

$$z - kci = kc(u - 1)e^{-3\pi i/2}, z + kci = kc(u + 1)e^{\pi i/2}, 1 < u < \infty \qquad (3.1.199)$$

and

$$\sqrt{z^2 + k^2c^2} = kc\sqrt{u^2 - 1}\, e^{-\pi i/2}. \qquad (3.1.200)$$

Finally, along C_4,

$$z - kci = kc(u - 1)e^{\pi i/2}, z + kci = kc(u + 1)e^{\pi i/2}, 1 < u < \infty \qquad (3.1.201)$$

and

$$\sqrt{z^2 + k^2c^2} = kc\sqrt{u^2 - 1}\, e^{\pi i/2}. \qquad (3.1.202)$$

Substituting these definitions into the contour integration,

$$f(t) = \frac{c}{2\pi ki} \int_\infty^1 \frac{\exp(-cktui + ikx\sqrt{u^2 - 1})}{1 - u^2}(-i\,du)$$
$$+ \frac{c}{2\pi ki} \int_1^\infty \frac{\exp(-cktui - ikx\sqrt{u^2 - 1})}{1 - u^2}(-i\,du)$$
$$+ \frac{c}{2\pi ki} \int_\infty^1 \frac{\exp(cktui + ikx\sqrt{u^2 - 1})}{1 - u^2}(i\,du)$$
$$+ \frac{c}{2\pi ki} \int_1^\infty \frac{\exp(cktui - ikx\sqrt{u^2 - 1})}{1 - u^2}(i\,du)$$
$$+ \frac{c}{k} \sin(ckt) \qquad (3.1.203)$$

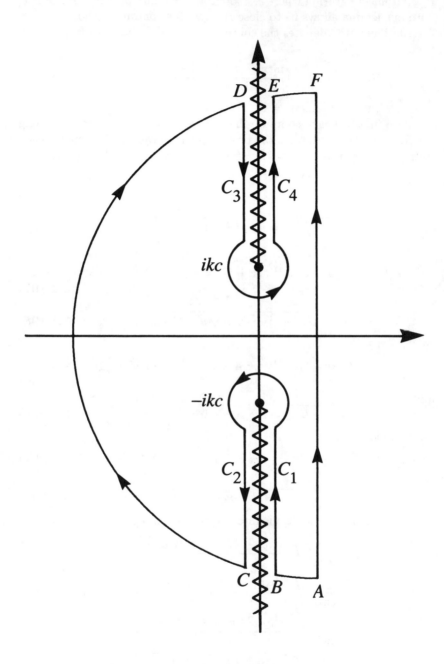

Fig. 3.1.10: Contour used in the integration of (3.1.194).

where the last term in (3.1.203) arises from the integration around the branch points $z = \pm kci + \epsilon e^{\theta i}$. Combining the exponentials,

$$f(t) = \frac{c}{k}\sin(ckt) - \frac{2c}{k\pi}\int_1^\infty \frac{\sin(kx\sqrt{u^2-1})}{u^2-1}\sin(cktu)\,du \quad \textbf{(3.1.204)}$$

$$= \frac{c}{k}\sin(ckt) - \frac{2c}{k\pi}\int_0^\infty \frac{\sin(ckt\sqrt{v^2+1})}{v\sqrt{v^2+1}}\sin(xkv)\,dv \quad \textbf{(3.1.205)}$$

$$= \frac{c}{k}\sin(ckt) - \frac{2c}{\pi}\int_0^x\left[\int_0^\infty \frac{\sin(ckt\sqrt{v^2+1})}{v\sqrt{v^2+1}}\cos(\eta kv)\,dv\right]d\eta$$

$$\textbf{(3.1.206)}$$

if $u^2 = v^2 + 1$. To obtain (3.1.206) we used the relationship that

$$\sin(xkv) = kv\int_0^x \cos(kv\eta)\,d\eta. \quad \textbf{(3.1.207)}$$

We have done this because we can evaluate the integral inside the large brackets exactly. From a table of integrals,

$$\int_0^\infty \frac{\sin(b\sqrt{a^2+x^2})\cos(xy)}{\sqrt{a^2+x^2}}\,dx = \frac{\pi}{2}H(b-y)J_0(a\sqrt{b^2-y^2}) \quad \textbf{(3.1.208)}$$

so that

$$f(t) = \frac{c}{k}\sin(kct) - c\int_0^x J_0(k\sqrt{c^2t^2-\eta^2})\,d\eta. \quad \textbf{(3.1.209)}$$

Finally, because

$$\sin(x) = \int_0^x J_0(\sqrt{x^2-\eta^2})\,d\eta, \quad \textbf{(3.1.210)}$$

our final result is

$$f(t) = cH(t-x/c)\int_x^{ct} J_0(k\sqrt{c^2t^2-\eta^2})\,d\eta. \quad \textbf{(3.1.211)}$$

Problems

1. Show that the Laplace transform

$$F(x,s) = \frac{\omega}{s^2+\omega^2}e^{-x\sqrt{s/\kappa}}$$

has the inverse

$$f(x,t) = \exp(-x\sqrt{\omega/2\kappa})\sin(\omega t - x\sqrt{\omega/2\kappa})$$
$$-\frac{2\kappa}{\pi}\int_0^\infty \frac{\omega}{\omega^2 + \kappa^2\eta^4}e^{-\kappa t\eta^2}\sin(x\eta)\eta\,d\eta.$$

Take the branch cut for $\sqrt{s}$ along the negative real axis of the s-plane.

2. Find the inverse of Laplace transform

$$F(s) = \frac{1}{s^2 + 1}\exp\left[-\frac{sx}{\sqrt{s+1}}\right]$$

by taking the branch cut associated with the branch point $s = -1$ along the negative real axis of the s-plane and using Bromwich's integral. You should find that

$$f(t) = \exp[-x\sin(\pi/8)/\sqrt{2}]\sin[t - x\cos(\pi/8)/\sqrt{2}]$$
$$-\frac{1}{\pi}\int_0^\infty \frac{\exp[-t(1+u)]}{u^2 + 2u + 2}\sin\left[\frac{(1+u)x}{\sqrt{u}}\right]du.$$

3. Show that the inverse[10] of the Laplace transform

$$F(s) = \frac{1}{s[a + b\log(1 + c/s)]}$$

is

$$f(t) = -\frac{\exp[-ct/(1 - e^{-a/b})]}{b(e^{a/b} - 1)} + b\int_0^1 \frac{\exp(-uct)}{u\{[a + b\ln(1/u - 1)]^2 + b^2\pi^2\}}du$$

if we take the branch cut along the negative real axis of the s-plane between its branch points at $s = -c$ and $s = 0$. The constants a, b and c are real and positive.

4. Find the inverse[11] of the Laplace transform

$$F(s) = \frac{1}{s}\exp\left[-r\sqrt{\frac{s(1+s)}{1+as}}\right]$$

[10] Reprinted from *Int. J. Solids Structures*, **29**, J. Lubliner and V. P. Panoskaltsis, The modified Kuhn model of linear viscoelasticity, pp. 3099–3112, ©1992, with kind permission from Pergamon Press Ltd., Headington Hill Hall, Oxford OX3 0BW, UK.

[11] Taken from Tanner, R. I., 1962: Note on the Rayleigh problem for a visco-elastic fluid. *Zeit. angew. Math. Phys.*, **13**, 573–580. Published by Birkhäuser Verlag, Basel, Switzerland.

where $0 \leq a \leq 1$ and $0 \leq r < \infty$ by taking the branch cuts associated with the branch points $s = 0$, $s = -1$ and $s = -1/a$ along the negative real axis of the s-plane. Then we may deform the Bromwich contour along the imaginary axis of the s-plane except for a small semicircle around $s = 0$. Using this contour you should find that

$$
f(t) = \frac{1}{2} + \frac{1}{\pi} \int_0^\infty \exp\left\{ -r\sqrt{\frac{u}{2}} \, M[\cos(\theta) - \sin(\theta)] \right\} \frac{du}{u}
$$

$$
\times \sin\left\{ tu - r\sqrt{\frac{u}{2}} \, M[\cos(\theta) + \sin(\theta)] \right\}
$$

where

$$
M = \left(\frac{1 + u^2}{1 + a^2 u^2} \right)^{1/4}
$$

and

$$
2\theta = \tan^{-1}(u) - \tan^{-1}(au).
$$

5. Show that the inverse of the Laplace transform

$$
F(s) = \frac{K_0(r\sqrt{s/\nu + 1/b^2})}{sK_0(a\sqrt{s/\nu + 1/b^2})}
$$

is

$$
f(t) = \frac{K_0(r/b)}{K_0(a/b)} + \frac{2}{\pi} e^{-\nu t/b^2} \int_0^\infty \frac{\exp(-\nu t\eta^2/a^2)}{\eta^2 + (a/b)^2} \eta \, d\eta
$$

$$
\times \frac{J_0(r\eta/a)Y_0(\eta) - Y_0(r\eta/a)J_0(\eta)}{J_0^2(\eta) + Y_0^2(\eta)}
$$

where a, b and ν are real, positive constants. The function $K_0(\ )$ is a zeroth-order, modified Bessel functions of the second kind while $J_0(\ )$ and $Y_0(\ )$ are zeroth-order Bessel functions of the first and second kind, respectively. Take the branch cut along the negative real axis of the s-plane from the branch point at $s = -\nu/b^2$.

6. Find the inverse[12] of the Laplace transform

$$
F(s) = \frac{K_0(r\sqrt{s})}{s^{3/2}K_1(a\sqrt{s})}
$$

---

[12] Taken from Goldstein, R. J., and D. G. Briggs, 1964: Transient free convection about vertical plates and circular cylinders. *J. Heat Transfer*, **86**, 490–500.

by first showing that

$$\mathcal{L}^{-1}\left[\frac{K_0(r\sqrt{s})}{\sqrt{s}K_1(a\sqrt{s})}\right] = -\frac{2}{\pi}\int_0^\infty e^{-\eta^2 t}\frac{J_0(\eta r)Y_1(\eta a) - Y_0(\eta r)J_1(\eta a)}{J_1^2(\eta a) + Y_1^2(\eta a)}\,d\eta$$

where $K_0(\ )$ and $K_1(\ )$ are zeroth-order and first-order, modified Bessel functions of the second kind, respectively, and $J_0(\ )$ and $Y_0(\ )$ are zeroth-order Bessel functions of the first and second kind, respectively. Take the branch cuts associated with $K_0(\ )$, $K_1(\ )$ and $\sqrt{s}$ along the negative real axis of the s-plane. Then

$$f(t) = \int_0^t \mathcal{L}^{-1}\left[\frac{K_0(r\sqrt{s})}{\sqrt{s}K_1(a\sqrt{s})}\right]d\tau$$

$$= -\frac{2}{\pi}\int_0^\infty \left(1 - e^{-\eta^2 t}\right)\frac{J_0(\eta r)Y_1(\eta a) - Y_0(\eta r)J_1(\eta a)}{J_1^2(\eta a) + Y_1^2(\eta a)}\frac{d\eta}{\eta^2}.$$

7. Find the inverse of the Laplace transform

$$F(s) = \frac{K_0(r\sqrt{s+1})}{sK_0(\sqrt{s+1})}$$

where $r > 0$ and $K_0(\ )$ is the zeroth-order, modified Bessel function of the second kind. If you take the branch cut associated with $K_0(\ )$ along the negative real axis of the s-plane, you should find that

$$f(t) = \frac{K_0(r)}{K_0(1)} - \frac{2}{\pi}e^{-t}\int_0^\infty \frac{u}{u^2+1}\frac{J_0(u)Y_0(ru) - J_0(ru)Y_0(u)}{J_0^2(u) + Y_0^2(u)}e^{-u^2 t}\,du$$

where $J_0(\ )$ and $Y_0(\ )$ are zeroth-order Bessel functions of the first and second kind, respectively.

8. Find the inverse[13] of the Laplace transform

$$F(s) = \frac{2\pi}{s^2 + 4\pi^2}\exp\left[-x\sqrt{\frac{s(1+as)}{1+bs}}\right]$$

where $b > a > 0$. If we take the branch cuts from the three branch points $s = 0$, $s = -1/b$ and $s = -1/a$ along the negative real axis, you should find that

$$f(t) = \sin[2\pi t - x\Delta\sin(\psi/2)]\exp[-x\Delta\cos(\psi/2)]$$

$$+ 2\int_0^{1/b} F(x,t,\eta)\,d\eta + 2\int_{1/a}^\infty F(x,t,\eta)\,d\eta$$

---

[13] Taken from Zheltov, Yu. P., and V. S. Kutlyarov, 1965: Transient motion of a liquid in a fissured porous stratum subject to periodic pressure variation at the boundary. *J. Appl. Mech. Tech. Phys.*, **6(6)**, 45–49.

where

$$F(x,t,\eta) = \frac{\exp(-\eta t)}{\eta^2 + 4\pi^2} \sin\left[x\sqrt{\frac{\eta(1-a\eta)}{1-b\eta}}\right],$$

$$\Delta = [(d_1^2 + d_2^2)d_3^2]^{1/4},$$

$$d_1 = 1 + 4\pi^2 ab, \qquad d_2 = 2\pi(b-a),$$

$$d_3 = \frac{2\pi}{1 + 4\pi^2 b^2} \quad \text{and} \quad \psi = \tan^{-1}(d_1/d_2).$$

9. Find the inverse[14] of the Laplace transform

$$F(s) = \frac{\omega}{s^2 + \omega^2} \frac{K_1(r\sqrt{s})}{K_1(\sqrt{s})}$$

where $K_1(\ )$ is the first-order, modified Bessel function of the second kind, r and ω are real and positive. If we take the branch cut associated with K_1 along the negative real axis of the s-plane, show that

$$f(t) = \frac{N_1(r\sqrt{\omega})}{N_1(\sqrt{\omega})} \sin[\omega t + \phi_1(r\sqrt{\omega}) - \phi_1(\sqrt{\omega})]$$
$$- \frac{2}{\pi} \int_0^\infty \frac{\eta}{\eta^4 + \omega^2} e^{-\eta^2 t} \frac{J_1(r\eta)Y_1(\eta) - Y_1(r\eta)J_1(\eta)}{J_1^2(\eta) + Y_1^2(\eta)} d\eta$$

where

$$N_1(z) = \sqrt{\ker_1^2(z) + \kei_1^2(z)}, \phi_1(z) = \tan^{-1}[\kei_1(z)/\ker_1(z)],$$

$\ker_1(z)$, $\kei_1(z)$ are Kelvin's ker and kei functions, respectively, and $J_1(\ )$, $Y_1(\ )$ are first-order Bessel functions of the first and second kind, respectively.

10. Find the inverse[15] of the Laplace transform

$$F(s) = \frac{1 + r\sqrt{s}}{s[1 + a\sqrt{s}]} e^{-(r-a)\sqrt{s}}$$

---

[14] Taken from Schwarz, W. H., 1963: The unsteady motion of an infinite oscillating cylinder in an incompressible Newtonian fluid at rest. *Appl. Sci. Res.*, **A11**, 115–124. Reprinted by permission of Kluwer Academic Publishers.

[15] Taken from Seth, S. S., 1977: Unsteady motion of viscoelastic fluid due to rotation of a sphere. *Indian J. Pure Appl. Math.*, **8**, 302–308.

where $r > a > 0$. If we take the branch cut associated with the square root along the negative real axis of the s-plane, show that

$$f(t) = 1 - \frac{2}{\pi} \int_0^\infty \frac{1+ar\eta^2}{1+a^2\eta^2} \sin[(r-a)\eta] e^{-\eta^2 t} \frac{d\eta}{\eta}$$
$$+ \frac{2}{\pi}(r-a) \int_0^\infty \frac{1}{1+a^2\eta^2} \cos[(r-a)\eta] e^{-\eta^2 t} d\eta.$$

11. Find the inverse[16] of the Laplace transform

$$F(s) = \frac{aM}{s} \frac{K_1(r\sqrt{s})}{(Is+2\pi\mu a^2)K_1(a\sqrt{s}) - 2\pi\mu a^3 \partial K_1(r\sqrt{s})/\partial r|_{r=a}}$$

where a, M, r, I, π and μ are constants and $K_1(\ )$ is a modified Bessel function of first order and second kind. If we take the branch cut associated with the Bessel function along the negative real axis of the s-plane, show that

$$f(t) = \frac{M}{4\pi\mu r} \left\{ 1 + \frac{4\chi r}{a\pi} \int_0^\infty \frac{P(\eta)}{\eta^2 Q(\eta)} e^{-t\eta^2/a^2} d\eta \right\}$$

where

$$P(\eta) = Y_1(r\eta/a)[\eta J_1(\eta) - \chi J_2(\eta)] - J_1(r\eta/a)[\eta Y_1(\eta) - \chi Y_2(\eta)],$$

$$Q(\eta) = [\eta J_1(\eta) - \chi J_2(\eta)]^2 + [\eta Y_1(\eta) - \chi Y_2(\eta)]^2,$$

$\chi = 2\pi\mu a^4/I$, $J_1(\ )$, $J_2(\ )$ are first and second-order Bessel functions of the first kind, respectively, and $Y_1(\ )$, $Y_2(\ )$ are first and second-order Bessel functions of the second kind, respectively.

12. Find the inverse of the Laplace transform

$$F(s) = \frac{1}{s} \exp\left(-x\sqrt{\frac{s}{s+a}}\right)$$

where a and $x > 0$. If we take the branch cuts associated with the $\sqrt{s}$ and $\sqrt{s+a}$ along the negative real axis of the s-plane, show that

$$f(t) = 1 - \frac{2}{\pi} \int_0^\infty \exp\left(-\frac{at\sigma^2}{1+\sigma^2}\right) \frac{\sin(\sigma x)}{\sigma(1+\sigma^2)} d\sigma.$$

[16] Taken from Sennitskii, 1981: Unsteady rotation of a cylinder in a viscous fluid. *J. Appl. Mech. Tech. Phys.*, **21**, 347–349.

13. Find the inverse[17] of the Laplace transform

$$F(s) = \frac{1}{s + b^2} K_0 \left(rc \sqrt{\frac{s + a^2}{s + b^2}} \right)$$

where a, b, c and r are real and positive, $b > a$ and $K_0(\ )$ is the zeroth-order, modified Bessel function of the second kind. If we take the branch cuts associated with $\sqrt{s + a^2}$ and $\sqrt{s + b^2}$ along the negative real axis of the s-plane, show that

$$f(t) = \int_{a^2}^{b^2} \frac{1}{b^2 - \rho} J_0 \left(rc \sqrt{\frac{\rho - a^2}{b^2 - \rho}} \right) e^{-\rho t} d\rho$$

where $J_0(\ )$ is the zeroth-order Bessel function of the first kind.

14. Find the inverse[18] of the Laplace transform

$$F(s) = \frac{\exp(-sr/c)}{sK_0(sa/c)}$$

where a, c and r are real and positive and $K_0(\ )$ is the zeroth-order, modified Bessel function of the second kind. If we take the branch cut associated with $K_0(\ )$ along the negative real axis of the s-plane, you should find that

$$f(t) = H(\tau) \int_0^\infty \frac{I_0(\eta)}{K_0^2(\eta) + \pi^2 I_0^2(\eta)} e^{-\eta \tau} \frac{d\eta}{\eta}$$

where $\tau = (ct - r)/a$ and $I_0(\ )$ is the zeroth-order, modified Bessel function of the first kind.

15. Find the inverse of the Laplace transform

$$F(s) = \frac{K_1(as)}{s\, K_1(bs)}, \qquad a > b > 0,$$

where $K_1(\ )$ is the first-order, modified Bessel function of the second kind. If we take the branch cut associated with the Bessel function along the negative real axis of the s-plane, show that

$$f(t) = 1 + \int_0^\infty \left[\frac{I_1(a\eta)K_1(b\eta) - K_1(a\eta)I_1(b\eta)}{K_1^2(b\eta) + \pi^2 I_1^2(b\eta)} \right] \frac{e^{-\eta t}}{\eta}\, d\eta$$

---

[17] Taken from Raichenko, L. M., 1976: Flow of liquid to an incomplete well in a bed of fissile-porous rocks. *Sov. Appl. Mech.*, **12**, 1196–1199.

[18] Taken from Pine, Z. L., and F. M. Tesche, 1973: Calculation of the early time radiated electric field from a linear antenna with a finite source gap. *IEEE Trans. Antennas Propagat.*, **AP-21**, 740–743. ©1973 IEEE.

where $I_1(\ )$ is a first-order, modified Bessel function of the first kind.

16. Find the inverse[19] of the Laplace transform

$$F(s) = \frac{\exp(-a\sqrt{s})}{(s-k)\sqrt{s}}$$

where $k > 0$. If we take the branch cut associated with $\sqrt{s}$ along the negative real axis of the s-plane, you should find that

$$f(t) = \frac{\exp(kt - a\sqrt{k})}{\sqrt{k}} - \frac{2}{\pi} \int_0^\infty \frac{\cos(a\eta)}{\eta^2 + k} e^{-\eta^2 t} d\eta.$$

17. Show that the inverse of the Laplace transform

$$F(s) = \frac{\sqrt{s}}{s\sqrt{s} + a^3}, \qquad a > 0,$$

is

$$f(t) = \frac{4}{3} e^{-a^2 t/2} \cos\left(\frac{\sqrt{3}}{2} a^2 t\right) - \frac{2a^3}{\pi} \int_0^\infty e^{-tx^2} \frac{x^2}{x^6 + a^6} dx.$$

Take the branch cut associated with the square root along the negative real axis of the s-plane.

18. Find the inverse[20] of the Laplace transform

$$F(s) = \frac{K_1(a\sqrt{s})}{s[bK_1(\sqrt{s}) + \sqrt{s}K_0(\sqrt{s})]}$$

where $K_0(\ )$ and $K_1(\ )$ are modified Bessel functions of zeroth and first order and second kind, respectively, $a > 0$ and $b > 0$. If you take the branch cuts associated with $K_0(\ )$, $K_1(\ )$ and $\sqrt{s}$ along the negative real axis of the s-plane, you should find that

$$f(t) = \frac{2}{\pi} \int_0^\infty \left(1 - e^{-\eta^2 t}\right) \frac{d\eta}{\eta}$$
$$\times \frac{[\eta Y_0(\eta) - bY_0(\eta)]J_1(\eta a) + Y_1(\eta a)[\eta J_0(\eta) - bJ_1(\eta)]}{[\eta J_0(\eta) - bJ_1(\eta)]^2 + [\eta Y_0(\eta) - bY_1(\eta)]^2}$$

[19] Taken from Duffy, P., 1989: The acceleration of cometary ions by Alfvén waves. *J. Plasma Phys.*, **42**, 13–25. Reprinted with the permission of Cambridge University Press.

[20] A more complicated version occurred in Das, A., A. Bello and R. A. Jishi, 1990: Theoretical considerations relating to the characteristic curves of the silver chalcogenide glass inorganic photoresists. *J. Appl. Phys.*, **68**, 3957–3963.

where $J_0(\ )$ and $Y_0(\ )$ are zeroth-order Bessel functions of the first and second kind, respectively.

19. Find the inverse[21] of the Laplace transform

$$F(s) = K_0\left[r\sqrt{\frac{s(s+b)}{c(s+a)}}\ \right]$$

where $b > a > 0$, c, r are real and positive and $K_0(\ )$ is the modified Bessel function of the second kind and zeroth order. If we take all of the branch cuts along the negative real axis of the s-plane, then

$$f(t) = \frac{1}{2}\int_0^a J_0\left[r\sqrt{\frac{x(b-x)}{c(a-x)}}\right]e^{-xt}dx + \frac{1}{2}\int_b^\infty J_0\left[r\sqrt{\frac{x(x-b)}{c(x-a)}}\right]e^{-xt}dx.$$

20. Find the inverse[22] of the Laplace transform

$$F(s) = \frac{1}{1+(as)^p}, \qquad 0 < p < 1.$$

If we take the branch cut associated with s^p along the negative real axis, show that

$$f(t) = \frac{\sin(p\pi)}{\pi}\int_0^\infty \frac{e^{-\eta t}}{a^p\eta^p + a^{-p}\eta^{-p} + 2\cos(p\pi)}\,d\eta.$$

21. Find the inverse[23] of the Laplace transform

$$F(s) = \frac{a}{z(a+s^{1/n})}$$

[21] Taken from Cooley, R. L., and C. M. Case, 1973: Effect of a water table aquitard on drawdown in an underlying pumped aquifer. *Water Resour. Res.*, **9**, 434–447. ©1973 American Geophysical Union.

[22] Taken from Meshkov, S. I., 1970: The integral representation of fractionally exponential functions and their application to dynamic problems of linear visco-elasticity. *J. Appl. Mech. Tech. Phys.*, **11**, 100–107.

[23] Reprinted from *Int. J. Heat Mass Transfer*, **21**, R. Ghez, Mass transport and surface reactions in Lévêque's approximation, pp. 745–750, ©1978, with kind permission from Pergamon Press Ltd., Headington Hill Hall, Oxford OX3 0BW, UK.

where $a > 0$ and $n > 1$. If we take the branch cut along the negative real axis of the s-plane so that $-\pi < \arg(z) < \pi$, show that

$$f(t) = 1 - \frac{n}{\pi} \sin\left(\frac{\pi}{n}\right) \int_0^\infty \frac{\exp(-a^n \eta^n t)}{\eta^2 + 2\eta \cos(\pi/n) + 1}\, d\eta.$$

22. Find the inverse[24] of the Laplace transform

$$F(s) = \frac{1}{s} \exp\left(-\lambda \sqrt{\frac{s^\alpha}{s^\alpha + a}}\right)$$

where $0 < \alpha < 1$, $0 < a < 1$ and $\lambda \geq 0$. If the branch cut associated with s^α lies along the negative real axis of the s-plane, show that the inverse is

$$f(t) = 1 - \frac{1}{\pi} \int_0^\infty e^{-t\eta} \exp\left[-\lambda \sqrt{\frac{\eta^\alpha}{\rho}} \cos\left(\frac{\alpha\pi - \phi}{2}\right)\right]$$
$$\times \sin\left[\lambda \sqrt{\frac{\eta^\alpha}{\rho}} \sin\left(\frac{\alpha\pi - \phi}{2}\right)\right] \frac{d\eta}{\eta}$$

where

$$\rho = \sqrt{\eta^{2\alpha} + a^2 + 2a\eta^\alpha \cos(\alpha\pi)}$$

and

$$\phi = \tan^{-1}\left[\frac{\eta^\alpha \sin(\alpha\pi)}{\eta^\alpha \cos(\alpha\pi) + a}\right], \qquad 0 < \phi < \pi.$$

23. Find the inverse[25] of the Laplace transform

$$F(s) = \frac{s\sqrt{1 + s^2/k^2}}{(s^2 + 2)^2 - 4\sqrt{(1 + s^2/k^2)(s^2 + 1)}}$$

where $k > 1$. The transform $F(s)$ has a simple pole at $s = 0$ because $F(s) \sim k^2/[2s(k^2 - 1)]$ as $s \to 0$, simple poles at $s = \pm i/\gamma$ where

---

[24] Reprinted from *Int. J. Solids Structures*, **25**, D. C. Lagoudas, C.-Y. Hui and S. L. Phoenix, Time evolution of overstress profiles near broken fibers in a composite with a viscoelastic matrix, pp. 45–66, ©1989, with kind permission from Pergamon Press Ltd., Headington Hill Hall, Oxford OX3 0BW, UK.

[25] Reprinted from *Int. J. Solids Structures*, **7**, Y. M. Tsai, Dynamic contact stresses produced by the impact of an axisymmetrical projectile on an elastic half-space, pp. 543–558, ©1971, with kind permission from Pergamon Press Ltd., Headington Hill Hall, Oxford OX3 0BW, UK.

$\gamma > 1$, and branch points at $s = \pm i$ and $s = \pm ki$. If we take the branch cuts associated with $\sqrt{s+i}$, $\sqrt{s-i}$, $\sqrt{s+ki}$ and $\sqrt{s-ki}$ along the imaginary axis of the s-plane as shown in Fig. 3.1.10, show that

$$f(t) = \frac{k^2}{2(k^2 - 1)} + \frac{2\sqrt{1 - 1/(\gamma k)^2}}{\gamma[-ig'(i/\gamma)]} \cos(t/\gamma)$$

$$- \frac{8}{\pi} \int_1^k \frac{(1 - \eta^2/k^2)\eta\sqrt{\eta^2 - 1}}{(2 - \eta^2)^4 + 16(1 - \eta^2/k^2)(\eta^2 - 1)} \cos(\eta t)\, d\eta$$

$$- \frac{2}{\pi} \int_k^\infty \frac{\eta\sqrt{\eta^2/k^2 - 1}}{(2 - \eta^2)^2 + 4\sqrt{(\eta^2/k^2 - 1)(\eta^2 - 1)}} \cos(\eta t)\, d\eta$$

where $g(s)$ is the denominator of the Laplace transform.

24. Find the inverse[26] of the Laplace transform

$$F(s) = \frac{a\exp(-x\sqrt{s})}{s[a + (2 - a)\sqrt{\pi s/2}]}$$

where both a and $x > 0$. If we take the branch cut associated with $\sqrt{s}$ along the negative real axis, show that the inverse is

$$f(t) = 1 - \frac{2ac}{\pi}\int_0^\infty \frac{\cos(x\eta)}{a^2 + c^2\eta^2}e^{-t\eta^2}\,d\eta - \frac{2a^2}{\pi}\int_0^\infty \frac{\sin(x\eta)}{\eta(a^2 + c^2\eta^2)}e^{-t\eta^2}\,d\eta$$

where $c^2 = \pi(2 - a)^2/2$.

25. Find the inverse[26] of the Laplace transform

$$F(s) = \frac{a\sqrt{s}\exp[-x\sqrt{s(s+1)}]}{s[a\sqrt{s+1} + (2 - a)\sqrt{\pi s/2}]}$$

where both a and $x > 0$. If we take the branch cuts associated with $\sqrt{s}$ and $\sqrt{s+1}$ along the negative real axis of the s-plane, show that the inverse is

$$f(t) = \left\{\frac{2a^2}{\pi}\int_0^1 \frac{\sqrt{1 - \eta^2}\cos(x\eta\sqrt{1 - \eta^2})}{a^2 + b^2\eta^2}e^{-t\eta^2}\,d\eta\right.$$

$$\left. - a(2 - a)\sqrt{\frac{2}{\pi}}\int_0^1 \frac{\eta\sin(x\eta\sqrt{1 - \eta^2})}{a^2 + b^2\eta^2}e^{-t\eta^2}\,d\eta\right\}H(t - x)$$

---

[26] Taken from Kuznetsov, M. M., 1975: Unsteady-state slip of a gas near an infinite plane with diffusion-mirror reflection of molecules. *J. Appl. Mech. Tech. Phys.*, **16**, 853–858.

where $b^2 = \pi(2-a)^2/2 - a^2$.

26. Find the inverse of the Laplace transform

$$F(s) = \frac{1}{s}\exp[-a\sqrt{s}(1 - ce^{-b\sqrt{s}})/(1 + ce^{-b\sqrt{s}})]$$

where $a > 0$ and b and c are real. If we take the branch cut associated with the square root along the negative real axis, then

$$f(t) = 1 - \frac{1}{\pi}\int_0^\infty e^{-\eta t + a\sqrt{\eta}\,u(\eta)}\sin[a\sqrt{\eta}\,v(\eta)]\frac{d\eta}{\eta}$$

where

$$u(\eta) = \frac{2c\sin(b\sqrt{\eta})}{1 + c^2 + 2c\cos(b\sqrt{\eta})}$$

and

$$v(\eta) = \frac{1 - c^2}{1 + c^2 + 2c\cos(b\sqrt{\eta})}.$$

27. Following the example 3.1.5, find the inverse[27] of the Laplace transform

$$F(s) = \frac{1}{s}\sqrt{\frac{v}{c}\left(s + \frac{\beta}{2}\right) + \sqrt{s(s+\beta) + \frac{v^2\beta^2}{4c^2}}}$$

where $0 < v/c < 1$ and $\beta > 0$. The transform has three branch points at $s = -\beta$ and

$$s = \alpha_{1,2} = \frac{\beta}{2}\left(-1 \pm \sqrt{1 - \frac{v^2}{c^2}}\right)$$

so that $-\beta < \alpha_2 < \alpha_1$. Taking the branch cut associated with each of these branch points along the negative real axis of the s-plane, show that the inverse is

$$f(t) = \sqrt{\frac{v\beta}{c}} - \int_0^{\beta\sqrt{1-(v/c)^2}} \frac{e^{(\alpha_1 - \eta)t}}{\alpha_1 - \eta}\sqrt{\frac{r - x}{2}}\,d\eta$$

$$- \int_{\beta\sqrt{1-(v/c)^2}}^\infty \frac{e^{(\alpha_1 - \eta)t}}{\alpha_1 - \eta}\,d\eta$$

$$\times \sqrt{\frac{v}{c}\left[\eta - \frac{\beta}{2}\sqrt{1 - (v/c)^2}\right] + \sqrt{\eta[\eta - \beta\sqrt{1 - (v/c)^2}]}}$$

---

[27] Taken from Mondal, S. C., and M. L. Ghosh, 1990: Moving punch on a viscoelastic semi-infinite medium. *Indian J. Pure Appl. Math.*, **21**, 847–864.

where

$$r = \sqrt{\eta[\beta\sqrt{1 - (v/c)^2} - \eta] + \frac{v^2}{c^2}\left[\frac{\beta}{2}\sqrt{1 - (v/c)^2} - \eta\right]^2}$$

and

$$x = \frac{v}{c}\left[\frac{\beta}{2}\sqrt{1 - (v/c)^2} - \eta\right].$$

28. Find the inverse[28] of the Laplace transform

$$F(s) = \frac{\exp[-(h - \delta)s/a]}{\sqrt{s}\,(s + a/2\delta)\sinh[(s + a/2\delta)(R\pi/c)]}.$$

Step 1. Use the properties of geometric series to show that

$$\frac{1}{\sinh[(s + a/2\delta)(R\pi/c)]} = \frac{2\exp[-(s + a/2\delta)(R\pi/c)]}{1 - \exp[-2(s + a/2\delta)(R\pi/c)]}$$

$$= 2\sum_{n=0}^{\infty}\exp\left[-(2n + 1)\left(\frac{aR\pi}{2\delta c}\right)\right]$$

$$\times \exp\left[-s(2n + 1)\frac{R\pi}{c}\right]$$

so that

$$F(s) = 2\sum_{n=0}^{\infty}\exp\left[-(2n + 1)\left(\frac{aR\pi}{2\delta c}\right)\right]f_n(s)$$

where

$$f_n(s) = \frac{\exp\{-s[(h - \delta)/a + (2n + 1)(R\pi/c)]\}}{\sqrt{s}\,(s + a/2\delta)}.$$

Step 2. Use the convolution theorem to show that

$$\mathcal{L}^{-1}\left[\frac{1}{\sqrt{s}\,(s + a/2\delta)}\right] = \frac{\exp(-at/2\delta)}{\sqrt{\pi}}\int_0^t\frac{\exp(au/2\delta)}{\sqrt{u}}\,du$$

$$= 2\sqrt{\frac{2\delta}{\pi a}}D\left(\sqrt{\frac{at}{2\delta}}\right)$$

---

[28] Taken from Kahalas, S. L., 1965: Excitation of extremely low frequency electromagnetic waves in the earth-ionosphere cavity by high-altitude nuclear detonations. *J. Geophys. Res.*, **70**, 3587–3594. ©1965 American Geophysical Union.

where

$$D(x) = e^{-x^2} \int_0^x e^{\eta^2} d\eta$$

is Dawson's function[29].

Step 3. Use the second shifting theorem to show that

$$f(t) = 4\sqrt{\frac{2\delta}{\pi a}} \sum_{n=0}^{\infty} \exp\left[-(2n+1)\left(\frac{aR\pi}{2\delta c}\right)\right]$$

$$\times D\left\{\sqrt{\frac{a}{2\delta}}\left[\tau - (2n+1)\left(\frac{\pi R}{c}\right)\right]\right\} H\left[\tau - (2n+1)\left(\frac{\pi R}{c}\right)\right]$$

where $\tau = t - (h - \delta)/a$.

3.2 INVERSION OF FOURIER TRANSFORMS BY CONTOUR INTEGRATION

In section 2.2 we showed that we may invert many Fourier transforms by using the inversion integral and complex variables. In that chapter we restricted ourselves to transforms which contained only single-valued functions. In many applications the transform will be multivalued. In this section we show how we may apply the techniques of sections 1.5–1.6 to the inversion of such transforms.

- **Example 3.2.1**

In this example we find the inverse[30] to the generalized Fourier transform

$$F(\omega) = \frac{-i \exp[-\sqrt{a^2 + \omega^2}]}{\omega\sqrt{a - i\omega}}. \tag{3.2.1}$$

We know that this is a generalized Fourier transform because of the singularity on the real axis at $\omega = 0$. Taking the contour of integration just below the real axis, Fig. 3.2.1 shows the closed contour. From the residue theorem,

$$\oint_C \frac{-i \exp[izt - \sqrt{a^2 + z^2}]}{z\sqrt{a - iz}} dz = \frac{2\pi}{\sqrt{a}} e^{-a}. \tag{3.2.2}$$

---

[29] Dawson, H. G., 1898: On the numerical value of $\int_0^h e^{x^2} dx$. *Proc. Lond. Math. Soc., Ser. 1*, **29**, 519–522.

[30] A similar Fourier transform arose in A. G. Azpeitia and G. F. Newell, 1958: Theory of oscillation type viscometers III: A thin disk. *Zeit. angew. Math. Phys.*, **9a**, 98–118. Published by Birkhäuser Verlag, Basel, Switzerland.

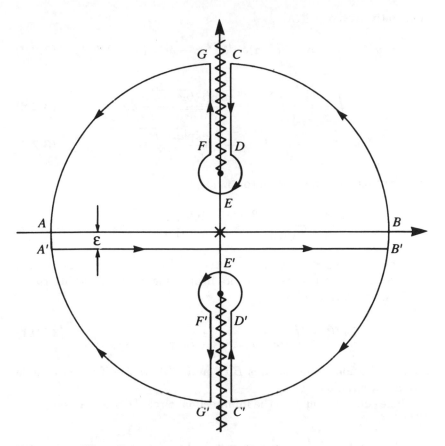

Fig. 3.2.1: The contour used to invert (3.2.1).

From Fig. 3.2.1 the closed contour for $t > 0$ may be broken down into several line integrations

$$2\pi f(t) + \int_{CD} + \int_{DEF} + \int_{FG} = \frac{2\pi}{\sqrt{a}}e^{-a} \qquad (3.2.3)$$

because the arcs BC and GA vanish as the radius of the semicircle becomes infinite.

Turning to the remaining integrals, the contribution from the infinitesimally small circle DEF vanishes. Along CD,

$$z = ue^{\pi i/2}, z - ia = (u - a)e^{\pi i/2}, z + ia = (u + a)e^{\pi i/2} \qquad (3.2.4)$$

$$\int_{CD} = -i \int_{\infty}^{a} \frac{e^{-ut} \exp[-i\sqrt{u^2 - a^2}]}{u\sqrt{u + a}} \, du \qquad (3.2.5)$$

$$= \frac{i}{\sqrt{a}} \int_{1}^{\infty} \frac{e^{-axt} \exp[-ai\sqrt{x^2 - 1}]}{x\sqrt{x + 1}} \, dx. \qquad (3.2.6)$$

Alternatively, along FG,

$$z = ue^{\pi i/2}, z - ia = (u - a)e^{-3\pi i/2}, z + ia = (u + a)e^{\pi i/2} \quad \text{(3.2.7)}$$

$$\int_{FG} = -i \int_a^\infty \frac{e^{-ut} \, \exp[i\sqrt{u^2 - a^2}]}{u\sqrt{u + a}} \, du \quad \text{(3.2.8)}$$

$$= -\frac{i}{\sqrt{a}} \int_1^\infty \frac{e^{-axt} \, \exp[ai\sqrt{x^2 - 1}]}{x\sqrt{x + 1}} \, dx. \quad \text{(3.2.9)}$$

Substituting these relations into (3.2.3),

$$f(t) = \frac{e^{-a}}{\sqrt{a}} - \frac{1}{\pi\sqrt{a}} \int_1^\infty \frac{e^{-axt} \, \sin[a\sqrt{x^2 - 1}]}{x\sqrt{x + 1}} \, dx. \quad \text{(3.2.10)}$$

For the case $t < 0$, we must use the contour in the lower half-plane. Because the contour does not contain any singularities, the inverse of the Fourier transform is

$$2\pi f(t) + \int_{C'D'} + \int_{D'E'F'} + \int_{F'G'} = 0; \quad \text{(3.2.11)}$$

the contributions from the arcs $B'C'$ and $G'A'$ vanish if the semicircle is of infinite radius.

The contribution from the infinitesimal circle $D'E'F'$ vanishes. For $C'D'$,

$$z = ue^{-\pi i/2}, z - ia = (u + a)e^{-\pi i/2}, z + ia = (u - a)e^{-\pi i/2} \quad \text{(3.2.12)}$$

$$\int_{C'D'} = \int_\infty^a \frac{e^{ut} \, \exp[i\sqrt{u^2 - a^2}]}{u\sqrt{u - a}} \, du \quad \text{(3.2.13)}$$

$$= -\frac{1}{\sqrt{a}} \int_1^\infty \frac{e^{axt} \, \exp[ai\sqrt{x^2 - 1}]}{x\sqrt{x - 1}} \, dx. \quad \text{(3.2.14)}$$

Alternatively, along $F'G'$,

$$z = ue^{-\pi i/2}, z - ia = (u + a)e^{-\pi i/2}, z + ia = (u - a)e^{3\pi i/2} \quad \text{(3.2.15)}$$

$$\int_{F'G'} = -\int_a^\infty \frac{e^{ut} \, \exp[-i\sqrt{u^2 - a^2}]}{u\sqrt{u - a}} \, du \quad \text{(3.2.16)}$$

$$= -\frac{1}{\sqrt{a}} \int_1^\infty \frac{e^{axt} \, \exp[-ai\sqrt{x^2 - 1}]}{x\sqrt{x - 1}} \, dx. \quad \text{(3.2.17)}$$

Consequently, the inverse of the Fourier transform (3.2.1) for $t < 0$ is

$$f(t) = \frac{1}{\pi\sqrt{a}} \int_1^\infty e^{axt} \frac{\cos[a\sqrt{x^2-1}]}{x\sqrt{x-1}}\,dx. \qquad (3.2.18)$$

- **Example 3.2.2**

We now find the inverse of the generalized Fourier transform[31]

$$F(\omega) = \frac{\exp[-\omega^2 + \omega\sqrt{\omega^2-1}]}{i\omega} \qquad (3.2.19)$$

or

$$f(t) = \frac{1}{2\pi} \int_{-\infty-\epsilon i}^{\infty-\epsilon i} \exp[-\omega^2 + \omega\sqrt{\omega^2-1} + i\omega t]\,\frac{d\omega}{i\omega}, 0 < \epsilon \ll 1. \qquad (3.2.20)$$

Because the singularities at $\omega = 0$ and $\omega = \pm 1$ lie above the contour, $f(t) = 0$ for $t < 0$. For $t > 0$ we convert the line integral into a closed contour integral in the upper half-plane shown in Fig. 3.2.2.

If our branch cut lies on the real axis from $z = -1$ to $z = 1$, the singularity at $z = 0$ also lies on the branch cut. This poses no difficulty but we can simplify the integration by a trick. Note that

$$\frac{df}{dt} = \frac{1}{2\pi} \int_{-\infty-\epsilon i}^{\infty-\epsilon i} \exp[-z^2 + z\sqrt{z^2-1} + izt]\,dz \qquad (3.2.21)$$

only has the branch points at $z = \pm 1$; there is no singularity at $z = 0$. If we introduce the closed contour shown in Fig. 3.2.2 with $C = C_1 + C_2 + C_3 + C_4 + C_5$,

$$\frac{df}{dt} + \frac{1}{2\pi} \int_C \exp[-z^2 + z\sqrt{z^2-1} + izt]\,dz = 0 \qquad (3.2.22)$$

because the function is analytic in the interior. As we expand the semicircle without limit, the integrations along C_1 and C_5 tend to zero if we choose the positive square root. Next, the line integral C_2 cancels the line integral C_4. This leaves only the path integral C_3. Along the top, to the right of the ordinate,

$$z = xe^{0i}, z - 1 = (1-x)e^{\pi i}, z + 1 = (1+x)e^{0i} \qquad (3.2.23)$$

---

[31] Taken from Menkes, J., 1972: The propagation of sound in the ionosphere. *J. Sound Vib.*, **20**, 311–319. Published by Academic Press Ltd., London, U.K.

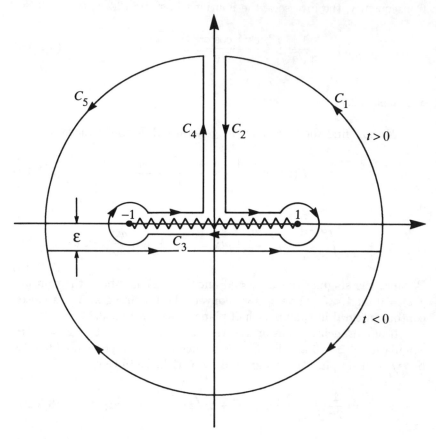

Fig. 3.2.2: The contour used to invert (3.2.19).

$$\sqrt{z^2 - 1} = \sqrt{1 - x^2}\, e^{\pi i/2} \tag{3.2.24}$$

with $0 < x < 1$. To the left of the ordinate,

$$z = xe^{\pi i}, z - 1 = (1 + x)e^{\pi i}, z + 1 = (1 - x)e^{0i} \tag{3.2.25}$$

$$\sqrt{z^2 - 1} = \sqrt{1 - x^2}\, e^{\pi i/2}. \tag{3.2.26}$$

Below the branch cut, to the right of the ordinate, with $0 < x < 1$

$$z = xe^{2\pi i}, z - 1 = (1 - x)e^{\pi i}, z + 1 = (1 + x)e^{2\pi i} \tag{3.2.27}$$

and

$$\sqrt{z^2 - 1} = \sqrt{1 - x^2}\, e^{3\pi i/2} \tag{3.2.28}$$

and to the left of the ordinate

$$z = xe^{\pi i}, z - 1 = (1 - x)e^{\pi i}, z + 1 = (1 + x)e^{2\pi i} \tag{3.2.29}$$

so that

$$\sqrt{z^2 - 1} = \sqrt{1 - x^2}\, e^{3\pi i/2}. \tag{3.2.30}$$

Upon substitution,

$$\frac{df}{dt} = -\frac{2}{\pi} \int_0^1 e^{-x^2} \sin[x\sqrt{1 - x^2}\,] \sin(xt)\, dx. \tag{3.2.31}$$

We obtain the final result by integrating Eq. (3.2.31),

$$f(t) = \frac{2}{\pi} \int_0^1 \frac{e^{-x^2} \sin[x\sqrt{1 - x^2}\,]\cos(xt)}{x}\, dx + C. \tag{3.2.32}$$

We specify the constant of integration from the initial value of $f(t)$.

- **Example 3.2.3**

For our next example, we will find the inverse[32] of

$$F(\omega, t) = 2\frac{\tanh(\omega b)}{\omega b} \cos\left(t\sqrt{\omega^2 + \frac{n^2\pi^2}{b^2}}\right) \tag{3.2.33}$$

or

$$f(x, t) = \frac{1}{\pi} \int_{-\infty}^{\infty} \frac{\tanh(\omega b)}{\omega b} \cos\left(t\sqrt{\omega^2 + \frac{n^2\pi^2}{b^2}}\right) e^{i\omega x}\, d\omega \tag{3.2.34}$$

where $n \geq 1$, $b > 0$ and $0 < x < t$. For the case $x > t$, we may invert (3.2.34) using the techniques shown earlier in this section. The case of $0 < x < t$ is much more complicated and we will now do it in detail.

First, we replace the cosine with its complex definition and obtain

$$f(x, t) = \frac{1}{2\pi} \int_{-\infty}^{\infty} \frac{\sinh(\omega b)}{\omega b \cosh(\omega b)} \left[e^{i\omega x + it\sqrt{\omega^2 + N_n^2}} + e^{i\omega x - it\sqrt{\omega^2 + N_n^2}} \right] d\omega \tag{3.2.35}$$

where $N_n = n\pi/b$. Next, we distort the original contour along the real axis into contours that include arcs with infinite radii. See Fig. 3.2.3. For $t > x$ the behavior of $\sqrt{\omega^2 + N_n^2}$ determines the choice of the contour rather than ω as $|\omega| \to \infty$. Let us define the phase of $\omega - iN_n$ as lying between $-3\pi/2$ and $\pi/2$ and the phase of $\omega + iN_n$ as lying

---

[32] A more complicated version appeared in Walker, J. S., 1978: Solitary fluid transients in rectangular ducts with transverse magnetic fields. *Zeit. angew. Math. Phys.*, **29**, 35–53. Published by Birkhäuser Verlag, Basel, Switzerland.

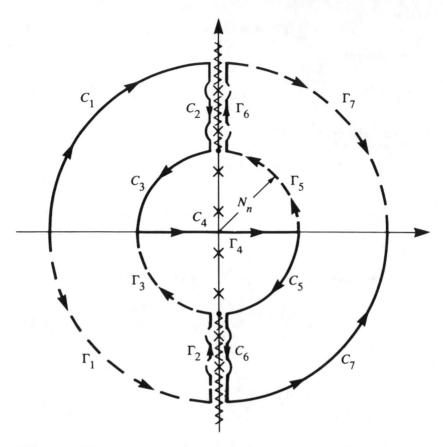

Fig. 3.2.3: The contour used in the evaluation of (3.2.33).

between $-\pi/2$ and $3\pi/2$. Consequently, $e^{it\sqrt{\omega^2+N_n^2}} \to 0$ as $|\omega| \to \infty$ in the first and third quadrants while $e^{-it\sqrt{\omega^2+N_n^2}} \to 0$ as $|\omega| \to \infty$ in the second and fourth quadrant. With these considerations, we may distort the line integration along the real axis as shown in Fig. 3.2.3. We use the Γ_i contours with $e^{it\sqrt{\omega^2+N_n^2}}$ while we employ the C_i contours with $e^{-it\sqrt{\omega^2+N_n^2}}$.

Turning now to the actual evaluation of the various contour integrals, those over C_1, C_7, Γ_1, Γ_7 vanish as $|\omega| \to \infty$. For the other contours,

$$
\int_{C_2} = \frac{1}{2\pi} \int_\infty^{N_n} \frac{\sin(\eta b)}{\eta b \cos(\eta b)} e^{-\eta x - t\sqrt{\eta^2-N_n^2}} \, i \, d\eta
$$
$$
+ \frac{1}{2b^2} \sum_{m=n}^\infty \frac{e^{-M_m x - t\sqrt{M_m^2-N_n^2}}}{M_m} \tag{3.2.36}
$$

$$\int_{C_6} = \frac{1}{2\pi} \int_{N_n}^{\infty} \frac{\sin(\eta b)}{\eta b \cos(\eta b)} e^{\eta x - t\sqrt{\eta^2 - N_n^2}} (-i\, d\eta)$$

$$+ \frac{1}{2b^2} \sum_{m=n}^{\infty} \frac{e^{M_m x - t\sqrt{M_m^2 - N_n^2}}}{M_m} \tag{3.2.37}$$

$$\int_{\Gamma_2} = \frac{1}{2\pi} \int_{\infty}^{N_n} \frac{\sin(\eta b)}{\eta b \cos(\eta b)} e^{\eta x - t\sqrt{\eta^2 - N_n^2}} (-i\, d\eta)$$

$$+ \frac{1}{2b^2} \sum_{m=n}^{\infty} \frac{e^{M_m x - t\sqrt{M_m^2 - N_n^2}}}{M_m} \tag{3.2.38}$$

$$\int_{\Gamma_6} = \frac{1}{2\pi} \int_{N_n}^{\infty} \frac{\sin(\eta b)}{\eta b \cos(\eta b)} e^{-\eta x - t\sqrt{\eta^2 - N_n^2}} i\, d\eta$$

$$+ \frac{1}{2b^2} \sum_{m=n}^{\infty} \frac{e^{-M_m x - t\sqrt{M_m^2 - N_n^2}}}{M_m} \tag{3.2.39}$$

$$\int_{C_4} = \frac{1}{2\pi} \int_{-N_n}^{N_n} \frac{\sinh(\eta b)}{\eta b \cosh(\eta b)} e^{i\eta x + it\sqrt{\eta^2 + N_n^2}}\, d\eta \tag{3.2.40}$$

$$\int_{\Gamma_4} = \frac{1}{2\pi} \int_{-N_n}^{N_n} \frac{\sinh(\eta b)}{\eta b \cosh(\eta b)} e^{i\eta x - it\sqrt{\eta^2 + N_n^2}}\, d\eta \tag{3.2.41}$$

where $M_n = (n + \frac{1}{2})\pi/b$.

The remaining contour integrals are more involved. Along C_3, from purely geometric considerations,

$$\omega = -N_n e^{-\theta i}, \quad \omega - iN_n = N_n \sqrt{2[1 - \sin(\theta)]} e^{-3\pi i/4 - \theta i/2}, \tag{3.2.42}$$

$$\omega + iN_n = N_n \sqrt{2[1 + \sin(\theta)]} e^{3\pi i/4 - \theta i/2}, \tag{3.2.43}$$

$$\sqrt{\omega^2 + N_n^2} = N_n \sqrt{2\cos(\theta)} e^{-\theta i/2} \tag{3.2.44}$$

where θ is clockwise from the negative real axis and

$$\int_{C_3} = \frac{1}{2\pi b} \int_{\pi/2}^{0} \frac{\sinh(-N_n b e^{-\theta i})}{-N_n e^{-\theta i} \cosh(-N_n b e^{-\theta i})} iN_n e^{-\theta i} d\theta$$

$$\times \exp\left[-iN_n x e^{-\theta i} - it N_n \sqrt{2\cos(\theta)} e^{-\theta i/2} \right] \tag{3.2.45}$$

$$= -\frac{i}{2\pi b} \int_0^{\pi/2} \frac{\exp\left[-N_n x \sin(\theta) - \xi \sin(\theta/2) \right]}{\cosh(\kappa) + \cos(\lambda)} d\theta$$

$$\times \Big\{ \cos[\zeta(-x)] \sinh(\kappa) - \sin(\lambda) \sin[\zeta(-x)]$$

$$- i\sin(\lambda) \cos[\zeta(-x)] - i\sinh(\kappa) \sin[\zeta(-x)] \Big\} \tag{3.2.46}$$

where $\zeta(x) = \xi \cos(\theta/2) - N_n x \cos(\theta)$, $\xi = N_n t \sqrt{2 \cos(\theta)}$, $\kappa = 2n\pi \cos(\theta)$ and $\lambda = 2n\pi \sin(\theta)$. Similarly, along C_5,

$$\omega = N_n e^{-\theta i}, \quad \omega - iN_n = N_n \sqrt{2[1 + \sin(\theta)]} e^{-\pi i/4 - \theta i/2}, \qquad (3.2.47)$$

$$\omega + iN_n = N_n \sqrt{2[1 - \sin(\theta)]} e^{\pi i/4 - \theta i/2}, \qquad (3.2.48)$$

$$\sqrt{\omega^2 + N_n^2} = N_n \sqrt{2 \cos(\theta)} e^{-\theta i/2} \qquad (3.2.49)$$

where θ is clockwise from the positive real axis and

$$\int_{C_5} = \frac{1}{2\pi b} \int_0^{\pi/2} \frac{\sinh(N_n b e^{-\theta i})}{N_n e^{-\theta i} \cosh(N_n b e^{-\theta i})} (-iN_n e^{-\theta i}) \, d\theta$$

$$\times \exp\left[iN_n x e^{-\theta i} - it N_n \sqrt{2 \cos(\theta)} e^{-\theta i/2}\right] \qquad (3.2.50)$$

$$= -\frac{i}{2\pi b} \int_0^{\pi/2} \frac{\exp\left[N_n x \sin(\theta) - \xi \sin(\theta/2)\right]}{\cosh(\kappa) + \cos(\lambda)} \, d\theta$$

$$\times \left\{ \cos[\zeta(x)] \sinh(\kappa) - \sin(\lambda) \sin[\zeta(x)] \right.$$

$$\left. - i \sin(\lambda) \cos[\zeta(x)] - i \sinh(\kappa) \sin[\zeta(x)] \right\}. \qquad (3.2.51)$$

Along Γ_3,

$$\omega = -N_n e^{-\theta i}, \quad \omega - iN_n = N_n \sqrt{2[1 + \sin(\theta)]} e^{-3\pi i/4 + \theta i/2}, \qquad (3.2.52)$$

$$\omega + iN_n = N_n \sqrt{2[1 - \sin(\theta)]} e^{3\pi i/4 + \theta i/2}, \qquad (3.2.53)$$

$$\sqrt{\omega^2 + N_n^2} = N_n \sqrt{2 \cos(\theta)} e^{\theta i/2} \qquad (3.2.54)$$

where θ measures counterclockwise from the negative axis. Therefore,

$$\int_{\Gamma_3} = \frac{1}{2\pi b} \int_{\pi/2}^0 \frac{\sinh(-N_n b e^{\theta i})}{-N_n e^{\theta i} \cosh(-N_n b e^{\theta i})} (-iN_n e^{\theta i}) \, d\theta$$

$$\times \exp\left[-iN_n x e^{\theta i} + it N_n \sqrt{2 \cos(\theta)} e^{\theta i/2}\right] \qquad (3.2.55)$$

$$= \frac{i}{2\pi b} \int_0^{\pi/2} \frac{\exp\left[N_n x \sin(\theta) - \xi \sin(\theta/2)\right]}{\cosh(\kappa) + \cos(\lambda)} \, d\theta$$

$$\times \left\{ \cos[\zeta(x)] \sinh(\kappa) - \sin(\lambda) \sin[\zeta(x)] \right.$$

$$\left. + i \sin(\lambda) \cos[\zeta(x)] + i \sinh(\kappa) \sin[\zeta(x)] \right\}. \qquad (3.2.56)$$

Finally, along Γ_5,

$$\omega = N_n e^{\theta i}, \quad \omega - iN_n = N_n \sqrt{2[1 - \sin(\theta)]} e^{-\pi i/4 + \theta i/2}, \qquad (3.2.57)$$

$$\omega + iN_n = N_n \sqrt{2[1 + \sin(\theta)]} e^{\pi i/4 + \theta i/2}, \qquad (3.2.58)$$

$$\sqrt{\omega^2 + N_n^2} = N_n \sqrt{2\cos(\theta)} e^{\theta i/2} \qquad (3.2.59)$$

where θ increases counterclockwise from the positive real axis. Therefore,

$$\int_{\Gamma_5} = \frac{1}{2\pi b} \int_0^{\pi/2} \frac{\sinh(N_n b e^{\theta i})}{N_n e^{\theta i} \cosh(N_n b e^{\theta i})} (iN_n e^{\theta i}) \, d\theta$$

$$\times \exp\left[iN_n x e^{\theta i} + it N_n \sqrt{2\cos(\theta)} e^{\theta i/2}\right] \qquad (3.2.60)$$

$$= \frac{i}{2\pi b} \int_0^{\pi/2} \frac{\exp\left[-N_n x \sin(\theta) - \xi \sin(\theta/2)\right]}{\cosh(\kappa) + \cos(\lambda)} \, d\theta$$

$$\times \left\{ \cos[\zeta(-x)] \sinh(\kappa) - \sin(\lambda)\sin[\zeta(-x)] \right.$$

$$\left. + i\sin(\lambda)\cos[\zeta(-x)] + i\sinh(\kappa)\sin[\zeta(-x)] \right\}. \qquad (3.2.61)$$

Summing the results from (3.2.36)–(3.2.41), (3.2.46), (3.2.51), (3.2.56) and (3.2.61), we obtain the final result that

$$f(x,t) = \frac{2}{b^2} \sum_{m=n}^{\infty} \frac{\cosh(M_n x)}{M_n} e^{-t\sqrt{M_n^2 - N_n^2}}$$

$$+ \frac{2}{\pi b} \int_0^{N_n} \frac{\sinh(\eta b)}{\eta \cosh(\eta b)} \cos[t\sqrt{\eta^2 + N_n^2}] \, d\eta$$

$$- \frac{1}{\pi b} \int_0^{\pi/2} \frac{\exp\left[-N_n x \sin(\theta) - \xi \sin(\theta/2)\right]}{\cosh(\kappa) + \cos(\lambda)} \, d\theta$$

$$\times \left\{ \sin(\lambda)\cos[\zeta(x)] + \sinh(\kappa)\sin[\zeta(x)] \right\}$$

$$- \frac{1}{\pi b} \int_0^{\pi/2} \frac{\exp\left[N_n x \sin(\theta) - \xi \sin(\theta/2)\right]}{\cosh(\kappa) + \cos(\lambda)} \, d\theta$$

$$\times \left\{ \sin(\lambda)\cos[\zeta(-x)] + \sinh(\kappa)\sin[\zeta(-x)] \right\}. \qquad (3.2.62)$$

• **Example 3.2.4**

For our final example, let us find the inverse[33]

$$f(x,y) = \frac{1}{2\pi} \int_{-\infty}^{\infty} \exp\left[ikx - y\sqrt{k^2 - \frac{ikV}{c}}\right] dk \qquad (3.2.63)$$

where the square root must have a positive real part. To recast (3.2.63) into a more symmetric form, we introduce $\alpha = k - iV/2c$ so that

$$f(x,y) = \frac{e^{-Vx/2c}}{2\pi} \int_{-\infty-iV/2c}^{\infty-iV/2c} \exp\left[i\alpha x - y\sqrt{\alpha^2 + \left(\frac{V}{2c}\right)^2}\right] d\alpha. \qquad (3.2.64)$$

In order that the radical has a positive real part, we define the function $\alpha - iV/2c$ with a phase between $-3\pi/2$ to $\pi/2$ and $\alpha + iV/2c$ with a phase $-\pi/2$ to $3\pi/2$. See Fig. 3.2.4. We will give the details for $x > 0$; the case $x < 0$ follows by analog.

One method of integrating (3.2.64) would be to deform the contour so that it consists of arcs at infinity in the first and second quadrants of the α-plane plus integrations along the branch cuts. Instead of proceeding along this line of attack, we will use another method. We will explore this new method in greater detail in Section 4.2 where we will use it to invert Laplace transforms.

Let us define

$$i\alpha x - y\sqrt{\alpha^2 + \left(\frac{V}{2c}\right)^2} = -s. \qquad (3.2.65)$$

A little algebra shows that we must define α by

$$\alpha_{\pm} = \frac{ixs}{r^2} \pm \frac{y}{r^2}\sqrt{s^2 - \left(\frac{rV}{2c}\right)^2}, \quad \frac{rV}{2c} \le s \le \infty, \qquad (3.2.66)$$

if we introduce the polar coordinates $x = r\cos(\theta)$ and $y = r\sin(\theta)$. We now deform our original contour so that it conforms to α_+ in the first quadrant and α_- in the second quadrant of the α-plane. This deformation is permissible because we do not cross any singularities and the integration along the arcs at infinity vanish by Jordan's lemma.

---

[33] Patterned after Rudnicki, J. W., and E. A. Roeloffs, 1990: Plane-strain shear dislocations moving steadily in linear elastic diffusive solids. *J. Appl. Mech.*, **57**, 32–38. See also Chowdhury, K. L., and P. G. Glockner, 1980: On a boundary value problem for an elastic dielectric half-space. *Acta Mech.*, **37**, 65–74.

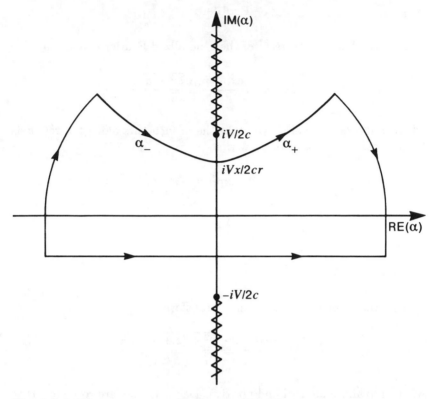

Fig. 3.2.4: The contour used to invert (3.2.64).

Our motivation lies in the simplicity that this transformation introduces in the exponential.

Upon introducing this new contour into (3.2.64),

$$f(x,y) = e^{-Vx/2c} \frac{y}{\pi r^2} \int_{rV/2c}^{\infty} \frac{se^{-s}}{\sqrt{s^2 - \left(\frac{rV}{2c}\right)^2}} \, ds \qquad (3.2.67)$$

$$= e^{-Vx/2c} \frac{Vy}{2\pi cr} \int_1^{\infty} \frac{\xi e^{-rV\xi/2c}}{\sqrt{\xi^2 - 1}} \, d\xi \qquad (3.2.68)$$

$$= e^{-Vx/2c} \frac{y}{\pi} \left(\frac{V}{2c}\right)^2 \int_1^{\infty} e^{-rV\xi/2c} \sqrt{\xi^2 - 1} \, d\xi \qquad (3.2.69)$$

after an integration by parts and introducing $s = rV\xi/2c$. We may tabulate (3.2.69) very simply by numerical quadrature.

Transform Methods for Solving Partial Differential Equations

Problems

1. Find the Fourier inverse[34] of the generalized Fourier transform

$$F(\omega) = -i\frac{\exp[-it_0\sqrt{\omega^2 - a^2}]}{\sqrt{\omega^2 - a^2}}$$

where a and t_0 are real. If you take the branch cuts along the real axis from $[a, \infty]$ and $[-\infty, -a]$, show that

$$f(t) = -\frac{i}{2\pi}\int_{-\infty-\epsilon i}^{\infty-\epsilon i} \frac{\exp[i\omega t - it_0\sqrt{\omega^2 - a^2}]}{\sqrt{\omega^2 - a^2}}\,d\omega$$

$$= J_0\left(a\sqrt{t^2 - t_0^2}\right)H(t - t_0)$$

where $0 < \epsilon << 1$.

2. Find the inverse[35] of the Fourier transform

$$F(\omega) = \frac{\exp[-i\omega\sqrt{r^2 + (z + a/\omega i)^2}]}{\sqrt{r^2 + (z + a/\omega i)^2}}$$

where a, r and z are real and positive. Define the square root such that

$$\mathrm{Im}\left[\omega\sqrt{r^2 + (z + a/\omega i)^2}\right] \le 0.$$

Show that

$$f(t) = \frac{2}{\pi r_i}\frac{d}{dt}\left[e^{-\alpha t}J_0\left(\beta\sqrt{t^2 - r_i^2}\right)H(t - r_i)\right]$$

where $J_0(\ )$ is the zeroth-order Bessel function of the first kind, $r_i = \sqrt{r^2 + z^2}$, $\alpha = az/r_i^2$ and $\beta = ar/r_i^2$.

---

[34] For a problem which uses this inverse, see Row, R. V., 1967: Acoustic-gravity waves in the upper atmosphere due to a nuclear detonation and an earthquake. *J. Geophys. Res.*, **72**, 1599–1610. ©1967 American Geophysical Union.

[35] Taken from Nikoskinen, K. I., and I. V. Lindell, 1990: Time-domain analysis of the Sommerfeld VMD problem based on the exact image theory. *IEEE Trans. Antennas Propagat.*, **AP-38**, 241–250. ©1990 IEEE.

3. One method[36] for dealing with Fourier transforms with singularities on the real axis which also have symmetry properties in t is to compute the inverse using the formula

$$f(t) = \frac{1}{2\pi} \left[\int_{-\infty-\epsilon i}^{\infty-\epsilon i} F(\omega)e^{i\omega t}d\omega + \int_{-\infty+\epsilon i}^{\infty+\epsilon i} F(\omega)e^{i\omega t}d\omega \right]$$

where $0 < \epsilon << 1$. Using this formula, find the inverse of

$$F(\omega) = \frac{e^{z\sqrt{\omega^2+k^2}}}{k - \sqrt{\omega^2 + k^2}}, \qquad z < 0$$

where k is real and greater than zero. If we take the branch cuts associated with $\sqrt{\omega^2 + k^2}$ along the imaginary axis from $[ki, \infty i]$ and $[-ki, -\infty i]$, show that the inverse is

$$f(t) = k|t|e^{kz} - \frac{1}{\pi} \int_k^\infty \frac{e^{-|t|u}}{u^2} \, du$$
$$\times [k\sin(z\sqrt{u^2 - k^2}) + \sqrt{u^2 - k^2}\cos(z\sqrt{u^2 - k^2})].$$

4. Show that the inverse[37] of

$$F(\omega) = -\frac{|\omega|\exp(-s|\omega|)}{|\omega| - a_1 + i\epsilon a_2 \mathrm{sgn}(\omega)}$$

where a_1, a_2 and s are greater than zero, $\epsilon \to 0$ and $\mathrm{sgn}(\omega) = 1$ if $\omega > 0$ and -1 if $\omega < 0$. By deforming the path of integration along the real ω axis so that it passes over the $\omega = \pm a_1$ and taking the branch cuts associated with $|\omega|$ along the imaginary axis from $[-\infty i, 0^- i]$ and $[0^+ i, \infty i]$, show that

$$f(t) = -2a_1 e^{-a_1 s}\sin(a_1 t)H(-t)$$
$$- \frac{1}{\pi}\int_0^\infty \frac{\eta e^{-t\eta}}{\eta^2 + a_1^2}[a_1\cos(s\eta) + \eta\sin(s\eta)]\,d\eta.$$

[Hint: Treat $|z| = \sqrt{(z - 0^+ i)(z - 0^- i)}$ where the real part of the square root must be always positive.]

---

[36] Taken from Haren, P., and C. C. Mei, 1981: Head-sea diffraction by a slender raft with application to wave-power absorption. *J. Fluid Mech.*, **104**, 505–526. Reprinted with the permission of Cambridge University Press.

[37] Taken from Savage, M. D., 1967: Stationary waves at a plasma-magnetic field interface. *J. Plasma Phys.*, **1**, 229–239. Reprinted with the permission of Cambridge University Press.

5. Find the inverse[38] of the Fourier transform

$$F(\omega) = \frac{V(\omega)i}{\omega[1 - cV(\omega)]} - \frac{V(\omega)\omega i}{(\omega^2 + \alpha^2)[1 - cV(\omega)]}$$

where

$$V(\omega) = \frac{1}{2i\omega} \log\left[\frac{1 + \omega i}{1 - i\omega}\right] = \frac{\tan^{-1}(\omega)}{\omega}$$

and $0 < \alpha < 1$. Take the branch cuts associated with the logarithms along the imaginary axis from $[-\infty i, -i]$ and $[i, \infty i]$. For $t > 0$ show that the inverse is

$$f(t) = \frac{V(\alpha i)e^{-\alpha t}}{2[1 - cV(\alpha i)]} - \frac{1}{1 - c} - \frac{\alpha^2(1 - \omega_0^2)e^{-\omega_0 t}}{c(\alpha^2 - \omega_0^2)(\omega_0^2 + c - 1)}$$
$$- \frac{\alpha^2}{2} \int_0^1 \frac{x^2}{1 - \alpha^2 x^2} e^{-t/x} \left\{ \left[1 + \frac{cx}{2} \ln\left(\frac{1 - x}{1 + x}\right) \right]^2 + \frac{\pi^2 c^2 x^2}{4} \right\}^{-1} dx$$

where $cV(i\omega_0) = 1$ and $0 < \omega_0 < 1$.

6. Find the inverse of the generalized Fourier transform

$$F(\omega) = \frac{\log(i\omega\tau)}{(\lambda + \sqrt{\lambda^2 + \omega i})^2}$$

where τ and λ are real and positive. If we take the branch cut associated with the square root along the positive imaginary axis of the ω-plane, from $\omega = \lambda^2 i$ to ∞i, and the branch cut of $\log(z)$ lies along the positive imaginary axis, then show that

$$f(t) = \frac{1}{\pi} \int_1^\infty e^{-\lambda^2 t\eta} \left[2\sqrt{\eta - 1} \ln(\lambda^2\tau\eta) - \pi(2 - \eta) \right] \frac{d\eta}{\eta^2}$$
$$- \int_0^1 e^{-\lambda^2 t\eta} \frac{d\eta}{(1 + \sqrt{1 - \eta})^2}$$

if $t > 0$; $f(t) = 0$ if $t < 0$.

7. Show that the inverse[39] of

$$F(\omega, y) = e^{-|k|y}$$

---

[38] Taken from Williams, M. M. R., 1965: Neutron transport in differentially heated media. *Brit. J. Appl. Phys.*, **16**, 1727–1732. See also Pearlstein, L. D., and G. W. Stuart, 1961: Effect of collisional energy loss on ionization growth in H_2. *Phys. Fluids*, **4**, 1293–1297.

[39] Taken from Chowdhury, K. L., and P. G. Glockner, 1980: On a boundary value problem for an elastic dielectric half-space. *Acta Mech.*, **37**, 65–74.

is

$$f(x, y) = \pi \frac{\partial [\ln(r)]}{\partial y}.$$

Step 1. Use polar coordinates to show that

$$f(x, y) = \frac{1}{2\pi} \int_{-\infty}^{\infty} e^{r[ik \cos(\theta) - |k| \sin(\theta)]} dk.$$

Step 2. Introduce the transformation $s = -ik \cos(\theta) + |k| \sin(\theta) = ke^{i(-\pi/2 \pm \theta)}$, $0 < s < \infty$ where we take the upper sign for $k > 0$ and the lower sign for $k < 0$. Then $k = se^{i(\pi/2 \mp \theta)}$ and $dk/ds = e^{i(\pi/2 \mp \theta)}$. By breaking the integral in step 1 into two parts, from $-\infty$ to 0 and 0 to ∞, show that

$$f(x, y) = \pi \sin(\theta) \int_0^{\infty} e^{-rs} ds = \pi \frac{\partial [\ln(r)]}{\partial y}.$$

8. Rework Example 3.2.4 with the transform

$$F(k, y) = \frac{\exp(-y\sqrt{k^2 - ikV/c})}{\sqrt{k^2 - ikV/c}}, \quad y > 0$$

and show that the inverse equals

$$f(x, y) = \frac{\exp(-Vx/2c)}{\pi} \int_1^{\infty} \frac{\exp(-rV\eta/2c)}{\sqrt{\eta^2 - 1}} d\eta$$

where $r = \sqrt{x^2 + y^2}$.

3.3 THE SOLUTION OF PARTIAL DIFFERENTIAL EQUATIONS BY LAPLACE TRANSFORMS

In section 2.3 we showed how we may successfully employ Laplace transforms in solving linear partial differential equations. However, at that time we restricted ourselves to situations where the Laplace transform was single-valued. In this section we extend our technique to multivalued transforms.

In their study of the seepage of a homogeneous fluid in fissured rocks, Barenblatt *et al.*[40] solved an equation similar to

$$\frac{\partial u}{\partial t} = \frac{\partial^2 u}{\partial x^2} + \frac{\partial^3 u}{\partial t \partial x^2} \tag{3.3.1}$$

---

[40] Reprinted from *J. Appl. Math. Mech.*, **24**, G. I. Barenblatt, Iu. P. Zheltov and I. N. Kochina, Basic concepts in the theory of seepage of homogeneous liquids in fissured rocks (strata), 1286–1303, ©1960, with kind permission from Pergamon Press Ltd., Headington Hill Hall, Oxford OX3 0BX, UK.

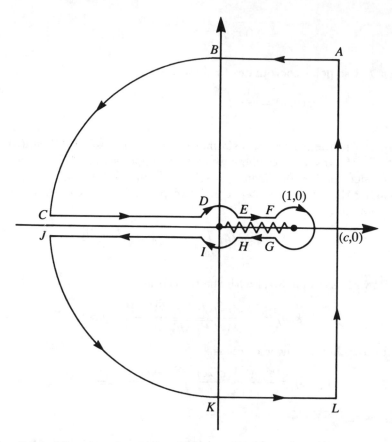

Fig. 3.3.1: The contour used to evaluate (3.3.9).

with the boundary conditions

$$u(0,t) = 1 - e^{-t} \qquad \text{and} \qquad u(\infty, t) = 0 \qquad (\mathbf{3.3.2})$$

and the initial condition

$$u(x,0) = 0. \qquad (\mathbf{3.3.3})$$

Upon applying the Laplace transform to (3.3.1) and using the boundary conditions and initial conditions, (3.3.1) becomes

$$\frac{d^2\overline{u}}{dx^2} - \frac{s}{s+1}\overline{u} = 0 \qquad (\mathbf{3.3.4})$$

with

$$\overline{u}(0,s) = \frac{1}{s(s+1)} \qquad (\mathbf{3.3.5})$$

$$\overline{u}(\infty, s) = 0. \qquad (\mathbf{3.3.6})$$

The general solution of (3.3.4) which satisfies the boundary conditions is

$$\overline{u}(x, s) = \frac{1}{s(s+1)} \exp\left[-x\sqrt{\frac{s}{s+1}}\right]. \tag{3.3.7}$$

The inversion of (3.3.7) is

$$u(x, t) = \frac{1}{2\pi i} \int_{c-\infty i}^{c+\infty i} \frac{e^{st}}{s(s+1)} \exp\left[-x\sqrt{\frac{s}{s+1}}\right] ds \tag{3.3.8}$$

where $c > 0$. We facilitate the evaluation of the integral (3.3.8) by the change of variables $s = z - 1$, so

$$u(x, t) = \frac{e^{-t}}{2\pi i} \int_{c-\infty i}^{c+\infty i} \frac{e^{zt}}{z(z-1)} \exp\left[-x\sqrt{\frac{z-1}{z}}\right] dz \tag{3.3.9}$$

where $c > 1$.

Because of the square root in the exponential in (3.3.9), the integrand is a multivalued function. A quick check shows that $z = 0$ and $z = 1$ are the branch points of the integrand. A convenient choice for the branch cut is the line segment lying along the real axis running between $z = 0$ and $z = 1$.

At this point we shall convert the line integral running from $c - \infty i$ to $c + \infty i$ into a closed contour so that we can apply the residue theorem. Figure 3.3.1 shows the contour that we will use. The contribution from the arcs ABC and JKL are negligibly small by Jordan's lemma (see Section 1.4). The contribution from the line segment CD cancels the contribution from IJ. Consequently the dumbbell-shaped contour integral shown in Fig. 3.3.2 is equivalent to the contour $ABCDEFGHIJKL$ shown in Fig. 3.3.1. Because there are no singularities inside the closed contour, the value given by the contour integral shown in Fig. 3.3.2 must equal the negative of the integral from $c - \infty i$ to $c + \infty i$.

Along C_1, $z = \epsilon e^{\theta i}$ and $dz = i\epsilon e^{\theta i} d\theta$ so that

$$\frac{e^{-t}}{2\pi i} \int_{C_1} \frac{e^{zt}}{z(z-1)} \exp\left[-x\sqrt{\frac{z-1}{z}}\right] dz$$

$$= \frac{e^{-t}}{2\pi} \lim_{\epsilon \to 0} \int_{2\pi}^{0} \frac{\exp(\epsilon t e^{\theta i})}{-1 + \epsilon e^{\theta i}} \exp\left[-x\sqrt{1 - e^{-\theta i}/\epsilon}\right] d\theta \tag{3.3.10}$$

$$= -\frac{e^{-t}}{2\pi} \lim_{\epsilon \to 0} \int_{2\pi}^{0} \exp\left[-x\frac{e^{-\theta i/2 + i\pi/2}}{\sqrt{\epsilon}}\right] d\theta = 0. \tag{3.3.11}$$

Along C_3, $z = 1 + \epsilon e^{\theta i}$, $dz = i\epsilon e^{\theta i} d\theta$ and

$$\frac{e^{-t}}{2\pi i} \int_{C_3} \frac{e^{zt}}{z(z-1)} \exp\left[-x\sqrt{\frac{z-1}{z}}\right] dz$$

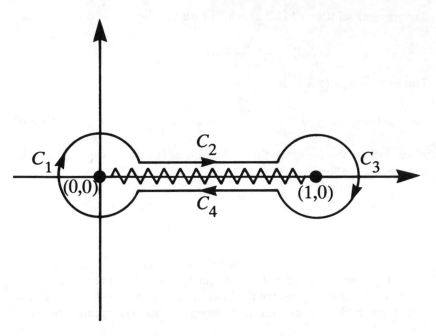

Fig. 3.3.2: Another contour used in the evaluation of (3.3.9).

$$= \frac{e^{-t}}{2\pi} \lim_{\epsilon \to 0} \int_{\pi}^{-\pi} \frac{\exp(\epsilon t e^{\theta i})}{1 + \epsilon e^{\theta i}} \exp\left[-x\sqrt{\frac{\epsilon e^{\theta i}}{1 + \epsilon e^{\theta i}}}\right] d\theta \quad \textbf{(3.3.12)}$$

$$= -1. \quad \textbf{(3.3.13)}$$

Along C_2, $z = \sigma e^{0i}$, $dz = d\sigma$, $z - 1 = (1-\sigma)e^{\pi i}$ and

$$\frac{e^{-t}}{2\pi i} \int_{C_2} \frac{e^{zt}}{z(z-1)} \exp\left[-x\sqrt{\frac{z-1}{z}}\right] dz$$

$$= \frac{e^{-t}}{2\pi i} \int_0^1 \frac{e^{\sigma t}}{\sigma(\sigma-1)} \exp\left[-ix\sqrt{\frac{1-\sigma}{\sigma}}\right] d\sigma \textbf{(3.3.14)}$$

Along C_4, $z = \sigma e^{0i}$, $dz = d\sigma$, $z - 1 = (1-\sigma)e^{-\pi i}$ and

$$\frac{e^{-t}}{2\pi i} \int_{C_4} \frac{e^{zt}}{z(z-1)} \exp\left[-x\sqrt{\frac{z-1}{z}}\right] dz$$

$$= \frac{e^{-t}}{2\pi i} \int_1^0 \frac{e^{\sigma t}}{\sigma(\sigma-1)} \exp\left[ix\sqrt{\frac{1-\sigma}{\sigma}}\right] d\sigma. \textbf{(3.3.15)}$$

Therefore,

$$u(x,t) = 1 + \frac{1}{\pi} \int_0^1 \frac{e^{(\sigma-1)t}}{\sigma(\sigma-1)} \sin\left[x\sqrt{\frac{1-\sigma}{\sigma}}\right] d\sigma \quad \textbf{(3.3.16)}$$

$$= 1 - \frac{1}{\pi} \int_0^1 \frac{e^{-\eta t}}{\eta(1-\eta)} \sin\left[x\sqrt{\frac{\eta}{1-\eta}}\,\right] d\eta \qquad (3.3.17)$$

$$= 1 - \frac{2}{\pi} \int_0^\infty \frac{1}{\nu} \sin(\nu x) \exp\left[-\frac{\nu^2 t}{1+\nu^2}\right] d\nu \qquad (3.3.18)$$

where we introduced the new variables $\eta = 1 - \sigma$ and $\nu^2 = \eta/(1-\eta)$.

For our second example, we find the sound waves that propagate away from an exponentially flared loud speaker[41]. To find the pressure field $p(x,t)$ at a distance x away from the throat of the speaker, we must find first the velocity potential $u(x,t)$ because $p(x,t) = \rho_0 u_t(x,t)$ where ρ_0 is the density of air. The wave equation

$$\frac{\partial^2 u}{\partial x^2} + \beta \frac{\partial u}{\partial x} = \frac{1}{c^2} \frac{\partial^2 u}{\partial t^2} \qquad (3.3.19)$$

governs the velocity potential where c is the speed of sound and β is the flaring index. At time $t = 0$, we subject the air at the throat ($x = 0$) to a forcing

$$\left.\frac{\partial u}{\partial x}\right|_{x=0} = -\frac{d\xi}{dt} = \omega \xi_0 \sin(\omega t) \qquad (3.3.20)$$

where $\omega > \beta c/2$. If $\omega < \beta c/2$, there are no wave solutions. Initially the air is at rest.

We begin our analysis by first taking the Laplace transform of (3.3.19) and find that

$$\frac{d^2 \bar{u}}{dx^2} + \beta \frac{d\bar{u}}{dx} - \frac{s^2}{c^2} \bar{u} = 0 \qquad (3.3.21)$$

where $\bar{u}(x,s)$ denotes the Laplace transform of $u(x,t)$. The general solution of (3.3.21) is

$$\bar{u}(x,s) = A(s) e^{-\beta x/2} e^{-x\sqrt{\beta^2 + 4s^2/c^2}/2} \qquad (3.3.22)$$

where we have discarded the exponentially growing solution because the solution must be finite in the limit of $x \to \infty$.

To evaluate $A(s)$, we take the Laplace transform of (3.3.20) which gives

$$-\left.\frac{d\bar{u}}{dx}\right|_{x=0} = \tfrac{1}{2} A(s)\left(\beta + \sqrt{\beta^2 + 4s^2/c^2}\,\right) = -\frac{\omega^2 \xi_0}{s^2 + \omega^2} \qquad (3.3.23)$$

---

[41] Taken from McLachlan, N. W., and A. T. McKay, 1936: Transient oscillations in a loud-speaker horn. *Proc. Camb. Phil. Soc.*, **32**, 265–275. Reprinted with the permission of Cambridge University Press.

or

$$A(s) = \frac{\omega^2 c^2 \xi_0 [\beta - \sqrt{\beta^2 + 4s^2/c^2}]}{2s^2(s^2 + \omega^2)}. \tag{3.3.24}$$

Thus, we may write the Laplace transform of the velocity potential

$$\bar{u}(x, s) = \frac{1}{2}\omega^2 c^2 \xi_0 e^{-\beta x/2}\left(\beta + 2\frac{\partial}{\partial x}\right)\left[\frac{e^{-x\sqrt{\beta^2 + 4s^2/c^2}/2}}{s^2(s^2 + \omega^2)}\right]. \tag{3.3.25}$$

Because $\bar{p}(x, s) = \rho_0 s \bar{u}(x, s)$,

$$p(x, t) = \frac{\rho_0 \omega^2 c^2 \xi_0}{2} e^{-\beta x/2}\left(\beta + 2\frac{\partial}{\partial x}\right)$$
$$\times \left[\frac{1}{2\pi i}\int_{c-\infty i}^{c+\infty i}\frac{e^{zt - x\sqrt{\beta^2 + 4z^2/c^2}/2}}{z(z^2 + \omega^2)}\,dz\right]. \tag{3.3.26}$$

We have written $p(x, t)$ in this form because it is easier to invert the transform inside the brackets rather than in the original expression.

We may evaluate Bromwich's integral in (3.3.26) using complex variables. The integrand has poles at $z = 0$ and $z = \pm\omega i$ and branch points at $z = \pm\beta ci/2 = \pm ai$. For $t < x/c$, Jordan's lemma dictates that we close the contour in the right half-plane and $p(x, t) = 0$ because there are no singularities. However, when $t > x/c$, we must close the contour in the left half-plane as shown in Fig. 3.3.3.

Turning our attention first to the poles,

$$\text{Res}(z = 0) = \lim_{z \to 0}\frac{e^{zt - x\sqrt{\beta^2 + 4z^2/c^2}/2}}{z^2 + \omega^2} = \frac{e^{-\beta x/2}}{\omega^2}, \tag{3.3.27}$$

$$\text{Res}(z = \omega i) = \lim_{z \to \omega i}\frac{e^{zt - x\sqrt{\beta^2 + 4z^2/c^2}/2}}{z(z + \omega i)} = -\frac{e^{i(\omega t - mx)}}{2\omega^2} \tag{3.3.28}$$

and

$$\text{Res}(z = -\omega i) = \lim_{z \to -\omega i}\frac{e^{zt - x\sqrt{\beta^2 + 4z^2/c^2}/2}}{z(z - \omega i)} = -\frac{e^{-i(\omega t - mx)}}{2\omega^2} \tag{3.3.29}$$

where $m = \sqrt{\omega^2/c^2 - \beta^2/4}$. Thus, the total contribution from the poles is

$$[e^{-\beta x/2} - \cos(\omega t - mx)]/\omega^2 \tag{3.3.30}$$

and the pressure contribution is

$$p_1(x, t) = -\frac{\rho_0 c^2 \xi_0}{2}e^{-\beta x/2}H\left(t - \frac{x}{c}\right)$$
$$\times [\beta\cos(\omega t - mx) + 2m\sin(\omega t - mx)]. \tag{3.3.31}$$

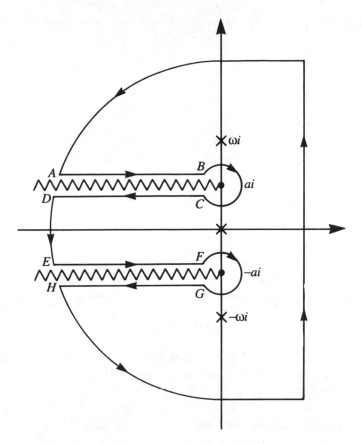

Fig. 3.3.3: The contour used to evaluate (3.3.26).

We now turn to the branch cut integrals. The contributions from the integrations around the branch points $z = \pm ai$ vanish. Patterning our analysis along the lines in Example 3.1.4, we have along AB

$$z = -\rho + ai, \quad \sqrt{z^2 + a^2} = -f_2 + f_1 i, \quad z(z^2 + \omega^2) = -f_3 + f_4 i \quad (\mathbf{3.3.32})$$

where

$$f_1 = \rho^{1/2}(\sqrt{a^2 + \rho^2/4} - \rho/2)^{1/2}, \qquad (\mathbf{3.3.33})$$

$$f_2 = \rho^{1/2}(\sqrt{a^2 + \rho^2/4} + \rho/2)^{1/2}, \qquad (\mathbf{3.3.34})$$

$$f_3 = \rho(\rho^2 + \omega^2 - 3a^2), \qquad (\mathbf{3.3.35})$$

$$f_4 = a(3\rho^2 + \omega^2 - a^2) \qquad (\mathbf{3.3.36})$$

and $0 < \rho < \infty$. Along CD, the only change is $\sqrt{z^2 + a^2} = f_2 - f_1 i$. Along the other branch cut,

$$z = -\rho - ai, \quad \sqrt{z^2 + a^2} = f_2 + f_1 i, \quad z(z^2 + \omega^2) = -f_3 - f_4 i \quad (\mathbf{3.3.37})$$

227

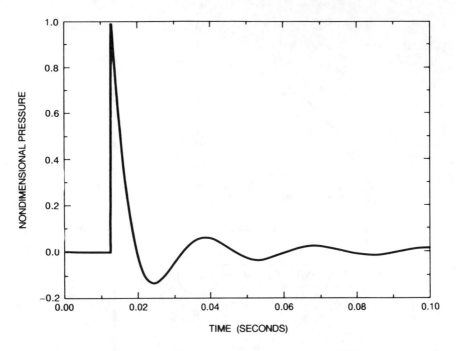

Fig. 3.3.4: The evolution of the transient pressure field $p_2(427\text{ cm}, t)$ generated by the sounding of a loud speaker with time for parameters stated in the text.

for EF. Finally, the only difference between EF and GH is

$$\sqrt{z^2 + a^2} = -f_2 - f_1 i. \tag{3.3.38}$$

Direct substitution into (3.3.26) yields

$$\int_{AB} = -\int_{\infty}^{0} \frac{e^{-\rho t} e^{iat} e^{-if_1 x/c} e^{f_2 x/c}}{-f_3 + f_4 i}\, d\rho, \tag{3.3.39}$$

$$\int_{CD} = -\int_{0}^{\infty} \frac{e^{-\rho t} e^{iat} e^{if_1 x/c} e^{-f_2 x/c}}{-f_3 + f_4 i}\, d\rho, \tag{3.3.40}$$

$$\int_{EF} = -\int_{\infty}^{0} \frac{e^{-\rho t} e^{-iat} e^{-if_1 x/c} e^{-f_2 x/c}}{-f_3 - f_4 i}\, d\rho \tag{3.3.41}$$

and

$$\int_{GH} = -\int_{0}^{\infty} \frac{e^{-\rho t} e^{-iat} e^{if_1 x/c} e^{f_2 x/c}}{-f_3 - f_4 i}\, d\rho. \tag{3.3.42}$$

We may combine (3.3.39)–(3.3.42) together to yield

$$
\frac{1}{2\pi i}\int_{ABCDEFGH}\frac{e^{zt-(x/c)\sqrt{z^2+a^2}}}{z(z^2+\omega^2)}\,dz =
$$
$$
-\frac{1}{\pi}\int_0^\infty \frac{f_3 e^{-\rho t}}{f_3^2+f_4^2}\,d\rho
$$
$$
\times\left[e^{f_2 x/c}\sin(at-f_1 x/c)-e^{-f_2 x/c}\sin(at+f_1 x/c)\right]
$$
$$
-\frac{1}{\pi}\int_0^\infty \frac{f_4 e^{-\rho t}}{f_3^2+f_4^2}\,d\rho
$$
$$
\times\left[e^{f_2 x/c}\cos(at-f_1 x/c)-e^{-f_2 x/c}\cos(at+f_1 x/c)\right]
\tag{3.3.43}
$$

and

$$
p_2(x,t) = \frac{\rho_0\omega^2 c^2\xi_0}{2\pi}e^{-\beta x/2}H\left(t-\frac{x}{c}\right)\int_0^\infty \frac{f_3 e^{-\rho t}}{f_3^2+f_4^2}\,d\rho
$$
$$
\times\left[\beta e^{f_2 x/c}\sin(at-f_1 x/c)-\beta e^{-f_2 x/c}\sin(at+f_1 x/c)\right.
$$
$$
+2f_2 e^{f_2 x/c}\sin(at-f_1 x/c)/c+2f_2 e^{-f_2 x/c}\sin(at+f_1 x/c)/c
$$
$$
\left.-2f_1 e^{f_2 x/c}\cos(at-f_1 x/c)/c-2f_1 e^{-f_2 x/c}\cos(at+f_1 x/c)/c\right]
$$
$$
+\frac{\rho_0\omega^2 c^2\xi_0}{2\pi}e^{-\beta x/2}H\left(t-\frac{x}{c}\right)\int_0^\infty \frac{f_4 e^{-\rho t}}{f_3^2+f_4^2}\,d\rho
$$
$$
\times\left[\beta e^{f_2 x/c}\cos(at-f_1 x/c)-\beta e^{-f_2 x/c}\cos(at+f_1 x/c)\right.
$$
$$
+2f_2 e^{f_2 x/c}\cos(at-f_1 x/c)/c+2f_2 e^{-f_2 x/c}\cos(at+f_1 x/c)/c
$$
$$
\left.+2f_1 e^{f_2 x/c}\sin(at-f_1 x/c)/c+2f_1 e^{-f_2 x/c}\sin(at+f_1 x/c)/c\right].
\tag{3.3.44}
$$

The sum of $p_1(x,t)$ and $p_2(x,t)$ gives the total radiation field. Fig. 3.3.4 presents $p_2(x,t)$ for $c = 3.43\times 10^4$ cm s^{-1}, $x = 427$ cm, $\beta = 0.0117$ cm^{-1} and $\omega = 256$ Hz.

Next let us find the heat flow in a two-layer earth[42] where

$$
\frac{\partial u_1}{\partial t} = a_1\frac{\partial^2 u_1}{\partial x^2}, \qquad 0\le x\le h, t>0
\tag{3.3.45}
$$

and

$$
\frac{\partial u_2}{\partial t} = a_2\frac{\partial^2 u_2}{\partial x^2}, \qquad x\ge h, t>0
\tag{3.3.46}
$$

---

[42] The steady-state solution was found by Parasnis, D. S., 1976: Effect of a uniform overburden on the passage of a thermal wave and the temperatures in the underlying rock. *Geophys. J. R. Astr. Soc.*, **46**, 189–192.

a_i denotes the diffusivity of the layer, u_i denotes the temperature and x denotes the distance down from the surface. At the earth's surface, there is a diurnal cycle

$$u_1(0, t) = T_0 \sin(\omega t) \tag{3.3.47}$$

while at the interface between the two layers

$$u_1(h, t) = u_2(h, t) \quad \text{and} \quad k_1 \frac{\partial u_1(h, t)}{\partial x} = k_2 \frac{\partial u_2(h, t)}{\partial x} \tag{3.3.48}$$

where k_i denotes the thermal conductivity. We also require that $\lim_{x \to \infty} u_2(x, t) \to 0$. The initial conditions are $u_1(x, 0) = u_2(x, 0) = 0$.

Taking the Laplace transform of (3.3.45)–(3.3.48),

$$\frac{d^2 \bar{u}_i}{dx^2} - \frac{s}{a_i} \bar{u}_i = 0, \qquad i = 1, 2 \tag{3.3.49}$$

while the boundary conditions are

$$\bar{u}_1(0, s) = \frac{T_0 \omega}{s^2 + \omega^2}, \qquad \bar{u}_1(h, s) = \bar{u}_2(h, s) \tag{3.3.50}$$

and

$$k_1 \frac{d\bar{u}_1(h, s)}{dx} = k_2 \frac{d\bar{u}_2(h, s)}{dx}, \qquad \lim_{x \to \infty} \bar{u}_2(x, s) \to 0. \tag{3.3.51}$$

The solutions to (3.3.49)–(3.3.51) are

$$\bar{u}_1(x, s) = \frac{T_0 \omega}{s^2 + \omega^2} \exp(x\sqrt{s/a_1}) - \frac{T_0 \omega}{s^2 + \omega^2} \frac{\exp(x\sqrt{s/a_1})}{1 + \beta \exp(-2h\sqrt{s/a_1})}$$
$$+ \frac{T_0 \omega}{s^2 + \omega^2} \frac{\exp(-x\sqrt{s/a_1})}{1 + \beta \exp(-2h\sqrt{s/a_1})} \tag{3.3.52}$$

and

$$\bar{u}_2(x, s) = \frac{T_0 \omega (1 + \beta)}{s^2 + \omega^2} \frac{\exp[h(\sqrt{s/a_2} - \sqrt{s/a_1}) - x\sqrt{s/a_2}]}{1 + \beta \exp(-2h\sqrt{s/a_1})} \tag{3.3.53}$$

where $\beta = (k_1\sqrt{a_2} - k_2\sqrt{a_1})/(k_1\sqrt{a_2} + k_2\sqrt{a_1})$. Except for the first term of (3.3.52), we cannot invert directly the remaining terms, as well as (3.3.53). To invert these terms, we note that we can express

$$[1 + \beta \exp(-2h\sqrt{s/a_1})]^{-1}$$

as the geometric series

$$\sum_{n=0}^{\infty}(-1)^n\beta^n\exp(-2nh\sqrt{s/a_1})$$

if $|\beta\exp(-2h\sqrt{s/a_1})| < 1$. Upon expanding the denominator of (3.3.52)–(3.3.53),

$$\bar{u}_1(x,s) = \frac{T_0\omega}{s^2+\omega^2}\exp(x\sqrt{s/a_1})$$

$$-\frac{T_0\omega}{s^2+\omega^2}\sum_{n=0}^{\infty}(-1)^n\beta^n\exp[-(2nh-x)\sqrt{s/a_1}]$$

$$+\frac{T_0\omega}{s^2+\omega^2}\sum_{n=0}^{\infty}(-1)^n\beta^n\exp[-(2nh+x)\sqrt{s/a_1}]\qquad\textbf{(3.3.54)}$$

and

$$\bar{u}_2(x,s) = \frac{T_0\omega(1+\beta)}{s^2+\omega^2}$$

$$\times\sum_{n=0}^{\infty}(-1)^n\beta^n\exp\{-[(2n+1)h+(x-h)\sqrt{a_1/a_2}]\sqrt{s/a_1}\}.$$

$$\textbf{(3.3.55)}$$

Using Bromwich's integral, we can show (see problem 1 at the end of section 3.1) that the inverse of

$$F(x,s) = \frac{\omega}{s^2+\omega^2}\exp(x\sqrt{s/\kappa})\qquad\textbf{(3.3.56)}$$

is

$$f(x,t) = \exp(x\sqrt{\omega/2\kappa})\sin(\omega t + x\sqrt{\omega/2\kappa})$$

$$+\frac{2\kappa}{\pi}\int_0^{\infty}\frac{\omega}{\omega^2+\kappa^2\eta^4}\exp(-\kappa t\eta^2)\sin(x\eta)\,\eta\,d\eta.\qquad\textbf{(3.3.57)}$$

At this point we note that we can express all of (3.3.54)–(3.3.55) in terms of $f(x,t)$. Inverting (3.3.54)–(3.3.55) term by term, the final solution is

$$u_1(x,t) = T_0 f(x,t) - T_0\sum_{n=0}^{\infty}(-1)^n\beta^n f(x-2nh,t)$$

$$+T_0\sum_{n=0}^{\infty}(-1)^n\beta^n f(-x-2nh,t)\qquad\textbf{(3.3.58)}$$

and

$$u_2(x,t) = (1+\beta)T_0 \sum_{n=0}^{\infty} (-1)^n \beta^n f[(h-x)\sqrt{a_1/a_2} - (2n+1)h, t].$$

$$(3.3.59)$$

For our final example[43], we solve the partial differential equation

$$\frac{\partial u}{\partial t} = \frac{\partial}{\partial x}\left[(1+x^2)\frac{\partial u}{\partial x}\right], \qquad 0 < x < \infty, t > 0 \qquad (3.3.60)$$

subject to the boundary conditions $\lim_{x\to\infty} |u_x(x,t)| < \infty$ and $u(0,t) = 0$ and the initial condition $u(x,0) = 1$.

If the Laplace transform of $u(x,t)$ is

$$\bar{u}(x,s) = \int_0^{\infty} u(x,t)e^{-st}dt, \qquad (3.3.61)$$

the partial differential equation and boundary conditions reduce to the ordinary differential equation

$$\frac{d}{dx}\left[(1+x^2)\frac{d\bar{u}}{dx}\right] - s\bar{u} = -1, \qquad 0 < x < \infty \qquad (3.3.62)$$

subject to the boundary conditions $\lim_{x\to\infty} |\bar{u}'(x,s)| < \infty$ and $\bar{u}(0,s) = 0$. The particular solution to (3.3.62) is $1/s$. If we introduce $z = xi$ into (3.3.62), the left side becomes Legendre's equation. Consequently, the homogeneous solution is

$$\bar{u}(x,s) = AP_\nu(xi) + BQ_\nu(xi) \qquad (3.3.63)$$

where $\nu = -1/2 + \sqrt{1/4 + s}$, $P_\nu(z)$ and $Q_\nu(z)$ are Legendre functions of order ν of the first and second kind[44], respectively. Because the power series for $P_\nu(z)$ becomes unbounded as $|z| \to \infty$ while $Q_\nu(z) \to 0$, we must discard the $P_\nu(z)$ solutions. Consequently the general solution is

$$\bar{u}(x,s) = \frac{1}{s} - \frac{Q_\nu(xi)}{sQ_\nu(0^+i)}. \qquad (3.3.64)$$

We have written 0^+i for the argument of $Q_\nu(z)$ because Q_ν is a multi-valued function and we must cut the x-plane with a branch cut along

---

[43] Taken from Moshinskii, A. I., 1987: Effective diffusion of a dynamically passive impurity in narrow channels. *J. Appl. Mech. Tech. Phys.*, **28**, 374–382.

[44] See Lebedev, N. N., 1972: *Special Functions and Their Applications*. New York: Dover Publications, Inc., Chapter 7.

the real axis from $-\infty$ to 1. Consequently, 0^+i denotes a point lying on the imaginary axis just above the branch cut.

From Bromwich's integral, the inverse of $\bar{u}(x,s)$ is

$$u(x,t) = \frac{1}{2\pi i} \int_{c-\infty i}^{c+\infty i} \left[\frac{1}{z} - \frac{Q_\nu(xi)}{zQ_\nu(0^+i)} \right] e^{tz} dz. \tag{3.3.65}$$

We close the line integral in (3.3.65) with an infinite semicircle on the left side of the complex z-plane excluding the branch cut along the negative real axis from $z = -\infty$ to $z = -1/4$ which is associated with the multivalued ν. We also have a simple pole at $z = 0$. The intrigue of this problem is the presence of a multivalued function in the order of the Legendre function.

Computing first the residue at $z = 0$,

$$\text{Res}(z = 0) = \left[1 - \frac{Q_0(xi)}{Q_0(0^+i)} \right] = 1 + \frac{2}{\pi} \tan^{-1}(x) \tag{3.3.66}$$

because $Q_0(z) = \tanh(z) = \frac{1}{2} \log[(z+1)/(z-1)]$. On the other hand, along the branch cut we have $z + 1/4 = \eta^2 e^{\pm \pi i}$ so that along the top of the branch cut,

$$\int_{top} = -2e^{-t/4} \int_0^\infty \frac{Q_{-1/2+\eta i}(xi)}{(1/4 + \eta^2)Q_{-1/2+\eta i}(0^+i)} e^{-t\eta^2} \eta \, d\eta \tag{3.3.67}$$

while just below the branch cut

$$\int_{bottom} = 2e^{-t/4} \int_0^\infty \frac{Q_{-1/2-\eta i}(xi)}{(1/4 - \eta^2)Q_{-1/2-\eta i}(0^+i)} e^{-t\eta^2} \eta \, d\eta. \tag{3.3.68}$$

Combining (3.3.67)–(3.3.68) yields

$$\int_{branch\ cut} = 2e^{-t/4} \int_0^\infty \left[\frac{Q_{-1/2-\eta i}(xi)}{Q_{-1/2-\eta i}(0^+i)} - \frac{Q_{-1/2+\eta i}(xi)}{Q_{-1/2+\eta i}(0^+i)} \right]$$
$$\times e^{-\eta^2 t} \frac{\eta}{1/4 + \eta^2} \, d\eta. \tag{3.3.69}$$

Let us introduce the relationships[45]

$$Q_\nu(z) - Q_{-\nu-1}(z) = \pi \cot(\nu\pi)P_\nu(z) \tag{3.3.70}$$

---

[45] Taken from Erdélyi, A., W. Magnus, F. Oberhettinger and F. G. Tricomi, 1953: *Higher Transcendental Functions*. New York: McGraw-Hill Book Co., Inc., Chapter 3.

233

$$\frac{2}{\pi}\sin(\nu\pi)Q_\nu(z) = P_\nu(z)e^{-\nu\pi i} - P_\nu(-z), \quad \text{Im}(z) > 0 \qquad (3.3.71)$$

$$Q_\nu(0^+i) = Q_\nu(0) - \pi i P_\nu(0)/2 \qquad (3.3.72)$$

$$P_\nu(0) = \frac{1}{\sqrt{\pi}}\cos\left(\frac{\pi\nu}{2}\right)\frac{\Gamma(1/2+\nu/2)}{\Gamma(1+\nu/2)} \qquad (3.3.73)$$

and

$$Q_\nu(0) = -\frac{\sqrt{\pi}}{2}\sin\left(\frac{\pi\nu}{2}\right)\frac{\Gamma(1/2+\nu/2)}{\Gamma(1+\nu/2)}, \qquad (3.3.74)$$

where $\Gamma(z)$ is the gamma function. Combining (3.3.72)–(3.3.74) yields

$$iQ_{\eta i-1/2}(0^+i) = \frac{\sqrt{\pi}}{2}\frac{\Gamma(1/4+\eta i/2)}{\Gamma(3/4+\eta i/2)}e^{\pi i/4}e^{\eta\pi/2}. \qquad (3.3.75)$$

Because $\Gamma(z)\Gamma(1-z) = \pi/\sin(\pi z)$,

$$iQ_{\eta i-1/2}(0^+i) = \frac{1}{2\sqrt{\pi}}\Gamma(1/4+\eta i/2)\Gamma(1/4-\eta i/2)$$

$$\times \cos\left(\frac{\eta\pi i}{2}+\frac{\pi}{4}\right)e^{\pi i/4}e^{\eta\pi/2}. \qquad (3.3.76)$$

From (3.3.70)–(3.3.71),

$$\frac{Q_{-1/2-\eta i}(xi)}{Q_{-1/2-\eta i}(0^+i)} - \frac{Q_{-1/2+\eta i}(xi)}{Q_{-1/2+\eta i}(0^+i)}$$

$$= \pi i \tanh(\eta\pi)\frac{P_{\eta i-1/2}(0)Q_{\eta i-1/2}(xi) - Q_{\eta i-1/2}(0^+i)P_{\eta i-1/2}(xi)}{Q_{\eta i-1/2}(0^+i)Q_{-\eta i-1/2}(0^+i)}$$

$$(3.3.77)$$

$$= -\frac{2\pi\sqrt{\pi}\tanh(\eta\pi)[P_{\eta i-1/2}(xi) - P_{\eta i-1/2}(-xi)]}{\Gamma(1/4+\eta i/2)\Gamma(1/4-\eta i/2)\cosh(\eta\pi)}. \qquad (3.3.78)$$

Consequently the final solution equals the residue minus the branch cut integral after we have divided the sum by $2\pi i$ or

$$u(x,t) = 1 + \frac{2}{\pi}\tan^{-1}(x) - 4\sqrt{\pi}\,e^{-t/4}$$

$$\times \int_0^\infty \frac{\eta\tanh(\eta\pi)}{\cosh(\eta\pi)\Gamma(1/4+\eta i/2)\Gamma(1/4-\eta i/2)(1/4+\eta^2)}$$

$$\times e^{-\eta^2 t}\frac{P_{\eta i-1/2}(xi) - P_{\eta i-1/2}(-xi)}{2i}\,d\eta. \qquad (3.3.79)$$

Problems

1. Solve the partial differential equation

$$\frac{\partial^2 u}{\partial t^2} = \frac{\partial^2 u}{\partial x^2} + \frac{\partial^3 u}{\partial t \partial x^2}, \qquad 0 < x < \infty, t > 0$$

with the boundary conditions $u(0, t) = \sin(t)H(t)$ and $\lim_{x \to \infty} u(x, t) \to 0$ and the initial conditions $u(x, 0) = u_t(x, 0) = 0$.

Step 1. Show that the Laplace transform of the partial differential equation yields the ordinary differential equation

$$\frac{d^2 \overline{u}(x, s)}{dx^2} - \frac{s^2}{s+1} \overline{u}(x, s) = 0$$

with

$$\overline{u}(0, s) = \frac{1}{s^2 + 1}, \quad \lim_{x \to \infty} \overline{u}(x, s) = 0.$$

Step 2. Show that the solution to step 1 is

$$\overline{u}(x, s) = \frac{1}{s^2 + 1} \exp\left[-\frac{sx}{\sqrt{1+s}}\right].$$

Step 3. Show that the singularities in the Laplace transform consist of simple poles at $s = \pm i$ and a branch point at $s = -1$.

Step 4. By defining the branch cut along the negative real axis, show that we may express the solution as a sum of the residues at the poles plus an integral along the branch cut:

$$u(x, t) = \exp[-x \sin(\pi/8)/\sqrt{2}] \sin[t - x \cos(\pi/8)/\sqrt{2}]$$
$$- \frac{1}{\pi} \int_0^\infty \frac{\exp[-t(1+u)]}{u^2 + 2u + 2} \sin\left[\frac{(1+u)x}{\sqrt{u}}\right] du.$$

2. Solve the partial differential equation[11]

$$\frac{\partial^2 u}{\partial t^2} + \frac{\partial u}{\partial t} = \frac{\partial^2 u}{\partial x^2} + a\frac{\partial^3 u}{\partial t \partial x^2}, \qquad a > 0, 0 < x < \infty, t > 0$$

with the boundary conditions $u(0, t) = H(t)$ and $\lim_{x \to \infty} u(x, t) \to 0$ and the initial conditions $u(x, 0) = u_t(x, 0) = 0$.

Transform Methods for Solving Partial Differential Equations

Step 1. Show that the Laplace transform of the partial differential equation yields the ordinary differential equation

$$(1 + as)\frac{d^2\bar{u}(x,s)}{dx^2} - s(1+s)\bar{u}(x,s) = 0$$

with

$$\bar{u}(0,s) = \frac{1}{s}, \qquad \lim_{x\to\infty} \bar{u}(x,s) \to 0.$$

Step 2. Show that the solution to step 1 is

$$\bar{u}(x,s) = \frac{1}{s}\exp\left[-x\sqrt{\frac{s(1+s)}{1+as}}\right].$$

Step 3. Show that the singularities in the Laplace transform consist of a simple pole at $s = 0$ and branch points at $s = 0$, $s = -1$ and $s = -1/a$.

Step 4. By defining all of the branch cuts along the negative real axis and deforming the Bromwich contour along the imaginary axis of the s-plane except for a small semicircle around $s = 0$, show that we may express the solution as

$$u(x,t) = \frac{1}{2} + \frac{1}{\pi}\int_0^\infty \exp\left\{-x\sqrt{\frac{y}{2}}\, M[\cos(\theta) - \sin(\theta)]\right\}\frac{dy}{y}$$

$$\times \sin\left\{ty - x\sqrt{\frac{y}{2}}\, M[\cos(\theta) + \sin(\theta)]\right\}$$

where

$$M = \left(\frac{1+y^2}{1+a^2y^2}\right)^{1/4}$$

and

$$2\theta = \tan^{-1}(y) - \tan^{-1}(ay).$$

3. Solve the partial differential equation

$$\frac{\partial^2 u}{\partial r^2} + \frac{1}{r}\frac{\partial u}{\partial r} - \frac{u}{b^2} = \frac{1}{\nu}\frac{\partial u}{\partial t}, \qquad a < r < \infty, t > 0$$

where a, b and ν are real, positive constants. This partial differential equation is subject to the boundary conditions

$$u(a,t) = H(t), \qquad \lim_{r\to\infty} u(r,t) \to 0$$

where $a > 0$ and the initial condition $u(r, 0) = 0$.

Step 1. By taking the Laplace transform of the partial differential equation, show that it reduces to the ordinary differential equation

$$\frac{d^2\bar{u}}{dr^2} + \frac{1}{r}\frac{d\bar{u}}{dr} - \left(\frac{1}{b^2} + \frac{s}{\nu}\right)\bar{u} = 0$$

with $\bar{u}(a, s) = 1/s$ and $\lim_{r\to\infty} \bar{u}(r, s) \to 0$.

Step 2. Show that the solution to step 1 is

$$\bar{u}(r, s) = \frac{K_0(r\sqrt{s/\nu + 1/b^2})}{sK_0(a\sqrt{s/\nu + 1/b^2})}$$

where $K_0(\ )$ is a zeroth-order, modified Bessel function of the second kind.

Step 3. Show that the singularities of the transform in step 2 are a simple pole at $s = 0$ and branch point located at $s = -\nu/b^2$.

Step 4. By taking the branch cut along the negative real axis from $s = -\nu/b^2$ to $s = -\infty$, show that Bromwich's integral gives the inverse of the transform in step 2 as

$$u(r, t) = \frac{K_0(r/b)}{K_0(a/b)} + \frac{2}{\pi}e^{-\nu t/b^2}\int_0^\infty \frac{\exp(-\nu t\eta^2/a^2)}{\eta^2 + (a/b)^2}\eta\, d\eta$$
$$\times \frac{J_0(r\eta/a)Y_0(\eta) - Y_0(r\eta/a)J_0(\eta)}{J_0^2(\eta) + Y_0^2(\eta)}$$

where $J_0(\ )$ and $Y_0(\ )$ are zeroth-order Bessel functions of the first and second kind, respectively.

4. Solve the partial differential equation[12]

$$\frac{\partial^2 u}{\partial r^2} + \frac{1}{r}\frac{\partial u}{\partial r} = \frac{\partial u}{\partial t}, \qquad a < r < \infty, t > 0$$

with the boundary conditions

$$\lim_{r\to\infty} u(r, t) \to 0, \qquad \frac{\partial u(a, t)}{\partial r} = -H(t)$$

and the initial condition $u(r, 0) = 0$.

Transform Methods for Solving Partial Differential Equations

Step 1. By taking the Laplace transform of the partial differential equation and boundary conditions, show that they reduce down to the ordinary differential equation

$$\frac{d^2\bar{u}}{dr^2} + \frac{1}{r}\frac{d\bar{u}}{dr} - s\bar{u} = 0, \qquad a < r < \infty$$

with the boundary conditions

$$\lim_{r\to\infty} \bar{u}(r,s) \to 0, \qquad \bar{u}'(a,s) = -\frac{1}{s}.$$

Step 2. Show that the solution to step 1 is

$$\bar{u}(r,s) = \frac{K_0(r\sqrt{s})}{s^{3/2}K_1(a\sqrt{s})}$$

where $K_0(\ )$ and $K_1(\ )$ are zeroth-order and first-order, modified Bessel functions of the second kind, respectively.

Step 3. Show that we can invert the Laplace transform in step 2 as follows:

$$u(r,t) = \int_0^t \mathcal{L}^{-1}\left[\frac{K_0(r\sqrt{s})}{\sqrt{s}K_1(a\sqrt{s})}\right] d\tau.$$

Step 4. Show that

$$\mathcal{L}^{-1}\left[\frac{K_0(r\sqrt{s})}{\sqrt{s}K_1(a\sqrt{s})}\right] = -\frac{2}{\pi}\int_0^\infty e^{-\eta^2\tau}\frac{J_0(\eta r)Y_1(\eta a) - Y_0(\eta r)J_1(\eta a)}{J_1^2(\eta a) + Y_1^2(\eta a)}d\eta$$

so that

$$u(r,t) = -\frac{2}{\pi}\int_0^\infty \left(1 - e^{-\eta^2 t}\right)\frac{J_0(\eta r)Y_1(\eta a) - Y_0(\eta r)J_1(\eta a)}{J_1^2(\eta a) + Y_1^2(\eta a)}\frac{d\eta}{\eta^2}$$

where $J_0(\ )$ and $Y_0(\ )$ are zeroth-order Bessel functions of the first and second kind, respectively. Take the branch cuts associated with the square root and $K_0(\ )$ along the negative real axis of the s-plane.

5. Solve the partial differential equation

$$\frac{\partial^2 u}{\partial r^2} + \frac{1}{r}\frac{\partial u}{\partial r} - u = \frac{\partial u}{\partial t}, \qquad 1 < r < \infty, t > 0$$

with the boundary conditions

$$\lim_{r\to\infty} u(r,t) \to 0, \qquad u(1,t) = 1$$

and the initial condition $u(r, 0) = 0$.

Step 1. By taking the Laplace transform of the partial differential equation and boundary conditions, show that they reduce down to the ordinary differential equation

$$\frac{d^2\bar{u}}{dr^2} + \frac{1}{r}\frac{d\bar{u}}{dr} - (s+1)\bar{u} = 0, \qquad 1 < r < \infty$$

with the boundary conditions

$$\lim_{r\to\infty} \bar{u}(r, s) \to 0, \qquad \bar{u}(1, s) = \frac{1}{s}.$$

Step 2. Show that the solution to step 1 is

$$\bar{u}(r, s) = \frac{K_0(r\sqrt{s+1})}{sK_0(\sqrt{s+1})}$$

where $K_0(\ )$ is the zeroth-order, modified Bessel functions of the second kind.

Step 3. Show that the Laplace transform in step 2 has a simple pole at $s = 0$ and a branch point at $s = -1$.

Step 4. Use Bromwich's integral to invert $\bar{u}(r, s)$. If you take the branch cut associated with $K_0(\ )$ along the negative real axis of the s-plane, you should find that

$$u(r, t) = \frac{K_0(r)}{K_0(1)} - \frac{2}{\pi}e^{-t}\int_0^\infty \frac{u}{u^2+1}e^{-u^2t}\frac{J_0(u)Y_0(ru) - J_0(ru)Y_0(u)}{J_0^2(u) + Y_0^2(u)}\,du$$

where $J_0(\ )$ and $Y_0(\ )$ are zeroth-order Bessel functions of the first and second kind, respectively.

6. Solve the partial differential equation[13]

$$\frac{\partial u}{\partial t} + a\frac{\partial^2 u}{\partial t^2} - b\frac{\partial^3 u}{\partial x^2 \partial t} - \frac{\partial^2 u}{\partial x^2} = 0, \qquad 0 < x < \infty, t > 0$$

where $b > a > 0$ with the boundary conditions that

$$\lim_{x\to\infty} u(x, t) \to 0 \qquad \text{and} \qquad u(0, t) = \sin(2\pi t)$$

and the initial conditions that

$$u(x, 0) = u_t(x, 0) = 0.$$

Transform Methods for Solving Partial Differential Equations

Step 1. By taking the Laplace transform of the partial differential equation, show that it reduces to the ordinary differential equation

$$\frac{d^2\bar{u}}{dx^2} - \frac{s(1+as)}{1+bs}\bar{u} = 0$$

with the boundary conditions

$$\bar{u}(0,s) = \frac{2\pi}{s^2 + 4\pi^2} \qquad \text{and} \qquad \lim_{x\to\infty} \bar{u}(x,s) \to 0.$$

Step 2. Show that the solution to step 1 is

$$\bar{u}(x,s) = \frac{2\pi}{s^2 + \pi^2} \exp\left[-x\sqrt{\frac{s(1+as)}{1+bs}}\right].$$

Step 3. Show that the Laplace transform in step 2 has simple poles at $s = \pm 2\pi i$ and branch points at $s = 0$, $s = -1/b$ and $s = -1/a$.

Step 4. By taking the branch cuts associated with each branch point along the negative real s axis, use Bromwich's integral and show that

$$u(x,t) = \sin[2\pi t - x\Delta\sin(\psi/2)]\exp[-x\Delta\cos(\psi/2)]$$

$$+ 2\int_0^{1/b} F(x,t,\eta)\,d\eta + 2\int_{1/a}^{\infty} F(x,t,\eta)\,d\eta$$

where

$$F(x,t,\eta) = \frac{\exp(-\eta t)}{\eta^2 + 4\pi^2}\sin\left[x\sqrt{\frac{\eta(1-a\eta)}{1-b\eta}}\right],$$

$$\Delta = [(d_1^2 + d_2^2)d_3^2]^{1/4},$$

$$d_1 = 1 + 4\pi^2 ab, \qquad d_2 = 2\pi(b-a),$$

$$d_3 = \frac{2\pi}{1 + 4\pi^2 b^2} \qquad \text{and} \qquad \psi = \tan^{-1}(d_1/d_2).$$

7. Solve the partial differential equation[14]

$$\frac{\partial u}{\partial t} = \frac{\partial^2 u}{\partial r^2} + \frac{1}{r}\frac{\partial u}{\partial r} - \frac{u}{r^2}, \qquad 1 \le r < \infty, t > 0$$

with the boundary conditions that $\lim_{r\to\infty} u(r,t) \to 0$, $u(1,t) = \sin(\omega t)$ and the initial condition that $u(r,0) = 0$.

Step 1. By taking the Laplace transform of the partial differential equation and boundary conditions, show that they reduce to the ordinary differential equation

$$\frac{d^2\bar{u}}{dr^2} + \frac{1}{r}\frac{d\bar{u}}{dr} - \left(s + \frac{1}{r^2}\right)\bar{u} = 0$$

with the boundary conditions $\lim_{r\to\infty}\bar{u}(r,s) \to 0$ and $\bar{u}(1,s) = \omega/(s^2 + \omega^2)$.

Step 2. Show that the solution to step 1 is

$$\bar{u}(r,s) = \frac{\omega}{s^2 + \omega^2}\frac{K_1(r\sqrt{s})}{K_1(\sqrt{s})}$$

where $K_1(\ )$ is a modified Bessel function of the first order and second kind.

Step 3. Show that the Laplace transform in step 2 has simple poles at $s = \pm\omega i$ and a branch point at $s = 0$.

Step 4. By taking the branch cut associated with $K_1(\ )$ along the negative real axis of the s-plane, use Bromwich's integral and show that

$$u(r,t) = \frac{N_1(r\sqrt{\omega})}{N_1(\sqrt{\omega})}\sin[\omega t + \phi_1(r\sqrt{\omega}) - \phi_1(\sqrt{\omega})]$$
$$- \frac{2}{\pi}\int_0^\infty \frac{\eta}{\eta^4 + \omega^2}e^{-\eta^2 t}\frac{J_1(r\eta)Y_1(\eta) - Y_1(r\eta)J_1(\eta)}{J_1^2(\eta) + Y_1^2(\eta)}\,d\eta$$

where

$$N_1(z) = \sqrt{\text{ker}_1^2(z) + \text{kei}_1^2(z)},\ \phi_1(z) = \tan^{-1}[\text{kei}_1(z)/\text{ker}_1(z)],$$

$\text{ker}_1(z)$, $\text{kei}_1(z)$ are Kelvin's ker and kei functions, respectively, and $J_1(\ )$, $Y_1(\ )$ are first-order Bessel functions of the first and second kind, respectively.

8. Solve the partial differential equation[15]

$$\frac{\partial u}{\partial t} = \frac{\partial^2 u}{\partial r^2} + \frac{1}{r}\frac{\partial u}{\partial r} - \frac{9u}{4r^2}, \qquad a < r < \infty, t > 0$$

with the boundary conditions that $u(a,t) = a^{3/2}H(t)$ and $\lim_{r\to\infty} u(r,t) \to 0$ and the initial condition that $u(r,0) = 0$.

Transform Methods for Solving Partial Differential Equations

Step 1. By taking the Laplace transform of the partial differential equation and boundary conditions, show that they reduce to the ordinary differential equation

$$\frac{d^2\bar{u}}{dr^2} + \frac{1}{r}\frac{d\bar{u}}{dr} - \left(s + \frac{9}{4r^2}\right)\bar{u} = 0$$

with the boundary conditions $\lim_{r\to\infty} \bar{u}(r,s) \to 0$ and $\bar{u}(a,s) = a^{3/2}/s$.

Step 2. Show that the solution to step 1 is

$$\bar{u}(r,s) = \frac{a^{3/2}K_{3/2}(r\sqrt{s})}{sK_{3/2}(a\sqrt{s})} = \frac{a^3[1 + r\sqrt{s}]}{sr^{3/2}[1 + a\sqrt{s}]}e^{-(r-a)\sqrt{s}}$$

where $K_{3/2}(\ )$ is a modified Bessel function of order 3/2 and the second kind.

Step 3. Show that the Laplace transform in step 2 has a branch point at $s = 0$.

Step 4. By taking the branch cut associated with the square root along the negative real axis of the s-plane, use Bromwich's integral and show that

$$u(r,t) = \frac{a^3}{r^{3/2}}\left\{1 - \frac{2}{\pi}\int_0^\infty \frac{1 + ar\eta^2}{1 + a^2\eta^2}\sin[(r-a)\eta]e^{-\eta^2 t}\frac{d\eta}{\eta}\right.$$
$$\left. + \frac{2}{\pi}(r-a)\int_0^\infty \frac{1}{1 + a^2\eta^2}\cos[(r-a)\eta]e^{-\eta^2 t}d\eta\right\}.$$

9. Solve the partial differential equation[16]

$$\frac{\partial u}{\partial t} = \frac{\partial^2 u}{\partial r^2} + \frac{1}{r}\frac{\partial u}{\partial r} - \frac{u}{r^2}, \qquad a < r < \infty, t > 0$$

with the boundary conditions that $\lim_{r\to\infty} u(r,t) \to 0$ and

$$u(a,t) = a\Omega(t); \qquad \frac{d\Omega(t)}{dt} = M + 2\pi\mu a^2\left[\frac{\partial u(a,t)}{\partial r} - \Omega(t)\right]$$

where I, M and μ are constants. Initially all systems are at rest so that $u(r,0) = 0$ and $\Omega(0) = 0$.

Step 1. By taking the Laplace transform of the partial differential equation and boundary conditions, show that they reduce to the ordinary differential equation

$$\frac{d^2\bar{u}}{dr^2} + \frac{1}{r}\frac{d\bar{u}}{dr} - \left(s + \frac{1}{r^2}\right)\bar{u} = 0$$

with the boundary conditions $\lim_{r \to \infty} \bar{u}(r, s) \to 0$ and

$$(Is + 2\pi\mu a^2)\bar{u}(a, s) - 2\pi\mu a^3 \frac{d\bar{u}(a, s)}{dr} = \frac{aM}{s}.$$

Step 2. Show that the solution to step 1 is

$$\bar{u}(r, s) = \frac{aM}{s} \frac{K_1(r\sqrt{s})}{(Is + 2\pi\mu a^2)K_1(a\sqrt{s}) - 2\pi\mu a^3 \partial K_1(r\sqrt{s})/\partial r|_{r=a}}.$$

where $K_1(\ )$ is a modified Bessel function of first order and second kind.

Step 3. Show that the Laplace transform in step 2 has a branch point at $s = 0$.

Step 4. By taking the branch cut associated with the square root along the negative real axis of the s-plane, use Bromwich's integral and show that

$$u(r, t) = \frac{M}{4\pi\mu r}\left\{1 + \frac{4\chi r}{a\pi} \int_0^\infty \frac{P(\eta)}{\eta^2 Q(\eta)} e^{-t\eta^2/a^2} d\eta\right\}$$

where

$$P(\eta) = Y_1(r\eta/a)[\eta J_1(\eta) - \chi J_2(\eta)] - J_1(r\eta/a)[\eta Y_1(\eta) - \chi Y_2(\eta)],$$

$$Q(\eta) = [\eta J_1(\eta) - \chi J_2(\eta)]^2 + [\eta Y_1(\eta) - \chi Y_2(\eta)]^2,$$

$\chi = 2\pi\mu a^4/I$, $J_1(\ )$, $J_2(\ )$ are first and second-order Bessel functions of the first kind, respectively, and $Y_1(\ )$, $Y_2(\ )$ are first and second-order Bessel functions of the second kind, respectively.

10. Solve the partial differential equation

$$\frac{\partial u}{\partial t} = \left(a + \frac{\partial}{\partial t}\right)\frac{\partial^2 u}{\partial x^2}, \qquad 0 < x < \infty, t > 0$$

where $a > 0$ with the boundary conditions $\lim_{x \to \infty} u(x, t) \to 0$ and $u(0, t) = H(t)$ and the initial condition $u(x, 0) = 0$.

Step 1. By taking the Laplace transform of the partial differential equation and boundary conditions, show that they reduce to the ordinary differential equation

$$\frac{d^2\bar{u}}{dx^2} + \frac{s}{s + a}\bar{u} = 0$$

with the boundary conditions $\lim_{x \to \infty} \bar{u}(x, s) \to 0$ and $\bar{u}(0, s) = 1/s$.

Step 2. Show that the solution to step 1 is

$$\overline{u}(r, s) = \frac{1}{s} \exp\left(-x\sqrt{\frac{s}{s+a}}\right).$$

Step 3. Show that the Laplace transform in step 2 has branch points at $s = 0$ and $s = -a$.

Step 4. By taking the branch cuts associated with the square roots $\sqrt{s}$ and $\sqrt{s+a}$ along the negative real axis of the s-plane, use Bromwich's integral and show that

$$u(r, t) = 1 - \frac{2a}{\pi} \int_0^\infty \exp\left(-\frac{a t \sigma^2}{1 + \sigma^2}\right) \frac{\sin(\sigma x)}{\sigma(1 + \sigma^2)} d\sigma.$$

11. Solve the partial differential equation[17]

$$\frac{\partial u}{\partial t} - \eta \frac{\partial}{\partial t}\left(\frac{\partial^2 u}{\partial r^2} + \frac{1}{r}\frac{\partial u}{\partial r} + \frac{\partial^2 u}{\partial z^2}\right) = \kappa\left(\frac{\partial^2 u}{\partial r^2} + \frac{1}{r}\frac{\partial u}{\partial r} + \frac{\partial^2 u}{\partial z^2}\right),$$

where $0 < r < \infty$, $0 < z < h$, $t > 0$ and η and κ are real and positive, subject to the boundary conditions

$$\lim_{r \to \infty} u(r, z, t) \to 0, \qquad \frac{\partial u(r, 0, t)}{\partial z} = \frac{\partial u(r, h, t)}{\partial z} = 0$$

and

$$\lim_{r \to 0} r\frac{\partial u(r, z, t)}{\partial r} = \begin{cases} \exp(-\kappa t/\eta) - 1 & \text{if } 0 \le z \le h \\ 0 & \text{if } l < z \le h. \end{cases}$$

and the initial condition $u(r, z, 0) = 0$.

Step 1. By taking the Laplace transform of the partial differential equation and boundary conditions, show that they reduce to the partial differential equation

$$s\overline{u} - (\kappa + \eta s)\left(\frac{\partial^2 \overline{u}}{\partial r^2} + \frac{1}{r}\frac{\partial \overline{u}}{\partial r} + \frac{\partial^2 \overline{u}}{\partial z^2}\right) = 0$$

with the boundary conditions

$$\lim_{r \to \infty} \overline{u}(r, z, s) \to 0, \qquad \frac{\partial \overline{u}(r, 0, s)}{\partial z} = \frac{\partial \overline{u}(r, h, s)}{\partial z} = 0$$

and

$$\lim_{r \to 0} r\frac{\partial \overline{u}(r, z, s)}{\partial r} = \begin{cases} -\kappa/[s(\kappa + \eta s)] & \text{if } 0 \le z \le h \\ 0 & \text{if } l < z \le h. \end{cases}$$

Step 2. Using the method of separation of variables, solve the partial differential equation in step 1 and show that

$$\bar{u}(r, z, s) = \frac{\kappa l}{h} \frac{K_0(r\sigma_0)}{s(\kappa + \eta s)} + \sum_{n=1}^{\infty} \frac{2\kappa K_0(r\sigma_n)}{n\pi s(\kappa + \eta s)} \sin(\lambda_n l) \cos(\lambda_n z)$$

$$= \frac{\kappa l}{h} \frac{c_0^2}{s(s + b^2)} K_0\left(rc_0\sqrt{\frac{s + a_0^2}{s + b^2}}\right)$$

$$+ \frac{2\kappa}{s\pi} \sum_{n=1}^{\infty} \frac{c_0^2}{n(s + b^2)} K_0\left(rc_n\sqrt{\frac{s + a_n^2}{s + b^2}}\right) \sin(\lambda_n l) \cos(\lambda_n z)$$

where $a_n^2 = b^2\lambda_n^2/c_n^2$, $b^2 = \kappa/\eta$, $c_n^2 = \lambda_n^2 + 1/\eta$, $\lambda_n = n\pi/h$, $\sigma_n^2 = s/(\kappa + \eta s) + \lambda^2$ and $K_0(\ )$ is a modified Bessel function of the second kind and zeroth order. Note that $b > a_n$.

Step 3. Taking the branch cuts associated with $\sqrt{s + a_n^2}$ and $\sqrt{s + b^2}$ along the negative real axis of the s-plane, show that

$$\mathcal{L}^{-1}\left[\frac{c_0^2}{s + b^2} K_0\left(rc_n\sqrt{\frac{s + a_n^2}{s + b^2}}\right)\right] = \int_{a_n^2}^{b^2} \frac{c_0^2}{b^2 - \rho} J_0\left(rc_n\sqrt{\frac{\rho - a_n^2}{b^2 - \rho}}\right) e^{-\rho t} d\rho$$

Step 4. Because $\mathcal{L}^{-1}[F(s)/s] = \int_0^t f(u)\, du$, show that the final solution is

$$u(r, z, t) = \frac{l\kappa}{2h} \int_0^{b^2} \frac{1 - \exp(-\rho t)}{\rho(\kappa - \eta\rho)} J_0\left(r\sqrt{\frac{\rho}{\kappa - \eta\rho}}\right) d\rho$$

$$+ \frac{\kappa}{\pi} \sum_{n=1}^{\infty} \frac{\sin(\lambda_n l) \cos(\lambda_n z)}{n} \int_{a_n^2}^{b^2} \frac{1 - \exp(-\rho t)}{\rho(\kappa - \eta\rho)}$$

$$J_0\left(r\sqrt{\frac{\rho}{\kappa - \eta\rho} - \lambda_n^2}\right) d\rho.$$

3.4 THE SOLUTION OF PARTIAL DIFFERENTIAL EQUATIONS BY FOURIER TRANSFORMS

Fourier transforms are often used to find the spatial dependence in problems where the domains are infinite or semi-infinite. To illustrate this technique we find the velocity potential[46] for an incompressible,

---

[46] Taken from Troesch, A. W., 1979: The diffraction forces for a ship moving in oblique seas. *J. Ship Res.*, **23**, 127–139.

irrotational, three-dimensional fluid of infinite depth. At the free surface, $z = 0$, we apply a specified pressure distribution. The governing equations are

$$\frac{\partial^2 \phi}{\partial y^2} + \frac{\partial^2 \phi}{\partial z^2} - \nu^2 \phi = 0, \qquad -\infty < z < 0 \qquad (3.4.1)$$

and

$$-\kappa\phi + \frac{\partial \phi}{\partial z} = p(y), \qquad z = 0 \qquad (3.4.2)$$

where we have factored out $\exp[i(\kappa x - \omega t)]$ and the usual dispersion relationship $\omega^2 = g\kappa$ holds. The relationship between κ and ν is $\nu = \kappa \cos(\chi)$ where χ is the heading angle. We take the pressure field to be symmetric

$$p(y) = p_0 e^{-\mu|y| - i|y|\sqrt{\gamma^2 - \nu^2}}. \qquad (3.4.3)$$

The presence of the term $e^{-\mu|y|}$ plays a very important role in our analysis. Mathematically it ensures that the Fourier transform for $p(y)$ exists. Physically, it introduces some friction into the system. In Section 2.4 we showed that a certain ambiguity exists in non-dissipative systems because incoming and outgoing waves are both present. We resolved this problem by requiring that energy radiate outward, away from the source of the disturbance, the so-called "Sommerfeld radiation condition". The addition of a small amount of friction is another method of imposing the same condition, because any energy propagating inwards from infinity would be dissipated before it reaches any areas of interest.

Let us now define the Fourier transform and inverse for the velocity potential as

$$\Phi(k, z) = \int_{-\infty}^{\infty} \phi(y, z) e^{-iky} dy \qquad (3.4.4)$$

and

$$\phi(y, z) = \frac{1}{2\pi} \int_{-\infty}^{\infty} \Phi(k, z) e^{iky} dk. \qquad (3.4.5)$$

Upon substituting into (3.4.1),

$$\Phi(k, z) = C(k) e^{z\sqrt{\nu^2 + k^2}} \qquad (3.4.6)$$

where $\mathrm{Re}[\sqrt{\nu^2 + k^2}] > 0$ so that the solution does not grow exponentially for $z < 0$. From the boundary condition at $z = 0$, (3.4.2),

$$-\kappa C(k) + C(k)\sqrt{\nu^2 + k^2} = P(k) \qquad (3.4.7)$$

where $P(k)$ is the Fourier transform of $p(y)$. From the definition of the Fourier transform, we may compute

$$P(k) = p_0 \int_{-\infty}^{\infty} e^{-\mu|y| - i|y|\sqrt{\gamma^2 - \nu^2}} e^{-iky} dy \tag{3.4.8}$$

$$= -p_0 \left[\frac{1}{-\mu + i(k - \sqrt{\gamma^2 - \nu^2})} + \frac{1}{-\mu - i(k + \sqrt{\gamma^2 - \nu^2})} \right] \tag{3.4.9}$$

and

$$C(k) = p_0 \frac{2\mu + 2i\sqrt{\gamma^2 - \nu^2}}{[i(k - k_1)][-i(k - k_2)][-\kappa + \sqrt{\nu^2 + k^2}]} \tag{3.4.10}$$

where

$$k_1 = \sqrt{\gamma^2 - \nu^2} - i\mu \tag{3.4.11}$$

$$k_2 = -\sqrt{\gamma^2 - \nu^2} + i\mu. \tag{3.4.12}$$

Then we may write $\phi(y, z)$ as

$$\phi(y, z) = \frac{p_0}{2\pi} \int_{-\infty}^{\infty} e^{iky + z\sqrt{\nu^2 + k^2}} \frac{2\mu + 2i\sqrt{\gamma^2 - \nu^2}}{(k - k_1)(k - k_2)(-\kappa + \sqrt{\nu^2 + k^2})} dk. \tag{3.4.13}$$

To invert (3.4.13) we employ the residue theorem where k is now treated as a complex variable. The integrand has branch points at $k = \pm i\nu$ and simple poles at $k = k_1$, $k = k_2$, $k = k_3 = \sqrt{\kappa^2 - \nu^2}$ and $k = k_4 = -\sqrt{\kappa^2 - \nu^2}$. In Fig. 3.4.1 we show that we have taken the branch cuts to lie along the imaginary axis. Note the important role that friction plays here; if it were not present we would have singularities along the path of integration for the inverse.

For $y > 0$, we close the contour with an infinite semicircle in the upper half k-plane as dictated by Jordan's lemma. Therefore, the contribution from contours C_1 and C_5 are zero. Further analysis shows that the contribution near the branch point $k = i\nu$ is also zero. However, the integration along the branch cuts yields

$$\int_{C_2} = \int_{\infty}^{\nu} e^{-ky + iz\sqrt{k^2 - \nu^2}} \frac{2i\sqrt{\gamma^2 - \nu^2}}{(ik - k_1)(ik - k_2)(-\kappa + i\sqrt{k^2 - \nu^2})} i \, dk \tag{3.4.14}$$

and

$$\int_{C_4} = \int_{\nu}^{\infty} e^{-ky - iz\sqrt{k^2 - \nu^2}} \frac{2i\sqrt{\gamma^2 - \nu^2} \, (i \, dk)}{(ik - k_1)(ik - k_2)(-\kappa - i\sqrt{k^2 - \nu^2})}. \tag{3.4.15}$$

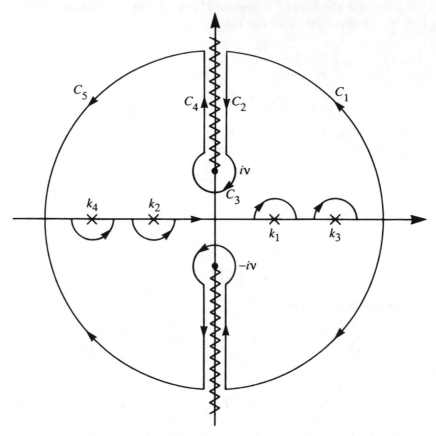

Fig. 3.4.1: The contour used to invert (3.4.1).

The residue at k_2 equals

$$\text{Res}(k_2) = e^{ik_2 y + z\sqrt{\nu^2 + k_2^2}} \frac{2i\sqrt{\gamma^2 - \nu^2}}{(k_2 - k_1)(-\kappa + \sqrt{\nu^2 + k_2^2})} \qquad (3.4.16)$$

$$= e^{-iy\sqrt{\gamma^2 - \nu^2} + \gamma z} \frac{2i\sqrt{\gamma^2 - \nu^2}}{(-2\sqrt{\gamma^2 - \nu^2})(\gamma - \kappa)}. \qquad (3.4.17)$$

An interesting case occurs when $\gamma \to \kappa$. The forcing produces a resonant solution. Expanding $\sqrt{\gamma^2 - \nu^2}$ with $\gamma = \kappa - \delta$ and $\delta/\kappa \ll 1$,

$$\sqrt{\gamma^2 - \nu^2} = \sqrt{\kappa^2 - \nu^2} - \frac{\kappa\delta}{\sqrt{\kappa^2 - \nu^2}} + O(\delta^2) \qquad (3.4.18)$$

so that

$$\text{Res}(k_2) = \lim_{\delta \to 0} i \left[\frac{1}{\delta} - z + \frac{i\kappa y}{\sqrt{\kappa^2 - \nu^2}} \right] e^{-iy\sqrt{\kappa^2 - \nu^2} + \kappa z} \qquad (3.4.19)$$

while at the other residue $k = k_4$

$$\text{Res}(k_4) = \lim_{\delta \to 0} \frac{-ie^{-iy\sqrt{\kappa^2 - \nu^2} + \kappa z}}{\delta}. \tag{3.4.20}$$

Consequently, the sum of the two residues is

$$\text{Res}(k_2) + \text{Res}(k_4) = -ie^{-iy\sqrt{\kappa^2 - \nu^2} + \kappa z}\left(z - \frac{i\kappa y}{\sqrt{\kappa^2 - \nu^2}}\right) \tag{3.4.21}$$

where the singular nature of k_2 and k_4 cancel out.

For $y < 0$, a similar analysis follows with the exception that we close the contour by introducing an infinite semicircle in the lower half of the k-plane. Combining our results yields the final result for the special case $\kappa = \gamma$:

$$\phi(y, z) = p_0 e^{-i|y|\sqrt{\kappa^2 - \nu^2} + \kappa z}\left(z - \frac{i\kappa|y|}{\sqrt{\kappa^2 - \nu^2}}\right)$$

$$- \frac{p_0\sqrt{\kappa^2 - \nu^2}}{\pi}\int_\nu^\infty \frac{e^{-k|y|}}{(ik - \sqrt{\kappa^2 - \nu^2})(ik + \sqrt{\kappa^2 - \nu^2})}$$

$$\times \left[\frac{e^{iz\sqrt{\kappa^2 - \nu^2}}}{-\kappa + i\sqrt{\kappa^2 - \nu^2}} + \frac{e^{-iz\sqrt{\nu^2 + k^2}}}{\kappa + i\sqrt{\nu^2 + k^2}}\right] dk. \tag{3.4.22}$$

For our second example, we solve a classic problem from seismology. In 1904 Lamb[47] modeled the seismic waves generated by an earthquake as the response of an infinite, elastic half-space to an oscillating, impulsive load. If we denote the y component of the displacement by u, then the governing equation is the Helmholtz equation

$$\frac{\partial^2 u}{\partial x^2} + \frac{\partial^2 u}{\partial z^2} + \frac{\omega^2}{V_S^2}u = 0, \quad 0 < z < \infty \tag{3.4.23}$$

where the forcing has a frequency ω and the phase speed of the shear waves is V_S. Assuming that the Fourier transform exists, we have

$$u(x, z, \omega) = \frac{1}{2\pi}\int_{-\infty}^{\infty} U(k, z, \omega)e^{-ikx} dk \tag{3.4.24}$$

and the transformed equation is

$$\frac{d^2 U}{dz^2} - \left(k^2 - \frac{\omega^2}{V_S^2}\right)U = 0. \tag{3.4.25}$$

---

[47] Lamb, H., 1904: The propagation of tremors over the surface of an elastic solid. *Phil. Trans. R. Soc. Lond.*, **A203**, 1–42.

Transform Methods for Solving Partial Differential Equations

The general solution to (3.4.25) is

$$U(k, z, \omega) = A(k, \omega)e^{-\nu z} + B(k, \omega)e^{\nu z} \qquad (3.4.26)$$

where $\nu = \sqrt{k^2 - \omega^2/V_S^2}$. At the surface of the half-plane, $z = 0$, the boundary condition relates the surface stress to the impulsive point load

$$\mu \frac{\partial u}{\partial z} = -P\delta(x), \qquad (3.4.27)$$

where μ is one of Lamé's constants. The transformed version of (3.4.27) is

$$\mu \frac{dU}{dz} = -P. \qquad (3.4.28)$$

Upon substituting (3.4.26) into (3.4.28), $A(k, \omega) = P/\mu\nu$ or

$$u(x, z, \omega) = \frac{P}{2\pi\mu} \int_{-\infty}^{\infty} \frac{e^{-\nu z - ikx}}{\nu} \, dk. \qquad (3.4.29)$$

In the derivation of (3.4.29), we have assumed that the $\text{Re}(\nu) \geq 0$. Therefore, $B(k, \omega) = 0$ so that the solution remains finite as $z \to \infty$.

At this point, we introduce polar coordinates $x = r\cos(\theta)$ and $y = r\sin(\theta)$ along with $k = \omega\cos(\zeta)/V_S$ and $\nu = i\omega\sin(\zeta)/V_S$. Equation (3.4.29) now becomes

$$u(x, z, \omega) = \frac{P}{2\pi\mu i} \int_{0-i\infty}^{\pi+i\infty} e^{-i\omega r \sin(\zeta+\theta)/V_S} \, d\zeta. \qquad (3.4.30)$$

Before reducing (3.4.30) to its final form, we must consider the integral of the form

$$I(\rho) = \frac{1}{\pi} \int_W e^{i\rho\cos(w)} e^{in(w-\pi/2)} \, dw. \qquad (3.4.31)$$

Sommerfeld[48] showed that for a wide class of contour integrals in the complex w-plane, $I(\rho)$ equals either the first or second Hankel functions[49], $H_n^{(1)}(\rho)$ and $H_n^{(2)}(\rho)$. Fig. 3.4.2 illustrates these so-called "*Sommerfeld contours*". We transform our integral into a contour integral similar to W_2 by defining $n = 0$, $w = \zeta + \theta + \pi/2$ because $-\pi/2 < \theta < \pi/2$ so that

$$u(x, z, \omega) = \frac{P}{2i\mu} H_0^{(2)}(\omega r/V_S). \qquad (3.4.32)$$

---

[48] Sommerfeld, A., 1949: *Partial Differential Equations in Physics*. New York: Academic Press, Inc., Section 19.

[49] Watson, G. N., 1966: *A Treatise on the Theory of Bessel Functions*. Cambridge (England): At the University Press, Section 3.6.

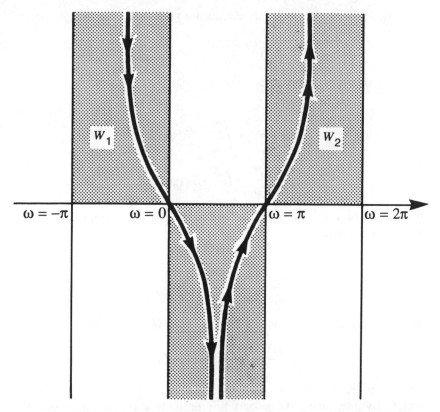

Fig. 3.4.2: The Sommerfeld contours associated with (3.4.31). Integration along W_1 gives $H_n^{(1)}(\rho)$ while integration along W_2 gives $H_n^{(2)}(\rho)$. The contours W_1 and W_2 may lie anywhere in the shaded area.

For large $\omega r/V_S$, we can use the asymptotic approximation for the Hankel functions (Watson's book, p. 196) and find that

$$u(x, z, \omega) = \frac{P}{2i\mu} \sqrt{\frac{2V_S}{\pi \omega r}} e^{-i\omega r/V_S + \pi i/4}. \qquad (3.4.33)$$

Problem

Find $\eta(x)$ defined by

$$\eta(x) = \frac{\partial u(x, y_0)}{\partial x}, \qquad -\infty < x < \infty$$

where the partial differential equation[37]

$$\frac{\partial^2 u}{\partial x^2} + \frac{\partial^2 u}{\partial y^2} = \delta'(x)\delta(y - a), \qquad -\infty < x < \infty, y_0 < y < \infty$$

Transform Methods for Solving Partial Differential Equations

gives $u(x, y)$ with the boundary conditions that $\lim_{y \to \infty} u(x, y) \to 0$ and

$$\frac{\partial^2 u(x, y_0)}{\partial x^2} - a_1 \frac{\partial u(x, y_0)}{\partial y} - \epsilon a_2 \frac{\partial u(x, y_0)}{\partial x} = 0$$

with $a > y_0 > 0$, a_1 and $a_2 > 0$ and $\epsilon \to 0$.

Step 1. If

$$u(x, y) = \frac{1}{2\pi} \int_{-\infty}^{\infty} U(k, y) e^{ikx} dk$$

and

$$\eta(x) = \frac{1}{2\pi} \int_{-\infty}^{\infty} H(k) e^{ikx} dk,$$

show that the partial differential equations reduce to the ordinary differential equations

$$H(k) = ikU(k, y_0)$$

and

$$\frac{d^2 U}{dy^2} - k^2 U = ik\delta(y - a)$$

with the boundary conditions that $\lim_{y \to \infty} U(k, y) \to 0$ and

$$a_1 \frac{dU(k, y_0)}{dy} + (k^2 + ik\epsilon a_2)U(k, y_0) = 0.$$

Step 2. By integrating the second differential equation in step 1 over the interval $[a^-, a^+]$ where a^+ and a^- are points just above and below a, respectively, show that we can replace the differential equation in step 1 by

$$\frac{d^2 U}{dy^2} - k^2 U = 0 \qquad \text{if} \qquad y \neq a$$

$$\lim_{y \to \infty} U(k, y) \to 0, \qquad U(k, a^+) = U(k, a^-),$$

$$\frac{dU(k, a^+)}{dy} - \frac{dU(k, a^-)}{dy} = ik,$$

and

$$a_1 \frac{dU(k, y_0)}{dy} + (k^2 + ik\epsilon a_2)U(k, y_0) = 0.$$

Step 3. Show that the solution to step 2 is

$$U(k, y) = \left\{ \frac{i}{2}\mathrm{sgn}(k) \left[\frac{|k| + a_1 + i\epsilon a_2 \mathrm{sgn}(k)}{|k| - a_1 + i\epsilon a_2 \mathrm{sgn}(k)} \right] e^{(2y_0 - a)|k|} \right.$$
$$\left. - \frac{i}{2}\mathrm{sgn}(k) e^{|k|a} \right\} e^{-|k|y}$$

if $y > a$ and

$$U(k, y) = \frac{i}{2}\text{sgn}(k)\left[\frac{|k| + a_1 + i\epsilon a_2\text{sgn}(k)}{|k| - a_1 + i\epsilon a_2\text{sgn}(k)}\right]e^{(2y_0-a)|k|}e^{-|k|y}$$
$$- \frac{i}{2}\text{sgn}(k)e^{-|k|a}e^{|k|y}$$

if $y_0 < y < a$ and

$$H(k) = -\frac{a_1|k|\exp[-(a-y_0)|k|]}{|k| - a_1 + i\epsilon a_2\text{sgn}(k)}$$

where $\text{sgn}(k) = 1$ if $k > 0$ and -1 if $k < 0$.

Step 4. Show that the inverse of $H(k)$ is

$$\eta(x) = -2a_1^2 e^{-a_1(a-y_0)}\sin(a_1 x)H(-x)$$
$$- \frac{a_1}{\pi}\int_0^\infty \frac{\eta e^{-x\eta}}{\eta^2 + a_1^2}\{a_1\cos[(a-y_0)\eta] + \eta\sin[(a-y_0)\eta]\}\,d\eta$$

by deforming the path of integration along the real k axis so that it passes over the $k = \pm a_1$ and taking the branch cuts associated with $|k|$ along the imaginary axis from $[-\infty i, 0^- i]$ and $[0^+ i, \infty i]$. [Hint: Treat $|z| = \sqrt{(z - 0^+ i)(z - 0^- i)}$ where the real part of the square root is always positive.]

Papers Using Laplace Transforms
to Solve Partial Differential Equations

Akkaş, N., and F. Erdoğan, 1989: The residual variable method applied to the diffusion equation in cylindrical coordinates. *Acta Mech.*, **79**, 207–219.

Balasubramaniam, T. A., and H. F. Bowman, 1974: Temperature field due to a time dependent heat source of spherical geometry in an infinite medium. *J. Heat Transfer*, **96**, 296–299.

Barakat, R., 1965: Diffraction of a plane step pulse by a perfectly conducting circular cylinder. *J. Opt. Soc. Am.*, **55**, 998–1002.

Barenblatt, G. I., Iu. P. Zheltov and I. N. Kochina, 1960: Basic concepts in the theory of seepage of homogeneous liquids in fissured rocks [strata]. *J. Appl. Math. Mech.*, **24**, 1286–1303.

Baron, M. L., 1961: Response of nonlinearly supported cylindrical boundaries to shock waves. *J. Appl. Mech.*, **28**, 135–136.

Baron, M. L., and R. Parnes, 1962: Displacements and velocities produced by the diffraction of a pressure wave by a cylindrical cavity in an elastic medium. *J. Appl. Mech.*, **29**, 385–395.

Belluigi, A., 1959: Aspetto elettrogeosmotico bidimensionale variabile col tempo. *Geofisica pura e appl.*, **43**, 182–194.

Bhattacharyya, S., 1972: Note on diffraction of waves incident obliquely on a half-plane. *Rev. Roum. Phys.*, **17**, 723–728.

Blackstock, D. T., 1967: Transient solution for sound radiated into a viscous fluid. *J. Acoust. Soc. Am.*, **41**, 1312–1319.

Blackwell, J. H., 1954: A transient-flow method for determination of thermal constants of insulating material in bulk. *J. Appl. Phys.*, **25**, 137–144.

Boerner, W. M., and Y. M. Antar, 1972: Aspects of electromagnetic pulse scattering from a grounded dielectric slab. *Arch. für Electronik Übertragungstechnik*, **26**, 14–21.

Boley, B. A., and C. C. Chao, 1955: Some solutions of the Timoshenko beam equation. *J. Appl. Mech.*, **22**, 579–586.

Carslaw, H. S., and J. C. Jaeger, 1938: Some problems in mathematical theory of the conduction of heat. *Phil. Mag.*, *Ser. 7*, **26**, 473–495.

Carslaw, H. S., and J. C. Jaeger, 1963: *Operational Methods in Applied Mathematics.* New York: Dover Publications, Inc., Chapters 5–10.

Chaudhuri, B. R., 1971: Dynamical problems of thermo-magnetoelasticity for cylindrical regions. *Indian J. Pure Appl. Math.*, **2**, 631–656.

Chen, C.-S., 1987: Analytical solutions for radial dispersion with Cauchy boundary at injection well. *Water Resour. Res.*, **23**, 1217–1224.

Chen, C.-S., and G. D. Woodside, 1988: Analytical solution for aquifer decontamination by pumping. *Water Resour. Res.*, **24**, 1329–1338.

Chen, Y. M., 1964: The transient behavior of diffraction of plane pulse by a circular cylinder. *Int. J. Engng. Sci.*, **2**, 417–429.

Chester W., R. Bobone and E. Brocher, 1984: Transient conduction through a two-layer medium. *Int. J. Heat Mass Transfer*, **27**, 2167–2170.

Ciric, I. R., and J. Ma, 1991: Transient currents on a cylinder excited by a parallel line-current. *Can. J. Phys.*, **69**, 1242–1248.

De Wiest, J. M. R., 1963: Flow to an eccentric well in a leaky circular aquifer with varied lateral replenishment. *Geofisica pura e appl.*, **54**, 87–102.

Dhaliwal, R. S., and K. L. Chowdhury, 1968: Dynamical problems of thermoelasticity for cylindrical regions. *Arch. Mech. Stos.*, **20**, 46–66.

Dillon, O. W., 1967: Coupled thermoelasticity of bars. *J. Appl. Mech.*, **34**, 137–145.

Dimaggio, F. L., 1956: Effect of an acoustic medium on the dynamic buckling of plates. *J. Appl. Mech.*, **23**, 201–206.

Duffy, P., 1989: The acceleration of cometary ions by Alfvén waves. *J. Plasma Phys.*, **42**, 13–25.

Eason, G., 1969: Wave propagation in inhomogeneous elastic media, solution in terms of Bessel functions. *Acta Mech.*, **7**, 137–160.

Eason, G., 1970: The displacement produced in a semi-infinite linear Cosserat continuum by an impulsive force. *Proc. Vibr. Prob.*, **11**, 199–221.

Ehlers, F. E., and T. Strand, 1958: The flow of a supersonic jet in a supersonic stream at an angle of attack. *J. Aero/space Sci.*, **25**, 497–506.

Engevik, L., 1971: A note on a stability problem in hydrodynamics. *Acta Mech.*, **12**, 143–153.

Florence, A. L., 1965: Traveling force on a Timoshenko beam. *J. Appl. Mech.*, **32**, 351–358.

Forrestal, M. J., and W. E. Alzheimer, 1968: Transient motion of a rigid cylinder produced by elastic and acoustic waves. *J. Appl. Mech.*, **55**, 134–138.

Forrestal, M. J., 1968: Response of an elastic cylindrical shell to a transverse, acoustic pulse. *J. Appl. Mech.*, **35**, 614–616.

Forrestal, M. J., 1968: Transient response at the boundary of a cylindrical cavity in an elastic medium. *Int. J. Solids Structures*, **4**, 391–395.

Forrestal, M. J., and M. J. Sagartz, 1971: Radiated pressure in an acoustic medium produced by pulsed cylindrical and spherical shells. *J. Appl. Mech.*, **38**, 1057–1060.

Forrestal, M. J., G. E. Sliter and M. J. Sagartz, 1972: Stresses emanating from the supports of a cylindrical shell produced by a lateral pressure pulse. *J. Appl. Mech.*, **39**, 124–128.

Geers, T. L., 1969: Excitation of an elastic cylindrical shell by a transient acoustic wave. *J. Appl. Mech.*, **36**, 459–469.

Ghosh, A. K., and L. Debnath, 1986: Hydromagnetic Stokes flow in a rotating fluid with suspended small particles. *Appl. Sci. Res.*, **43**, 165–192.

Goldenberg, H., 1952: Heat flow in an infinite medium heated by a sphere. *Brit. J. Appl. Phys.*, **3**, 296–298.

Goldstein, R. J., and D. G. Briggs, 1964: Transient free convection about vertical plates and circular cylinders. *J. Heat Transfer*, **86**, 490–500.

Gonsovskii, V. L., S. I. Meshkov and Yu. A. Rossikhin, 1972: Impact of a viscoelastic rod onto a rigid target. *Sov. Appl. Mech.*, **8**, 1109–1113.

Gonsovskii, V. L., and Yu. A. Rossikhin, 1973: Stress waves in a viscoelastic medium with a singular hereditary kernel. *J. Appl. Mech. Tech. Phys.*, **14**, 595–597.

Hantush, M. S., 1959: Nonsteady flow to flowing wells in leaky aquifers. *J. Geophys. Res.*, **64**, 1043–1052.

Holt, M., and S. L. Strack, 1961: Supersonic panel flutter of a cylindrical shell of finite length. *J. Aerospace Sci.*, **28**, 197–208.

Iben, H., 1974: Ebene, kreissymmetrische Rand-Anfangswertprobleme zäher Flüssigkeiten. *Zeit. angew. Math. Mech.*, **54**, 213–224.

Jaeger, J. C., 1941: Heat conduction in composite circular cylinders. *Phil. Mag., Ser. 7*, **32**, 324–335.

Jaeger, J. C., 1944: Some problems involving line sources in conduction of heat. *Phil. Mag., Ser. 7*, **35**, 169–179.

Johri, A. K., and R. K. Singhal, 1980: Unsteady flow of visco-elastic fluid. *Indian J. Theoret. Phys.*, **28**, 197–203.

Kildal, A., 1970: On the motion generated by a plate vibrating in a stratified fluid. *Acta Mech.*, **9**, 78–104.

Koizamu, T., 1970: Transient thermal stresses in a semi-infinite body heated axi-symmetrically. *Zeit. angew. Math. Mech.*, **50**, 747–757.

Kramer, B. M., and N. P. Suh, 1973: Plane strain pulse propagation in a semi-infinite viscoelastic Maxwell solid. *J. Sound Vib.*, **29**, 435–442.

Kuznetsov, M. M., 1975: Unsteady-state slip of a gas near an infinite plane with diffusion-mirror reflection of molecules. *J. Appl. Mech. Tech. Phys.*, **16**, 853–858.

Lou, Y. K., and J. M. Klosner, 1973: Transient response of a point-excited submerged spherical shell. *J. Appl. Mech.*, **40**, 1078–1084.

Lubliner, J., and V. P. Panoskaltsis, 1992: The modified Kuhn model of linear viscoelasticity. *Int. J. Solids Structures*, **29**, 3099–3112.

Ma, J., and I. R. Ciric, 1992: Transient response of a circular cylinder to an electromagnetic pulse. *Radio Sci.*, **27**, 561–567.

Mæland, E., 1983: On the response of a wind-driven current over a continental shelf. *J. Geophys. Res.*, **88**, 4534–4538.

Majumdar, S. R., 1969: On the motion of an infinite circular cylinder in a rotating viscous liquid. *Rev. Roum. Sci. Tech. - Méc. Appl.*, **14**, 701–719.

Mallick, D. D., 1957: Nonuniform rotation of an infinite circular cylinder in an infinite viscous liquid. *Zeit. angew. Math. Mech.*, **37**, 385–392.

Mareschal, J. C., and A. F. Gangi, 1977: A linear approximation to the solution of a one-dimensional Stefan problem and its geophysical implications. *Geophys. J. R. Astr. Soc.*, **49**, 443–458.

Matsumoto, H., E. Tsuchida, S. Miyao and N. Tsunadu, 1976: Torsional stress wave propagation in a semi-infinite conical bar. *Bull. JSME*, **19**, 8–14.

McLachlan, N. W., and A. T. McKay, 1936: Transient oscillations in a loud-speaker horn. *Proc. Camb. Phil. Soc.*, **32**, 265–275.

McLachlan, N. W., 1939: *Complex Variables & Operational Calculus with Technical Applications.* Cambridge: At the University Press, Chapters 13–15.

Miklowitz, J., 1960: Plane-stress unloading waves emanating from a suddenly punched hole in a stretched elastic plate. *J. Appl. Mech.*, **27**, 165–171.

Miklowitz, J., 1960: Flexural stress waves in an infinite elastic plate due to a suddenly applied concentrated transverse load. *J. Appl. Mech.*, **27**, 681–689.

Miles, J. W., 1951: On virtual mass and transient motion in subsonic compressible flow. *Quart. J. Mech. Appl. Math.*, **4**, 388–400.

Minster, J. B., 1978: Transient and impulsive responses of a one-dimensional linearly attenuating medium – I. Analytical results. *Geophys. J. R. Astr. Soc.*, **52**, 479–501.

Misiak, P., A. Papliński and E. Włodarczyk, 1975: Oblique stress bi-waves generated in elastic medium by a moving axisymmetrical concentrated load. *J. Tech. Phys.*, **16**, 303–314.

Mittal, M. L., 1963: Unsteady hydrodynamic viscous flow in an annular channel. *Appl. Sci. Res.*, **B10**, 86–90.

Moshinskii, A. I., 1987: Effective diffusion of a dynamically passive impurity in narrow channels. *J. Appl. Mech. Tech. Phys.*, **28**, 374–382.

Nanda, R. S., 1960: Unsteady circulatory flow about a circular cylinder with suction. *Appl. Sci. Res.*, **A9**, 85–92.

Odaka, T., and I. Nakahara, 1967: Stresses in an infinite beam impacted by an elastic bar. *Bull. JSME*, **10**, 863–872.

Pack, D. C., 1956: The oscillations of a supersonic gas jet embedded in a supersonic stream. *J. Aeronaut. Sci.*, **23**, 747–753.

Pal, S. C., M. L. Ghosh and P. K. Chowdhuri, 1985: Spectral representation of a certain class of self-adjoint differential operators and its application to axisymmetric boundary value problems in electrodynamics. *J. Tech. Phys.*, **26**, 97–115.

Papadopulos, I. S., and H. H. Cooper, 1967: Drawdown in a well of large diameter. *Water Resour. Res.*, **3**, 241–244.

Parnes, R., 1969: Response of an infinite elastic medium to traveling loads in a cylindrical bore. *J. Appl. Mech.*, **36**, 51–58.

Parnes, R., 1980: Progressing torsional loads along a bore in an elastic medium. *Int. J. Solids Structures*, **16**, 653–670.

Parnes, R., and L. Banks-Sills, 1983: Transient response of an elastic medium to torsional loads on a cylindrical cavity. *J. Appl. Mech.*, **50**, 397–404.

Puri, P., and P. K. Kythe, 1988: Some inverse Laplace transforms of exponential form. *Zeit. angew. Math. Phys.*, **39**, 150–156.

Raichenko, L. M., 1973: The problem of the influx of liquid to a perfect well in a layer of fissured-porous rock in the presence of an end-face zone. *Sov. Appl. Mech.*, **9**, 1225–1228.

Raichenko, L. M., 1976: Flow of liquid to an incomplete well in a bed of fissile-porous rocks. *Sov. Appl. Mech.*, **12**, 1196–1200.

Rushchitskii, Ya. Ya., and B. B. Érgashev, 1987: Nonsteady effects in a massive sample of fiber composite under the short-term action of a harmonic pulse. *Sov. Appl. Mech.*, **23**, 1177–1181.

Sagartz, M. J., and M. J. Forrestal, 1969: Transient stresses at a clamped support of a circular cylindrical shell. *J. Appl. Mech.*, **36**, 367–369.

Schubert, G. U., 1950: Über eine in der Theorie der elektrischen Schmelz-sicherungen auftretende Lösung der Wärmeleitungsgleichung. *Zeit. angew. Phys.*, **2**, 174–179.

Schwarz, W. H., 1963: The unsteady motion of an infinite oscillating cylinder in an incompressible Newtonian fluid at rest. *Appl. Sci. Res.*, **A11**, 115–124.

Selberg, H. L., 1952: Transient compression waves from spherical and cylindrical cavities. *Ark. för Fys.*, **5**, 97–108.

Sennitskii, V. L., 1981: Unsteady rotation of a cylinder in a viscous fluid. *J. Appl. Phys. Tech. Phys.*, **21**, 347–349.

Seth, S. S., 1977: Unsteady motion of viscoelastic fluid due to rotation of a sphere. *Indian J. Pure Appl. Math.*, **8**, 302–308.

Shibuya, T., I. Nakahara, T. Koizumi and K. Kaibara, 1975: Impact stress analysis of a semi-infinite plate by the finite difference method. *Bull. JSME*, **18**, 649–655.

Shvidler, M. I., 1965: Sorption in a plane-radial filtration flow. *J. Appl. Mech. Tech. Phys.*, **6(3)**, 77–79.

Smith, F. B., 1957: The diffusion of smoke from a continuous elevated point-source into a turbulent atmosphere. *J. Fluid Mech.*, **2**, 49–76.

Standing, R. G., 1971: The Rayleigh problem for a slightly diffusive density-stratified fluid. *J. Fluid Mech.*, **48**, 673–688.

Tanaka, K., and T. Kurokawa, 1972: Thermoelastic effect caused by longitudinal collision of bars. *Bull. JSME*, **15**, 816–821.

Tanner, R. I., 1962: Note on the Rayleigh problem for a visco-elastic fluid. *Zeit. angew. Math. Phys.*, **13**, 573–580.

Thomson, W. T., 1950: *Laplace Transforms*. Englewood Cliffs, NJ: Prentice-Hall, Inc., Chapter 8.

Tsai, Y. M., 1971: Dynamic contact stresses produced by the impact of an axisymmetrical projectile on an elastic half-space. *Int. J. Solids Structures*, **7**, 543–558.

Uflyand, Ya. S., 1948: The propagation of waves in the transverse vibration of bars and plates (in Russian). *Prikl. Mat. Mek.*, **12**, 287–300.

Vimala, C. S., and G. Nath, 1976: Unsteady motion of a slightly rarefied gas about an infinite circular cylinder. *Rev. Roum. Sci. Tech. - Méc. Appl.*, **21**, 219–227.

Walker, J. S., 1978: Solitary fluid transients in rectangular ducts with transverse magnetic fields. *Zeit. angew. Math. Phys.*, **29**, 35–53.

Watson, K., 1973: Periodic heating of a layer over a semi-infinite solid. *J. Geophys. Res.*, **78**, 5904–5910.

Wikramaratna, R. S., 1984: An analytical solution for the effects of abstraction from a multi-layered confined aquifer with no cross flow. *Water Resour. Res.*, **20**, 1067–1074.

Wilks, G., 1969: The flow around a semi-infinite oscillating plate and the skin friction on arbitrarily cross-sectioned infinite cylinders oscillating parallel to their length. *Proc. Camb. Phil. Soc.*, **66**, 163–187.

Wilms, E. V., 1966: Temperature induced in a medium due to a suddenly applied pressure inside a spherical cavity. *J. Appl. Mech.*, **33**, 941–943.

Widnall, S. E., and E. H. Dowell, 1967: Aerodynamic forces on an oscillating cylindrical duct with an internal flow. *J. Sound Vib.*, **6**, 71–85.

Yamashita, T., 1977: Dependence of source time function on tectonic field. *J. Phys. Earth*, **25**, 419–445.

Yang, H. T., and J. V. Healy, 1973: The Stokes problems for a conducting fluid with a suspension of particles. *Appl. Sci. Res.*, **27**, 387–397.

Yates, S. R., 1990: An analytical solution for one-dimensional transport in heterogeneous porous media. *Water Resour. Res.*, **26**, 2331–2338.

Valathur, M., 1972: Wave propagation in a truncated conical shell. *Int. J. Solids Structures*, **8**, 1223–1233.

Volkov, F. G., and A. M. Golovin, 1970: Thermal and diffusive relaxation of an evaporating drop with internal heat liberation. *J. Appl. Mech. Tech. Phys.*, **11**, 76–85.

Zaitsev, A. S., 1974: Dynamics of a rod with an elastic shock absorber. *Sov. Appl. Mech.*, **10**, 1003–1008.

Zheltov, Yu. P., and V. S. Kutlyarov, 1965: Transient motion of a liquid in a fissured porous stratum subject to periodic pressure variation at the boundary. *J. Appl. Mech. Tech. Phys.*, **6(6)**, 69–76.

Papers Using Fourier Transforms
to Solve Partial Differential Equations

Azpeitia, A. G., and G. F. Newell, 1958: Theory of oscillatory type viscometers. III: A thin disk. *Zeit. angew. Math. Phys.*, **9a**, 97–118.

Baron, M. L., and A. T. Matthews, 1961: Diffraction of a pressure wave by a cylindrical cavity in an elastic medium. *J. Appl. Mech.*, **28**, 347–354.

Bhutani, O. M., and K. D. Nanda, 1968: A general theory of thin airfoils in nonequilibrium magnetogasdynamics. Part I: Aligned magnetic field. *AIAA J.*, **6**, 1757–1762.

Carrier, G. F., and R. C. DiPrima, 1957: On the unsteady motion of a viscous fluid past a semi-infinite flat plate. *J. Math. and Phys.*, **35**, 359–383.

Chian, C. T., and F. C. Moon, 1981: Magnetically induced cylindrical stress waves in a thermoelastic conductor. *Int. J. Solids Structures*, **17**, 1021–1035.

Chowdhury, K. L., and P. G. Glockner, 1980: On a boundary value problem for an elastic dielectric half-plane. *Acta Mech.*, **37**, 65–74.

Goldstein, M. E., 1975: Cascade with subsonic leading-edge locus. *AIAA J.*, **13**, 1117–1119.

Goswami, S. K., 1982: A note on the problem of scattering of surface waves by a submerged fixed vertical barrier. *Zeit. angew. Math. Mech.*, **62**, 637–639.

Haren, P., and C. C. Mei, 1981: Head-sea diffraction by a slender raft with application to wave-power absorption. *J. Fluid Mech.*, **104**, 505–526.

Karasudhi, P., L. M. Keer and S. L. Lee, 1968: Vibration motion of a body on an elastic half space. *J. Appl. Mech.*, **35**, 697–705.

Lee, T. J., 1980/81: Transient electromagnetic response of a sphere in a layered medium. *PAGEOPH*, **119**, 309–338.

Menkes, J., 1972: The propagation of sound in the ionosphere. *J. Sound Vib.*, **20**, 311–319.

Nagaya, K., and Y. Hirano, 1976: Responses of an infinite medium with cavities to an impact load at one of the cavities. *Bull. JSME*, **19**, 1430–1434.

Nakano, H., 1925: On Rayleigh waves. *Jap. J. Astron. Geophys.*, **2**, 233–326.

Norwood, F. R., and J. Miklowitz, 1967: Diffraction of transient elastic waves by a spherical cavity. *J. Appl. Mech.*, **34**, 735–744.

Papliński, A., and E. Włodarczyk, 1977: Propagation in acoustic medium of two-dimensional, cylindrical pressure waves excited by a moving load. *J. Tech. Phys.*, **18**, 81–97.

Papliński, A., and E. Włodarczyk, 1980: Response of elastic medium to a traveling line load applied in a cylindrical bore. *J. Tech. Phys.*, **21**,

313–335.

Rudnicki, J. W., and E. A. Roeloffs, 1990: Plane-strain shear dislocations moving steadily in linear elastic diffusive solids. *J. Appl. Mech.*, **57**, 32–39.

Savage, M. D., 1967: Stationary waves at a plasma-magnetic field interface. *J. Plasma Phys.*, **1**, 229–239.

Sezawa, K., and K. Kanai, 1932: Possibility of free oscillations of strata excited by seismic waves. Part III. *Bull. Earthq. Res. Inst.*, **10**, 1–19.

Singh, S. K., 1973: Electromagnetic transient response of a conducting sphere embedded in a conductive medium. *Geophysics*, **38**, 864–893.

Skalak, R., and M. B. Friedman, 1958: Reflection of an acoustic step wave from an elastic cylinder. *J. Appl. Mech.*, **25**, 103–108.

Weaver, A. J., L. A. Mysak and A. F. Bennett, 1988: The steady state response of the atmosphere to midlatitude heating with various zonal structures. *Geophys. Astrophys. Fluid Dynamics*, **41**, 1–44.

Williams, M. M. R., 1965: Neutron transport in differentially heated media. *Brit. J. Appl. Phys.*, **16**, 1727–1732.

Wimp, J., C. Rorres and R. F. Wayland, 1992: Acoustic impulse responses for nonuniform media. *J. Comput. Appl. Math.*, **42**, 89–107.

Chapter 4

The Joint Transform Method

In the previous two chapters, we have shown that we can use Laplace and Fourier transforms separately to solve partial differential equations. In section 4.1 we shall show how we can apply them together to the same end. In general, this requires a double inversion of the transformed version of the solution–often a formidable task. However, over the last sixty years, several analytic techniques, in particular, the Cagniard-de Hoop method, have been developed for the joint inversion problem. We will explore these methods in sections 4.2–4.3.

4.1 THE SOLUTION OF PARTIAL DIFFERENTIAL EQUATIONS USING THE JOINT TRANSFORM METHOD

A popular method for solving linear partial differential equations that have an infinite or semi-infinite spatial domain and specified initial conditions is the joint transform method. The general procedure is as follows: We use the Laplace transform to eliminate the temporal dependence while we apply a Fourier or Hankel transform in the spatial dimension. This results in an algebraic or ordinary differential equation which we solve to obtain the joint transform. We then compute the inverses. Whether we invert the Laplace or the spatial transform first is usually dictated by the nature of the joint transform.

Transform Methods for Solving Partial Differential Equations

To illustrate this technique, we use it to find the shallow-water gravity waves excited on an infinite, one-dimensional, flat earth. Here we neglect variations of the Coriolis parameter with latitude. The nondimensional x-momentum, y-momentum and continuity equations are

$$\frac{\partial u}{\partial t} + \frac{\partial h}{\partial x} - v = 0 \tag{4.1.1}$$

$$\frac{\partial v}{\partial t} + u = 0 \tag{4.1.2}$$

$$\frac{\partial h}{\partial t} + c^2 \frac{\partial u}{\partial x} = 0 \tag{4.1.3}$$

where c is the nondimensional phase speed of the shallow water waves.

To solve (4.1.1)–(4.1.3), we assume that a Fourier transform exists for each of the dependent variables. For example,

$$u(x,t) = \frac{1}{2\pi} \int_{-\infty}^{\infty} U(k,t) e^{ikx} dk \tag{4.1.4}$$

with similar expressions for $v(x,t)$ and $h(x,t)$. If all of the perturbations vanish at infinity, (4.1.1)–(4.1.3) reduce to the set of ordinary differential equations:

$$\frac{dU}{dt} + ik\Theta - V = 0 \tag{4.1.5}$$

$$\frac{dV}{dt} + U = 0 \tag{4.1.6}$$

$$\frac{d\Theta}{dt} + ikc^2 U = 0. \tag{4.1.7}$$

To solve the system (4.1.5)–(4.1.7) we take the Laplace transform of (4.1.5)–(4.1.7) by defining

$$\overline{U}(k,s) = \int_{0}^{\infty} U(k,t) e^{-st} dt \tag{4.1.8}$$

with similar expressions for $\overline{V}(k,s)$ and $\overline{\Theta}(k,s)$.

At this point we must specify the initial conditions. In this problem we choose to find the solution if the initial height field is $\theta(x,0) = H(x+a) - H(x-a)$ where $H(\ )$ is the Heaviside step function. Taking the Laplace transform of (4.1.5)–(4.1.7),

$$s\overline{U} + ik\overline{\Theta} - \overline{V} = 0 \tag{4.1.9}$$

$$s\overline{V} + \overline{U} = 0 \tag{4.1.10}$$

$$s\overline{\Theta} + ikc^2 \overline{U} = \Theta(k,0). \tag{4.1.11}$$

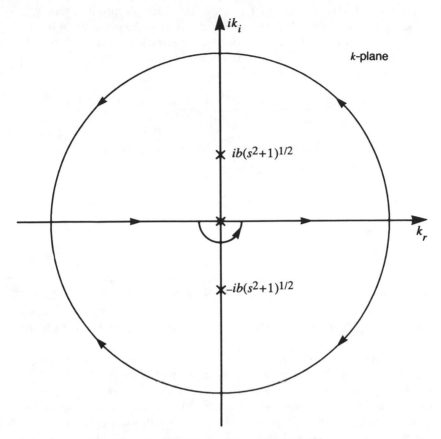

Fig. 4.1.1: Contours used in the inversion of (4.1.15).

Because $\Theta(k,0) = \sin(ka)/k$, we obtain the following transformed solution:

$$\overline{U} = \frac{-ik}{c^2 k^2 + s^2 + 1} \frac{\sin(ka)}{k} \tag{4.1.12}$$

$$\overline{V} = \frac{i}{c^2 k^2 + s^2 + 1} \frac{\sin(ka)}{s} \tag{4.1.13}$$

$$\overline{\Theta} = \frac{s^2 + 1}{c^2 k^2 + s^2 + 1} \frac{\sin(ka)}{ks}. \tag{4.1.14}$$

Because the transform $\overline{V}$ is the most involved, we present a detailed analysis for this case; the remaining transforms follow by analogy.

If we rewrite $\sin(ka)$ in terms of complex exponentials, then the inverse Fourier transform is

$$\overline{v}(x,s) = \frac{b^2}{2\pi s} \int_{-\infty}^{\infty} \frac{e^{ik(x+a)} - e^{-ik(x-a)}}{k^2 + b^2 s^2 + b^2} \, dk \tag{4.1.15}$$

Transform Methods for Solving Partial Differential Equations

where $b = 1/c$. To evaluate (4.1.15), we introduce an infinite semicircle in the upper or lower half-plane (see Fig. 4.1.1) as dictated by Jordan's lemma and evaluate the closed contour by Cauchy's residue theorem. The integrand possesses simple poles at $k = \pm ib(s^2 + 1)^{1/2}$ and an application of the residue theorem yields

$$\overline{v}(x, s) = \frac{b}{2} \frac{\exp[-b|x + a|(1 + s^2)^{1/2}] - \exp[-b|x - a|(1 + s^2)^{1/2}]}{s(s^2 + 1)^{1/2}}.$$

(4.1.16)

We invert the Laplace transform $\overline{v}(x, s)$ by applying Bromwich's integral on the complex plane, i.e.,

$$v(x, t) = \frac{b}{4\pi i} \int_C \frac{1}{s(s^2 + 1)^{1/2}} \, ds$$
$$\times \left\{ \exp[st - b|x + a|(1 + s^2)^{1/2}] - \exp[st - b|x - a|(1 + s^2)^{1/2}] \right\}$$

(4.1.17)

where the contour C runs to the right of any singularities. Treating each term in (4.1.17) separately, we convert each line integral into a closed contour and then use the residue theorem. Because the analysis is essentially identical for both terms, we shall only give the details for the first term.

For the first term, Jordan's lemma requires that we close the contour with an infinite semicircle in the right half-plane for $t < b|x + a|$. Because there are no singularities within the contour, the integral vanishes as we would expect from causality. When $t > b|x + a|$, we close the contour with an infinite semicircle in the left half-plane. However, we must be careful because the integrand is multivalued with branch points at $s = \pm i$. There are several possible choices and Fig. 4.1.2 shows the one that we shall use.

Because the argument of the square root must be positive for large positive s, we define the amplitude and phase as follows along each contour:

$$C_1: \quad s = \rho e^{-\pi i/2}, \quad s - i = (\rho + 1)e^{-\pi i/2}, \quad s + i = (\rho - 1)e^{-\pi i/2};$$

(4.1.18)

$$C_2: \quad s = \rho e^{-i\pi/2}, \quad s - i = (\rho + 1)e^{-\pi i/2}, \quad s + i = (\rho - 1)e^{3\pi i/2};$$

(4.1.19)

$$C_3: \quad s = \rho e^{\pi i/2}, \quad s - i = (\rho - 1)e^{-3\pi i/2}, \quad s + i = (\rho + 1)e^{\pi i/2};$$

(4.1.20)

and

$$C_4: \quad s = \rho e^{\pi i/2}, \quad s - i = (\rho - 1)e^{\pi i/2}, \quad s + i = (\rho + 1)e^{\pi i/2}$$

(4.1.21)

where $1 \leq \rho \leq \infty$. The contribution from the arcs AB, CD and EF at infinity vanish.

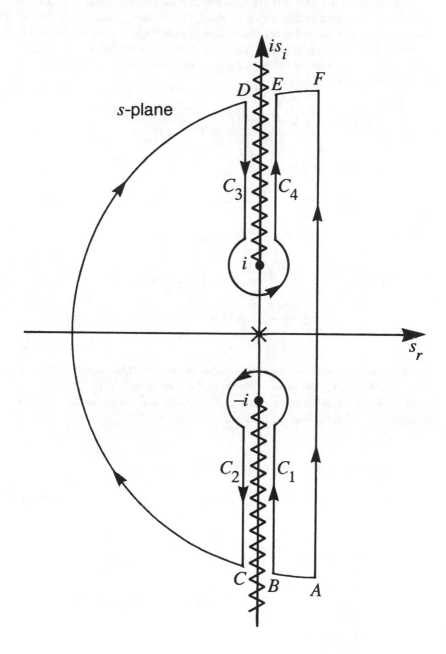

Fig. 4.1.2: The contours used in the inversion of (4.1.17).

Transform Methods for Solving Partial Differential Equations

In addition to line integrals, contributions come from the simple pole at $s = 0$, and integrations around the branch points at $s = \pm i$. We compute the contribution from the simple pole $s = 0$ in the usual manner. For the branch points we introduce the variable $s = \pm i + \epsilon e^{\theta i}$ and perform an integration around the infinitesimally small circles at the branch points as $\epsilon \to 0$.

Thus,

$$\frac{b}{4\pi i} \int_C \frac{\exp[st - b|x + a|(s^2 + 1)^{1/2}]}{s(s^2 + 1)^{1/2}} \, ds$$

$$= \frac{b}{2} \exp[-b|x + a|]$$

$$+ \frac{b}{4\pi i} \int_\infty^1 \frac{e^{-i\rho t}(i \, d\rho)}{(-i\rho)[-i(\rho^2 - 1)^{1/2}]} \exp[-b|x + a|(-i)(\rho^2 - 1)^{1/2}]$$

$$+ \frac{b}{4\pi i} \int_1^\infty \frac{e^{-i\rho t}(-i \, d\rho)}{(-i\rho)[i(\rho^2 - 1)^{1/2}]} \exp[-b|x + a|i(\rho^2 - 1)^{1/2}]$$

$$+ \frac{b}{4\pi i} \int_\infty^1 \frac{e^{i\rho t} i \, d\rho}{i\rho[-i(\rho^2 - 1)^{1/2}]} \exp[-b|x + a|(-i)(\rho^2 - 1)^{1/2}]$$

$$+ \frac{b}{4\pi i} \int_1^\infty \frac{e^{i\rho t} i \, d\rho}{i\rho[i(\rho^2 - 1)^{1/2}]} \exp[-b|x + a|i(\rho^2 - 1)^{1/2}].$$

$$(4.1.22)$$

In this case, the contribution from the branch points vanish although this is not always the case. We may further simplify (4.1.22) by introducing the transformation $\eta^2 = \rho^2 - 1$. When we carry out similar analyses for (4.1.12) and (4.1.14), we find that

$$u(x, t) = \frac{b}{2} J_0\left\{[t^2 - b^2|x - a|^2]^{1/2}\right\} H(t - b|x - a|)$$

$$- \frac{b}{2} J_0\left\{[t^2 - b^2|x + a|^2]^{1/2}\right\} H(t - b|x + a|), \qquad (4.1.23)$$

$$v(x, t) = \left\{ \frac{b}{2} e^{-b|x+a|} \right.$$

$$- \frac{b}{\pi} \int_0^\infty \frac{\cos[t(1 + \eta^2)^{1/2}]}{1 + \eta^2} \cos[b|x + a|\eta] \, d\eta \left. \right\} H(t - b|x + a|)$$

$$- \left\{ \frac{b}{2} e^{-b|x-a|} \right.$$

$$- \frac{b}{\pi} \int_0^\infty \frac{\cos[t(1 + \eta^2)^{1/2}]}{1 + \eta^2} \cos[b|x - a|\eta] \, d\eta \left. \right\} H(t - b|x - a|)$$

$$(4.1.24)$$

and

$$
\begin{aligned}
h(x,t) = &\, H(a - |x|) \\
&- \text{sgn}(x + a)H(t - b|x + a|) \\
&\times \left\{ \frac{1}{2}e^{-b|x+a|} - \frac{1}{\pi}\int_0^\infty \frac{\eta\cos[t(1+\eta^2)^{1/2}]}{1+\eta^2}\sin[b|x + a|\eta]\,d\eta \right\} \\
&+ \text{sgn}(x - a)H(t - b|x - a|) \\
&\times \left\{ \frac{1}{2}e^{-b|x-a|} - \frac{1}{\pi}\int_0^\infty \frac{\eta\cos[t(1+\eta^2)^{1/2}]}{1+\eta^2}\sin[b|x - a|\eta]\,d\eta \right\}
\end{aligned}
$$

(4.1.25)

where $J_0(\ )$ is the Bessel function of the first kind of order zero and sgn$(\ )$ is the sign operator.

When there is axial symmetry, we replace the Fourier transform by the Hankel transform. To illustrate this, let us find the response[1] $u(r, z, t)$ of a quiescent, compressible ocean when we force it impulsively at $r = 0$ and $z = d$. The corresponding governing equation is

$$
\frac{1}{V_0^2}\frac{\partial^2 u}{\partial t^2} - \frac{\partial^2 u}{\partial z^2} - \frac{\partial^2 u}{\partial r^2} - \frac{1}{r}\frac{\partial u}{\partial r} = 4\pi\delta(r)\delta(z - d)H(t) \quad (4.1.26)
$$

where V_0 is the (nondimensional) speed of sound in the ocean. The boundary condition at the free surface is

$$
\frac{\partial^2 u(r,0,t)}{\partial t^2} = \frac{\partial u(r,0,t)}{\partial z} \quad (4.1.27)
$$

while at the bottom

$$
\frac{\partial u(r,1,t)}{\partial z} = 0. \quad (4.1.28)
$$

We solve (4.1.26)–(4.1.28) by a joint application of Laplace and Hankel transforms

$$
u(r,z,t) = \frac{1}{2\pi i}\int_{c-\infty i}^{c+\infty i} e^{st}\int_0^\infty \overline{U}(k,z,s)J_0(kr)\,k\,dk\,ds. \quad (4.1.29)
$$

The application of these transforms yields the system of equations

$$
\frac{d^2\overline{U}}{dz^2} - m^2\overline{U} = 0, \qquad 0 \le z \le 1 \quad (4.1.30)
$$

---

[1] Taken from Duffy, D. G., 1992: On the generation of oceanic surface waves by underwater volcanic explosions. *J. Volcanol. Geotherm. Res.*, **50**, 323–344.

$$s^2 \overline{U}(r, 0, s) = \frac{d\overline{U}(k, 0, s)}{dz} \qquad (4.1.31)$$

$$\overline{U}(k, d^+, s) = \overline{U}(k, d^-, s) \qquad (4.1.32)$$

$$\frac{d\overline{U}(k, d^+, s)}{dz} - \frac{d\overline{U}(k, d^-, s)}{dz} = -\frac{2}{s} \qquad (4.1.33)$$

$$\frac{d\overline{U}(k, 1, s)}{dz} = 0 \qquad (4.1.34)$$

where $m^2 = k^2 + s^2/V_0^2$ and d^+ and d^- are points slightly greater than and less than $z = d$, respectively. We obtain (4.1.33) by integrating the transformed form of (4.1.26) across a narrow strip between d^+ and d^-.

The solution which satisfies (4.1.30) along with (4.1.31)–(4.1. 34) is

$$\overline{U}(k, z, s) = \frac{2 \cosh[m(1 - d)] \cosh(mz)}{s[s^2 \cosh(m) + m \sinh(m)]}$$
$$+ \frac{2s \cosh[m(1 - d)] \sinh(mz)}{m[s^2 \cosh(m) + m \sinh(m)]} \qquad (4.1.35)$$

for $0 \leq z \leq d$ and

$$\overline{U}(k, z, s) = \frac{2m \cosh(md) + 2s^2 \sinh(md)}{sm[s^2 \cosh(m) + m \sinh(m)]} \cosh[m(1 - z)] \qquad (4.1.36)$$

for $d \leq z \leq 1$. Because $m = \sqrt{k^2 + s^2/V_0^2}$, it appears that we must deal with a branch cut to invert (4.1.35)–(4.1.36). However, when we expand the hyperbolic functions in (4.1.35)–(4.1.36), the transforms are purely even functions of m and thus no branch points or cuts exist. Unlike our previous example, we first invert the Laplace transform by converting the line integral into a closed contour by introducing an infinite semi-circle in the left side of the complex s-plane. By Jordan's lemma the contribution from the semicircle at infinity is zero. For this reason the inversion follows as a straightforward application of the residue theorem. This transform has singularities at $s = 0$ and

$$m \tanh(m) + s^2 = 0. \qquad (4.1.37)$$

The solutions of this transcendental equation (4.1.37) lie along the imaginary axis. Although we must compute the precise solutions numerically, an asymptotic expansion in powers of the small quantity $1/V_0^2$ gives

$$s_0 = i\omega_0 = i\sqrt{k \tanh(k)} \left\{ 1 - \frac{1}{4V_0^2} \left[\text{sech}^2(k) + \frac{\tanh(k)}{k} \right] + \cdots \right\} \qquad (4.1.38)$$

$$s_n = i\omega_n$$

$$= iV_0\sqrt{(2n-1)^2\pi^2/4 + k^2}\left\{1 + \frac{(2n-1)^2\pi^2/4}{V_0^2[(2n-1)^2\pi^2/4 + k^2]^2}\right.$$
$$\left. + \frac{4\pi^2(2n-1)^2[12k^2 - 7(2n-1)^2\pi^2]}{V_0^4[(2n-1)^2\pi^2/4 + k^2]^4} + \cdots\right\},$$

$$(4.1.39)$$

where $n = 1, 2, 3, \ldots$ We observe that (4.1.38) is the external surface gravity wave with a correction due to the compressibility of the water. The solutions expressed by (4.1.39) correspond to sound waves within the ocean.

Using these poles we can invert the Laplace transform and then the Hankel transform. The solution $u(r, z, t)$ is

$$u(r, z, t) = \int_0^\infty J_0(kr) k\, dk$$

$$\times \left\{ \frac{\cosh[k(1-z-d)] + \cosh[k(1-|z-d|)]}{k\sinh(k)}\right.$$

$$- 2V_0^2 \frac{m_0\{\cosh[m_0(1-z-d)] + \cosh[m_0(1+z-d)]\}\cos(\omega_0 t)}{\omega_0^2\{m_0(1+2V_0^2)\cosh(m_0) + (1-\omega_0^2)\sinh(m_0)\}}$$

$$- 2V_0^2 \frac{\{\sinh[m_0(1-z-d)] - \sinh[m_0(1-|z-d|)]\}\cos(\omega_0 t)}{m_0(1+2V_0^2)\cosh(m_0) + (1-\omega_0^2)\sinh(m_0)}$$

$$- 2V_0^2 \sum_{n=1}^\infty \frac{m_n\{\cos[m_n(1-z-d)] + \cos[m_n(1+z-d)]\}\cos(\omega_n t)}{\omega_n^2\{m_n(1+2V_0^2)\cos(m_n) + (1-\omega_n^2)\sin(m_n)\}}$$

$$\left.- 2V_0^2 \sum_{n=1}^\infty \frac{\{\sin[m_n(1-z-d)] - \sin[m_n(1-|z-d|)]\}\cos(\omega_n t)}{m_n(1+2V_0^2)\cos(m_n) + (1-\omega_n^2)\sin(m_n)}\right\}$$

$$(4.1.40)$$

where $m_0 = \sqrt{k^2 - \omega_0^2/V_0^2}$ and $m_n = \sqrt{\omega_n^2/V_0^2 - k^2}$. We can compute values of $u(r, z, t)$ by numerical quadrature.

For our next example[2] with Hankel transforms, let us solve the wave equation in cylindrical coordinates

$$\frac{\partial^2 u}{\partial r^2} + \frac{1}{r}\frac{\partial u}{\partial r} + \frac{\partial^2 u}{\partial z^2} - \frac{u}{r^2} = \frac{1}{c^2}\frac{\partial^2 u}{\partial t^2} \qquad (4.1.41)$$

for $0 < r < \infty$, $0 < z < \infty$ and $t > 0$. Assuming that the system is initially at rest, $u(r, z, 0) = 0$ and $u_t(r, z, 0) = 0$, we have the following boundary conditions:

$$|u(0, z, t)| < \infty, \quad \lim_{r\to\infty} u(r, z, t) \to 0, \quad \lim_{z\to\infty} u(r, z, t) \to 0 \quad (4.1.42)$$

---

[2] Taken from Sarkar, G., 1966: Wave motion due to impulsive twist on the surface. *PAGEOPH*, **65**, 43–47. Published by Birkhäuser Verlag, Basel, Switzerland.

and

$$\mu\frac{\partial u(r,0,t)}{\partial z} = \begin{cases} 0 & \text{if } 0 < r < a, \\ Pa\delta(t)/[r\sqrt{r^2 - a^2}] & \text{if } r > a. \end{cases} \tag{4.1.43}$$

We begin by defining the Laplace transform

$$\overline{u}(r,z,s) = \int_0^\infty u(r,z,t)e^{-st}\,dt \tag{4.1.44}$$

so that (4.1.41) becomes

$$\frac{\partial^2 \overline{u}}{\partial r^2} + \frac{1}{r}\frac{\partial \overline{u}}{\partial r} + \frac{\partial^2 \overline{u}}{\partial z^2} - \frac{\overline{u}}{r^2} - \frac{s^2}{c^2}\overline{u} = 0. \tag{4.1.45}$$

Next, we introduce the Hankel transform

$$\overline{U}(k,z,t) = \int_0^\infty \overline{u}(r,z,s)J_1(kr)r\,dr \tag{4.1.46}$$

so that (4.1.45) becomes

$$\frac{d^2\overline{U}}{dz^2} - \left(k^2 + \frac{s^2}{c^2}\right)\overline{U} = 0 \tag{4.1.47}$$

which has the solution

$$\overline{U}(k,z,s) = A(k,s)\exp[-z\sqrt{k^2 + (s/c)^2}]. \tag{4.1.48}$$

We have assumed that $\text{Re}[\sqrt{k^2 + (s/c)^2}] \geq 0$ and also discarded the exponentially growing solution.

To compute $A(k,s)$, we take the Hankel and Laplace transforms of the boundary condition (4.1.43) and find that

$$\mu\frac{d\overline{U}(k,0,s)}{dz} = P\frac{\sin(ka)}{k} \tag{4.1.49}$$

so that

$$A(k,s) = -\frac{P}{\mu}\frac{\sin(ka)}{\sqrt{k^2 + (s/c)^2}}. \tag{4.1.50}$$

Therefore, taking the inverse of the Hankel transform,

$$\overline{u}(r,z,s) = -\frac{P}{\mu}\int_0^\infty \frac{\sin(ka)\exp[-z\sqrt{k^2 + (s/c)^2}]}{\sqrt{k^2 + (s/c)^2}}J_1(kr)\,dk. \tag{4.1.51}$$

Consider now the integral

$$\int_{C_1} H_1^{(1)}(\zeta r)\frac{\sin(\zeta a)\exp[-z\sqrt{\zeta^2 + (s/c)^2}]}{\sqrt{\zeta^2 + (s/c)^2}}\,d\zeta \qquad (4.1.52)$$

where $H_1^{(1)}(\ )$ is the Hankel function of the first kind and order. The contour C_1 is pie-shaped and includes the positive real axis ($\zeta = k$), the positive imaginary axis ($\zeta = ki$) and an arc at infinity in the first quadrant of the ζ-plane. By Jordan's lemma, the contribution from the arc vanishes. Therefore,

$$\int_0^\infty H_1^{(1)}(kr)\frac{\sin(ka)\exp[-z\sqrt{k^2 + (s/c)^2}]}{\sqrt{k^2 + (s/c)^2}}\,dk$$

$$= \int_0^\infty H_1^{(1)}(ikr)\frac{\sinh(ka)\exp[-z\sqrt{(s/c)^2 - k^2}]}{\sqrt{(s/c)^2 - k^2}}\,dk. \quad (4.1.53)$$

Similarly, consider the integral

$$\int_{C_2} H_1^{(2)}(\zeta r)\frac{\sin(\zeta a)\exp[-z\sqrt{\zeta^2 + (s/c)^2}]}{\sqrt{\zeta^2 + (s/c)^2}}\,d\zeta \qquad (4.1.54)$$

where $H_1^{(2)}(\ )$ is the Hankel function of the second kind and first order. The contour C_2 is pie-shaped and includes the positive real axis ($\zeta = k$), the negative imaginary axis ($\zeta = -ki$) and an arc at infinity in the fourth quadrant of the ζ-plane. By Jordan's lemma, the contribution from the arc vanishes. Therefore,

$$\int_0^\infty H_1^{(2)}(kr)\frac{\sin(ka)\exp[-z\sqrt{k^2 + (s/c)^2}]}{\sqrt{k^2 + (s/c)^2}}\,dk$$

$$= \int_0^\infty H_1^{(2)}(-ikr)\frac{\sinh(ka)\exp[-z\sqrt{(s/c)^2 - k^2}]}{\sqrt{(s/c)^2 - k^2}}\,dk. \quad (4.1.55)$$

Combining (4.1.53) and (4.1.55),

$$\bar{u}(r, z, s) = \frac{2P}{\mu\pi}\int_0^\infty K_1(kr)\frac{\sinh(ka)\exp[-z\sqrt{(s/c)^2 - k^2}]}{\sqrt{(s/c)^2 - k^2}}\,dk$$

$$(4.1.56)$$

because $K_1(z) = -\pi H_1^{(1)}(iz)/2$ if $-\pi < \arg(z) < \pi/2$, $H_1^{(2)}(-z) = H_1^{(1)}(z)$ and $J_1(x) = [H_1^{(1)}(x) + H_1^{(2)}(x)]/2$. Replacing the hyperbolic sine with its exponential equivalent,

$$\bar{u}(r, z, s) = \frac{P}{\mu\pi}\int_0^1 K_1\left(\frac{srp}{c}\right)[e^{pas/c} - e^{-pas/c}]\frac{\exp[-sz\sqrt{1 - p^2}/c]}{\sqrt{1 - p^2}}\,dp.$$

$$(4.1.57)$$

Using the second shifting theorem, we can take the inverse Laplace transform of (4.1.57) and obtain the final solution

$$
u(r, z, t) = -\frac{Pc}{\mu\pi} \int_0^{ct/r} \left[\frac{(t - z\sqrt{1 - p^2}/c + ap/c)H(t - z/c + ap/c)}{\sqrt{(t - z\sqrt{1 - p^2}/c + ap/c)^2 - (rp/c)^2}} \right.
$$

$$
\left. - \frac{(t - z\sqrt{1 - p^2}/c - ap/c)H(t - z/c - ap/c)}{\sqrt{(t - z\sqrt{1 - p^2}/c - ap/c)^2 - (rp/c)^2}} \right] \frac{dp}{rp\sqrt{1 - p^2}}.
$$

$$(4.1.58)$$

For our final problem, let us solve again the wave equation in cylindrical coordinates

$$
\frac{1}{r}\frac{\partial}{\partial r}\left(r\frac{\partial u}{\partial r} \right) + \frac{\partial^2 u}{\partial z^2} - \frac{1}{c^2}\frac{\partial^2 u}{\partial t^2} = -4\pi\delta(r)\delta(z - h)\delta(t) \qquad (4.1.59)
$$

subject to the boundary conditions $|u(0, z, t)| < \infty$, $\lim_{r\to\infty} u(r, z, t) \to 0$, $u_{rz}(r, 0, t) = u_{rz}(r, H, t) = 0$ and the initial conditions $u(r, z, 0) = u_t(r, z, 0) = 0$ where $0 < h < H$.

We begin by taking the Laplace transform

$$
\bar{u}(r, z, s) = \int_0^\infty u(r, z, t)e^{-st}dt \qquad (4.1.60)
$$

and the Hankel transform

$$
\overline{U}(k, z, s) = \int_0^\infty \bar{u}(r, z, s)J_0(kr)r\, dr \qquad (4.1.61)
$$

of (4.1.59) and the boundary conditions. This results in the ordinary differential equation

$$
\frac{d^2\overline{U}}{dz^2} - m^2\overline{U} = -2\delta(z - h) \qquad (4.1.62)
$$

and the boundary conditions

$$
\overline{U}'(k, 0, s) = \overline{U}'(k, H, s) = 0 \qquad (4.1.63)
$$

where $m^2 = k^2 + s^2/c^2$. We construct solutions to (4.1.62)–(4.1.63) in the regions $0 < z < h$ and $h < z < H$ and then use the conditions that $\overline{U}(k, h^+, s) = \overline{U}(k, h^-, s)$ and $\overline{U}'(k, h^+, s) - \overline{U}'(k, h^-, s) = -2$ to piece the two parts together. After some algebra,

$$
\bar{u}(x, z, s) = 2 \int_0^\infty \frac{\cosh(ma)\cosh(mb)}{m\sinh(mH)} J_0(kr)k\, dk \qquad (4.1.64)
$$

where we have taken the inverse Hankel transform, $a = z$, $b = H - h$ for $0 \leq z \leq h$ and $a = h$, $b = H - z$ for $h \leq z \leq H$.

At this point, we replace $J_0(kr)$ with its representation in terms of Hankel functions of the first and second kind, $2J_0(kr) = H_0^{(1)}(kr) + H_0^{(2)}(kr)$. Therefore, we can write (4.1.64) as

$$\bar{u}(x, z, s) = \int_0^\infty \frac{\cosh(ma)\cosh(mb)}{m\sinh(mH)} H_0^{(1)}(kr)k\,dk$$
$$+ \int_0^\infty \frac{\cosh(ma)\cosh(mb)}{m\sinh(mH)} H_0^{(2)}(kr)k\,dk. \qquad (4.1.65)$$

At this point we consider the integral

$$\oint_{C_1} \frac{\cosh(ma)\cosh(mb)}{m\sinh(mH)} H_0^{(1)}(rz)z\,dz = 2\pi i \sum_{n=0}^\infty \text{Res}(z = iZ_n) \quad (4.1.66)$$

and

$$\oint_{C_2} \frac{\cosh(ma)\cosh(mb)}{m\sinh(mH)} H_0^{(2)}(rz)z\,dz = -2\pi i \sum_{n=0}^\infty \text{Res}(z = -iZ_n)$$
$$(4.1.67)$$

where $Z_n = \sqrt{s^2/c^2 + n^2\pi^2/H^2}$. C_1 is a pie-shaped closed contour in the first quadrant of the z-plane and consists of the positive real and imaginary axes and a circular arc at infinity that connects them. C_2 is the mirror image of C_1 in the fourth quadrant. Because of the behavior of $H_0^{(1)}$ and $H_0^{(2)}$, the contribution from the arcs at infinity vanish. Furthermore, the integration along the positive imaginary axis in (4.1.66) gives the negative of the integration along the negative imaginary axis in (4.1.67) because $H_0^{(1)}(ikr) = -H_0^{(2)}(-ikr)$. Consequently,

$$\bar{u}(x, z, s) = 2\pi i \left\{ \sum_{n=0}^\infty \left[\text{Res}(z = iZ_n) - \text{Res}(z = -iZ_n) \right] \right\}. \qquad (4.1.68)$$

We now compute the residues and they equal

$$\text{Res}(k = is/c) = \frac{1}{H\pi i} K_0\left(\frac{rs}{c}\right) \qquad (4.1.69)$$

$$\text{Res}(k = -is/c) = -\frac{1}{H\pi i} K_0\left(\frac{rs}{c}\right) \qquad (4.1.70)$$

$$\text{Res}(k = iZ_n) = \frac{2(-1)^n}{H\pi i} \cos\left(\frac{n\pi a}{H}\right) \cos\left(\frac{n\pi b}{H}\right) K_0(rZ_n) \qquad (4.1.71)$$

$$\text{Res}(k = -iZ_n) = -\frac{2(-1)^n}{H\pi i} \cos\left(\frac{n\pi a}{H}\right) \cos\left(\frac{n\pi b}{H}\right) K_0(rZ_n) \qquad (4.1.72)$$

because $H_0^{(1)}(ri) = 2K_0(r)/\pi i$ and $H_0^{(2)}(ri) = -2K_0(r)/\pi i$. A summation of the residues gives

$$\overline{u}(x,z,s) = \frac{2}{H}\left[K_0\left(\frac{rs}{c}\right) + 2\sum_{n=1}^{\infty}(-1)^n\cos\left(\frac{n\pi a}{H}\right)\cos\left(\frac{n\pi b}{H}\right)K_0\left(rZ_n\right)\right].$$

(4.1.73)

Finally, taking the inverse of the Laplace transform, we obtain the desired solution

$$u(r,z,t) = \frac{2}{H\sqrt{t^2 - r^2/c^2}}\left[1 + 2\sum_{n=1}^{\infty}(-1)^n\cos\left(\frac{n\pi a}{H}\right)\cos\left(\frac{n\pi b}{H}\right)\right.$$
$$\left. \times \cos\left(\frac{n\pi c}{H}\sqrt{t^2 - r^2/c^2}\right)\right]$$

(4.1.74)

if $t > r/c$; otherwise, $u(r,z,t) = 0$.

Problems

1. Solve the partial differential equation[3]

$$EI\frac{\partial^4 u}{\partial x^4} + ku + m\frac{\partial^2 u}{\partial t^2} = P_0\delta(x)\delta(t), \qquad -\infty < x < \infty, t > 0$$

with the initial conditions $u(x,0) = u_t(x,0) = 0$ by the joint transform method.

Step 1. If we define the Fourier transform

$$U(k,t) = \int_{-\infty}^{\infty}u(x,t)e^{-ikx}dx$$

and the Laplace transform by

$$\overline{U}(k,s) = \int_0^{\infty}U(k,t)e^{-st}dt,$$

show that the joint transform of our system is

$$\overline{U}(k,s) = \frac{P_0/m}{s^2 + a^2k^4 + \omega^2}$$

---

[3] Taken from Stadler, W., and R. W. Shreever, 1970: The transient and steady-state response of an infinite Bernoulli-Euler beam with damping and an elastic foundation. *Quart. J. Mech. Appl. Math.*, **23**, 197–208. By permission of Oxford University Press.

with $a^2 = EI/m$ and $\omega^2 = k/m$.

Step 2. Find the inverse Laplace transform of $\overline{U}(k, s)$ and show that

$$U(k, t) = \frac{P_0}{m} \frac{\sin(t\sqrt{a^2 k^4 + \omega^2})}{\sqrt{a^2 k^4 + \omega^2}}.$$

Step 3. Use the fact that $U(k, t)$ is an even function in k to show that the inverse of the Fourier transform is

$$u(x, t) = \frac{P_0}{m\pi} \int_0^\infty \frac{\sin(t\sqrt{a^2 k^4 + \omega^2})}{\sqrt{a^2 k^4 + \omega^2}} \cos(kx) \, dk.$$

2. Solve the partial differential equation[4]

$$\frac{\partial^2 u}{\partial x^2} - \frac{1}{c^2} \frac{\partial^2 u}{\partial t^2} = P \cos(\omega t) \delta[x - X(t)], \qquad -\infty < x < \infty, t > 0$$

with the boundary conditions that $\lim_{|x| \to \infty} u(x, t) \to 0$ and the initial conditions $u(x, 0) = u_t(x, 0) = 0$ by the joint transform method.

Step 1. By taking the Laplace transform in t and the Fourier transform in x, show that the partial differential equation and boundary conditions reduce to the algebraic equation

$$\overline{U}(k, s) = -\frac{c^2 P \overline{F}(s)}{k^2 c^2 + s^2}$$

where

$$\overline{F}(s) = \mathcal{L}\left[\cos(\omega t) e^{-ikX(t)}\right].$$

Step 2. Use the convolution theorem as it applies to Laplace transforms and show that

$$U(k, t) = -cP \int_0^t \frac{\sin[kc(t - \eta)]}{k} \cos(\omega \eta) e^{-ikX(\eta)} \, d\eta$$

and

$$u(x, t) = -\frac{cP}{2\pi} \int_{-\infty}^\infty \int_0^t \frac{\sin[kc(t - \eta)]}{k} \cos(\omega \eta) e^{ik[x - X(\eta)]} \, d\eta \, dk.$$

---

[4] Taken from Knowles, J. K., 1968: Propagation of one-dimensional waves from a source in random motion. *J. Acoust. Soc. Am.*, **43**, 948–957.

Step 3. By reversing the order of integration, show that

$$u(x,t) = -\frac{cP}{2} \int_0^t H\left[t - \eta - \frac{|X(\eta) - x|}{c}\right] \cos(\omega\eta)\, d\eta.$$

3. Solve the partial differential equation[5]

$$\frac{\partial^2 u}{\partial x^2} + \frac{\partial^2 u}{\partial z^2} = 0, \qquad -h < z < 0, -\infty < x < \infty, t > 0$$

with the boundary conditions

$$\frac{\partial u(x, -h, t)}{\partial z} = \xi_0(1 - e^{-\alpha t})H(b^2 - x^2)H(t)$$

and

$$\frac{\partial^2 u(x, 0, t)}{\partial t^2} + g\frac{\partial u(x, 0, t)}{\partial z} = 0$$

and the initial conditions $u(x, z, 0) = u_t(x, z, 0) = 0$ by the joint transform method.

Step 1. If we define the Fourier transform

$$U(k, z, t) = \int_{-\infty}^{\infty} u(x, z, t)e^{-ikx}\, dx$$

and the Laplace transform by

$$\overline{U}(k, z, s) = \int_0^\infty U(k, z, t)e^{-st}\, dt,$$

show that the joint transform of our system is

$$\frac{d^2\overline{U}}{dz^2} - k^2\overline{U} = 0, \qquad -h < z < 0$$

with

$$\frac{d\overline{U}(k, -h, s)}{dz} = \frac{2\xi_0\alpha \sin(kb)}{sk(s + \alpha)}$$

and

$$\frac{d\overline{U}(k, 0, s)}{dz} + \frac{s^2}{g}\overline{U}(k, 0, s) = 0.$$

---

[5] Modeled after a problem that occurred in Hammack, J. L., 1973: A note on tsunamis: Their generation and propagation in an ocean of uniform depth. *J. Fluid Mech.*, **60**, 769–799.

Step 2. Show that the solution to step 1 is

$$\overline{U}(k, z, s) = -\frac{2g\alpha\xi_0 \sin(kb)}{ks(s+\alpha)(s^2+\omega^2)\cosh(kh)}\left[\cosh(kz) - \frac{s^2}{gk}\sinh(kz)\right]$$

where $\omega^2 = gk\tanh(kh)$.

Step 3. Find the inverse Laplace transform of $\overline{U}(k, z, s)$ and show that

$$U(k, z, t) = \frac{2g\xi_0 \sin(kb)}{k\cosh(kh)}\left[\frac{\sinh(kz)}{gk}\left\{\frac{\alpha^2}{\omega^2+\alpha^2}\cos(\omega t)\right.\right.$$
$$-\frac{\alpha^2}{\omega^2+\alpha^2}e^{-\alpha t} + \frac{\alpha\omega}{\omega^2+\alpha^2}\sin(\omega t)\bigg\}$$
$$-\cosh(kz)\left\{\frac{1-\cos(\omega t)}{\omega^2}\right.$$
$$\left.\left.-\frac{1}{\omega^2+\alpha^2}\left[e^{-\alpha t} - \cos(\omega t) + \frac{\alpha}{\omega}\sin(\omega t)\right]\right\}\right].$$

Step 4. Use the fact that $U(k, z, t)$ is an even function in k to show that the inverse of the Fourier transform is

$$u(x, z, t) = \frac{2g\xi_0}{\pi}\int_0^\infty \frac{\sin(kb)\cos(kx)}{k\cosh(kh)}dk\left[\frac{\sinh(kz)}{gk}\left\{\frac{\alpha^2}{\omega^2+\alpha^2}\right.\right.$$
$$\cos(\omega t) - \frac{\alpha^2}{\omega^2+\alpha^2}e^{-\alpha t} + \frac{\alpha\omega}{\omega^2+\alpha^2}\sin(\omega t)\bigg\}$$
$$-\cosh(kz)\left\{\frac{1-\cos(\omega t)}{\omega^2}\right.$$
$$\left.\left.-\frac{1}{\omega^2+\alpha^2}\left[e^{-\alpha t} - \cos(\omega t) + \frac{\alpha}{\omega}\sin(\omega t)\right]\right\}\right].$$

4. Solve the partial differential equation[6]

$$x\frac{\partial u}{\partial t} + \frac{\partial u}{\partial x} = f(x), \qquad -\infty < x < \infty, t > 0$$

with the boundary conditions $\lim_{x\to\pm\infty} u(x, t) \to 0$ and the initial condition $u(x, 0) = 0$ by the joint transform method. The function $f(x)$ is an odd function.

---

[6] Suggested by a considerably more complicated problem given by Shankar, P. N., 1981: On the evolution of disturbances at an inviscid interface. *J. Fluid Mech.*, **108**, 159–170. Reprinted with the permission of Cambridge University Press.

Transform Methods for Solving Partial Differential Equations

Step 1. If we define the Laplace transform by

$$\overline{u}(x, s) = \int_0^\infty u(x, t)e^{-st}dt,$$

show that the Laplace transform of our system is

$$\frac{d\overline{u}}{dx} + sx\overline{u} = \frac{f(x)}{s}.$$

Step 2. If we define the Fourier transform by

$$\overline{U}(k, s) = \int_{-\infty}^\infty \overline{u}(x, s)e^{-ikx}dx$$

and assume that $\overline{u}(x, s)$ goes to zero sufficient rapidly as $x \to \infty$, show that the joint transform of our system is

$$\frac{d\overline{U}}{dk} + \frac{k}{s}\overline{U} = \frac{F(k)}{is^2}.$$

Step 3. Show that the solution to step 2 is

$$\overline{U}(k, t) = \int_0^k \frac{F(\xi)\exp[-(k^2 - \xi^2)/(2s)]}{s^2 i} d\xi.$$

Step 4. Find the inverse Laplace transform of $\overline{U}(k, s)$ and show that

$$U(k, t) = \sqrt{2t} \int_0^k \frac{F(\xi)J_1[\sqrt{2(k^2 - \xi^2)t}]}{i\sqrt{k^2 - \xi^2}} d\xi.$$

Step 5. Find the inverse Fourier transform of $U(k, t)$ and show that

$$u(x, t) = \frac{\sqrt{2t}}{\pi} \int_0^\infty \left\{ \int_0^k \frac{F(\xi)J_1[\sqrt{2(k^2 - \xi^2)t}]}{\sqrt{k^2 - \xi^2}} d\xi \right\} \sin(kx) \, dk.$$

5. Solve the partial differential equations[7]

$$\frac{\partial^2 u}{\partial z^2} + \frac{\partial^2 u}{\partial x^2} - \frac{1}{c^2}\frac{\partial^2 u}{\partial t^2} = 0, \qquad z > 0$$

---

[7] Taken from Lee, K. H., G. Lin and H. F. Morrison, 1989: A new approach to modeling the electromagnetic response of conductive media. *Geophysics*, **54**, 1180–1192.

and

$$\frac{\partial^2 u}{\partial z^2} + \frac{\partial^2 u}{\partial x^2} = 0, \qquad z < 0$$

with the boundary conditions

$$\lim_{z \to \pm\infty} u(x, z, t) \to 0, \qquad u(x, 0^+, t) = u(x, 0^-, t)$$

and

$$u_z(x, 0^+, t) - u_z(x, 0^-, t) = -\delta(x)\delta(t).$$

Assume that the initial conditions are $u(x, z, 0) = u_t(x, z, 0) = 0$.

Step 1. By taking the Laplace and Fourier transforms of the partial differential equations and boundary conditions, show that

$$\frac{d^2\overline{U}}{dz^2} - \left(k^2 + \frac{s^2}{c^2}\right)\overline{U} = 0, \qquad z > 0$$

and

$$\frac{d^2\overline{U}}{dz^2} - k^2\overline{U} = 0, \qquad z < 0$$

where $\overline{U}(k, z, s)$ is the joint transform of $u(x, z, t)$. The boundary conditions become

$$\lim_{z \to \pm\infty} \overline{U}(k, z, s) \to 0, \qquad \overline{U}(k, 0^+, s) = \overline{U}(k, 0^-, s)$$

and

$$\overline{U}'(k, 0^+, s) - \overline{U}'(k, 0^-, s) = -1.$$

Step 2. Show that the solution to the transform equations is

$$\overline{U}(k, z, s) = \frac{c^2(|k| + m)}{s^2} e^{|k|z}, \qquad z < 0$$

and

$$\overline{U}(k, z, s) = \frac{c^2(|k| + m)}{s^2} e^{-mz}, \qquad z > 0$$

where $m = \sqrt{k^2 + s^2/c^2}$.

Step 3. For $z > 0$, show that we may write the inverse of the joint transform

$$u(x, z, t) = I_1 + I_2$$

where

$$I_1 = \frac{c^2}{4\pi^2 i} \int_{c-\infty i}^{c+\infty i} e^{st} \left[\int_{-\infty}^{\infty} \frac{|k|}{s^2} e^{-mz} e^{ikx} \, dk \right] ds$$

and

$$I_2 = \frac{c^2}{4\pi^2 i} \int_{c-\infty i}^{c+\infty i} e^{st} \left[\int_{-\infty}^{\infty} \frac{m}{s^2} e^{-mz} e^{ikx} dk \right] ds.$$

Step 4. Using tables of Laplace transforms and integrals, verify the following evaluations of I_1 and I_2:

$$I_1 = \frac{c^2}{\pi} \frac{\partial^2}{\partial x \partial z} \int_0^{\infty} \mathcal{L}^{-1}\left(\frac{1}{s^2}\right) * \mathcal{L}^{-1}\left[\frac{\exp(-z\sqrt{k^2 + s^2/c^2})}{\sqrt{k^2 + s^2/c^2}}\right] \sin(kx)\, dk$$

$$= \begin{cases} 0 & 0 < t < z/c \\ \frac{c^3}{\pi} \frac{\partial^2}{\partial x \partial z} \int_0^{\infty} t * J_0(ck\sqrt{t^2 - z^2/c^2})\sin(kx)\, dk & t > z/c \end{cases}$$

$$= \begin{cases} 0 & 0 < t < z/c \\ \frac{c^2}{\pi} \frac{\partial^2}{\partial x \partial z} \int_{z/c}^{t} (t-y)/\sqrt{r^2/c^2 - y^2}\, dy & z/c < t < r/c \\ \frac{c^2}{\pi} \frac{\partial^2}{\partial x \partial z} \int_{z/c}^{r/c} (y-t)/\sqrt{r^2/c^2 - y^2}\, dy & t > r/c \end{cases}$$

$$= \begin{cases} 0 & 0 < t < z/c \\ c^2[(x^2 - z^2)t \\ \quad + |x|z(2t^2 - r^2/c^2)/\sqrt{r^2/c^2 - t^2}]/\pi r^4 & z/c < t < r/c \\ c^2(x^2 - z^2)t/\pi r^4 & t > r/c \end{cases}$$

where $r^2 = x^2 + z^2$ and

$$I_2 = \frac{c^2}{2\pi i} \frac{\partial^2}{\partial z^2}\left\{ \int_{c-\infty i}^{c+\infty i} \frac{e^{st}}{s^2}\left[\frac{1}{\pi} \int_0^{\infty} \frac{e^{-mz}}{m} \cos(kx)\, dk \right] ds \right\}$$

$$= \frac{c^2}{2\pi i} \frac{\partial^2}{\partial z^2}\left[\int_{c-\infty i}^{c+\infty i} K_0(rs/c)\frac{e^{st}}{s^2}\, ds \right]$$

$$= \begin{cases} 0 & 0 < t < r/c \\ \frac{c^2}{\pi} \frac{\partial^2}{\partial z^2} \int_{r/c}^{t} (t-y)/\sqrt{y^2 - r^2/c^2}\, dy & t > r/c \end{cases}$$

$$= \begin{cases} 0 & 0 < t < r/c \\ c^2[-x^2\sqrt{t^2 - r^2/c^2} \\ \quad + z^2 t^2/\sqrt{t^2 - r^2/c^2}]/\pi r^4 & t > r/c \end{cases}$$

Thus, the solution for $z > 0$ is

$$
u(x, z, t) = \begin{cases}
0 & 0 < t < z/c \\
\begin{aligned} & c^2[(x^2 - z^2)t + |x|z(2t^2 - r^2/c^2) \\ & /\sqrt{r^2/c^2 - t^2}]/\pi r^4 \end{aligned} & z/c < t < r/c \\
\begin{aligned} & c^2[(x^2 - z^2)t - x^2\sqrt{t^2 - r^2/c^2} \\ & + z^2 t^2/\sqrt{t^2 - r^2/c^2}/\pi r^4 \end{aligned} & t > r/c.
\end{cases}
$$

6. Solve the partial differential equation[8]

$$
\frac{\partial^2 u}{\partial r^2} + \frac{1}{r}\frac{\partial u}{\partial r} + \frac{\partial^2 u}{\partial z^2} + \nu\frac{\partial u}{\partial z} = -2I\rho_0\delta(r)\delta(z)
$$

where $0 < r < \infty$ and $0 < z < \infty$ with the boundary conditions $\lim_{r\to\infty} u(r,z) \to 0$, $\lim_{z\to\infty} u(r,z) \to 0$, $|u(0,z)| < \infty$ and $u_z(r,0) = -I\rho_0\delta(r)$.

Step 1. By defining the Hankel transform

$$
U(k, z) = \int_0^\infty u(r, z)J_0(kr)r\,dr,
$$

show that the partial differential equation becomes the ordinary differential equation

$$
\frac{d^2 U}{dz^2} + \nu\frac{dU}{dz} - k^2 U = -\frac{I\rho_0}{\pi}\delta(z)
$$

with the boundary conditions

$$
\lim_{z\to\infty} U(k, z) \to 0 \qquad \text{and} \qquad U'(k, 0) = -\frac{I\rho_0}{2\pi}.
$$

Step 2. By taking the Fourier transform of the ordinary differential equation in step 1, show that the particular solution reduces to the algebraic equation

$$
\overline{U}_p(k, n) = \frac{I\rho_0}{\pi}\frac{1}{n^2 - i\nu n + k^2}
$$

---

[8] Taken from Paul, M. K., and B. Banerjee, 1970: Electrical potentials due to a point source upon models of continuously varying conductivity. *PAGEOPH*, **80**, 218–237. Published by Birkhäuser Verlag, Basel, Switzerland.

where

$$\overline{U}(k,n) = \int_{-\infty}^{\infty} U(k,z)e^{-inz}dz.$$

Step 3. Use the residue theorem to invert $\overline{U}_p(k,n)$ and show that

$$U_p(k,z) = \frac{I\rho_0}{\pi} \frac{\exp[-(\nu + \sqrt{\nu^2 + 4k^2})z/2]}{\sqrt{\nu^2 + 4k^2}}$$

while the general solution is

$$U(k,z) = \frac{I\rho_0}{\pi} \frac{\exp[-(\nu + \sqrt{\nu^2 + 4k^2})z/2]}{\sqrt{\nu^2 + 4k^2}}$$
$$+ A(k)\exp[-(\nu + \sqrt{\nu^2 + 4k^2})z/2]$$

because $\lim_{z\to\infty} U(k,z) \to 0$.

Step 4. Using the boundary condition $U'(k,0) = -I\rho_0/2\pi$, show that

$$u(r,z) = \frac{I\rho_0}{\pi} \int_0^\infty \left[\frac{\exp[-(\nu + \sqrt{\nu^2 + 4k^2})z/2]}{\sqrt{\nu^2 + 4k^2}} \right.$$
$$\left. - \frac{\nu\exp[-(\nu + \sqrt{\nu^2 + 4k^2})z/2]}{\sqrt{\nu^2 + 4k^2}(\nu + \sqrt{\nu^2 + 4k^2})} \right] J_0(kr)k\, dk$$

or

$$u(r,z) = \frac{I\rho_0}{2\pi} \left[\frac{\exp[-\nu(z + \sqrt{r^2 + z^2})/2]}{\sqrt{r^2 + z^2}} \right.$$
$$\left. - 2\nu \int_0^\infty \frac{\exp[-(\nu + \sqrt{\nu^2 + 4k^2})z/2]}{\sqrt{\nu^2 + 4k^2}(\nu + \sqrt{\nu^2 + 4k^2})} J_0(kr)k\, dk \right].$$

7. Solve the partial differential equation

$$\frac{\partial^2 u}{\partial r^2} + \frac{1}{r}\frac{\partial u}{\partial r} + \frac{\partial^2 u}{\partial z^2} = \frac{\partial u}{\partial t}, \qquad 0 \le r < \infty, 0 \le z < \infty, t > 0$$

subject to the boundary conditions

$$\frac{\partial u(r,0,t)}{\partial z} = \begin{cases} -1/\sqrt{a^2 - r^2} & \text{if} \quad 0 \le r < a, \\ 0 & \text{if} \quad r > a, \end{cases}$$

$$\frac{\partial u(0,z,t)}{\partial r} = 0, \qquad 0 \le z < \infty$$

$$\lim_{r\to\infty} u(r,z,t) \to 0, \qquad \lim_{z\to\infty} u(r,z,t) \to 0$$

and the initial condition $u(r, z, 0) = 0$.

Step 1. By defining the Laplace transform

$$\overline{u}(x, z, s) = \int_0^\infty u(x, z, t)e^{-st}dt$$

and the Hankel transform

$$\overline{U}(k, z, s) = \int_0^\infty \overline{u}(x, z, s)J_0(kr)r\,dr,$$

show that the partial differential equation and boundary conditions reduce to the ordinary differential equation

$$\frac{d^2\overline{U}}{dz^2} - (k^2 + s)\overline{U} = 0$$

with $\lim_{z\to\infty} \overline{U}(k, z, s) \to 0$ and

$$\frac{d\overline{U}(k, 0, s)}{dz} = -\frac{\sin(ka)}{ks}.$$

Step 2. Show that the solution to step 1 is

$$\overline{U}(k, z, s) = \frac{\sin(ka)}{ks\sqrt{k^2 + s}}e^{-z\sqrt{k^2+s}} = \frac{\sin(ka)}{k(s' - k^2)\sqrt{s'}}e^{-z\sqrt{s'}}$$

$$= \frac{\sin(ka)}{2k^2\sqrt{s'}}e^{-z\sqrt{s'}}\left[\frac{1}{\sqrt{s'} - k} - \frac{1}{\sqrt{s'} + k}\right]$$

where $s' = s + k^2$.

Step 3. Using the first shifting theorem and tables, invert the Laplace transform and show that

$$U(k, z, t) = \frac{\sin(ka)}{2k^2}\left[e^{-kz}\text{erfc}\left(\frac{z}{2\sqrt{t}} - k\sqrt{t}\right) - e^{kz}\text{erfc}\left(\frac{z}{2\sqrt{t}} + k\sqrt{t}\right)\right].$$

Step 4. Complete the problem by inverting the Hankel transform and showing that

$$u(x, z, t) = \int_0^\infty \frac{\sin(ka)}{2k}J_0(kr)\,dk$$

$$\times \left[e^{-kz}\text{erfc}\left(\frac{z}{2\sqrt{t}} - k\sqrt{t}\right) - e^{kz}\text{erfc}\left(\frac{z}{2\sqrt{t}} + k\sqrt{t}\right)\right].$$

8. Solve the partial differential equation[9]

$$\frac{\partial^2 u}{\partial x^2} + \frac{\partial^2 u}{\partial z^2} = \frac{\partial^2 u}{\partial t^2}, \qquad -\infty < x < \infty, 0 \le z < \infty, t > 0$$

with the boundary conditions $\lim_{|x| \to \infty} u(x, z, t) \to 0$, $\lim_{z \to \infty} u(x, z, t) \to 0$ and $u(x, 0, t) = H(x + t) - H(x - t)$ with the initial conditions $u(x, z, 0) = u_t(x, z, 0) = 0$.

Step 1. Define the Laplace transform by

$$\bar{u}(x, z, s) = \int_0^\infty u(x, z, t) e^{-st} dt$$

and the Fourier transform by

$$\overline{U}(k, z, s) = \int_{-\infty}^\infty \bar{u}(x, z, s) e^{-ikx} dx.$$

Show that the partial differential equation and boundary conditions reduce to the ordinary differential equation

$$\frac{d^2 \overline{U}}{dz^2} - (s^2 + k^2) \overline{U} = 0, \qquad 0 \le z < \infty$$

with the boundary conditions $\lim_{z \to \infty} \overline{U}(k, z, s) \to 0$ and

$$\overline{U}(k, 0, s) = \frac{2}{s^2 + k^2}.$$

Step 2. Show that the solution to step 1 is

$$\overline{U}(k, z, s) = \frac{2}{s^2 + k^2} \exp[-z\sqrt{s^2 + k^2}].$$

Step 3. Using the relationship[10] that

$$\mathcal{L}^{-1} \left[\frac{F(\sqrt{s^2 + k^2})}{\sqrt{s^2 + k^2}} \right] = \int_0^t J_0[k\sqrt{t^2 - \eta^2}] f(\eta) \, d\eta$$

---

[9] Reprinted from *Int. J. Mech. Sci.*, **11**, W. R. Spillers and A. Callegari, Impact of two elastic cylinders: Short-time solution, pp. 846–851, ©1969, with kind permission from Pergamon Press Ltd., Headington Hill Hall, Oxford OX3 0BW, UK.

[10] From Erdélyi, A., W. Magnus, F. Oberhettinger and F. G. Tricomi, 1954: *Table of Integral Transforms. Volume 1.* New York: McGraw-Hill Book Co., Inc., p. 133, we have

$$\mathcal{L} \left\{ \int_0^t \left(\frac{t - \eta}{t + \eta} \right)^\nu J_{2\nu}(\sqrt{t^2 - \eta^2}) f(\eta) \, d\eta \right\}$$

$$= \frac{[\sqrt{s^2 + 1} - s]^{-2\nu} F(\sqrt{s^2 + 1})}{\sqrt{s^2 + 1}}.$$

if $\mathcal{L}[f(t)] = F(s)$, show that

$$\mathcal{F}^{-1}\left\{\mathcal{L}^{-1}\left[\frac{F(\sqrt{s^2 + k^2})}{\sqrt{s^2 + k^2}}\right]\right\} = \frac{1}{\pi}\int_0^{\sqrt{t^2 - x^2}} \frac{f(\eta)}{\sqrt{t^2 - x^2 - \eta^2}}\, d\eta.$$

Step 4. If we choose $F(s) = 2e^{-zs}/s$, show that

$$u(x, z, t) = \frac{2}{\pi}\int_0^{\sqrt{t^2 - x^2}} \frac{H(t - z)}{\sqrt{t^2 - x^2 - \eta^2}}\, d\eta$$

$$= \frac{2}{\pi}\left[\frac{\pi}{2} - \sin^{-1}\left(\frac{z}{\sqrt{t^2 - x^2}}\right)\right]H(t^2 - x^2 - z^2).$$

9. Solve the partial differential equation[11]

$$\frac{\partial^2 u}{\partial z^2} + \frac{\partial^2 u}{\partial r^2} + \frac{1}{r}\frac{\partial u}{\partial r} = \frac{\partial^2 u}{\partial t^2}, \qquad 0 < r < \infty, 0 < z < \infty, t > 0$$

with the boundary conditions $|u(0, z, t)| < \infty$, $\lim_{r\to\infty} u(r, z, t) \to 0$, $\lim_{z\to\infty} u(r, z, t) \to 0$ and

$$\frac{\partial u(r, 0, t)}{\partial z} = H(1 - r)H(t)$$

with the initial conditions $u(r, z, 0) = u_t(r, z, 0) = 0$.

Step 1. By defining the Laplace transform

$$\overline{u}(r, z, s) = \int_0^\infty u(r, z, t)e^{-st}\, dt$$

and the Hankel transform

$$\overline{U}(k, z, s) = \int_0^\infty \overline{u}(r, z, s)J_0(kr)r\, dr,$$

show that the partial differential equation and boundary conditions reduce to

$$\frac{d^2\overline{U}}{dz^2} - (s^2 + k^2)\overline{U} = 0, \qquad 0 < z < \infty$$

---

[11] Taken from Miles, J. W., 1953: Transient loading of a baffled piston. *J. Acoust. Soc. Am.*, **25**, 200–203.

with the boundary conditions $\lim_{z\to\infty} \overline{U}(k,z,s) \to 0$ and

$$\overline{U}'(k,0,s) = \frac{J_1(k)}{ks}.$$

Step 2. Show that the solution to step 1 is

$$\overline{U}(k,z,s) = -\frac{J_1(k)\exp[-z\sqrt{k^2+s^2}]}{sk\sqrt{k^2+s^2}}.$$

Step 3. Take the inverse Laplace transform and show that

$$U_t(k,z,t) = -\frac{J_1(k)}{k}J_0(k\sqrt{t^2-z^2})H(t-z).$$

Step 4. Take the inverse Hankel transform and show that

$$u_t(r,z,t) = -H(t-z)\int_0^\infty J_1(k)J_0(k\sqrt{t^2-z^2})J_0(kr)\,dk.$$

Step 5. Finally use the relationship[12]

$$\int_0^\infty J_\mu(at)J_\nu(bt)J_\nu(ct)t^{1-\mu}\,dt$$

$$= \begin{cases} 0 & \text{if } a^2 < (b+c)^2, a^2 < (b-c)^2 \\ \cos^{-1}\left(\frac{b^2+c^2-a^2}{2bc}\right) & \text{if } (b-c)^2 < a^2 < (b+c)^2 \\ \pi & \text{if } a^2 > (b+c)^2, a^2 > (b-c)^2, \end{cases}$$

and show that

$$u_t(r,z,t) = \begin{cases} 0 & \text{if } |R_-| > 1 \text{ or } t < z \\ -\frac{1}{\pi}\cos^{-1}\left(\frac{t^2+r^2-z^2-1}{2r\sqrt{t^2-z^2}}\right) & \text{if } |R_-| < 1 < R_+, t > z \\ -1 & \text{if } R_+ > 1, t > z \end{cases}$$

where $R_- = r - \sqrt{t^2-z^2}$ and $R_+ = r + \sqrt{t^2-z^2}$.

---

[12] Taken from Watson, G. N., 1966: *A Treatise on the Theory of Bessel Functions*. Second Edition. Cambridge: At the University Press, p. 411.

10. Solve the partial differential equation[13]

$$\frac{\partial^2 u}{\partial r^2} + \frac{1}{r}\frac{\partial u}{\partial r} - \frac{\partial^2 u}{\partial t^2} - 2\epsilon\frac{\partial u}{\partial t} = -2\pi\delta(r)\delta(t), \qquad 0 < r < \infty, t > 0$$

with the boundary conditions $|u(0,t)| < \infty$, $\lim_{r\to\infty} u(r,t) \to 0$ and the initial conditions $u(r,0) = u_t(r,0) = 0$ where $\epsilon \geq 0$.

Step 1. By defining the Laplace transform

$$\bar{u}(r,s) = \int_0^\infty u(r,t)e^{-st}\,dt$$

and the Hankel transform

$$\bar{U}(k,s) = \int_0^\infty \bar{u}(r,s)J_0(kr)r\,dr,$$

show that the partial differential equation and boundary conditions reduce to the algebraic equation

$$\bar{U}(k,s) = \frac{1}{k^2 + s(s + 2\epsilon)}.$$

Step 2. Using a table of integrals[14], invert the Hankel transform and show that

$$\mathcal{L}\left[e^{\epsilon t}u(r,t)\right] = \int_0^\infty \frac{1}{k^2 + s^2 - \epsilon^2}J_0(kr)k\,dk = K_0(r\sqrt{s^2 - \epsilon^2})$$

where $\text{Re}(\sqrt{s^2 - \epsilon^2}) > 0$ and K_0 is a zeroth-order, modified Bessel function of the second kind.

Step 3. By using a table of Laplace transform[15] or by direct evaluation using Bromwich's integral, invert the Laplace transform in step 2 and show that

$$u(r,t) = e^{-\epsilon t}\frac{\cosh(\epsilon\sqrt{t^2 - r^2})}{\sqrt{t^2 - r^2}}H(t - r).$$

---

[13] This result was derived in a different manner by Sezginer, A., and W. C. Chew, 1984: Closed form expression of the Green's function for the time-domain wave equation for a lossy two-dimensional medium. *IEEE Trans. Antennas Propagat.*, **AP-32**, 527–528. ©1984 IEEE.

[14] For example, Gradshteyn, I. S., and I. M. Ryzhik, 1965: *Table of Integrals, Series, and Products.* New York: Academic Press, 1086 pp.

[15] For example, Erdélyi, A., W. Magnus, F. Oberhettinger and F. G. Tricomi, 1954: *Table of Integral Transforms. Vol. 1.* New York: McGraw-Hill Book Co., Inc., 391 pp.

Fig. 4.2.1: Louis Paul Emile Cagniard (1900–1971).

4.2 INVERSION OF THE JOINT TRANSFORM BY CAGNIARD'S METHOD

Although the vast majority of joint transform problems are solved along the lines given in section 4.1, this direct assault is by no means the only method. In 1939 Cagniard published a book[16] in which he suggested that by using a series of transformations he could convert the joint transform into the integral definition of the Laplace transform. In this new form, he could then extract the inversion by inspection.

Consider the anisotropic wave equation[17]

$$\rho\frac{\partial^2 u}{\partial t^2} = N\frac{\partial^2 u}{\partial x^2} + L\frac{\partial^2 u}{\partial z^2}, \quad -\infty < x < \infty, -\infty < z < \infty, t > 0 \quad (4.2.1)$$

---

[16] The most accessible version is Cagniard, L., E. A. Flinn and C. H. Dix, 1962: *Reflection and Refraction of Progressive Seismic Waves.* New York: McGraw-Hill, 282 pp.

[17] Taken from Sakai, Y., and I. Kawasaki, 1990: Analytic waveforms for a line source in a transversely isotropic medium. *J. Geophys. Res.*, **95**, 11333–11344. ©1990 American Geophysical Union.

where N, L and ρ are constants. Assuming that the system is initially quiescent, the Laplace transform of (4.2.1) gives

$$N\frac{\partial^2\overline{u}}{\partial x^2} + L\frac{\partial^2\overline{u}}{\partial z^2} = \rho s^2\overline{u}, \quad -\infty < x < \infty, -\infty < z < \infty \qquad (4.2.2)$$

while the Fourier transform of (4.2.2) yields

$$\frac{d^2\overline{U}}{dz^2} - \left(\frac{\rho s^2 + Nk^2}{L}\right)\overline{U} = 0 \qquad (4.2.3)$$

where $\overline{U}(k, z, s)$ is the joint Fourier-Laplace transform of $u(x, z, t)$. The solution of (4.2.3) is

$$\overline{U}(k, z, s) = \begin{cases} a_+ e^{-\beta z} & \text{if} \quad z > 0, \\ a_- e^{+\beta z}, & \text{if} \quad z < 0, \end{cases} \qquad (4.2.4)$$

where

$$\beta = \sqrt{\frac{\rho s^2 + Nk^2}{L}}, \qquad \text{Re}(\beta) \geq 0. \qquad (4.2.5)$$

Let us now assume that there is an interface located at $z = 0$ and the boundary conditions across this interface are

$$u(x, 0^+, t) = u(x, 0^-, t) = 0 \qquad (4.2.6)$$

and

$$L\frac{\partial u(x, 0^+, t)}{\partial z} - L\frac{\partial u(x, 0^-, t)}{\partial z} = M\delta(t)\delta(x). \qquad (4.2.7)$$

The transformed form of (4.2.6)–(4.2.7) is then

$$\overline{U}(k, 0^+, s) = \overline{U}(k, 0^-, s) = 0 \qquad (4.2.8)$$

and

$$L\frac{d\overline{U}(k, 0^+, s)}{dz} - L\frac{d\overline{U}(k, 0^-, s)}{dz} = M. \qquad (4.2.9)$$

From these boundary conditions, $a_\pm = -M/(2L\beta)$ and

$$\overline{u}(x, z, s) = -\frac{M}{4\pi L}\int_{-\infty}^{\infty}\frac{e^{-\beta z + ikx}}{\beta}\,dk, \qquad z > 0 \qquad (4.2.10)$$

and

$$\overline{u}(x, z, s) = -\frac{M}{4\pi L}\int_{-\infty}^{\infty}\frac{e^{\beta z + ikx}}{\beta}\,dk, \qquad z < 0. \qquad (4.2.11)$$

From this point forward, we will only treat the case $z > 0$; the case of $z < 0$ follows by analogy.

Let us now introduce $k = -sp/V_{SH}$ which results in $\beta = s\nu/V_{SV}$ where $\nu = \sqrt{1+p^2}$, $V_{SH} = \sqrt{N/\rho}$ and $V_{SV} = \sqrt{L/\rho}$. Then, (4.2.10) becomes

$$\bar{u}(x,z,s) = -\frac{M\zeta}{4\pi L} \int_{-\infty}^{\infty} \frac{\exp[-s(\nu z + i\zeta px)/V_{SV}]}{\nu} \, dp, \qquad (4.2.12)$$

where $\zeta = \sqrt{L/N}$. Because the integrand in the region $p < 0$ is the complex conjugate to that in $p > 0$,

$$\bar{u}(x,z,s) = -\frac{M\zeta}{2\pi L} \mathrm{Re}\left\{ \int_0^{\infty} \frac{\exp[-s(\nu z + i\zeta px)/V_{SV}]}{\nu} \, dp \right\}. \qquad (4.2.13)$$

We now transform the variable of integration from p to a *complex* variable t defined by

$$t = \frac{\nu z + i\zeta px}{V_{SV}}. \qquad (4.2.14)$$

Solving for p,

$$p = \frac{-i\zeta V_{SV} xt \pm z\sqrt{V_{SV}^2 t^2 - \zeta^2 x^2 - z^2}}{\zeta^2 x^2 + z^2}. \qquad (4.2.15)$$

One of the roots given by (4.2.15) corresponds to an upward propagating wave in the upper half-space while the other solution is spurious. A check of (4.2.10) shows that we should choose the top sign. With that choice,

$$\frac{1}{\nu}\frac{\partial p}{\partial t} = -\frac{1}{\sqrt{t^2 - t_{SH}^2}} \qquad (4.2.16)$$

where $t_{SH}^2 = (\zeta^2 x^2 + z^2)/V_{SV}^2$. Consider now the change of limits. As p increases from 0 to ∞, t changes from z/V_{SV} to $\infty + \infty i$ along a contour AB in the first quadrant of the complex t-plane shown in Fig. 4.2.2. Then

$$\bar{u}(x,z,s) = -\frac{M\zeta}{2\pi L} \mathrm{Re}\left\{ \int_{z/V_{SV}}^{\infty+\infty i} \frac{1}{\nu}\frac{\partial p}{\partial t} e^{-st} dt \right\}. \qquad (4.2.17)$$

Finally, because there is no singularity inside the region enclosed by the loop ABC, we may deform the contour AB to AC provided we pass above the branch point at $t = t_{SH}$. Therefore,

$$\bar{u}(x,z,s) = -\frac{M\zeta}{2\pi L} \mathrm{Re}\left\{ \int_{z/V_{SV}}^{\infty} \frac{1}{\nu}\frac{\partial p}{\partial t} e^{-st} dt \right\}. \qquad (4.2.18)$$

Because the integrand is purely imaginary for $t < t_{SH}$, we can modify (4.2.18) so that

$$\bar{u}(x,z,s) = \int_0^{\infty} u(x,z,t) e^{-st} dt = -\frac{M\zeta}{2\pi L} \mathrm{Re}\left\{ \int_0^{\infty} \frac{1}{\nu}\frac{\partial p}{\partial t} e^{-st} dt \right\}. \qquad (4.2.19)$$

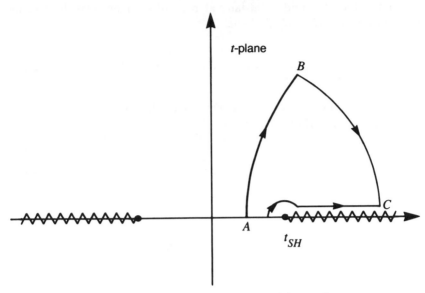

Fig. 4.2.2: Contours used in the integration of (4.2.17).

Consequently, by inspection,

$$u(x, z, t) = \frac{M\zeta}{2\pi L} \frac{H(t - t_{SH})}{\sqrt{t^2 - t_{SH}^2}}. \qquad (4.2.20)$$

Our next example is much more complicated. We find the velocity potential $u(x, z, t)$ that results from an explosion[18] located at $z = h$ within a compressible ocean having the speed of sound c. The governing equation is

$$\frac{\partial^2 u}{\partial t^2} - c^2 \left(\frac{\partial^2 u}{\partial x^2} + \frac{\partial^2 u}{\partial z^2} \right) - g \frac{\partial u}{\partial z} = -c^2 \delta(x)\delta(z - h)\delta(t) \qquad (4.2.21)$$

where g denotes the gravitational acceleration and $\delta(\ )$ is the Dirac delta function. The boundary condition at the free surface $z = 0$ is

$$\frac{\partial^2 u}{\partial t^2} = g \frac{\partial u}{\partial z}. \qquad (4.2.22)$$

Taking the Laplace transform of (4.2.21),

$$c^2 \left(\frac{\partial^2 \overline{u}}{\partial x^2} + \frac{\partial^2 \overline{u}}{\partial z^2} \right) + g \frac{\partial \overline{u}}{\partial z} - s^2 \overline{u} = c^2 \delta(x)\delta(z - h) \qquad (4.2.23)$$

---

[18] Taken from Ross, R. A., 1961: The effect of an explosion in a compressible fluid under gravity. *Can. J. Phys.*, **39**, 1330–1346.

assuming that the fluid is initially at rest. If we now take the Fourier transform of (4.2.23), we find that

$$c^2 \frac{d^2 \overline{U}}{dz^2} + g \frac{d\overline{U}}{dz} - (k^2 c^2 + s^2)\overline{U} = c^2 \delta(z - h) \tag{4.2.24}$$

where $\overline{U}(k, z, s)$ denotes the joint Fourier-Laplace transform.

The solution of (4.2.24) is

$$\overline{U}(k, z, s) = \frac{1}{2} \exp\left[-\frac{g(z - h)}{2c^2}\right] \frac{e^{-|z-h|\lambda}}{\lambda} \tag{4.2.25}$$

where

$$\lambda = \sqrt{k^2 + \frac{s^2}{c^2} + \frac{g^2}{4c^4}} \tag{4.2.26}$$

and the inverse Fourier transform is

$$\overline{u}(x, z, s) = \frac{1}{4\pi} \exp\left[-\frac{g(z - h)}{2c^2}\right] \int_{-\infty}^{\infty} \frac{e^{ikx - |z-h|\lambda}}{\lambda} \, dk. \tag{4.2.27}$$

The parameter λ is a multivalued function with branch points at

$$k = \pm i\sqrt{\frac{s^2}{c^2} + \frac{g^2}{4c^4}}. \tag{4.2.28}$$

In order for (4.2.27) to be finite as $|z - h| \to \infty$, $\mathrm{Re}(\lambda) \geq 0$. We ensure this if one branch cut lies between $i\sqrt{s^2/c^2 + g^2/4c^4}$ and ∞i while the other lies between $-i\sqrt{s^2/c^2 + g^2/4c^4}$ and $-\infty i$.

Equation (4.2.27) is the particular solution of (4.2.23). Consequently, the most general solution is this particular solution plus the homogeneous solution or

$$\overline{u}(x, z, s) = \frac{1}{4\pi} \exp\left[-\frac{g(z - h)}{2c^2}\right] \int_{-\infty}^{\infty} \frac{e^{ikx - |z-h|\lambda}}{\lambda} dk$$

$$+ \frac{1}{4\pi} \exp\left[-\frac{gz}{2c^2}\right] \int_{-\infty}^{\infty} A(k)e^{ikx - \lambda z} \, dk \tag{4.2.29}$$

where $A(k)$ is an arbitrary function of k. We determine this function by satisfying the Laplace transform of (4.2.22)

$$g \frac{\partial \overline{u}}{\partial z} = s^2 \overline{u} \tag{4.2.30}$$

at $z = 0$. Upon substituting (4.2.29) into (4.2.30), we find the general solution

$$
\bar{u}(x, z, s) = \frac{1}{2\pi} \exp\left[-\frac{g(z-h)}{2c^2}\right] \mathrm{Re}\left\{\int_0^\infty \frac{e^{ikx-|z-h|\lambda}}{\lambda} dk\right\}
$$
$$
- \frac{1}{2\pi} \exp\left[-\frac{g(z-h)}{2c^2}\right] \mathrm{Re}\left\{\int_0^\infty \frac{e^{ikx-|z+h|\lambda}}{\lambda} dk\right\}
$$
$$
+ \frac{1}{\pi} \exp\left[-\frac{g(z-h)}{2c^2}\right] \mathrm{Re}\left\{\int_0^\infty \frac{e^{ikx-(z+h)\lambda}}{s^2/g + g/2c^2 + \lambda} dk\right\}.
$$
$$(4.2.31)$$

Our next step is to write the right side of (4.2.31) as the forward Laplace transform

$$
\bar{u}(x, z, s) = \int_0^\infty u(x, z, t) e^{-st} dt. \tag{4.2.32}
$$

To this end we deform the path of integration in the second and third integrals of (4.2.31) into a new path given by

$$
k = \frac{[gx/2c^2 - (z+h)s/c]\sqrt{c^2p^2 - x^2 - (z+h)^2}}{x^2 + (z+h)^2} + \frac{ip[g(z+h)/2c + xs]}{x^2 + (z+h)^2} \tag{4.2.33}
$$

where p is a real parameter varying from 0 to ∞. We deform the first integral into the path given by (4.2.33) where we replace $z+h$ with $|z-h|$. These deformations are permissible because there are no contributions from poles (Cauchy's theorem) and the contribution from the arcs at infinity vanish (Jordan's lemma).

Because $p = 0$ corresponds to

$$
k = i\frac{gx/2c^2 - s(z+h)/c}{\sqrt{x^2 + (z+h)^2}}, \tag{4.2.34}
$$

a portion of the new path lies along the negative imaginary axis. However, because the integrands are purely imaginary along this segment, there is no contribution from this portion of the contour integration. Consequently, the first two integrals in (4.2.31) become

$$
\bar{u}_1(x, z, s) = \frac{1}{2\pi} \exp\left[-\frac{g(z-h)}{2c^2}\right] \mathrm{Re}\left\{\int_0^\infty \frac{e^{-sp+gi\sqrt{p^2-k_1^2}/2c}}{\sqrt{p^2 - k_1^2}} dp\right\}
$$
$$
- \frac{1}{2\pi} \exp\left[-\frac{g(z-h)}{2c^2}\right] \mathrm{Re}\left\{\int_0^\infty \frac{e^{-sp+gi\sqrt{p^2-k_2^2}/2c}}{\sqrt{p^2 - k_2^2}} dp\right\}
$$
$$(4.2.35)$$

where

$$k_1 = \frac{1}{c}\sqrt{x^2 + (z+h)^2} \qquad (4.2.36)$$

and

$$k_2 = \frac{1}{c}\sqrt{x^2 + (z-h)^2}. \qquad (4.2.37)$$

Because the first term in (4.2.35) vanishes if $0 < p < k_1$ (because the integral is imaginary) and the second one vanishes if $0 < p < k_2$, the contribution to $u(x, z, t)$ is

$$u_1(x, z, t) = \frac{1}{2\pi} \frac{\exp[-g(z-h)/2c^2]\cos[g\sqrt{t^2 - k_1^2}/2c]}{\sqrt{t^2 - k_1^2}} H(t - k_1)$$

$$- \frac{1}{2\pi} \frac{\exp[-g(z-h)/2c^2]\cos[g\sqrt{t^2 - k_2^2}/2c]}{\sqrt{t^2 - k_2^2}} H(t - k_2).$$

$$(4.2.38)$$

The third integral in (4.2.31) becomes

$$\bar{u}_2(x, z, s) = \frac{1}{2\pi} \exp\left[-\frac{g(z-h)}{2c^2}\right]$$

$$\times \mathrm{Re}\left\{ \int_{k_1}^{\infty} \frac{F_1(p)\exp[-sp - gi\sqrt{p^2 - k_1^2}/2c]}{[s + G_1(p)]\sqrt{p^2 - k_1^2}} \, dp \right\}$$

$$- \frac{1}{2\pi} \exp\left[-\frac{g(z-h)}{2c^2}\right]$$

$$\times \mathrm{Re}\left\{ \int_{k_1}^{\infty} \frac{F_2(p)\exp[-sp - gi\sqrt{p^2 - k_1^2}/2c]}{[s + G_2(p)]\sqrt{p^2 - k_1^2}} \, dp \right\}$$

$$(4.2.39)$$

where

$$F_{1,2}(p) = \frac{1}{2}\left[\sqrt{\omega^2 + \frac{1}{c^2}} \pm \frac{\omega^2 + 1/c^2 - i\omega/c}{\omega - i/c}\right] \qquad (4.2.40)$$

$$G_{1,2}(p) = \frac{g}{2}\left[\sqrt{\omega^2 + \frac{1}{c^2}} \pm \left(\omega - \frac{i}{c}\right)\right] \qquad (4.2.41)$$

and

$$\omega = \frac{ixp - (z+h)\sqrt{p^2 - k_1^2}}{c^2 k_1^2}. \qquad (4.2.42)$$

We now note that

$$\int_p^{\infty} e^{-G_{1,2}(p)(t-p)} e^{-st} dt = \frac{e^{-sp}}{s + G_{1,2}(p)}. \qquad (4.2.43)$$

Upon substituting this into (4.2.39) and interchanging the order of integration, we obtain a forward Laplace transform. Thus, the complete expression for $u(x, z, t)$ is

$$
u(x, z, t) = \frac{1}{2\pi} \frac{\exp[-g(z-h)/2c^2]\cos[g\sqrt{t^2-k_1^2}/2c]}{\sqrt{t^2-k_1^2}} H(t - k_1)
$$
$$
- \frac{1}{2\pi} \frac{\exp[-g(z-h)/2c^2]\cos[g\sqrt{t^2-k_2^2}/2c]}{\sqrt{t^2-k_2^2}} H(t - k_2)
$$
$$
- \frac{g}{\pi} \exp\left[\frac{-g(z-h)}{2c^2}\right] H(t - k_1)
$$
$$
\times \operatorname{Re}\left\{ \int_{k_1}^{t} \frac{F_1(p)\exp[gi\sqrt{p^2-k_1^2}/2c - (t-p)G_1(p)]}{\sqrt{p^2-k_1^2}}\,dp \right\}
$$
$$
- \frac{g}{\pi} \exp\left[\frac{-g(z-h)}{2c^2}\right] H(t - k_1)
$$
$$
\times \operatorname{Re}\left\{ \int_{k_1}^{t} \frac{F_2(p)\exp[gi\sqrt{p^2-k_1^2}/2c - (t-p)G_2(p)]}{\sqrt{p^2-k_1^2}}\,dp \right\}.
$$

$$(4.2.44)$$

The first two terms represent the source and image while the integrals are the surface effects.

Finally, let us solve the wave equation[19]

$$
\frac{\partial^2 u}{\partial x^2} + \frac{\partial^2 u}{\partial y^2} = \frac{1}{c^2}\frac{\partial^2 u}{\partial t^2} \qquad (4.2.45)
$$

where c denotes the phase speed. The medium is initially at rest so that $u(x, y, 0) = u_t(x, y, 0) = 0$. At infinity, the disturbances die away, $\lim_{y \to \pm\infty} u(x, y, t) \to 0$, while at $y = 0$

$$
u(x, 0^+, t) - u(x, 0^-, t) = \begin{cases} vt - |x| & \text{if } 0 < |x| < vt, \\ 0 & \text{if } |x| > vt \end{cases} \qquad (4.2.46)
$$

and

$$
\frac{\partial u(x, 0^+, t)}{\partial y} = \frac{\partial u(x, 0^-, t)}{\partial y}. \qquad (4.2.47)
$$

We begin by first taking the Laplace transform of (4.2.45), followed by taking the Fourier transform. If

$$
\bar{u}(x, y, s) = \int_0^\infty u(x, y, t)e^{-st}dt \qquad (4.2.48)
$$

[19] Taken from Bakhshi, V. S., 1965: Propagation of a fracture in an infinite medium. *PAGEOPH*, **62**, 23–34. Published by Birkhäuser Verlag, Basel, Switzerland.

and

$$\overline{U}(k, y, s) = \int_{-\infty}^{\infty} \overline{u}(x, y, s) e^{-ikx} dx, \qquad (4.2.49)$$

then (4.2.45) becomes

$$\frac{d^2 \overline{U}(k, y, s)}{dy^2} - \left(k^2 + \frac{s^2}{c^2}\right) \overline{U}(k, y, s) = 0 \qquad (4.2.50)$$

with $\lim_{y \to \pm\infty} \overline{U}(k, y, s) \to 0$,

$$\overline{U}(k, 0^+, s) - \overline{U}(k, 0^-, s) = \frac{2}{s} \left[\frac{1}{k^2 + (s/v)^2}\right] \qquad (4.2.51)$$

and

$$\frac{d\overline{U}(k, 0^+, s)}{dy} = \frac{d\overline{U}(k, 0^-, s)}{dy}. \qquad (4.2.52)$$

The solution to (4.2.50)–(4.2.52) is

$$\overline{U}(k, y, s) = \operatorname{sgn}(y) \frac{v^2}{s(s^2 + v^2 k^2)} e^{-|y|\sqrt{k^2 + s^2/c^2}} \qquad (4.2.53)$$

where $\operatorname{sgn}(y) = 1$ if $y > 0$ and $\operatorname{sgn}(y) = -1$ if $y < 0$. Therefore, taking the inverse Fourier transform,

$$\overline{u}(x, y, s) = \operatorname{sgn}(y) \frac{1}{2\pi s} \int_{-\infty}^{\infty} \frac{v^2}{s^2 + v^2 k^2} e^{ikx - |y|\sqrt{k^2 + s^2/c^2}} dk \qquad (4.2.54)$$

$$= \operatorname{sgn}(y) \frac{1}{\pi s} \int_{0}^{\infty} \frac{v^2}{s^2 + v^2 k^2} \cos(kx) e^{-|y|\sqrt{k^2 + s^2/c^2}} dk \qquad (4.2.55)$$

$$= \operatorname{sgn}(y) \frac{1}{\pi s^2} \operatorname{Re}\left\{ \int_{0}^{\infty} \frac{1}{w^2 + 1/v^2} \right.$$
$$\left. \times \exp[-s(iwx + |y|\sqrt{w^2 + 1/c^2})] \, dw \right\} \qquad (4.2.56)$$

where $k = s/w$.

We now invert $\overline{u}(x, y, s)$ by Cagniard's method. Consider the case of $y > 0$; the case for $y < 0$ follows by analog. Let us introduce the transformation

$$t = iwx + y\sqrt{w^2 + 1/c^2} \qquad (4.2.57)$$

where $\operatorname{Re}(\sqrt{w^2 + 1/c^2}) \geq 0$. We define the multivalued square root by introducing branch points at $w = \pm i/c$ and a branch cut along the imaginary axis of the w-plane. Our definition of the phases is

$$w - i/c = re^{\theta i}, \qquad -\pi/2 \leq \theta < 3\pi/2 \qquad (4.2.58)$$

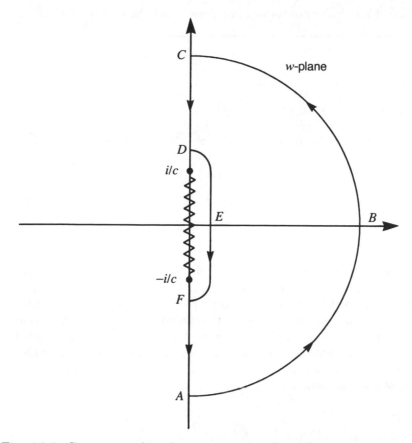

Fig. 4.2.3: Contour used in the integration of (4.2.56).

and

$$w + i/c = re^{\theta i}, \qquad -\pi/2 \leq \theta < 3\pi/2. \tag{4.2.59}$$

Consider the path $ABCDEFA$ shown in Fig. 4.2.3. The semicircle ABC has an infinite radius. Along this semicircle $t = iwx + yx + O(1/w)$. Along the imaginary axis, $CDEFA$, we set $w = li$ and give the values of t in Table 4.2.1. Fig. 4.2.4 shows the closed contour in the t-plane with corresponding points marked with primes. Note that as we move from E to F, the corresponding curve in the t-plane moves along the real axis until $t(l_0)$, marked by T', and then back-tracks to F'. To maintain uniqueness, we introduce a cut along the real axis of the t-plane from $E'T'$ and then take that k in (4.2.57) whose modulus is smaller.

Carrying out the change of variables,

$$\overline{u}(x,y,s) = \frac{1}{\pi s^2} \mathrm{Re}\left[\int_\Gamma f(x,y,t)\frac{\partial w}{\partial t}e^{-st}dt\right] \tag{4.2.60}$$

Transform Methods for Solving Partial Differential Equations

Table 4.2.1. Values of various functions along the contour $CDEFA$.

w	l	$\sqrt{w^2 + 1/c^2}$	t
C to D	∞ to $1/c$	$i\sqrt{l^2 - 1/c^2}$	$-lx + iy\sqrt{l^2 - 1/c^2}$
D	$1/c$	0	$-x/c$
D to E	$1/c$ to 0	$\sqrt{1/c^2 - l^2}$	$-lx + y\sqrt{1/c^2 - l^2}$
E	0	$1/c$	y/c
E to F	0 to $-1/c$	$\sqrt{1/c^2 - l^2}$	$-lx + y\sqrt{1/c^2 - l^2}$
F	$-1/c$	0	x/c
F to A	$-1/c$ to $-\infty$	$-i\sqrt{l^2 - 1/c^2}$	$-lx - iy\sqrt{l^2 - 1/c^2}$

where $f(x, y, t)$ is the expression for $[w^2 + 1/v^2]^{-1}$ after we transform to t. The path of integration Γ starts at the point y/c and extends out to infinity in the first quadrant along the curve shown in Fig. 4.2.4.

We would now like to deform Γ to an integration along the real axis so that the analogy with the forward Laplace transform is complete. This is permissible because the contribution from the arc $S'B'$ vanishes as the arc approaches infinite radius. If $v < c$ so that the branch points are of no importance,

$$\bar{u}(x, y, s) = \frac{1}{\pi s^2} \mathrm{Re}\left[\int_{E'S'} f(x, y, t) \frac{\partial w}{\partial t} e^{-st} dt\right] \tag{4.2.61}$$

$$= \frac{1}{\pi s^2} \mathrm{Re}\left[\int_{y/c}^{\infty} f(x, y, t) \frac{\partial w}{\partial t} e^{-st} dt\right] \tag{4.2.62}$$

$$= \frac{1}{\pi s^2} \mathrm{Re}\left[\int_{0}^{\infty} f(x, y, t) \frac{\partial w}{\partial t} e^{-st} H(t - y/c) \, dt\right]. \tag{4.2.63}$$

Because the value of w along the real axis is

$$w = \frac{-ixt + k}{x^2 + y^2}$$

where $k = \sqrt{y^2 t^2 - x^2 y^2 s^2 - y^4 s^2}$ and $p = 1/c$, we can write down the inverse of (4.2.63) as a convolution

$$u(x, y, t) = \frac{t}{\pi} * \frac{(x^2 + y^2)[2kx^2t + y^2t/k\{L^2(x^2 + y^2)^2 + k^2 - x^2t^2\}]}{[L^2(x^2 + y^2)^2 + k^2 - x^2t^2]^2 + 4x^2k^2t^2}$$

$$\times H(t - \sqrt{x^2 + y^2}/c) \tag{4.2.64}$$

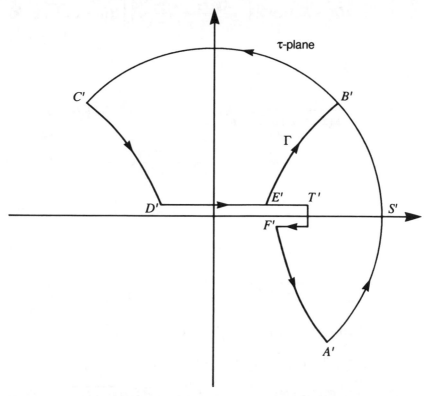

Fig. 4.2.4: New contour used in the integration of (4.2.56).

where $L^2 = 1/v^2$ or

$$u(x, y, t) = \frac{x^2 + y^2}{\pi} H(t - \sqrt{x^2 + y^2}/c)$$
$$\int_{\sqrt{x^2+y^2}/c}^{t} \frac{[2kx^2\eta + y^2\eta/k\{L^2(x^2 + y^2)^2 + k^2 - x^2\eta^2\}](t - \eta)}{[L^2(x^2 + y^2)^2 + k^2 - x^2\eta^2]^2 + 4x^2k^2\eta^2} \, d\eta.$$

$$(4.2.65)$$

Evaluating the integral in (4.2.65),

$$u(x, y, t) = B\left[2M \tan^{-1}\left\{ \frac{Ly(H - G)}{L^2y^2 + HG} \right\} + A\ln\left\{ \frac{L^2y^2 + H^2}{L^2y^2 + G^2} \right\} \right.$$
$$+ \frac{1}{2}(p\xi - Q\eta)\ln\left\{ \frac{(x_1 - x_2)^2 + (y_1 - y_2)^2}{(x_1 + x_2)^2 + (y_1 + y_2)^2} \right\}$$
$$- 4Q\ln\left\{ \frac{t + \sqrt{t^2 - C}}{\sqrt{C}} \right\} - \pi(Q\xi + P\eta)$$

$$+ (Q\xi + P\eta) \tan^{-1} \left\{ \frac{2(y_1 x_2 - x_1 y_2)}{x_2^2 + y_2^2 - x_1^2 - y_1^2} \right\} \Big] H(t - \sqrt{x^2 + y^2}/c)$$

$$(4.2.66)$$

where

$$B = \frac{1}{8\pi x L(x^2 + y^2)\sqrt{L^2 - p^2}}, \qquad P = 2x^2(x^2 + y^2)(L^2 - p^2),$$

$$(4.2.67)$$

$$Q = 2xyL\sqrt{L^2 - p^2}, \qquad \xi = \frac{p^2 xy}{L^2 y^2 + x^2(L^2 - p^2)}, \qquad (4.2.68)$$

$$\eta = \frac{L(x^2 + y^2)\sqrt{L^2 - x^2}}{L^2 y^2 + x^2(L^2 - p^2)}, \qquad F = 2Lxy\sqrt{L^2 - p^2}, \qquad (4.2.69)$$

$$G = x\sqrt{L^2 - p^2} - \sqrt{t^2 - C}, \qquad H = x\sqrt{L^2 - p^2} + \sqrt{t^2 - C}, \quad (4.2.70)$$

$$A = \frac{t(PLy - Qx\sqrt{L^2 - p^2})}{L^2 y^2 + x^2(L^2 - p^2)}, \qquad M = \frac{t(QLy + Px\sqrt{L^2 - p^2})}{L^2 y^2 + x^2(L^2 - p^2)},$$

$$(4.2.71)$$

$$J = C(t^2 - L^2 x^2) - Cy^2(p^2 - L^2), \qquad k = 2CLxy\sqrt{L^2 - p^2}, \quad (4.2.72)$$

$$x_1 = F(2t^2 - C), \qquad x_2 = 2txy(2L^2 - p^2)\sqrt{t^2 - C}, \qquad (4.2.73)$$

$$y_1 = Ct^2 + E(C - 2t^2), \qquad y_2 = 2Lt(y^2 - x^2)\sqrt{(t^2 - C)(L^2 - p^2)},$$

$$(4.2.74)$$

$$E = L^2 x^2 + y^2(p^2 - L^2) \qquad (4.2.75)$$

and $C = (x^2 + y^2)/c^2$.

Problems

1. Given the Laplace transform

$$F(s) = \int_0^\infty e^{-a\sqrt{s^2 + k^2}} \frac{[1 + \delta\sqrt{(s^2 + k^2)/(\gamma s^2 + k^2)}]^n s^{2n-1}}{(s^2 + \alpha^2 k^2)^n (s^2 + k^2)^{(n+1)/2}} k \, dk,$$

where a, α, δ and γ are real and positive and $n > 1$, use the Cagniard technique to find the inverse $f(t)$.

Step 1. Introducing $k = sz$, show that we can rewrite $F(s)$

$$F(s) = \int_0^\infty ze^{-as\sqrt{z^2 + 1}} \frac{[1 + \delta\sqrt{(1 + z^2)/(\gamma + z^2)}]^n}{s^n(1 + \alpha^2 z^2)^n(1 + z^2)^{(n+1)/2}} \, dz.$$

Step 2. Introducing $a\sqrt{z^2+1} = t$, show that we can rewrite $F(s)$

$$F(s) = \frac{1}{as^n} \int_0^\infty H(t-a)e^{-st}\frac{(a/t + \delta a/\sqrt{\gamma a^2 - a^2 + t^2})^n}{[1 + \alpha^2(t^2/a^2 - 1)]^n}\,dt.$$

Step 3. Using the convolution theorem, show that

$$f(t) = \frac{1}{a(n-1)!}\int_a^t \frac{(t-\tau)^{n-1}(a/\tau + \delta a/\sqrt{\gamma a^2 - a^2 + \tau^2})^n}{[1 + \alpha^2(\tau^2/a^2 - 1)]^n}\,d\tau.$$

2. Solve the ideal string problem[20]

$$\frac{\partial^2 u}{\partial t^2} = c^2\frac{\partial^2 u}{\partial x^2} + \frac{F}{\rho}\delta[t - f(x)]H(x)$$

by Cagniard's method where $u(x,t)$ is the displacement, c is the phase speed, F is the amplitude of the load, $\delta(\ )$ is the Dirac delta function, ρ is the density of the string and $f(0) = 0$. Assume $u(x,0) = u_t(x,0) = 0$ and $\lim_{x\to\pm\infty} u(x,t) \to 0$.

Step 1. Show that the Laplace transform of the partial differential equation is

$$s^2\bar{u} = c^2\frac{d^2\bar{u}}{dx^2} + \frac{F}{\rho}e^{-sf(x)}H(x).$$

Step 2. Take the Fourier transform of the ordinary differential equation in step 1 and show that the joint transform is

$$\frac{\rho\bar{U}(k,s)}{F} = \frac{1}{s^2 + c^2k^2}\int_0^\infty e^{-sf(\xi)-ik\xi}\,d\xi.$$

Step 3. Show that we can write Fourier inverse (after reversing the order of integration) of step 2 as

$$\frac{\rho s\bar{u}(x,s)}{F} = \frac{-1}{2\pi i}\int_0^\infty \int_{-\infty i}^{\infty i} \frac{e^{-s[f(\xi)+p(\xi-x)]}}{c^2(p - 1/c)(p + 1/c)}\,dp\,d\xi$$

after setting $p = ik/s$.

---

[20] Reprinted from *Int. J. Solids Structures*, **4**, F. T. Flaherty, Transient resonance of an ideal string under a load moving with varying speed, pp. 1221–1231, ©1968, with kind permission from Pergamon Press Ltd., Headington Hill Hall, Oxford OX3 0BW, UK.

Transform Methods for Solving Partial Differential Equations

Step 4. Use the residue theorem to evaluate the p integration by closing the contour on the left or right side of the p-plane, as dictated by Jordan's lemma. Keeping the sign of $\xi - x$ in mind, show that

$$\frac{2c\rho s\bar{u}(x,s)}{F} = \int_0^x e^{-s[f(\xi)-(\xi-x)/c]}\,d\xi + \int_x^\infty e^{-s[f(\xi)+(\xi-x)/c]}\,d\xi, \quad x > 0$$

and

$$\frac{2c\rho s\bar{u}(x,s)}{F} = \int_0^\infty e^{-s[f(\xi)+(\xi-x)/c]}\,d\xi, \quad x < 0.$$

Step 5. Define the Cagniard contours

$$f(\xi) \pm (\xi - x)/c = t$$

so that you can write the integrals in step 4 as

$$\frac{2c\rho s\bar{u}(x,s)}{F} = \int_{x/c}^{f(x)} \xi_1'(t)e^{-st}\,dt + \int_{f(x)}^\infty \xi_2'(t)e^{-st}\,dt, \quad x > 0$$

and

$$\frac{2c\rho s\bar{u}(x,s)}{F} = \int_{-x/c}^\infty \xi_2'(t)e^{-st}\,dt, \quad x < 0$$

where ξ_1 and ξ_2 are

$$t = f(\xi_1) - (\xi_1 - x)/c$$
$$t = f(\xi_2) + (\xi_2 - x)/c.$$

Step 6. Obtain the inverse of $s\bar{u}(x,s)$ by using the definition of the forward Laplace transform. Show that for $x > 0$,

$$\frac{2c\rho u_t(x,t)}{F} = \begin{cases} \xi_1'(t) & \text{if} \quad x/c < t < f(x), \\ \xi_2'(t) & \text{if} \qquad\quad t > f(x), \\ 0, & \text{otherwise} \end{cases}$$

while for $x < 0$

$$\frac{2c\rho u_t(x,t)}{F} = \begin{cases} \xi_2'(t) & \text{if} \quad t > -x/c, \\ 0, & \text{otherwise.} \end{cases}$$

4.3 THE MODIFICATION OF CAGNIARD'S METHOD BY De HOOP

Although the Cagniard technique is very clever, its use in more complicated problems becomes problematic because we must use series of transformation[21]. In 1960 De Hoop[22] suggested a modification which simplified matters by playing the Fourier transform off against the Laplace transform. It is this modification that has had the greatest acceptance.

To illustrate this technique, consider the two-dimensional wave equation[23]

$$\frac{\partial^2 u}{\partial x^2} + \frac{\partial^2 u}{\partial y^2} = \frac{1}{c^2}\frac{\partial^2 u}{\partial t^2} - M\delta(x)\delta(y)\delta(t) \tag{4.3.1}$$

where $-\infty < x < \infty$, $-\infty < y < \infty$ and $t > 0$. If all of the initial conditions are zero, the transformed form of (4.3.1) is

$$\frac{d^2\overline{U}}{dy^2} - s^2\left(\alpha^2 + \frac{1}{c^2}\right)\overline{U} = -M\delta(y) \tag{4.3.2}$$

where we have taken the Laplace transform with respect to time and the Fourier transform with respect to the x-direction. We have also replaced the Fourier transform parameter k with αs. The solution of (4.3.2) that tends to zero as $|y| \to \infty$ is

$$\overline{U}(\alpha, y, s) = \frac{M}{2s\beta}e^{-s\beta|y|} \tag{4.3.3}$$

where

$$\beta(\alpha) = \left(\alpha^2 + \frac{1}{c^2}\right)^{1/2} \quad \text{with } \mathrm{Re}(\beta) \geq 0. \tag{4.3.4}$$

We first invert with the Fourier transform. This yields

$$\overline{u}(x, y, s) = \frac{M}{2\pi}\int_{-\infty}^{\infty} \frac{e^{-s(\beta|y|-i\alpha x)}}{2\beta}\, d\alpha \tag{4.3.5}$$

where $-\infty < x < \infty$ and $-\infty < y < \infty$. Introducing $\alpha = iw$, (4.3.5) becomes

$$\overline{u}(x, y, s) = \frac{M}{2\pi i}\int_{-\infty i}^{\infty i} \frac{e^{-s(wx+\beta|y|)}}{2\beta}\, dw \tag{4.3.6}$$

[21] Dix, C. H., 1954: The method of Cagniard in seismic pulse problems. *Geophysics*, **19**, 722–738.

[22] De Hoop, A. T., 1960: A modification of Cagniard's method for solving seismic pulse problems. *Appl. Sci. Res.*, **B8**, 349–356.

[23] Patterned after Gopalsamy, K., and B. D. Aggarwala, 1972: Propagation of disturbances from randomly moving sources. *Zeit. angew. Math. Mech.*, **52**, 31–35.

with

$$\beta = \left(\frac{1}{c^2} - w^2\right)^{1/2} \quad \text{with } \text{Re}(\beta) \geq 0. \qquad (4.3.7)$$

As before, we note that choosing a path of integration so that

$$wx + \beta|y| = t \qquad (4.3.8)$$

is real and positive results in a forward Laplace transform. In this case, however, we must deform our integral to the new Cagniard contour with greater care. There are branch points at $w = 1/c$ and $w = -1/c$. To ensure that $\text{Re}(\beta) \geq 0$, the branch cuts lie along the real axis from $1/c$ to ∞ and from $-1/c$ to $-\infty$.

From (4.3.8), a little algebra gives

$$w = \frac{xt}{r^2} \pm \frac{i|y|}{r^2}\sqrt{t^2 - \frac{r^2}{c^2}}, \qquad \frac{r}{c} < t < \infty \qquad (4.3.9)$$

where $r^2 = x^2 + y^2$. Furthermore, along this hyperbola

$$\beta = \frac{|y|t}{r^2} \mp \frac{xi}{r^2}\sqrt{t^2 - \frac{r^2}{c^2}} \qquad (4.3.10)$$

and

$$\frac{\partial w}{\partial t} = \pm\frac{i\beta}{\sqrt{t^2 - r^2/c^2}}. \qquad (4.3.11)$$

Upon using the symmetry of the path of integration, a substitution to an integration in t yields

$$\bar{u}(x, y, s) = \frac{M}{2\pi}\int_{r/c}^{\infty} \frac{e^{-st}}{\sqrt{t^2 - r^2/c^2}}\, dt. \qquad (4.3.12)$$

Consequently, by inspection,

$$u(x, y, t) = \frac{M}{2\pi}\frac{H(t - r/c)}{\sqrt{t^2 - r^2/c^2}}. \qquad (4.3.13)$$

Our solution $u(x, y, t)$ gives the Green's function of the two-dimensional wave equation.

For our second example, we find the (SH) wave motion[24] in a two layer, semi-infinite elastic medium that we subject to a stress-discontinuity of magnitude P that grows in areal coverage $a < x < a+Vt$

---

[24] Taken from Nag, K. R., 1962: Disturbance due to shearing-stress discontinuity in a semi-infinite elastic medium. *Geophys. J.*, **6**, 468–478.

and occurs at the depth h from the free surface. The governing equation is

$$\frac{\partial^2 u}{\partial x^2} + \frac{\partial^2 u}{\partial z^2} = \frac{1}{V_s^2}\frac{\partial^2 u}{\partial t^2} \tag{4.3.14}$$

in the domain $-\infty < x < \infty$, $z > -h$ and $t > 0$. At the interfaces, the wave solution must satisfy the boundary conditions

$$u_1(x, 0, t) = u_2(x, 0, t), \tag{4.3.15}$$

$$\frac{\partial u_1(x, -h, t)}{\partial z} = 0 \tag{4.3.16}$$

and

$$\frac{\partial u_1(x, 0, t)}{\partial z} - \frac{\partial u_2(x, 0, t)}{\partial z} = \begin{cases} P, & a \le x \le a + Vt, \\ 0, & \text{otherwise} \end{cases} \tag{4.3.17}$$

for $-\infty < x < \infty$ and $t > 0$. We add the restriction that $V < V_s$.

The Laplace transform of (4.3.14) yields

$$\frac{\partial^2 \overline{u}_i}{\partial x^2} + \frac{\partial^2 \overline{u}_i}{\partial z^2} - \frac{s^2}{V_s^2}\overline{u}_i = 0 \tag{4.3.18}$$

if the medium is initially at rest while the Fourier transform of (4.3.18) gives

$$\frac{d^2 \overline{U}_i}{dz^2} - \nu^2 \overline{U}_i = 0, \qquad \nu = \sqrt{k^2 + s^2/V_s^2}. \tag{4.3.19}$$

The solution to (4.3.19) for the two regions is

$$\overline{U}_1(k, z, s) = A_1(k, s)\cosh(\nu z) + B_1(k, s)\sinh(\nu z), \quad -h < z < 0 \tag{4.3.20}$$

and

$$\overline{U}_2(k, z, s) = A_2(k, s)e^{-\nu z}, \qquad z > 0 \tag{4.3.21}$$

with $\text{Re}(\nu) \ge 0$.

If we substitute (4.3.20)–(4.3.21) into the boundary conditions (4.3.15)–(4.3.17) after we have taken the Laplace and Fourier transforms, then $\overline{u}_1(x, -h, s)$ is

$$\overline{u}_1(x, -h, s) = \frac{P}{2\pi s}\int_{-\infty}^{\infty} \frac{e^{-\nu h - ika + ikx}}{\nu(ik + s/V)}\,dk \tag{4.3.22}$$

$$= \frac{P}{\pi s}\int_0^{\infty} \frac{e^{-\nu h}}{\nu(ik + s/V)}\cos(kx_1)\,dk \tag{4.3.23}$$

$$= \frac{P}{\pi s}\text{Re}\left\{ \int_0^{\infty} \frac{e^{ikx_1 - \nu h}}{\nu(ik + s/V)}\,dk \right\} \tag{4.3.24}$$

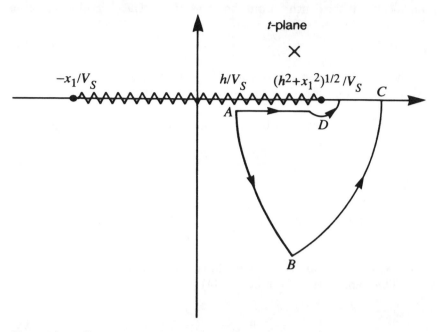

Fig. 4.3.1: Contours used in the inversion of (4.3.27).

where $x_1 = x - a$.

Let us change the integration variable from k to $p = kV_s/s$. Then

$$\bar{u}_1(x, -h, s) = \frac{PV_s}{\pi s^2} \mathrm{Re}\left\{\int_0^\infty \frac{e^{-st}}{(V_s/V + ip)\sqrt{p^2 + 1}}\, dp\right\} \qquad (4.3.25)$$

where $t = [-ipx_1 + h\sqrt{p^2 + 1}]/V_s$ or

$$p(t) = \frac{V_s}{x_1^2 + h^2}\left[itx_1 + h\sqrt{t^2 - (x_1^2 + h^2)/V_s^2}\right]. \qquad (4.3.26)$$

The integrand of (4.3.25) has branch points at $p = \pm i$ and a simple pole at $p = iV_s/V$.

From the definition of t, we see that as p varies from 0 to ∞ in the p-plane, t starts at h/V_s, moves into the fourth quadrant of the t-plane and approaches the limit $t \to phe^{-\theta i}/V_s$ as $p \to \infty$ where $\theta = \arctan(x_1/h)$. Fig. 4.3.1 illustrates this new contour by the curve AB. Because there are no poles in the fourth quadrant, we may deform the contour from AB to the real axis. Hence,

$$\int_0^\infty (\,)\, dp = \int_A^C (\,)\frac{dp}{dt}\, dt = \int_{h/V_s}^\infty (\,)\frac{dp}{dt}\, dt \qquad (4.3.27)$$

or

$$\bar{u}_1(x, -h, s) = \frac{PV_s}{\pi s^2} \text{Re}\left\{ \int_{h/V_s}^{\infty} F_1[p(t)] \frac{dp}{dt} e^{-st} dt \right\} \qquad (4.3.28)$$

where

$$F_1[p(t)] = \frac{1}{[ip(t) + V_s/V]\sqrt{p(t)^2 + 1}}. \qquad (4.3.29)$$

Now because $F_1[p(t)]$ is real and dp/dt is imaginary for $h/V_s < t < \sqrt{x_1^2 + h^2}/V_s$, we can replace the lower limit of h/V_s by $\sqrt{x_1^2 + h^2}/V_s$ so that

$$\bar{u}_1(x, -h, s) = \frac{1}{s^2} \frac{PV_s}{\pi} \text{Re}\left\{ \int_{\sqrt{x_1^2 + h^2}/V_s}^{\infty} F_1[p(t)] \frac{dp}{dt} e^{-st} dt \right\}. \qquad (4.3.30)$$

We recognize that $u_1(x, -h, t)$ equals the convolution of $tH(t)$ with a function $G(x, t)$ whose Laplace transform is

$$\bar{G}(x, s) = \text{Re}\left\{ \int_{\sqrt{x_1^2 + h^2}/V_s}^{\infty} F_1[p(t)] \frac{dp}{dt} e^{-st} dt \right\}. \qquad (4.3.31)$$

Because (4.3.31) is in the form of a forward Laplace transform, we immediately have

$$G[x, p(t)] = \text{Re}\left\{ F_1[p(t)] \frac{dp}{dt} \right\} H(t - \sqrt{x_1^2 + h^2}/V_s) \qquad (4.3.32)$$

$$- \frac{V(x_1^2 + h^2)(h^2 + x_1^2 - Vtx_1)}{V_s\sqrt{t^2 - (x_1^2 + h^2)/V_s^2}} \qquad (4.3.33)$$

$$\times \frac{H(t - \sqrt{x_1^2 + h^2}/V_s)}{\{(h^2 + x_1^2 - Vtx_1)^2 + V^2h^2[t^2 - (x_1^2 + h^2)/V_s^2]\}}$$

and

$$u_1(x, -h, t) = \frac{PV_s}{\pi} \int_{\sqrt{x_1^2 + h^2}/V_s}^{t} (t - \lambda)G[x, p(\lambda)] \, d\lambda. \qquad (4.3.34)$$

Problems

1. Use the Cagniard-de Hoop method to invert the transform[25]

$$\bar{u}(x, z, s) = \int_0^{\infty} J_0(kx) \frac{\exp(-z\sqrt{k^2 + s^2/c^2})}{\sqrt{k^2 + s^2/c^2}} k \, dk$$

---

[25] Taken from Wiggins, R. A., and D. V. Helmberger, 1974: Synthetic seismogram computation by expansion in generalized rays. *Geophys. J. R. Astr. Soc.*, **37**, 73–90.

for large x and $z > 0$.

Step 1. By changing variables via $k = -isp$, convert the original transform to

$$\bar{u}(x, z, s) = \frac{2s}{\pi} \text{Im}\left[\int_0^{\infty i} \frac{K_0(spx)\exp(-s\eta z)}{\eta} p\,dp\right]$$

where $\eta = \sqrt{1/c^2 - p^2}$. Fig. 4.3.2 shows the branch cuts for η.

Step 2. Using the leading term from the asymptotic expansion for $K_0(\ )$, show that for $x \gg 1$

$$\bar{u}(x, z, s) = \left(\frac{2s}{\pi x}\right)^{1/2} \text{Im}\left[\int_0^{\infty i} \frac{\sqrt{p}}{\eta} \exp[-s(px + \tau)]\,dp\right]$$

where $\tau = z\eta$.

Step 3. By setting $t = px + \tau$, we can deform the original p contour into one where t is real as shown in Fig. 4.3.2. Show then that

$$\bar{u}(x, z, s) = \left(\frac{2s}{\pi x}\right)^{1/2} \int_0^{\infty} \text{Im}\left[\frac{\sqrt{p(t)}}{\eta(t)}\frac{dp}{dt}\right]e^{-st}\,dt$$

or

$$u(x, z, t) = \left(\frac{2}{\pi x}\right)^{1/2} \frac{d}{dt}\left[\frac{H(t)}{\sqrt{t}}\right] * f(t)$$

where

$$f(t) = \text{Im}\left[\frac{\sqrt{p(t)}}{\eta(t)}\frac{dp}{dt}\right],$$

$$p(t) = \begin{cases} xt/r^2 - z\sqrt{r^2/c^2 - t^2}/r^2 & \text{if } t \leq r/c \\ xt/r^2 + iz\sqrt{t^2 - r^2/c^2}/r^2 & \text{if } t \geq r/c \end{cases}$$

and $r = \sqrt{x^2 + z^2}$.

2. We have shown how we may use the Cagniard-de Hoop method to solve the wave equation. Consider now the following system of diffusion equations[26]:

$$\frac{\partial^2 u_1}{\partial x^2} + \frac{\partial^2 u_1}{\partial z^2} = \frac{\partial u_1}{\partial t}, \qquad -\infty < x < \infty, 0 < z < \infty, t > 0$$

and

$$\frac{\partial^2 u_2}{\partial x^2} + \frac{\partial^2 u_2}{\partial z^2} = \frac{\partial u_2}{\partial t}, \qquad -\infty < x < \infty, -\infty < z < 0, t > 0$$

---

[26] Patterned after De Hoop, A. T., and M. L. Oristaglio, 1988: Application of the modified Cagniard technique to transient electromagnetic diffusion problems. *Geophys. J.*, **94**, 387–397.

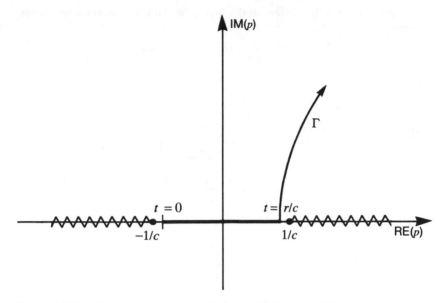

Fig. 4.3.2: Contour arising in the use of Cagniard-de Hoop to solve problem 1.

with the boundary conditions

$$\lim_{z \to \infty} u_1(x, z, t) \to 0, \qquad \lim_{z \to -\infty} u_2(x, z, t) \to 0,$$

$$u_1(x, 0^+, t) = u_2(x, 0^-, t)$$

and

$$\frac{\partial u_1(x, 0^+, t)}{\partial z} - \frac{\partial u_2(x, 0^-, t)}{\partial z} = \delta(x)\delta(t)$$

and the initial conditions

$$u_1(x, z, 0) = u_2(x, z, 0) = 0.$$

Assume $a_2 > a_1$.

Step 1. By defining the Laplace and Fourier transforms as follows:

$$\bar{u}_i(x, z, s) = \int_0^\infty u_i(x, z, t)e^{-st} dt$$

and

$$\overline{U}_i(k, z, s) = \int_{-\infty}^\infty \bar{u}_i(x, z, s)e^{-i\sqrt{s}kx} dx,$$

show that the partial differential equations and boundary conditions become

$$\frac{d^2\overline{U}_1}{dz^2} - s\left(k^2 - \frac{1}{a_1^2}\right)\overline{U}_1 = 0, \qquad 0 < z < \infty$$

and

$$\frac{d^2\overline{U}_2}{dz^2} - s\left(k^2 - \frac{1}{a_2^2}\right)\overline{U}_2 = 0, \qquad -\infty < z < 0$$

with

$$\lim_{z\to\infty}\overline{U}_1(k, z, s) \to 0, \qquad \lim_{z\to-\infty}\overline{U}_2(k, z, s) \to 0,$$

$$U_1(k, 0^+, s) = U_2(k, 0^-, s), \text{ and } \frac{dU_1(k, 0^+, s)}{dz} - \frac{dU_2(k, 0^-, s)}{dz} = 1.$$

Step 2. Show that the solution to step 1 is

$$\overline{U}_1(k, z, s) = -\frac{\exp(-\sqrt{s}\gamma_1 z)}{\sqrt{s}\,(\gamma_1 + \gamma_2)}, \qquad 0 < z < \infty$$

and

$$\overline{U}_2(k, z, s) = -\frac{\exp(\sqrt{s}\gamma_2 z)}{\sqrt{s}\,(\gamma_1 + \gamma_2)}, \qquad -\infty < z < 0$$

where $\gamma_i = \sqrt{k^2 + 1/a_i^2}$ with $\mathrm{Re}(\gamma_i) > 0$.

Step 3. Show that

$$\overline{u}_1(x, y, s) = \frac{\sqrt{s}}{2\pi}\int_{-\infty}^{\infty}\overline{U}_1(k, z, s)e^{i\sqrt{s}kx}\,dk$$

$$= -\frac{1}{2\pi}\int_{-\infty}^{\infty}\frac{\exp[-\sqrt{s}(-ikx + \gamma_1 z)]}{\gamma_1 + \gamma_2}\,dk$$

$$= -\frac{1}{2\pi i}\int_{-\infty i}^{\infty i}\frac{\exp[-\sqrt{s}(px + \overline{\gamma}_1 z)]}{\overline{\gamma}_1 + \overline{\gamma}_2}\,dp$$

where $\overline{\gamma}_i = \sqrt{1/a_i^2 - p^2}$.

Step 4. Let us choose the branch cut of $\overline{\gamma}_i$ to be along the real axis of the p-plane between the branch points $[-\infty, -1/a_i]$ and $[1/a_i, \infty]$. Then $\mathrm{Re}(\gamma_i) > 0$. Next, introduce the real, positive variable $\kappa = px + \overline{\gamma}_1|z|$. For $z > 0$,

$$p_1 = \frac{\kappa x}{r^2} + i\frac{z}{r^2}\sqrt{\kappa^2 - \kappa_1^2}, \qquad \kappa_1 < \kappa < \infty$$

where $r = \sqrt{x^2 + z^2}$ and $\kappa_1 = r/a_1$. Show that we can deform the integration in step 3 to this new path, yielding

$$\overline{u}_1(x, z, s) = -\frac{1}{\pi}\int_{\kappa_1}^{\infty}\mathrm{Im}\left(\frac{\partial_\kappa p_1}{\overline{\gamma}_1 + \overline{\gamma}_2}\right)e^{-\sqrt{s}\kappa}\,d\kappa.$$

See Example 3.2.4.

Step 5. Finish the problem by noting the the only quantity in the integral in step 4 that contains s is $\exp(-\sqrt{s}\kappa)$. Therefore, by inspection,

$$u_1(x, z, t) = -\frac{1}{\pi} \int_{\kappa_1}^{\infty} \text{Im}\left(\frac{\partial_\kappa p_1}{\overline{\gamma}_1 + \overline{\gamma}_2}\right) G(t, \kappa)\, d\kappa$$

where

$$\partial_\kappa p_1 = \frac{i\overline{\gamma}_1}{\sqrt{\kappa^2 - \kappa_1^2}}$$

and

$$G(t, \kappa) = \frac{\kappa}{\sqrt{4\pi t^3}} \exp(-\kappa^2/4t) H(t).$$

4.4 EXPANSIONS IN TERMS OF LEAKY MODES.

In section 4.1 we showed that the solution to a joint Laplace-Fourier transform problem often consists of two parts: normal modes due to the poles on the proper Riemann surface, which satisfy the Sommerfeld radiation condition, and an integration along the branch cut(s) which gives the continuous spectrum. In addition to these proper modes, we may also have improper modes that reside on the other, aphysical Riemann surface(s). Because these additional modes violate the Sommerfeld radiation condition, we may discard them from further considerations[27].

During the 1950's, before the advent of powerful digital computers, the only viable method of evaluating the branch cut integral was the asymptotic method of steepest descent[28]. During the process of deforming the branch cut integral to the path of steepest descent, the contour often crossed onto the improper Riemann surface. Thus, the branch cut integral sometimes contained two parts: a contribution from the path of steepest descent plus the contribution from various improper modes. Because these improper modes represent the radiation or leakage of energy away from the source[29], the term "leaky mode" was coined.

At approximately the same time, geophysicists found leaky modes in waveguides such as an oceanic layer overlying an infinite elastic half-space. One of the intriguing aspects of this problem was the migration

---

[27] For a simple example, see Tai, C., 1951: The effect of a grounded slab on the radiation from a line source. *J. Appl. Phys.*, **23**, 405–414.

[28] See, for example, Bleistein, N., and R. A. Handelsman, 1986: *Asymptotic Expansions of Integrals.* New York: Dover Publications, Inc., Chapter 7.

[29] Tamir, T., and A. A. Oliner, 1963: Guided complex waves. Part 1. Fields at an interface. *Proc. IEE*, **110**, 310–324.

and subsequent appearance on the proper Riemann surface of some of the leaky modes during the inversion of the Laplace transform. An unfortunate by-product was the discontinuous behavior in the variation of the branch cut integral as each leaky mode popped onto the proper Riemann surface.

During the late 1980's, Haddon[30,31,32,33,34], in a series of papers building upon an original idea by Rosenbaum[35], showed a way around this difficulty. He suggested that if we deform the *Bromwich contour* so that the evolution of the leaky modes from one Riemann surface to the other is continuous, then the discontinuity in the branch cut will not only vanish but so will the branch cut integral. Thus, we can express the continuous spectrum completely in terms of an expansion of leaky modes. It is this derivation that we explore in this section. Others[36,37,38,39] have suggested similar ideas but Haddon's are the most well developed.

---

[30] Haddon, R. A. W., 1984: Computation of synthetic seismograms in layered earth models using leaky modes. *Bull. Seism. Soc. Am.*, **74**, 1225–1248.

[31] Haddon, R. A. W., 1986: Exact evaluation of the response of a layered elastic medium to an explosive point source using leaky modes. *Bull. Seism. Soc. Am.*, **76**, 1755–1775.

[32] Haddon, R. A. W., 1987: Numerical evaluation of Green's function for axisymmetric boreholes using leaky modes. *Geophysics*, **52**, 1099–1105.

[33] Haddon, R. A. W., 1987: Response of an oceanic wave guide to an explosive point source using leaky modes. *Bull. Seism. Soc. Am.*, **77**, 1804–1822.

[34] Haddon, R. A. W., 1989: Exact Green's functions using leaky modes for axisymmetric boreholes in solid elastic media. *Geophysics*, **54**, 609–620.

[35] Rosenbaum, J. H., 1965: Transmission of explosive sound through a liquid layer over an infinitely deep elastic bottom. *Bull. Seism. Soc. Am.*, **55**, 879–901.

[36] Shatrov, A. D., 1970: Discrete representation of the field in the problem of excitation of a dielectric plate. *Radio Engng. Electron. Phys.*, **15**, 1564–1573.

[37] Shatrov, A. D., 1972: Expansions of fields in open wave guides and resonators. *Radio Engng. Electron. Phys.*, **17**, 896–902.

[38] Tesler, M. H., and G. Eichmann, 1978: Non-spectral field representations in dielectric fibre guides. *J. Inst. Math. Applics.*, **21**, 315–330.

[39] Carpentier, M. P., and A. F. dos Santos, 1985: Non-spectral complete field expansion in two-dimensional structures. *IMA J. Appl. Math.*, **35**, 1–12.

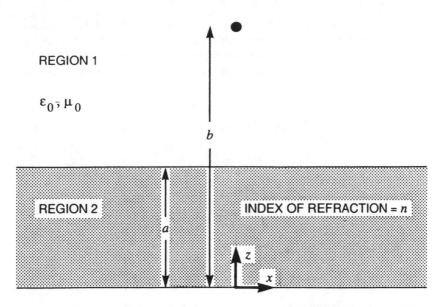

Fig. 4.4.1: A dielectric slab on a perfectly conducting half-space with a line source of current above it.

Consider[40] a dielectric slab of depth a and index of refraction n which we place on a perfectly conducting plane as shown in Fig. 4.4.1. We locate a filament of current at $x = 0$ and $z = b$. Assuming that the magnitude of the current element is such that it is a unit source function, the electric fields in regions 1 and 2 satisfy the wave equations

$$\frac{\partial^2 E_1}{\partial x^2} + \frac{\partial^2 E_1}{\partial z^2} - \frac{1}{c^2}\frac{\partial^2 E_1}{\partial t^2} = -\delta(z - b)\delta(x)\delta(t), \qquad z > a, \qquad (4.4.1)$$

and

$$\frac{\partial^2 E_2}{\partial x^2} + \frac{\partial^2 E_2}{\partial z^2} - \frac{n^2}{c^2}\frac{\partial^2 E_2}{\partial t^2} = 0, \qquad 0 < z < a, \qquad (4.4.2)$$

where c is the speed of light in the free space, x is the horizontal distance, z is the vertical distance and t is time. Because the dielectric lies on a conducting half-plane, $E_2(x, 0, t) = 0$ at $z = 0$. At infinity, the radiation condition requires that $\lim_{z\to\infty} E_1(x, z, t) \to 0$. Continuity in the electric and magnetic fields at the interface $z = a$ implies

$$E_1(x, a, t) = E_2(x, a, t) \quad \text{and} \quad \frac{\partial E_1(x, a, t)}{\partial z} = \frac{\partial E_2(x, a, t)}{\partial z}. \qquad (4.4.3)$$

---

[40] Most of the following material appeared in Duffy, D. G., 1994: Response of a grounded dielectric slab to an impulse line source using leaky modes. *IEEE Trans. Antenna Propagat.*, **42**, 340–346. ©1994 IEEE.

Transform Methods for Solving Partial Differential Equations

We solve (4.4.1)–(4.4.3) by the joint application of Laplace and Fourier transforms. We denote this joint transform of $E_j(x, z, t)$ by $\overline{\mathcal{E}}_j(k, z, s)$ with $j = 1, 2$ where s and k are the transform variables of the Laplace and Fourier transforms, respectively. Assuming that the system is initially at rest,

$$\frac{d^2\overline{\mathcal{E}}_1}{dz^2} - \nu_1^2\overline{\mathcal{E}}_1 = -\delta(z - b), \qquad z > a, \tag{4.4.4}$$

and

$$\frac{d^2\overline{\mathcal{E}}_2}{dz^2} - \nu_2^2\overline{\mathcal{E}}_2 = 0, \qquad 0 < z < a, \tag{4.4.5}$$

along with the boundary conditions $\overline{\mathcal{E}}_2(k, 0, s) = 0$, $\lim_{z \to \infty} \overline{\mathcal{E}}_1(k, z, s) \to 0$,

$$\overline{\mathcal{E}}_1(k, a, s) = \overline{\mathcal{E}}_2(k, a, s) \qquad \text{and} \qquad \frac{d\overline{\mathcal{E}}_1(k, a, s)}{dz} = \frac{d\overline{\mathcal{E}}_2(k, a, s)}{dz} \tag{4.4.6}$$

where $\nu_1 = \sqrt{k^2 + s^2/c^2}$ and $\nu_2 = \sqrt{k^2 + n^2s^2/c^2}$. Solutions which satisfy (4.4.4)–(4.4.6) are

$$\overline{\mathcal{E}}_1 = \frac{e^{-\nu_1|z-b|}}{2\nu_1} - \frac{e^{-\nu_1(z+b-2a)}}{2\nu_1}\left[\frac{\nu_2\cosh(\nu_2 a) - \nu_1\sinh(\nu_2 a)}{\nu_2\cosh(\nu_2 a) + \nu_1\sinh(\nu_2 a)}\right], z > a, \tag{4.4.7}$$

and

$$\overline{\mathcal{E}}_2 = \frac{\sinh(\nu_2 z)e^{-\nu_1(b-a)}}{\nu_2\cosh(\nu_2 a) + \nu_1\sinh(\nu_2 a)}, \qquad 0 < z < a. \tag{4.4.8}$$

Because (4.4.7)–(4.4.8) are even functions of ν_2, no branch points are associated with ν_2 but there are branch points and cuts with ν_1. In order to satisfy the radiation condition as $z \to \infty$, $\operatorname{Re}(\nu_1) \geq 0$.

With (4.4.7)–(4.4.8), we can find $E_j(x, z, t)$ via the inversion integrals

$$E_j(x, z, t) = \frac{1}{4\pi^2 i}\int_{\gamma-\infty i}^{\gamma+\infty i} e^{st}\left\{\int_{-\infty}^{\infty} \overline{\mathcal{E}}_j(k, z, s)e^{ikx}dk\right\}ds \tag{4.4.9}$$

where we choose $\gamma > 0$ so that the Laplace transform converges. If we introduce $s = \omega i$, we may rewrite (4.4.9) as

$$E_j(x, z, t) = \frac{1}{2\pi^2}\operatorname{Re}\left\{\int_{0-\gamma i}^{\infty-\gamma i}\left[\int_{-\infty}^{\infty} \overline{\mathcal{E}}_j(k, z, \omega)e^{i(kx+\omega t)}dk\right]d\omega\right\}. \tag{4.4.10}$$

The fields $E_j(x, z, t)$ vanish identically for $t < \tau(x, z)$ where $\tau(x, z)$ denotes the minimum travel-time for any disturbance originating at the

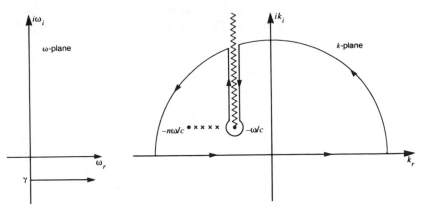

Fig. 4.4.2: Paths of integration, poles and branch cuts. The poles and branch cut occurring in the fourth quadrant have been suppressed.

source to reach the point (x, z). We shall now find the solution for $t \geq \tau(x, z)$.

The solution in region 1 consists of two parts: the first term is the direct wave from the source while the second is the reflection from the dielectric slab which appears to emanate from a source located at $z = 2a - b$. We can invert the first term of (4.4.7) exactly using the Cagniard-de Hoop method[41] and this inversion yields

$$E_1^d(x, z, t) = \frac{1}{2\pi c} \frac{H(t - r/c)}{\sqrt{c^2 t^2 - r^2}} \qquad (4.4.11)$$

where $r = \sqrt{x^2 + (z - b)^2}$ and $H(\ )$ denotes the Heaviside step function. We cannot invert the second term in (4.4.7) and $\overline{\mathcal{E}}_2(k, z, s)$ directly so we must perform the evaluation numerically.

Because of symmetry about the $x = 0$ plane, we need only consider the case $x > 0$. The conventional technique for evaluating (4.4.10) is to close the line integral along the real k axis as shown in Fig. 4.4.2. The arcs at infinity vanish by Jordan's lemma so that the inverse Fourier transform consists of two parts: poles arising from the zeros of

$$\Delta(k, \omega) := \nu_2 \cosh(\nu_2 a) + \nu_1 \sinh(\nu_2 a) = 0 \qquad (4.4.12)$$

where $\nu_1 = \sqrt{k^2 - \omega^2/c^2}$ and $\nu_2 = \sqrt{k^2 - n^2\omega^2/c^2}$ which lie in the upper half of the k-plane between $-\omega/c$ and $-n\omega/c$ in the second quadrant and a branch cut integral. Therefore, in principle, we can find the electric field first by computing the inverse Fourier transform in terms

[41] De Hoop, A. T., 1979: Pulsed electromagnetic radiation from a line source in a two-media configuration. *Radio Sci.*, **14**, 253–268.

of the residues and a branch cut integral for a specific (complex) value of ω along the contour shown in Fig. 4.4.2. Then by summing up the contributions we invert the Laplace transform.

Neglecting for the moment any questions concerning how we would actually compute numerically the branch cut integral, another, more pressing problem appears as ω varies from $-\gamma i$ to $\infty - \gamma i$. Some of the poles on the improper (non-spectral) Riemann surface migrate until they cross the branch cut at the cutoff frequency and "pop" onto the proper Riemann surface. A consequence of this migration of poles is a discontinuous change in the branch cut integral at the cutoff frequency. Because of the computational difficulty arising from this phenomena, we must find another method to invert the Fourier transform.

One approach is to deform the contour associated with the branch cut integral onto the improper Riemann surface so that we pick up the various improper modes that eventually become normal modes (see Fig. 4.4.3). This is permissible only if the poles on the improper Riemann surface are analytically continuous to its counterpart on the proper Riemann surface. If we remain with our original contour, this is not the case. Consequently, we must find a new ω contour which does produce an orderly, continuous evolution of poles from one Riemann surface to another. This is the essence of Haddon's technique: a deformation of *both* the k and ω contour. To sculpt such a contour, we must consider various singularities on the ω-plane.

First, to avoid any mathematical difficulties associated with a pass through the branch point $k = \omega/c$, our new ω contour must produce a $k_m(\omega)$ which avoids this point. We find this ω numerically by solving the equation $\Delta(\omega_m^{(C)}/c, \omega_m^{(C)}) = 0$ where we denote this singularity by $\omega_m^{(C)}$ and where we have used a counter m to count the poles. For our problem, we find

$$k_m^{(C)} = \frac{\omega_m^{(C)}}{c} = \frac{(2m-1)\pi}{2a\sqrt{n^2-1}}, \qquad m = 1, 2, 3, \ldots \qquad (4.4.13)$$

because $\nu_1 = 0$. Next, we must choose whether the contour passes above or below this singularity. Because $\Delta(k, \omega)$ is regular at this point, it makes no difference analytically which choice we make.

Next, we avoid points where two poles, say k_m and k_{m+1}, coalesce to form a second-order pole. This is purely for computational convenience so that we are always dealing with a simple pole. We could find these points numerically by solving the simultaneous equations $\Delta(k_m^{(G)}, \omega_m^{(G)}) = \Delta_k(k_m^{(G)}, \omega_m^{(G)}) = 0$ where the singularity is denoted by $\omega_m^{(G)}$. However, a little algebra shows that $\omega_m^{(G)}$ does not exist in the present problem. We avoid higher order poles in a similar manner.

The final singularities are branch points where any contour around it results in an interchange of $k_m(\omega)$ for $k_m^*(\omega)$, the complex conjugate

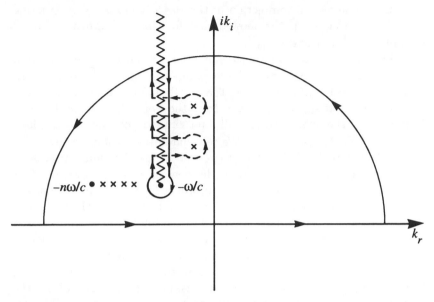

Fig. 4.4.3: Deformation of the original branch cut integral so that it captures those improper (non-spectral) modes which will eventually become normal modes.

of $k_m(\omega)$. In this case, it is very important to choose correctly whether to go over or under such branch points. These branch points will allow us to choose a $k_m(\omega)$ which has its origins in the upper half of the k-plane. Numerical experimentation reveals that the new ω contour must go above the singularity for this to occur. If our improper modes did not have their origin in the upper half plane, then $\text{Im}[k_m(\omega)] \to -\infty$ as $m \to \infty$ and the e^{ikx} factor would grow exponentially.

Locating these branch points is easy because they represent points where two solutions k_m and k_m^* merge. Consequently, the solution of the system of equations $\Delta(k,\omega) = \Delta_k(k,\omega) = 0$ gives us the ω's. We denote these branch points as $\omega_m^{(B)}$; the corresponding $k_m^{(B)}$ lies on the improper Riemann surface.

Another branch point occurs for $k = 0$. When we take a path around $k = 0$, the root $k_m(\omega)$ is replaced by $-k_m(\omega)$. In this case, the corresponding $\omega_m^{(D)}$ is

$$\omega_m^{(D)} = \frac{c}{2na}\left[(2m-1)\pi + i\ln\left(\frac{n+1}{n-1}\right)\right], \qquad m = 1, 2, 3, \ldots \quad (4.4.14)$$

Because we wish our sum of residues to be a regular analytic function of ω (for reasons soon to be explained), any closed path in the ω-plane must result in a root in the second quadrant of the k-plane returning to

319

its initial value at the completion of the closed contour. Constructing several simple closed paths showed that for this to happen, we must pass *over* $\omega_m^{(B)}$ and *under* $\omega_m^{(D)}$.

When additional multivalued functions are present, we must find additional singularities. They include points similar to $\omega_m^{(B)}$ and $\omega_m^{(D)}$ so that the procession of $k_m(\omega)$ from the one Riemann surface to the next is continuous and $k_m(\omega)$ originates from the upper half of the k-plane. Furthermore, the avoidance of additional branch points on the k-plane is necessary as it was for $\omega_m^{(C)}$. We refer interested reader to several of Haddon's papers[32,33,34] in which there were three Riemann surfaces.

It remains for us to deal with the new branch cut integral. Our deformation of the contour onto the improper Riemann surface guarantees that the branch cut integral will vary continuously with ω. We will now show that our deformation results in the unexpected windfall that this branch cut integral vanishes; we may express the Green's function solely as a superposition of improper and proper modes.

The first point in the proof is to show that the branch cut integral is a regular analytic function along *any* contour that passes above the branch point $\omega_m^{(B)}$ and remains in the first quadrant of the ω-plane. To establish this, we first note that the inverse Fourier transform, with its integration along the real k axis, is a regular analytic function. This is still true even after we have deformed the contour onto the improper Riemann surface because we have maintained analytic continuity. Finally, the sum of the residues is a regular analytic function of ω in the cut right half of the ω-plane. Thus, the branch cut integration must be a regular analytic function of ω. The point $\omega_m^{(D)}$ is of no concern here because as long as we pass above $\omega_m^{(B)}$, no root can cross the path of the branch cut integration and $\omega_m^{(D)}$ cannot be a singularity of the branch cut integration.

Let us now return to our deformed contour which passes above $\omega_m^{(B)}$ and below $\omega_m^{(D)}$ and close it with vertical and horizontal segments along the lines $\text{Re}(\omega) = 0$, $\text{Re}(\omega) = W_1$ and $\text{Im}(\omega) = W_2$. By Cauchy's theorem,

$$\text{Re}\left[\oint_\Omega L_j(\omega)e^{i\omega t}d\omega\right] = 0 \tag{4.4.15}$$

where Ω is the closed contour described above and $L_j(\omega)$ denotes our branch cut integral. The contribution to the closed integral along the imaginary axis is zero because the integral is pure imaginary. The contribution from the portion of Ω along the lines $\text{Re}(\omega) = W_1$ also vanishes as we take $W_1 \to \infty$. Thus, we must show that

$$\lim_{W_2 \to \infty} \int_{W_2 i}^{\infty + W_2 i} L_j(\omega)e^{i\omega t}d\omega = 0. \tag{4.4.16}$$

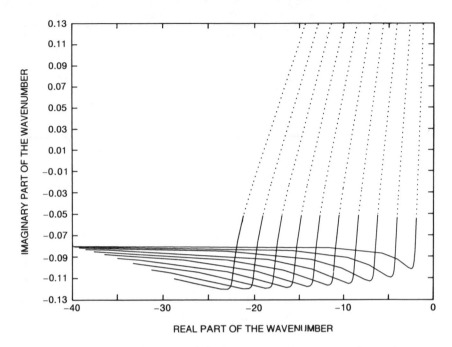

Fig. 4.4.4: The values of $k_m(\omega)$ for $m = 3, 6, 9, \ldots, 30$ along the new ω contour. As ω increases a pole $k_m(\omega)$ migrates from the improper Riemann surface (dotted line) to the proper Riemann surface (solid line). The units are cm^{-1}.

Following Haddon[31], we do this by first expanding the denominator of the integrand of $L_j(\omega)$ as a geometric series of exponentials. Because we can integrate each term separately and each term does not contain any poles, we can evaluate each integral by the method of steepest descent. Finally, each asymptotic result vanishes in the limit of $W_2 \to \infty$.

In summary, by a suitable modification of the original ω contour (in the present case, over $\omega_m^{(B)}$ and $\omega_m^{(C)}$ and under $\omega_m^{(D)}$), each pole initially in the second quadrant of the improper (non-spectral) Riemann surface follows a trajectory which leads around $k_m^{(B)}$ and $k_m^{(C)}$ onto the proper Riemann surface. Each corresponding residue varies with analytical continuity along this new contour. Most importantly, this process results in the entire solution being solely composed both of normal and of non-spectral modes.

Let us now detail how we can apply these theoretical considerations to numerically evaluate the Green's function for a grounded dielectric slab subjected to a impulse line source. Before we deform the ω contour from the fourth quadrant of the ω-plane to the first quadrant, we must calculate $\omega_m^{(B)}$, $\omega_m^{(C)}$ and $\omega_m^{(D)}$. For a specified a, c and n we obtain

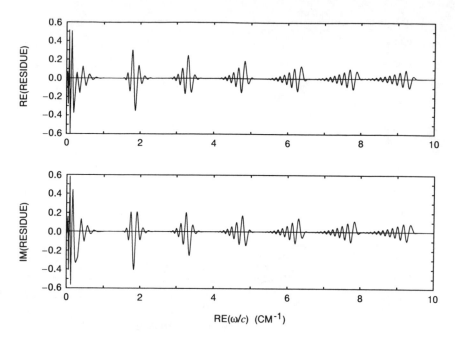

Fig. 4.4.5: The real and imaginary part of the residues given by (4.4.20) for $m = 1, 3, 5, \ldots, 13$ as a function of the real part of ω/c and the imaginary part of ω/c equaling 0.05 cm^{-1}. We scaled the $m = 1$ mode by dividing its original value by 5.

these points either analytically from (4.4.13) and (4.4.14) or numerically using the IMSL routine NEQNF which solves nonlinear, simultaneous equations.

Once we find these singularities, our new ω contour must pass over $\omega_m^{(B)}$ and under $\omega_m^{(D)}$ and avoid passing through $\omega_m^{(C)}$. Because $\omega_m^{(B)}$ and $\omega_m^{(C)}$ lie on the real k axis while $\text{Im}(\omega_m^{(D)}/c) = 0.13761$ cm^{-1}, our ω contour will pass over $\omega_m^{(B)}$ and under $\omega_m^{(D)}$ if we choose a line parallel to the real ω axis with $0 < \text{Im}(\omega/c) < 0.13761$ cm^{-1}. For convenience we choose the same line for all of the $k_m(\omega)$ poles, namely $W_s/c = \text{Im}(\omega/c) = 0.05$ cm^{-1}. A practical consideration is to pass sufficiently far from the singularities on the real ω axis. We note that we could have chosen different contours for each m. Furthermore, the contour line need not be a straight line.

For each ω contour, we compute $k_m(\omega)$ by using Newton's method. The definition of changing from one Riemann surface to another is $|\text{Re}(\nu_1)| < 5 \times 10^{-3}$ cm^{-1}. A portion of each $k_m(\omega)$ path lies on the improper Riemann surface while the remaining portion lies on the proper Riemann surface. Fig. 4.4.4 displays the $k_m(\omega)$ contours for the

$m = 3, 6, 9, \ldots, 30$ modes for $0 < \text{Re}(\omega/c) < 25$ cm^{-1}. We use a dotted line when the mode lies on the improper Riemann surface. Note that Fig. 4.4.4 does not show most of the trajectory on the improper mode because it has an imaginary part greater than 0.13 cm^{-1}.

For reasons that will become clear, we reverse the order of summation and integration so that, for example,

$$E_1(x, z, t) = \sum_m R_m \tag{4.4.17}$$

where

$$R_m = \text{Re}\left[\frac{i}{\pi} e^{i\omega_m^{(C)} t} \int_{W_s i}^{\infty + W_s i} [\text{Residue at } k = k_m \text{ of } \overline{\mathcal{E}}_1(k, z, \omega)]\right.$$
$$\left. \times e^{i[k_m x + (\omega - \omega_m^{(C)})t]} d\omega\right]. \tag{4.4.18}$$

We integrate (4.4.18) using a Cote's sixth-order formula with $\text{Re}(\Delta\omega /c) = 0.005$ cm^{-1} and $0 < \text{Re}(\omega/c) < 50$ cm^{-1}. Then we multiply the result of the integration by $ie^{i\omega_m^{(C)} t}/\pi$ before taking the real part. In order to obtain finite results in the following calculations, it is necessary to choose a well-behaved forcing function. To model our impulse forcing, we adopted the following:

$$f(t) = \frac{1}{\sqrt{\pi\epsilon}} e^{-t^2/\epsilon}, \qquad F(\omega) = \frac{1}{2\pi} e^{-\epsilon\omega^2/4} \tag{4.4.19}$$

where $\epsilon c^2 = 0.01$ cm^{-2}. This choice of forcing is convenient because it does not introduce further singularities.

In Fig. 4.4.5 we show the real and imaginary part of the residues:

$$\text{Residue}(k_m, z, \omega) = -\frac{e^{-\nu_1(z+b-2a)}}{2\nu_1}\left[\frac{\nu_2 \cosh(\nu_2 a) - \nu_1 \sinh(\nu_2 a)}{\Delta_k}\right]$$
$$\times e^{ik_m x} \tag{4.4.20}$$

for the modes $m = 1, 3, \ldots, 13$ when $x = 20$ cm and $z = 5$ cm. As the figure shows, the residues are small except in the vicinity of $\omega_m^{(B)}$ and $\omega_m^{(C)}$. Consequently, by performing the algebraic trick given in (4.4.18), we can make the integrand a slowly varying function of ω where the integrand is most significant and we achieve considerable accuracy.

Finally we illustrate in Fig. 4.4.6 the temporal evolution of the various modes at the sample point $x = 20$ cm and $z = 5$ cm. The results are strictly valid only for $ct > 20.277$ cm because that is the least amount of time before the first reflected wave would reach the observer. After the first pulse we see subsequently smaller pulses as the multiple reflections that occur within the dielectric reach the point.

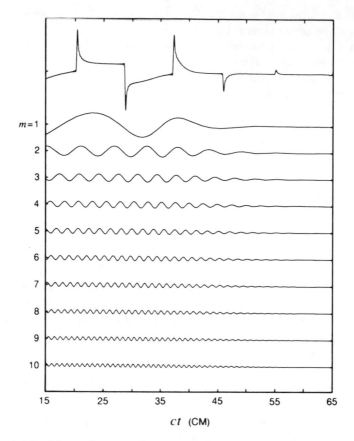

Fig. 4.4.6: Along the top, the temporal evolution of the total reflected wave field (consisting of *all* 30 modes) for the sample point $x = 20$ cm and $z = 5$ cm. Below this trace, the temporal evolution of the first ten individual modes at the same point.

Papers Using the Joint Transform Technique

Achari, R. M., 1975: A quasi-static thermoelastic problem for a semi-space. *Zeit. angew. Math. Mech.*, **55**, 688–692.

Aggarwal, H. R., and C. M. Ablow, 1965: Disturbance from a circularly symmetric load spreading over an acoustic half-space. *Bull. Seism. Soc. Am.*, **55**, 673–691.

Akulenko, L. D., S. A. Mikhailov and S. V. Nesterov, 1990: The oscillations of an oscillator near the interface between two liquids. *J. Appl. Mech. Math.*, **54**, 30–38.

Aoki, K., 1990: Theory of current distribution at a conical electrode under diffusion control with time dependence. *J. Electroanal. Chem.*, **281**, 29–40.

Ashpis, D. E., and E. Reshotko, 1990: The vibrating ribbon problem revisited. *J. Fluid Mech.*, **213**, 531–547.

Ben-Menahem, A., and A. G. Sena, 1990: Seismic source theory in stratified anisotropic media. *J. Geophys. Res.*, **95**, 15395–15427.

Bhattacharyya, P., 1968: Note on the rotatory motion set up in a semi-infinite viscous fluid by a certain impulsive velocity prescribed over a circular area on the surface. *Rev. Roum. Sci. Tech. - Méc. Appl.*, **13**, 259–263.

Boulton, N. S., and T. D. Streltsova, 1978: Unsteady flow to a pumped well in a fissured aquifer with a free surface level maintained constant. *Water Resour. Res.*, **14**, 527–532.

Bukreev, V. I., A. V. Gusev and I. V. Sturova, 1983: Unsteady motion of a circular cylinder in a two-layer liquid. *J. Appl. Mech. Tech. Phys.*, **24**, 856–861.

Burku, A. L., 1968: Unsteady radiative heating of a cylindrical body. *J. Appl. Mech. Tech. Phys.*, **9**, 184–186.

Carrier, G. F., and H. P. Greenspan, 1958: Water waves of finite amplitude on a sloping beach. *J. Fluid Mech.*, **4**, 97–109.

Chakraborty, M., 1985: Disturbance of SH type due to body forces and due to shearing stress-discontinuity in a pre-stressed semi-infinite viscoelastic medium. *Indian J. Pure Appl. Math.*, **16**, 309–322.

Chakravarti, S., 1977: One dimensional thermo-elastic wave in a non-simple medium. *Bull. Cal. Math. Soc.*, **64**, 129–135.

Chao, C. C., 1960: Dynamical response of an elastic half-space to tangential surface loadings. *J. Appl. Mech.*, **27**, 559–567.

Chao, C. C., H. H. Bleich and J. Sackman, 1961: Surface waves in an elastic half space. *J. Appl. Mech.*, **28**, 300–301.

Chaudhuri, K., 1975: Effect of viscosity on the surface waves produced by an explosion above a liquid. *Indian J. Pure Appl. Math.*, **6**, 49–68.

Chee-Seng, L., 1974: Unsteady current-induced perturbation of a magnetically contained magnetohydrodynamic flow. *J. Fluid Mech.*, **63**, 273–299.

Chen, K. C., 1987: Transient response of an infinite cylindrical antenna is a dissipative medium. *IEEE Trans. Antennas Propagat.*, **AP-35**, 562–573.

Cherkesov, L. V., 1966: The influence of viscosity on the propagation of tsunami type waves. *Izv. Acad. Sci. USSR, Atmos. Ocean. Phys.*, **2**, 793–797.

Cherkesov, L. V., 1970: Influence of viscosity on tsunami waves in a basin of variable depth. *Izv. Acad. Sci. USSR, Atmos. Ocean. Phys.*, **6**, 46–48.

Cole, J. D., and C. Greifinger, 1969: Acoustic-gravity waves from an energy source at the ground in an isothermal atmosphere. *J. Geophys. Res.*, **74**, 3693–3703.

Cooley, R. L., and C. M. Case, 1973: Effect of a water table aquitard on drawdown in an underlying pumped aquifer. *Water Resour. Res.*, **9**, 434–447.

Das, D. K., 1977: Motion of a viscous incompressible liquid due to a surface pressure. *Rev. Roum. Sci. Tech. - Méc. Appl.*, **22**, 819–828.

Das, K. P., 1971: Hydromagnetic disturbance produced by a momentary current in a semi-infinite fluid of finite electrical conductivity. *Indian J. Theoret. Phys.*, **19**, 131–139.

Das Gupta, S. C., 1959: On coda waves of earthquakes. *Geofisica pura e appl.*, **43**, 45–74.

Duffy, D. G., 1990: Geostrophic adjustment in a baroclinic atmosphere. *J. Atmos. Sci.*, **47**, 457–473.

Duffy, D. G., 1990: The response of an infinite railroad track to a moving, vibrating mass. *J. Appl. Mech.*, **57**, 66–73.

Duffy, D. G., 1992: On the generation of oceanic surface waves by underwater volcanic explosions. *J. Volcanol. Geotherm. Res.*, **50**, 323–344.

Eason, G., J. Fulton and I. N. Sneddon, 1956: The generation of waves in an infinite elastic solid by variable body forces. *Phil. Trans. R. Soc. Lond.*, **A248**, 575–607.

Eason, G., 1964: On the torsional impulsive loading of an elastic half space. *Quart. J. Mech. Appl. Math.*, **17**, 279–292.

Eason, G., 1965: The stresses produced in an semi-infinite solid by a moving surface load. *Int. J. Engng Sci.*, **2**, 581–609.

Eason, G., 1966: The displacements produced in an elastic half-space by a suddenly applied surface force. *J. Inst. Math. Appl.*, **2**, 299–326.

Eraslan, A. H., 1967: Oscillatory characteristics of unsteady MHD channel flows under Heaviside-type applied magnetic fields. *J. Appl. Mech.*, **34**, 854–859.

Estrin, T. A., and T. J. Higgins, 1951: The solution of boundary value problems by multiple Laplace transformations. *J. Franklin Inst.*, **252**, 153–167.

Filippov, I. G., and K. N. Smamutov, 1975: Axisymmetric nonstationary problem for an anisotropic inhomogeneous half-space and a layer. *Sov. Appl. Mech.*, **11**, 614–618.

Flinn, E. A., 1961: Exact transient solution to some elementary problems of elastic wave propagation. *J. Acoust. Soc. Am.*, **33**, 623–627.

Ghosh, M., and M. L. Ghosh, 1985: Harmonic rocking of a rigid strip on a semi-infinite elastic medium. *Indian J. Pure Appl. Math.*, **16**, 938–955.

Ghosh, S. S., 1969: On the disturbances in a thin elastic circular plate resting on a viscoelastic foundation of Pasternak type. *PAGEOPH*, **75**, 88–92.

Gilbert, F., and L. Knopoff, 1959: Scattering of impulsive elastic waves by a rigid cylinder. *J. Acoust. Soc. Am.*, **31**, 1169–1175.

Gilbert, F., 1960: Scattering of impulsive elastic waves by a smooth convex cylinder. *J. Acoust. Soc. Am.*, **32**, 841–857.

Gilbert, F., and L. Knopoff, 1960: Seismic scattering from topographic irregularities. *J. Geophys. Res.*, **65**, 3437–3444.

Greenspan, H. P., 1956: The generation of edge waves by moving pressure distributions. *J. Fluid Mech.*, **1**, 574–592.

Grigor'ev, G. I., N. G. Denisov and O. N. Savina, 1987: Emission of acoustic-gravity waves and a Lamb surface wave in an isothermal atmosphere. *Radiophys. Quantum Electron.*, **30**, 207–212.

Grigor'ev, G. I., N. G. Denisov and O. N. Savina, 1989: Transition radiation of acoustic waves by sources moving in the atmosphere above the earth's surface. *Radiophys. Quantum Electron.*, **32**, 110–115.

Gupta, P. C., 1983: Unsteady magneto-hydrodynamic flow through porous media in a channel whose cross-section is a circular section. *Bull. Cal. Math. Soc.*, **75**, 264–270.

Gupta, P. M., and V. P. Saxena, 1971: Heat conduction in a moving anisotropic rectangular slab. *Indian J. Pure Appl. Math.*, **2**, 290–296.

Gupta, S. C., 1979: Unsteady flow of a dusty gas in a channel whose cross-section is an annular section. *Indian J. Pure Appl. Math.*, **10**, 704–714.

Halepaska, J. C., 1972: Drawdown distribution around wells partially penetrating thick leaky artesian aquifers. *Water Resour. Res.*, **8**, 1332–1337.

Han, C. D., 1970: The effect of radial diffusion on the performance of a liquid-liquid displacement process. *Appl. Sci. Res.*, **22**, 223–238.

Hantush, M. S., 1967: Flow of groundwater in relatively thick leaky aquifers. *Water Resour. Res.*, **3**, 583–590.

Hantush, M. S., 1967: Flow of wells in aquifers separated by a semipervious layer. *J. Geophys. Res.*, **72**, 1709–1720.

Hibiya, T., 1986: Generation mechanism of internal waves by tidal flow over a sill. *J. Geophys. Res.*, **91**, 7697–7708.

Hudson, J. A., 1963: SH waves in a wedge-shaped medium. *Geophys. J. Astr. R. Soc.*, **7**, 517–546.

Imberger, J., and C. Fandry, 1975: Withdrawal of a stratified fluid from a vertical two-dimensional duct. *J. Fluid Mech.*, **70**, 321–332.

Jain, N. K., 1967: Torsional damped vibrations of an elastic half-space subjected to an impulsive radial stress distribution. *Indian J. Pure Appl. Phys.*, **5**, 599–601.

Javandel, I., and P. A. Witherspoon, 1983: Analytical solution of a partially penetrating well in a two-layer aquifer. *Water Resour. Res.*, **19**, 567–578.

Johnson, E. R., and R. Parnes, 1973: Propagation of axisymmetric waves in an elastic half-space containing a cylindrical inclusion. Part I: Formulation and general integral solution. *Quart. J. Mech. Appl. Math.*,

30, 235–253.

Jones, J. R., and T. S. Walters, 1966: Oscillatory motion of an elastico-viscous liquid contained in a cylindrical cup. I. Theoretical. *Brit. J. Appl. Phys.*, **17**, 937–944.

Joshi, B. K., and R. S. Sharma, 1976: Unsteady flow of visco-elastic fluid through long circular ducts. *Indian J. Pure Appl. Math.*, **7**, 1081–1090.

Kanwal, R. P., 1959: Impulsive rotatory motion of a circular disk in a viscous fluid. *Zeit. angew. Math. Phys.*, **10**, 552–557.

Kobayashi, N., and H. Takeuchi, 1955: Propagation of tremors over the sphere of an elastic solid. *J. Phys. Earth*, **3**, 17–22.

Kobayashi, N., and H. Takeuchi, 1957: Wave generation from line sources within the ground. *J. Phys. Earth*, **5**, 25–32.

Koizumi, T., 1970: Thermal stresses in a semi-infinite body with an instantaneous heat source on its surface. *Bull. JSME*, **13**, 26–33.

Kolyano, Yu. M., and E. A. Pakala, 1967: Temperature stresses in a heat-source heated plate strip with heat transfer. *Sov. Appl. Mech.*, **3(3)**, 48–54.

Kolyano, Yu. M., 1967: Temperature stresses in an orthotropic plate strip with heat outflow. *Sov. Appl. Mech.*, **3(6)**, 23–27.

Kolyano, Yu. M., and E. A. Pakula, 1970: Two-dimensional dynamic problem of thermoelasticity for heated thin plates. *Sov. Appl. Mech.*, **6**, 174–179.

Kubenko, V. D., and Yu. B. Moseenkov, 1987: Interaction of unsteady acoustic waves with plates and hollow shells in a fluid. *Sov. Appl. Mech.*, **23**, 951–957.

Lehner, F. K., V. C. Li and J. R. Rice, 1981: Stress diffusion along rupturing plate boundaries. *J. Geophys. Res.*, **86**, 6155–6169.

Leij, F. J., and J. H. Dane, 1990: Analytical solutions of the one-dimensional advection equation and two- or three-dimensional dispersion equation. *Water Resour. Res.*, **26**, 1475–1482.

Leij, F. J., T. H. Skaggs and M. Th. van Genuchten, 1991: Analytical solutions for solute transport in three-dimensional semi-infinite porous media. *Water Resour. Res.*, **27**, 2719–2733.

Levine, H., 1973: A note on problems of wave generation in semi-infinite media by surface forces. *Appl. Sci. Res.*, **28**, 207–222.

Liu, D. T., 1959: Wave propagation in a liquid layer. *Geophysics*, **24**, 658–666.

Lowndes, J. S., 1957: A transient magnetic dipole source above a two-layer earth. *Quart. J. Mech. Appl. Math.*, **10**, 79–89.

MacDonald, D. A., 1966: The Rayleigh problem for two-layer fluid. *Quart. J. Mech. Appl. Math.*, **19**, 198–215.

Matsumoto, H., I. Nakahara and Y. Matsuoko, 1978: The impact of an elastic cylinder on an elastic solid. *Bull. JSME*, **21**, 579–586.

Miles, J. W., 1958: On the disturbed motion of a plane vortex sheet. *J. Fluid Mech.*, **4**, 538–552.

Miles, J. W., 1968: The Cauchy-Poisson problem for a viscous liquid. *J. Fluid Mech.*, **34**, 359–370.

Milgram, J. H., 1969: The motion of a fluid in a cylindrical container with a free surface following vertical impact. *J. Fluid Mech.*, **37**, 435–448.

Mishra, S. K., 1964: Propagation of sound pulses in a semi-infinite stratified medium. *Proc. Indian Acad. Sci.*, **A59**, 21–48.

Misra, J. C., and R. M. Achari, 1980: Temperature stresses in an infinite disk having a heat source in the vicinity. *J. Therm. Stresses*, **3**, 57–66.

Mitra, A., 1966: Mixed boundary value problem in an elastic quarter-space. *PAGEOPH*, **63**, 60–67.

Mitra, M., 1958: Disturbance produced in an elastic half-space by an impulsive twisting moment applied to an attached rigid circular disc. *Zeit. angew. Math. Mech.*, **38**, 40–43.

Miyoshi, H., 1954: Generation of the tsunami in compressible water. *J. Oceanogr. Soc. Japan*, **10**, 1–9.

Miyoshi, H., 1955: Generation of the tsunami in compressible water (Part II). *Rec. Oceanogr. Works Japan*, **2**, 49–56.

Mohandis, M. G. S. el, 1969: Magnetohydrodynamic disturbances in the earth's core. I, II, III. *PAGEOPH*, **72**, 155–192.

Mohandis, M. G. S. el, 1969: Magnetohydrodynamic disturbances in the earth's core. IV. *PAGEOPH*, **74**, 45–56.

Morioka, S., and G. Matsui, 1975: Pressure-wave propagation through a separated gas-liquid layer in a duct. *J. Fluid Mech.*, **70**, 721–731.

Mucichescu, D., 1973: Response of infinite, mass-attached string on elastic foundation to an impact point load. *Rev. Roum. Sci. Tech. - Méc. Appl.*, **18**, 651–662.

Muller, P., 1976: Propagation d'ondes de flexion dans une tige micropolaire élastique tendue. *J. Mécanique*, **15**, 493–520.

Mukhopadhyay, A., 1966: Disturbances produced by variable body forces in a layer of finite thickness resting on a rigid foundation. *PAGEOPH*, **65**, 29–36.

Mukhopadhyay, A., 1967: Disturbances produced by a visco-elastic medium by transient torsional body forces. *PAGEOPH*, **67**, 43–53.

Neuman, S. P., and P. A. Witherspoon, 1969: Theory of flow in a confined two aquifer system. *Water Resour. Res.*, **5**, 803–816.

Neuman, S. P., 1972: Theory of flow in unconfined aquifers considering delayed response of the water table. *Water Resour. Res.*, **8**, 1031–1045.

Neuman, S. P., 1974: Effect of partial penetration in flow in unconfined aquifers considering delayed gravity response. *Water Resour. Res.*, **10**, 303–312.

Noda, T., 1988: On a certain inverse problem of coupled thermal stress fields in a thick plate. *Zeit. angew. Math. Mech.*, **68**, 411–415.

Norwood, F. R., 1968: Propagation of transient sound signals into a viscous fluid. *J. Acoust. Soc. Am.*, **44**, 450–457.

Oi, M., 1976: Some theoretical considerations upon the transient topographical perturbation of a zonal current. *J. Geophys. Res.*, **81**, 1084–1094.

Okeke, E. O., 1977: The acoustic vibrations induced in a homogeneous fluid by gravity waves. *Geophys. Astrophys. Fluid Dynamics*, **8**, 155–161.

Ölçer, N. Y., 1965: On the solar heating of rotating space vehicles. *Acta Mech.*, **1**, 148–170.

Parnes, R., 1978: Dynamic surface stresses on an elastic half-space containing a cylindrical inclusion. *J. Sound Vib.*, **58**, 167–178.

Parton, V. Z., 1972: The axially symmetric temperature problem for the space with a disk-shaped crack. *J. Appl. Math. Mech.*, **36**, 104–111.

Patil, S. P., 1988: Response of infinite railroad track to vibrating mass. *J. Eng. Mech.*, **114**, 688–703.

Pavlov, V. A., 1979: Effect of earthquakes and volcanic eruptions on the ionospheric plasma. *Radiophys. Quantum Electron.*, **22**, 10–23.

Payton, R. G., 1962: Initial bending stresses in elastic shells impacting into compressible fluids. *Quart. J. Mech. Appl. Math.*, **15**, 77–90.

Payton, R. G., 1965: Dynamic bond stress in a composite structure subjected to a sudden pressure rise. *J. Appl. Mech.*, **32**, 643–650.

Payton, R. G., 1968: Epicenter motion of an elastic half-space due to buried stationary and moving line sources. *Int. J. Solids Structures*, **4**, 287–300.

Payton, R. G., 1969: Two-dimensional pulse propagation in a two-parameter anisotropic elastic solid. *Q. Appl. Math.*, **27**, 147–160.

Peck, J. C., and J. Miklowitz, 1969: Shadow-zone response in the diffraction of a plane compressional pulse by a circular cavity. *Int. J. Solids Structures*, **5**, 437–454.

Pekeris, C. L., 1955: The seismic surface pulse. *Proc. Nat. Acad. Sci.*, **41**, 469–480.

Pekeris, C. L., 1955: The seismic buried pulse. *Proc. Nat. Acad. Sci.*, **41**, 629–639.

Pekeris, C. L., 1956: Solution of an integral equation occurring in impulsive wave propagation problems. *Proc. Nat. Acad. Sci.*, **42**, 439–443.

Pekeris, C. L., and H. Lifson, 1957: Motion of the surface of a uniform elastic half-space produced by a buried pulse. *J. Acoust. Soc. Am.*, **29**, 1233–1238.

Pekeris, C. L., and Z. Alterman, 1957: Radiation resulting from an impulsive current in a vertical antenna placed on a dielectric ground. *J. Appl. Phys.*, **28**, 1317–1323.

Pekeris, C. L., and I. M. Longman, 1958: The motion of the surface of a uniform elastic half-space produced by a buried torque-pulse. *Geophys. J. R. Astr. Soc.*, **1**, 146–153.

Pekeris, C. L., and I. M. Longman, 1958: Ray-theory solution of the problem of propagation of explosive sound in a layered liquid. *J. Acoust. Soc. Am.*, **30**, 323–328.

Pekeris, C. L., Z. Alterman and F. Abramovici, 1963: Propagation of an SH-torque pulse in a layered solid. *Bull. Seism. Soc. Am.*, **53**, 39–57.

Pekeris, C. L., Z. Alterman, F. Abramovici and A. Jarosch, 1965: Propagation of a compressional pulse in a layered solid. *Rev. Geophys.*, **3**, 25–47.

Peralta, L. A., and S. Raynor, 1964: Initial response of a fluid-filled, elastic, circular, cylindrical shell to a shock wave in acoustic medium. *J. Acoust. Soc. Am.*, **36**, 476–488.

Peterson, E. W., 1974: Acoustic wave propagation along a fluid-filled cylinder. *J. Appl. Phys.*, **45**, 3340–3350.

Phinney, R. A., 1961: Leaky modes in the crustal waveguide. Part 1. The oceanic PL waves. *J. Geophys. Res.*, **66**, 1445–1469.

Pinney, E., 1954: Surface motion due to a point source in a semi-infinite elastic medium. *Bull. Seism. Soc. Am.*, **44**, 571–596.

Podstrigach, Ya. S., and B. I. Kolodii, 1970: Two-dimensional temperature and stress field in induction heating of an elastic half-space. *Sov. Appl. Mech.*, **6**, 1329–1333.

Poritsky, H., 1955: Propagation of transient fields from dipoles near the ground. *Brit. J. Appl. Phys.*, **6**, 421–426.

Pramanik, A. K., 1972: The effect of viscosity on the waves due to a moving oscillating surface pressure. *Zeit. angew. Math. Phys.*, **23**, 85–95.

Rafalski, P., 1965: Dynamic thermal stresses in viscoelastic slab. *Arch. Mech. Stos.*, **17**, 617–631.

Raichenko, L. M., 1977: Fluid inflow to an imperfect borehole in an inhomogeneous medium. *Sov. Appl. Mech.*, **13**, 937–942.

Rawat, M. L., 1970: Flow of viscous incompressible fluids through a tube with sector of a circle as cross-section. *Bull. Cal. Math. Soc.*, **62**, 149–156.

Rosenbaum, J. H., 1959: A note on the propagation of a sound pulse in a two-layer liquid medium. *J. Geophys. Res.*, **64**, 95–102.

Rosenbaum, J. H., 1960: The long-time response of a layered elastic medium to explosive sound. *J. Geophys. Res.*, **65**, 1577–1613.

Roulet, C., 1972: Solution de l'équation de Fick avec source appliquée à la diffusion d'adatomes d'argent sur le cuivre. *Zeit. angew. Math. Phys.*, **23**, 412–419.

Roy, A. B., 1964: Disturbances in a viscoelastic medium due to a twist of finite duration on the surface of a spherical cavity. *PAGEOPH*, **59**,

21 26.

Roy, P. C., 1978: Unsteady flow of a Maxwell fluid past a flat plate. *Indian J. Pure Appl. Math.*, **9**, 157–166.

Rożnowski, T., 1969: The plane problem of thermoelasticity with a moving boundary condition. *Arch. Mech. Stos.*, **21**, 657–677.

Sabherwal, K. C., 1965: An inverse problem in transient heat conduction. *Indian J. Pure Appl. Phys.*, **3**, 397–398.

Sanyal, D. C., and S. K. Samanta, 1989: Unsteady motion of a semi-infinite and conducting liquid by a suddenly applied velocity on its surface. *Indian J. Pure Appl. Math.*, **20**, 1146–1156.

Sarkar, G., 1966: Wave motion due to impulsive twist on the surface. *PAGEOPH*, **65**, 43–47.

Sarkar, G., 1966: Displacement due to impulsive loadings on the surface of elastic half-space. *PAGEOPH*, **66**, 37–47.

Sarker, A. K., 1963: On SH type of motion due to body forces and due to stress-discontinuity in a semi-infinite viscoelastic medium. *Geofisica pura e appl.*, **55**, 42–52.

Scavuzzo, R. J., J. L. Bailey and D. D. Raftopoulos, 1971: Lateral structure interaction with seismic waves. *J. Appl. Mech.*, **38**, 125–134.

Schimmerl, J., 1965: Setzung einer Tonschicht endlicher Mächtigkeit bei zylindersymmetrischer Belastung. *Zeit. angew. Math. Mech.*, **45**, 553–562.

Sengupta, P. R., and J. Raymahrpatra, 1975: On the motion set up in a semi-infinite viscous conducting fluid. *Bull. Cal. Math. Soc.*, **67**, 29–35.

Sheehan, J. P., and L. Debnath, 1972: On the dynamic response of an infinite Bernoulli-Euler beam. *PAGEOPH*, **97**, 100–110.

Sherwood, J. W. C., 1958: Elastic wave propagation in a semi-infinite solid medium. *Proc. Phys. Soc. Lond.*, **71**, 207–219.

Shibuya, T., H. Matsumoto and I. Nakahara, 1968: The semi-infinite plate subjected to impact load on the free boundary. *Bull. JSME*, **11**, 203–210.

Shibuya, T., and I. Nakahara, 1968: The semi-infinite body subjected to a concentrated impact load on the surface. *Bull. JSME*, **11**, 983–992.

Shindo, Y., 1981: Sudden twisting of a flat annular crack. *Int. J. Solids Structures*, **17**, 1103–1112.

Shindo, Y., H. Nozaki and H. Higaki, 1986: Impact response of a finite crack in an orthotropic strip. *Acta Mech.*, **62**, 87–104.

Shindo, Y., 1988: Impact response of a crack in a semi-infinite body with a surface layer under longitudinal shear. *Acta Mech.*, **73**, 147–162.

Sidhu, R. S., 1972: Disturbances in semi-infinite heterogeneous media generated by torsional sources. I. *Bull. Seism. Soc. Am.*, **62**, 541–550.

Spence, D. A., and D. B. R. Kenning, 1978: Unsteady heat conduction in a quarter plate, with an application to bubble growth models. *Int. J. Heat Mass Transfer*, **21**, 719–724.

Spillers, W. R., and A. Callegari, 1969: Impact of two elastic cylinders: Short-time solution. *Int. J. Mech. Sci.*, **11**, 846–851.

Srivastava, K. N., R. M. Palaiya and D. S. Karaulia, 1981: Diffraction of SH waves by two coplanar Griffith cracks at the interface of two bonded dissimilar elastic half-spaces. *Indian J. Pure Appl. Math.*, **12**, 242–252.

Stadler, W., and R. W. Shreeves, 1970: The transient and steady-state response of the infinite Bernoulli-Euler beam with damping and an elastic foundation. *Quart. J. Mech. Appl. Math.*, **23**, 197–208.

Stick, E., 1959: Propagation of elastic wave motion from an impulsive source along a fluid/solid interface. Part II. Theoretical pressure response. *Phil. Trans. R. Soc. Lond.*, **A251**, 465–488.

Stick, E., 1959: Propagation of elastic wave motion from an impulsive source along a fluid/solid interface. Part III. The pseudo-Rayleigh wave. *Phil. Trans. R. Soc. Lond.*, **A251**, 488–523.

Stronge, W. J., 1970: A load accelerating on the surface of an acoustic half space. *J. Appl. Mech.*, **37**, 1077–1082.

Sve, C., and J. Miklowitz, 1973: Thermally induced stress waves in an elastic layer. *J. Appl. Mech.*, **40**, 161–167.

Takeuchi, H., and N. Kobayashi, 1955: Wave generations from line sources within the ground. *J. Phys. Earth*, **3**, 7–15.

Takeuchi, H., and N. Kobayachi, 1956: Wave generations in a superficial layer resting on a semi-infinite lower layer. *J. Phys. Earth*, **4**, 21–30.

Tawari, G., 1971: Effect of couple-stresses in a semi-infinite elastic medium due to impulsive twist over the surface. *PAGEOPH*, **91**, 71–75.

Thomson, R. E., 1970: On the generation of Kelvin-type waves by atmospheric disturbances. *J. Fluid Mech.*, **42**, 657–670.

Trogdon, S. A., and M. T. Farmer, 1991: Unsteady axisymmetric creeping flow from an orifice. *Acta Mech.*, **88**, 61–75.

Vasudevaiah, M., and K. V. S. N. Prasad, 1990: Viscous impulsive rotation of an annular disk. *Indian J. Pure Appl. Math.*, **21**, 1125–1136.

Watanabe, K., 1981: Response of an elastic plate on a Pasternak foundation to a moving load. *Bull. JSME*, **24**, 775–780.

Watanabe, K., 1981: Transient response of an inhomogeneous elastic half space to a torsional load. *Bull. JSME*, **24**, 1537–1542.

Watanabe, K., 1982: Transient response of an inhomogeneous elastic solid to an impulse SH-source (Variable SH-wave velocity). *Bull. JSME*, **25**, 315–320.

Watanabe, K., 1984: Transient response of an inhomogeneous elastic solid to a moving torsional load in a cylindrical bore. *Int. J. Solids Structures*, **20**, 359–376.

Yen, D. H. Y., 1968: Dynamic response of an infinite plate subjected to a steadily moving transverse force. *Zeit. angew. Math. Phys.*, **19**, 257–270.

Yen, D. H. Y., and C. C. Chou, 1974: An addendum to the paper 'Dynamic response of an infinite plate subjected to a steadily moving transverse force'. *Zeit. angew. Math. Phys.*, **25**, 463–485.

Youngdahl, C. K., and E. Sternberg, 1961: Transient thermal stresses in a circular cylinder. *J. Appl. Mech.*, **28**, 25–34.

Żórawski, M., 1961: Moving dynamic heat sources in a visco-elastic space and corresponding basic solutions for moving sources. *Arch. Mech. Stos.*, **13**, 257–274.

Papers Using the Cagniard Technique

Abubakar, I., 1962: Disturbance due to a line source in a semi-infinite transversely isotropic elastic medium. *Geophys. J.*, **6**, 337–359.

Abubakar, I., 1964: Magneto-elastic SH-type of motion. *PAGEOPH*, **59**, 10–20.

Ang, D. D., 1960: Elastic waves generated by a force moving along a crack. *J. Math. and Phys.*, **38**, 246–256.

Bakhski, V. S., 1965: Propagation of a fracture in an infinite medium. *PAGEOPH*, **62**, 23–34.

Ben-Menahem, A., and M. Vered, 1973: Extension and interpretation of the Cagniard-Pekeris method for dislocation sources. *Bull. Seism. Soc. Am.*, **63**, 1611–1636.

Bhattacharyya, S., 1973: SH waves in a semi-infinite elastic space due to torsional disturbance. *PAGEOPH*, **111**, 2216–2222.

Blowers, R. M., 1969: On the response of an elastic solid to droplet impact. *J. Inst. Math. Appl.*, **5**, 167–193.

Bortfeld, R., 1962: Exact solution of the reflection and refraction of arbitrary spherical compressional waves at liquid-liquid interfaces and at solid-solid interfaces with equal shear velocities and equal densities. *Geophys. Prospecting*, **10**, 35–67.

Bortfeld, R., 1962: Reflection and refraction of spherical compressional waves at arbitrary plane interfaces. *Geophys. Prospecting*, **10**, 517–538.

Broberg, K. B., 1960: The propagation of a brittle crack. *Ark. för Fyk.*, **18**, 159–192.

Bykovtsev, A. S., and D. B. Kramarovskii, 1989: Non-stationary super-sonic motion of a complex discontinuity. *J. Appl. Math. Mech.*, **53**, 779–786.

Chandra, U., 1967: Propagation of an SH-torque pulse in a three layered solid half-space. *PAGEOPH*, **67**, 54–64.

Chwalczyk, F., J. Rafa and E. Włodarczyk, 1972: Propagation of two-dimensional non-stationary stress waves in a semi-infinite viscoelastic body, produced by a normal load moving over the surface with sub-seismic velocity. *Proc. Vibr. Prob.*, **13**, 241–257.

Chwalczyk, F., J. Rafa and E. Włodarczyk, 1973: Propagation of two-dimensional non-stationary pressure waves in a layer of perfect compressible liquid. *Proc. Vibr. Prob.*, **14**, 245–256.

Chwalczyk, F., J. Rafa and E. Włodarczyk, 1984: Propagation of non-stationary elastic waves in an anisotropic half-space. Part I. Analytic solution. *J. Tech. Phys.*, **25**, 23–34.

Crosson, R. S., and C.-C. Chao, 1970: Radiation of Rayleigh waves from a vertical directional charge. *Geophysics*, **35**, 45–56.

Eason, G., and R. R. M. Wilson, 1971: The displacements produced in a composite infinite solid by an impulsive torsional body force. *Quart.*

J. Mech. Appl. Math., **24**, 169–185.

Emura, K., 1960: Propagation of the disturbances in the medium consisting of semi-infinite liquid and solid. *Sci. Rep. Tohoku Univ., Ser. 5 Geophys.*, **12**, 63–100.

Flaherty, F. T., 1968: Transient resonance of an ideal string under a load moving with varying speed. *Int. J. Solids Structures*, **4**, 1221–1231.

Garvin, W. W., 1956: Exact transient solution of the buried line source problem. *Proc. R. Soc. Lond.*, **A234**, 528–541.

Ghosh, M. L., 1964/65: Disturbance in an elastic half space due to an impulsive twisting moment applied to an attached rigid circular disc. *Appl. Sci. Res.*, **A14**, 31–42.

Ghosh, S. C., 1970: Disturbance produced in an elastic half-space by impulsive normal pressure. *PAGEOPH*, **80**, 71–83.

Ghosh, S. K., 1973: The transient disturbance produced in an elastic layer by a buried spherical source. *PAGEOPH*, **105**, 781–801.

Gopalsamy, K., 1974: Response of an acoustic half-space excited by a randomly moving pressure pulse on the surface. *PAGEOPH*, **112**, 240–252.

Jingu, T., H. Matsumoto and K. Nezu, 1985: Transient stress of an elastic half space subjected to a uniform impulsive load in a rectangular region of its surface. *Bull. JSME*, **28**, 2881–2889.

Kawasaki, I., Y. Suzuki and R. Sato, 1973: Seismic waves due to a shear fault in a semi-infinite medium. Part I: Point source. *J. Phys. Earth*, **21**, 251–284.

Knopoff, L., 1958: The surface motions of a thick plate. *J. Appl. Phys.*, **29**, 661–670.

Knopoff, L., F. Gilbert and W. L. Pilant, 1960: Wave propagation in a medium with a single layer. *J. Geophys. Res.*, **65**, 265–278.

Longman, I. M., 1961: Solution of an integral equation occurring in the study of certain wave-propagation problems in layered media. *J. Acoust. Soc. Am.*, **33**, 954–958.

Maiti, N. C., and L. Debnath, 1990: Transient wave motions due to an asymmetric shear stress discontinuity in a layered elastic medium. *Zeit. angew. Math. Mech.*, **70**, 35–40.

Mandal, S. B., 1972: Diffraction of elastic waves by a rigid half-space. *Proc. Vibr. Prob.*, **13**, 331–343.

Mandal, S. B., 1974: Diffraction of elastic pulses by a rigid half-plane. *Indian J. Pure Appl. Math.*, **5**, 594–600.

Mencher, A. G., 1953: Epicentral displacement caused by elastic waves in an elastic slab. *J. Appl. Phys.*, **24**, 1240–1246.

Miklowitz, J., 1982: Wavefront analysis in the nonseparable elastodynamic quarter-plate problems. Part 1: The general method. *J. Appl. Mech.*, **49**, 797–807.

Miklowitz, J., 1982: Wavefront analysis in the nonseparable elastodynamic quarter-plate problems. Part 2: Wavefront events in the edge uniform pressure problem. *J. Appl. Mech.*, **49**, 808–815.

Mitra, M., 1958: Exact transient solution of the buried line source problem for an asymmetric source. *Zeit. angew. Math. Phys.*, **9a**, 322–331.

Mitra, M., 1960: On the application of Cagniard's method to dynamical problems of elasticity. *Gerland's Beit. Geophys.*, **69**, 73–86.

Mitra, M., 1963: Exact solution of the source problem for a finite three-dimensional source. *Geofisica pura e appl.*, **56**, 31–38.

Mitra, M., 1964: Propagation of explosive sound in a layered liquid. *J. Acoust. Soc. Am.*, **36**, 1145–1149.

Mitra, M., 1970: On a finite SH type source in a layered half-space. II. *PAGEOPH*, **80**, 147–151.

Mittal, J. P., and R. S. Sidhu, 1982: Generation of SH waves from a nonuniformly moving stress discontinuity in a layered half space. *Indian J. Pure Appl. Math.*, **13**, 682–695.

Müller, G., 1967: Theoretical seismograms for some types of point-sources in layered media. Part I: Theory. *Zeit. Geophys.*, **33**, 15–35.

Nag, K. R., 1961: On "SH" type of motion due to body forces in a semi-infinite elastic medium. *Gerland's Beitr. Geophys.*, **70**, 221–232.

Nag, K. R., 1962: Disturbance due to shearing-stress discontinuity in a semi-infinite elastic medium. *Geophys. J.*, **6**, 468–478.

Nag, K. R., 1963: Generation of Love waves due to a point source in a layered medium. *Indian J. Theoret. Phys.*, **11**, 105–118.

Nag, K. R., 1963: Generation of 'SH' type of waves due to stress-discontinuity in a layered medium. *Quart. J. Mech. Appl. Math.*, **16**, 293–303.

Nag, K. R., 1965: Disturbance due to a point source in a transversely isotropic half-space. *Bull. Cal. Math. Soc.*, **57**, 16–24.

Nag, K. R., 1972: Disturbance due to a point source in a layered half-space. *PAGEOPH*, **98**, 72–86.

Niazy, A., 1973: Elastic displacements caused by a propagating crack in an infinite medium: An exact solution. *Bull. Seism. Soc. Am.*, **63**, 357–379.

Niazy, A., 1975: An exact solution for a finite, two-dimensional moving dislocation in an elastic half-space with application to San Fernando earthquake of 1971. *Bull. Seism. Soc. Am.*, **65**, 1797–1826.

Norwood, F. R., 1973: Similarity solutions in plane elastodynamics. *Int. J. Solid Structures*, **9**, 789–803.

Norwood, F. R., 1975: Transient response of an elastic plate to loads with finite characteristic dimensions. *Int. J. Solid Structures*, **11**, 33–51.

Pao, Y.-H., and F. Ziegler, 1982: Transient SH-waves in a wedge-shaped layer. *Geophys. J. R. Astr. Soc.*, **71**, 57–77.

Papadopoulos, M., 1963: The reflexion and refraction of point source fields. *Proc. R. Soc. Lond.*, **A273**, 198–221.

Papadopoulos, M., 1963: The use of singular integrals in wave propagation problems; with application to the point source in a semi-infinite elastic medium. *Proc. R. Soc. Lond.*, **A276**, 204–237.

Paul, S., 1976: On the displacements produced in a porous elastic half-space by an impulsive line load. (Non-dissipative case.) *PAGEOPH*, **114**, 605–614.

Paul, S., 1976: On the disturbance produced in a semi-infinite poroelastic medium by a surface load. *PAGEOPH*, **114**, 615–627.

Paul, S., 1976: Lamb's line load problem for a porous elastic half-space: Non-dissipative case. *Indian J. Pure Appl. Math.*, **7**, 854–867.

Paul, S., 1978: On the disturbance in a poro-elastic medium by an expanding ring and disk load. *Indian J. Pure Appl. Math.*, **9**, 324–331.

Payton, R. G., 1988: Stresses in a constrained transversely isotropic elastic solid caused by a moving dislocation. *Acta Mech.*, **74**, 35–49.

Rajhans, B. K., and P. Kesari, 1986: Scattering of compressional waves by a cylindrical cavity. *J. Math. Phys. Sci.*, **20**, 429–444.

Rajhans, B. K., and P. Kesari, 1988: Diffraction of elastic spherical P waves by a cylindrical cavity. *Acta Mech.*, **72**, 309–325.

Rajhans, B. K., and S. K. Samal, 1992: Diffraction of compressional waves by a fluid cylinder in a homogeneous medium. *Indian J. Pure Appl. Math.*, **23**, 603–616.

Ravera, R. J., and G. C. Sih, 1970: Transient analysis of stress waves around cracks under antiplane strain. *J. Acoust. Soc. Am.*, **47**, 875–881.

Ross, R. A., 1961: The effect of an explosion in a compressible fluid under gravity. *Can. J. Phys.*, **39**, 1330–1346.

Roy, A., 1974: Surface displacements in an elastic half space due to a buried moving point source. *Geophys. J. R. Astr. Soc.*, **40**, 289–304.

Roy, A., 1974: Exact transient response of an elastic half space to a non-uniformly expanding circular ring. *Indian J. Pure Appl. Math.*, **5**, 1063–1080.

Sato, R., 1972: Seismic waves in the mean field. *J. Phys. Earth*, **20**, 357–375.

Towne, D. H., 1968: Pulse shapes of spherical waves reflected and refracted at a plane interface separating two homogeneous fluids. *J. Acoust. Soc. Am.*, **44**, 65–76.

Vlaar, N. J., 1964: The transient electromagnetic field from an antenna near the plane boundary between two dielectric halfspaces. II. A closer investigation of the field. *Appl. Sci. Res.*, **B11**, 49–66.

Vlaar, N. J., 1964: The seismic pulse in a semi-infinite medium. *Appl. Sci. Res.*, **B11**, 67–83.

Watanabe, K., 1977: Transient response of a layered elastic half space subjected to a reciprocating anti-plane shear load. *Int. J. Solids Structures*, **13**, 63–74.

Wright, T. W., 1969: Impact on an elastic quarter space. *J. Acoust. Soc. Am.*, **45**, 935–943.

Papers Using the Cagniard-de Hoop Technique

Abramovici, F., 1978: A generalization of the Cagniard method. *J. Comp. Phys.*, **29**, 328–343.

Abramovici, F., L. H. T. Le and E. R. Kanasewich, 1989: The evanescent wave in Cagniard's problem for a line source generating SH waves. *Bull. Seism. Soc. Am.*, **79**, 1941–1955.

Achenbach, J. D., 1968: Longitudinal force on an embedded filament. *Appl. Sci. Res.*, **19**, 412–425.

Aggarwal, H. R., and C. M. Ablow, 1967: Solution to a class of three-dimensional pulse propagation problems in an elastic half-space. *Int. J. Engng. Sci.*, **5**, 663–679.

Ang, D. D., 1960: Transient motion of a line load on the surface of an elastic half-space. *Quart. Appl. Math.*, **18**, 251–256.

Barclay, D. W., 1990: Isochromatic curves for a dynamically loaded elastic plate. *Acta Mech.*, **84**, 127–137.

Barr, G. E., 1967: On the diffraction of a cylindrical pulse by a half-plane. *Quart. Appl. Math.*, **25**, 193–204.

Bennett, B. E., and G. Herrmann, 1976: The dynamic response of an elastic half space with an overlying acoustic fluid. *J. Appl. Mech.*, **43**, 39–42.

Ben-Zion, Y., 1989: The response of two joined quarter spaces to SH line sources located at the material discontinuity interface. *Geophys. J. Int.*, **98**, 213–222.

Ben-Zion, Y., 1990: The response of two half spaces to point dislocations at the material interface. *Geophys. J. Int.*, **101**, 507–528.

Bleistein, N., and J. K. Cohen, 1992: The Cagniard method in complex time revisited. *Geophys. Prospecting*, **40**, 619–649.

Brock, L. M., 1978: The effect of a thin layer surface inhomogeneity on dynamic surface response. *J. Appl. Mech.*, **45**, 95–99.

Brock, L. M., 1978: The non-uniform motion of a thin smooth rigid inclusion through an elastic solid. *Quart. Appl. Math.*, **36**, 269–277.

Brock, L. M., 1979: Two basic wave propagation problems for the non-uniform motion of displacement discontinuities in a half-plane. *Int. J. Engng. Sci.*, **17**, 1211–1223.

Brock, L. M., 1979: Two basic wave propagation problems for the non-uniform motion of displacement discontinuities across a bimaterial interface. *Int. J. Engng. Sci.*, **17**, 1289–1302.

Brock, L. M., 1980: The non-uniform motion of a thin smooth rigid wedge into an elastic half-plane. *Quart. Appl. Math.*, **38**, 209–223.

Brock, L. M., 1980: Exact dynamic surface response for sub and through-surface slip. *J. Appl. Mech.*, **47**, 525–530.

Brock, L. M., 1981: A wave propagation problem for non-uniform screw dislocation motion in a viscoelastic half-plane. *J. Elasticity*, **11**, 187–195.

Brock, L. M., 1982: Dynamic solutions for the non-uniform motion of an edge dislocation. *Int. J. Engng. Sci.*, **20**, 113–118.

Brock, L. M., 1983: The dynamic stress intensity factor due to arbitrary screw dislocation motion. *J. Appl. Mech.*, **50**, 383–389.

Brock, L. M., 1983: The dynamic stress intensity factor for a crack due to arbitrary rectilinear screw dislocation motion. *J. Elasticity*, **13**, 429–439.

Brock, L. M., 1986: The transient field under a point force acting on an infinite strip. *J. Appl. Mech.*, **53**, 321–325.

Brock, L. M., 1989: Transient analyses of dislocation emission in the three modes of fracture. *Int. J. Engng. Sci.*, **27**, 1479–1495.

Burridge, R., 1971: Lamb's problem for an isotropic half-space. *Quart. J. Mech. Appl. Math.*, **24**, 81–98.

Bykovtsev, A. S., and D. B. Kramarovskii, 1987: The propagation of a complex fracture area. The exact three-dimensional solution. *J. Appl. Math. Mech.*, **51**, 89–98.

Chandra, U., 1967: Propagation of an SH-torque pulse in a three layered solid half-space. *PAGEOPH*, **67**, 54–64.

Chapman, C. H., 1974: Generalized ray theory for an inhomogeneous medium. *Geophys. J. R. Astr. Soc.*, **36**, 673–704.

Chapman, C. H., 1976: Exact and approximate generalized ray theory in vertically inhomogeneous media. *Geophys. J. R. Astr. Soc.*, **46**, 201–233.

Chapman, C. H., and J. A. Orcutt, 1985: The computation of body wave synthetic seismograms in laterally homogeneous media. *Rev. Geophys.*, **23**, 105–163.

Chen, P., and Y.-H., Pao, 1977: The diffraction of sound pulses by a circular cylinder. *J. Math. Phys.*, **18**, 2397–2406.

Chwalczyk, F., J. Rafa and E. Włodarczyk, 1978: Propagation of three component non-stationary acoustic waves in layer of ideal liquid. *J. Tech. Phys.*, **19**, 467–480.

De Hoop, A. T., and H. J. Frankena, 1960: Radiation of pulses generated by a vertical electric dipole above a plane, non-conducting, earth. *Appl. Sci. Res.*, **B8**, 369–377.

De Hoop, A. T., 1979: Pulsed electromagnetic radiation from a line source in a two-media configuration. *Radio Sci.*, **14**, 253–268.

De Hoop, A. T., and J. H. M. T. van der Hijden, 1983: Generation of acoustic waves by an impulsive line source in a fluid/solid configuration with a plane boundary. *J. Acoust. Soc. Am.*, **74**, 333–342.

De Hoop, A. T., and J. H. M. T. van der Hijden, 1984: Generation of acoustic waves by an impulsive point source in a fluid/solid configuration with a plane boundary. *J. Acoust. Soc. Am.*, **75**, 1709–1715.

De Hoop, A. T., and J. H. M. T. van der Hijden, 1985: Seismic waves generated by an impulsive point source in a solid/fluid configuration with a plane boundary. *Geophysics*, **50**, 1083–1090.

De Hoop, A. T., and M. L. Oristaglio, 1988: Application of the modified Cagniard technique to transient electromagnetic diffusion problems. *Geophys. J.*, **94**, 387–397.

De Hoop, A. T., 1990: Acoustic radiation from an impulsive point source in a continuously layered fluid – An analysis based on the Cagniard method. *J. Acoust. Soc. Am.*, **88**, 2376–2388.

Drijkoningen, G. G., and J. T. Fokkema, 1987: The exact seismic response of an ocean and a n-layer configuration. *Geophys. Prospecting*, **35**, 33–61.

Drijkoningen, G. G., and C. H. Chapman, 1988: Tunneling rays using the Cagniard-de Hoop method. *Bull. Seism. Soc. Am.*, **78**, 898–907.

Drijkoningen, G. G., 1991: Tunneling and the generalized ray method in piecewise homogeneous media. *Geophys. Prospecting*, **39**, 757–781.

Du Cloux, R., 1984: Pulsed electromagnetic radiation from a line source in the presence of a semi-infinite screen in the plane interface of two different media. *Wave Motion*, **6**, 459–476.

Ezzeddine, A., J. A. Kong and L. Tsang, 1981: Transient fields of a vertical electric dipole over a two-layer nondispersive dielectric. *J. Appl. Phys.*, **52**, 1202–1208.

Felsen, L. B., 1965: Transient solutions for a class of diffraction problems. *Quart. Appl. Math.*, **23**, 151–169.

Finkler, H., and K. J. Langenberg, 1975: Das Einschwingverhalten der elektrischen Feldstärke in einem atmosphärischen Bodenwellenleiter bei beliebiger Empfängerentfernung in Abhängigkeit von der Trägerfrequenz. *Archiv für Electronik und Übertragungstechnik*, **29**, 37–45.

Fokkema, J. T., and P. M. van den Berg, 1989: Seismic inversion by a RMS Born approximation in the space-time domain. *Geophys. Prospecting*, **37**, 53–72.

Frankena, H. J., 1960: Transient phenomena associated with Sommerfeld's horizontal dipole problem. *Appl. Sci. Res.*, **B8**, 357–368.

Freund, L. B., 1972: Wave motion in an elastic solid due to a nonuniformly moving line load. *Quart. Appl. Math.*, **30**, 271–281.

Freund, L. B., 1987: The stress intensity factor history due to three-dimensional transient loading on the faces of a crack. *J. Mech. Phys. Solids*, **35**, 61–72.

Friedland, A. B., and A. D. Pierce, 1969: Reflection of acoustic pulses from stable and instable interfaces between moving fluids. *Phys. Fluids*, **12**, 1148–1159.

Gakenheimer, D. C., and J. Miklowitz, 1969: Transient excitation of an elastic half space by a point load traveling on the surface. *J. Appl. Mech.*, **36**, 505–515.

Gakenheimer, D. C., 1971: Response of an elastic half space to expanding surface loads. *J. Appl. Mech.*, **38**, 99–110.

Ghosh, M. L., 1971: The axisymmetric problem of propagation of a normal stress discontinuity in a semi-infinite elastic medium. *Appl. Sci. Res.*, **24**, 149–167.

Ghosh, M., 1980/81: Displacement produced in an elastic half-space by the impulsive torsional motion of a circular ring source. *PAGEOPH*, **119**, 102–117.

Ghosh, M., and M. L. Ghosh, 1983: Torsional response of an elastic half space to a nonuniformly expanding ring source. *Zeit. angew. Math. Mech.*, **63**, 621–629.

Gilbert, F., and L. Knopoff, 1961: The directivity problem for a buried line source. *Geophysics*, **26**, 626–634.

Gopalsamy, K., and B. D. Aggarwala, 1972: Propagation of disturbances from randomly moving sources. *Zeit. angew. Math. Mech.*, **52**, 31–35.

Harkrider, D. G., and D. V. Helmberger, 1978: A note of nonequivalent quadrupole source cylindrical shear potentials which give equal displacements. *Bull. Seism. Soc. Am.*, **68**, 125–132.

Harris, J. G., 1980: Diffraction by a crack of a cylindrical longitudinal pulse. *Zeit. angew. Math. Phys.*, **31**, 367–383.

Helmberger, D. V., 1968: The crust-mantle transition in the Bering sea. *Bull. Seism. Soc. Am.*, **58**, 179–214.

Heyman, E., and L. B. Felsen, 1985: Non-dispersive closed form approximations for transient propagation and scattering of ray fields. *Wave Motion*, **7**, 335–358.

Hwang, L.-F., J. T. Kuo and Y.-C. Teng, 1982: Three-dimensional elastic wave scattering and diffraction due to a rigid cylinder embedded in an elastic medium by a point source. *PAGEOPH*, **120**, 548–576.

Jingu, T., H. Matsumoto and K. Nezu, 1986: The transient stress in an elastic half-space subjected to a semi-infinite line load varying as unit step function on its surface. *Bull. JSME*, **29**, 35–43.

Jingu, T., H. Matsumoto and K. Nezu, 1986: The transient stress in an elastic half-space excited by impulsive loading over one quarter of its surface. *Bull. JSME*, **29**, 44–51.

Johnson, L. R., 1974: Green's function for Lamb's problem. *Geophys. J. R. Astr. Soc.*, **37**, 99–131.

Kennedy, T. C., and G. Herrmann, 1972: Moving load on a solid-solid interface: Supersonic region. *Arch. Mech.*, **24**, 1023–1028.

Kennedy, T. C., and G. Herrmann, 1973: Moving load on a fluid-solid interface: Supersonic region. *J. Appl. Mech.*, **40**, 137–142.

Kennedy, T. C., and G. Herrmann, 1974: The response of a fluid-solid interface to a moving disturbance. *J. Appl. Mech.*, **41**, 287–288.

Knopoff, L., F. Gilbert and W. L. Pilant, 1960: Wave propagation in a medium with a single layer. *J. Geophys. Res.*, **65**, 265–278.

Kooij, B. J., and D. Quak, 1988: Three-dimensional scattering of impulsive acoustic waves by a semi-infinite crack in the plane interface of a half-space and a layer. *J. Math. Phys.*, **29**, 1712–1721.

Kooij, B. J., 1990: Transient electromagnetic field of a vertical magnetic dipole above a plane conducting Earth. *Radio Sci.*, **25**, 349–356.

Kooij, B. J., 1991: The transient electromagnetic field of an electric line source above a plane conducting earth. *IEEE Trans. Electromagn. Compat.*, **EMC-33**, 19–24.

Kraut, E. A., 1965/66: The effect of anisotropy on the integral representation of a cylindrical pulse. *Appl. Sci. Res.*, **B12**, 308–314.

Kraut, E. A., 1963: Advances in the theory of anisotropic elastic wave propagation. *Rev. Geophys.*, **1**, 401–448.

Kraut, E. A., 1968: Diffraction of elastic waves by a rigid 90° wedge. Part II. *Bull. Seism. Soc. Am.*, **58**, 1097–1115.

Kuester, E. F., 1984: The transient electromagnetic field of a pulsed line source located above a dispersively reflecting surface. *IEEE Trans. Antennas Propagat.*, **AP-32**, 1154–1162.

Kunin, V. V., B. E. Nemtsov and B. Ya. Éidman, 1985: Precursor and lateral waves during pulse reflection from the separation boundary of two media. *Sov. Phys. Usp.*, **28**, 827–841.

Langenberg, K. J., 1974: The transient response of a dielectric layer. *Appl. Phys.*, **3**, 179–188.

Laturelle, F. G., 1990: The stresses produced in an elastic half-space by a normal step loading over a circular area: Analytical and numerical results. *Wave Motion*, **12**, 107–127.

Le, L. H. T., 1993: On Cagniard's problem for a qSH line source in transversely isotropic media. *Bull. Seism. Soc. Am.*, **83**, 529–541.

Markenscoff, X., and R. J. Clifton, 1981: The nonuniformly moving edge dislocation. *J. Mech. Phys. Solids*, **29**, 253–262.

Markenscoff, X., and L. Ni, 1984: Nonuniform motion of an edge dislocation in an anisotropic solid. I. *Quart. Appl. Math.*, **41**, 475–494.

Mitra, M., 1963: An SH-point source in a half-space with a layer. *Bull. Seism. Soc. Am.*, **53**, 1031–1037.

Mitra, M., 1964: Disturbance produced in an elastic half-space by impulsive normal pressure. *Proc. Camb. Phil. Soc.*, **60**, 683–696.

Norwood, F. R., 1969: Exact transient response of an elastic half space loaded over a rectangular region of its surface. *J. Appl. Mech.*, **36**, 516–522.

Nuismer, R. J., and J. D. Achenbach, 1972: Dynamically induced fracture. *J. Mech. Phys. Solids*, **20**, 203–222.

Olsen, D., 1974: Transient waves in two semi infinite viscoelastic media separated by a plane interface. *Int. J. Engng. Sci.*, **12**, 691–712.

Pal, P. C., 1983: Generation of SH-type waves due to non-uniformly moving stress-discontinuity in layered anisotropic elastic half-space. *Acta Mech.*, **49**, 209–220.

Pal, P. C., 1985: On the disturbance produced by an impulsive torsional motion of a circular ring source in a semi-infinite transversely isotropic medium. *Indian J. Pure Appl. Math.*, **16**, 179–188.

Pal, S. C., and M. L. Ghosh, 1987: Waves in a semi-infinite elastic medium due to an expanding elliptic ring source on the free surface. *Indian J. Pure Appl. Math.*, **18**, 648–674.

Pao, Y.-H., F. Ziegler and Y.-S. Wang, 1989: Acoustic waves generated by a point source in a sloping fluid layer. *J. Acoust. Soc. Am.*, **85**, 1414–1426.

Payton, R. G., 1967: Transient motion of an elastic half-space due to a moving surface line load. *Int. J. Engng. Sci.*, **5**, 49–79.

Ravera, R. J., and G. C. Sih, 1970: Transient analysis of stress waves around cracks under antiplane strain. *J. Acoust. Soc. Am.*, **47**, 875–881.

Richards, P. G., 1971: A theory for pressure radiation from ocean-bottom earthquakes. *Bull. Seism. Soc. Am.*, **61**, 707–721.

Richards, P. G., 1973: The dynamic field of a growing plane elliptical shear crack. *Int. J. Solid Structures*, **9**, 843–861.

Rose, L. R. F., 1984: Point-source representation for lasar-generated ultrasound. *J. Acoust. Soc. Am.*, **75**, 723–732.

Roy, A., 1975: Pulse generation in an elastic half space by normal pressure. *Int. J. Engng. Sci.*, **13**, 641–651.

Ryan, R. L., 1971: Pulse propagation in a transversely isotropic half-space. *J. Sound Vib.*, **14**, 511–524.

Sakai, Y., and I. Kawasaki, 1990: Analytic waveforms for a line source in a transversely isotropic medium. *J. Geophys. Res.*, **95**, 11333–11344.

Salvado, C., and J. B. Minster, 1980: Slipping interfaces: A possible source of S radiation from explosive sources. *Bull. Seism. Soc. Am.*, **70**, 659–670.

Sato, R., 1972: Seismic waves in the near field. *J. Phys. Earth*, **20**, 357–375.

Scott, R. A., and J. Miklowitz, 1969: Transient non-axisymmetric wave propagation in an infinite isotropic elastic plate. *Int. J. Solids Structure*, **5**, 65–79.

Scott, R. A., and J. Miklowitz, 1969: Near-field transient waves in an anisotropic elastic plates for two and three dimensional problems. *Int. J. Solids Structure*, **5**, 1059–1075.

Shmuely, M., 1974: Response of plates to transient sources. *J. Sound Vib.*, **32**, 491–506.

Suh, S. L., W. Goldsmith, J. L. Sackman and R. L. Taylor, 1974: Impact on a transversely anisotropic half-space. *Int. J. Rock Mech. Min. Sci. & Geomech. Abstr.*, **11**, 413–421.

Tanimoto, T., 1982: Cagniard-de Hoop method for a Haskell type vertical fault. *Geophys. J. R. Astr. Soc.*, **70**, 639–646.

Teng, Y. C., J. T. Kuo and C. Gong, 1975: Three-dimensional acoustic wave scattering and diffraction by an open-ended vertical soft cylinder in a half-space. *J. Acoust. Soc. Am.*, **57**, 782–790.

Tsai, C.-H., and C.-C. Ma, 1991: Exact transient solutions of buried dynamic point forces and displacement jumps for an elastic half space. *Int. J. Solids Structure*, **28**, 955–974.

Tsai, Y. M., 1968: Stress waves produced by impact on the surface of a plastic medium. *J. Franklin Inst.*, **285**, 204–221.

Van der Hijden, J. H. M. T., 1984: Quantitative analysis of the pseudo-Rayleigh phenomena. *J. Acoust. Soc. Am.*, **75**, 1041–1047.

Van der Hijden, J. H. M. T., 1987: Radiation from an impulsive line source in an unbounded homogeneous anisotropic medium. *Geophys. J. R. Astr. Soc.*, **91**, 355–372.

Verweij, M. D., and A. T. de Hoop, 1990: Determination of seismic wavefields in arbitrarily continuously layered media using the modified Cagniard method. *Geophys. J. Int.*, **103**, 731–754.

Vlaar, N. J., 1963/64: The transient electromagnetic field from an antenna near the plane boundary of two dielectric halfspaces. *Appl. Sci. Res.*, **B10**, 353–384.

Watanabe, K., 1978: Transient response of an acoustic half-space to a rotating point load. *Quart. Appl. Math.*, **36**, 39–48.

Watanabe, K., 1981: Transient response of an infinite elastic solid to a moving point load. *Bull. JSME*, **24**, 1115–1122.

Watanabe, K., 1984: Transient response of an elastic solid to a moving torsional load in a cylindrical bore: An approximate solution. *Int. J. Engng. Sci.*, **22**, 277–284.

Zemell, S. H., 1976: New derivation of the exact solution for the diffraction of a cylindrical or spherical pulse on a wedge. *Int. J. Engng. Sci.*, **14**, 845–851.

Ziegler, F., and Y.-H. Pao, 1990: Die Phasenfunktion kugeliger Wellen in der keilförmigen Oberflächenschicht. *Zeit. angew. Math. Mech.*, **70**, T222–T223.

Chapter 5
The Wiener-Hopf Technique

One of the difficulties in solving partial differential equations by Fourier transforms is finding a general transform that applies over the entire spatial domain. The Wiener-Hopf technique is a popular method that avoids this problem by defining Fourier transforms over certain regions and then uses function-theoretic analysis to piece together the complete solution.

Although Wiener and Hopf first devised this method to solve certain singular integral equations[1], it has been in radiation and diffraction problems that it has found its greatest applicability. For example, this technique reduces the problem of diffraction by a semi-infinite plate to the solution of a singular integral equation[2]. The Wiener-Hopf technique[3] then yields the classic result given by Sommerfeld[4].

---

[1] Wiener, N., and E. Hopf, 1931: Über eine Klasse singulärer Integralgleichungen. *Sitz. Ber. Preuss. Akad. Wiss., Phys.-Math. Kl.*, 696–706.

[2] Magnus, W., 1941: Über die Beugung electromagnetische Wellen an einer Halbebene. *Zeit. Phys.*, **117**, 168–179.

[3] Copson, E. T., 1946: On an integral equation arising in the theory of diffraction. *Quart. J. Math.*, **17**, 19–34.

[4] Sommerfeld, A., 1896: Mathematische Theorie der Diffraction. *Math. Ann.*, **47**, 317–374.

Since its original formulation, the Wiener-Hopf technique has undergone simplification by formulating the problem in terms of dual integral equations[5,6,7]. The essence of this technique is the process of *factorization* of the Fourier transform of the kernel function into the product of two other Fourier transforms which are regular and nonzero in certain half planes. In the first two sections, we will show various problems which illustrate this factorization. It is this simplified form that we commonly refer to as the Wiener-Hopf technique. In Section 5.3 we illustrate how to apply this technique with finite-sized scatterers.

5.1 THE WIENER-HOPF TECHNIQUE WHEN THE FACTORIZATION CONTAINS NO BRANCH POINTS

The Wiener-Hopf technique is most often used in solving partial differential equations where some discontinuity or obstacle generates reflected and transmitted waves. To illustrate this technique in its simplest form, consider an infinitely long channel $-\infty < x < \infty$, $0 < y < a$ rotating on a flat plate[8] filled with an inviscid, homogeneous fluid of uniform depth H. Within the channel, we have another plate of infinitesimal thickness located at $x < 0$ and $y = b$. See Fig. 5.1.1. The shallow-water equations govern the motion of the fluid:

$$-i\omega u - fv = -\frac{\partial h}{\partial x} \tag{5.1.1}$$

$$-i\omega v + fu = -\frac{\partial h}{\partial y} \tag{5.1.2}$$

$$-i\omega h + gH\left(\frac{\partial u}{\partial x} + \frac{\partial v}{\partial y}\right) = 0 \tag{5.1.3}$$

where u and v are the velocities in the x and y directions, respectively, h is the deviation of the free surface from its average height H, g is the gravitational acceleration and f is one half of the angular velocity at which it rotates. All motions within the fluid behave as $e^{-i\omega t}$.

---

[5] Kaup, S. N., 1950: Wiener-Hopf techniques and mixed boundary value problems. *Comm. Pure Appl. Math.*, **3**, 411–426.

[6] Clemmow, P. C., 1951: A method for the exact solution of a class of two-dimensional diffraction problems. *Proc. R. Soc. Lond.*, **A205**, 286–308.

[7] See Noble, B., 1958: *Methods Based on the Wiener-Hopf Technique for the Solution of Partial Differential Equations*. London: Pergamon Press, 246 pp.

[8] Taken from Kapoulitsas, G. M., 1980: Scattering of long waves in a rotating bifurcated channel. *Int. J. Theoret. Phys.*, **19**, 773–788.

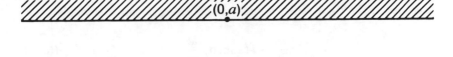

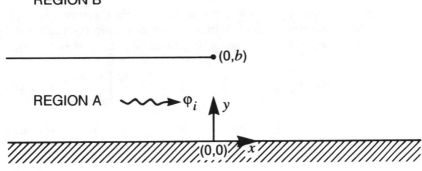

Fig. 5.1.1: Schematic of the rotating channel in which Kelvin waves are diffracted.

A little algebra shows that we may eliminate u and v and obtain the Helmholtz equation:

$$\frac{\partial^2 h}{\partial x^2} + \frac{\partial^2 h}{\partial y^2} + k^2 h = 0 \tag{5.1.4}$$

where

$$(\omega^2 - f^2)u = -i\omega \frac{\partial h}{\partial x} + f \frac{\partial h}{\partial y}, \tag{5.1.5}$$

$$(\omega^2 - f^2)v = -i\omega \frac{\partial h}{\partial y} - f \frac{\partial h}{\partial x} \tag{5.1.6}$$

and $k^2 = (\omega^2 - f^2)/gH$ assuming that $\omega > f$. We solve the problem when an incident wave of the form

$$h = \phi_i = \exp[(i\omega x - fy)/c] \tag{5.1.7}$$

$$= \exp\{k[ix\cosh(\beta) - y\sinh(\beta)]\}, \tag{5.1.8}$$

a so-called "Kelvin wave", propagates towards the origin from $-\infty$ within the lower channel A. See Fig. 5.1.1. We have introduced β such that $f = kc\sinh(\beta)$, $\omega = kc\cosh(\beta)$ and $c^2 = gH$.

The first step in the Wiener-Hopf technique is to write the solution as a sum of the incident wave plus some correction ϕ that represents the reflected and transmitted waves. For example, in region A, h consists of $\phi_i + \phi$ whereas in region B we have only ϕ. Because ϕ_i satisfies (5.1.4),

349

so must ϕ. Furthermore, ϕ must satisfy certain boundary conditions. Because of the rigid walls, v must vanish along them; this yields

$$\frac{\partial \phi}{\partial y} - i \tanh(\beta) \frac{\partial \phi}{\partial x} = 0 \tag{5.1.9}$$

along $-\infty < x < \infty$, $y = 0, a$ and $x < 0$, $y = b^{\pm}$. Furthermore, because the partition separating region A from region B is infinitesimally thin, we must have continuity of v across that boundary; this gives

$$\left[\frac{\partial \phi}{\partial y} - i \tanh(\beta) \frac{\partial \phi}{\partial x} \right]_{y=b^-} = \left[\frac{\partial \phi}{\partial y} - i \tanh(\beta) \frac{\partial \phi}{\partial x} \right]_{y=b^+} \tag{5.1.10}$$

for $-\infty < x < \infty$. Finally, to prevent infinite velocities in the right half of the channel, h must be continuous at $z = b$ or

$$\phi(x, b^-) + \phi_i(x, b) = \phi(x, b^+) \tag{5.1.11}$$

for $x > 0$.

An important assumption in the Wiener-Hopf technique concerns the so-called "edge conditions" at the edge point $(0, b)$, namely that

$$\phi = O(1) \quad \text{as} \quad x \to 0^{\pm} \quad \text{and} \quad y = b \tag{5.1.12}$$

and

$$\frac{\partial \phi}{\partial y} = O(x^{-1/2}) \quad \text{as} \quad x \to 0^{\pm} \quad \text{and} \quad y = b. \tag{5.1.13}$$

These conditions are necessary to guarantee the uniqueness of the solution because the edge point is a geometric singularity. Another assumption is the introduction of dissipation by allowing ω to have a small, positive imaginary part. We may either view this as merely reflecting reality or ensuring that we satisfy the Sommerfeld radiation condition that energy must radiate to infinity. As a result of this complex form of ω, k must also be complex with a small, positive imaginary part k_2.

We solve (5.1.4), as well as (5.1.9)–(5.1.11), by Fourier transforms. Let us define the double-sided Fourier transform of $\phi(x, y)$ by

$$\Phi(\alpha, y) = \int_{-\infty}^{\infty} \phi(x, y) e^{i\alpha x} dx, \qquad |\tau| < \tau_0, \tag{5.1.14}$$

along with the one-sided Fourier transforms

$$\Phi_-(\alpha, y) = \int_{-\infty}^{0} \phi(x, y) e^{i\alpha x} dx, \qquad \tau < \tau_0, \tag{5.1.15}$$

and

$$\Phi_+(\alpha, y) = \int_0^\infty \phi(x, y)e^{i\alpha x}\, dx, \qquad \tau > -\tau_0, \tag{5.1.16}$$

where $\alpha = \sigma + i\tau$ and σ, τ are real. Clearly,

$$\Phi_-(\alpha, y) + \Phi_+(\alpha, y) = \Phi(\alpha, y). \tag{5.1.17}$$

Note that (5.1.14)–(5.1.16) is regular in a common strip in the complex α-plane.

Taking the double-sided Fourier transform of (5.1.4),

$$\frac{d^2\Phi}{dy^2} - \gamma^2\Phi = 0 \tag{5.1.18}$$

where $\gamma = \sqrt{\alpha^2 - k^2}$. In general,

$$\Phi(\alpha, y) = \begin{cases} A(\alpha)e^{-\gamma y} + B(\alpha)e^{\gamma y} & \text{if } 0 \le y \le b, \\ C(\alpha)e^{-\gamma y} + D(\alpha)e^{\gamma y} & \text{if } b \le y \le a. \end{cases} \tag{5.1.19}$$

Taking the double-sided Fourier transform of (5.1.9) for the conditions on $y = 0, a$,

$$\Phi'(\alpha, 0) - \alpha\tanh(\beta)\Phi(\alpha, 0) = 0 \tag{5.1.20}$$

and

$$\Phi'(\alpha, a) - \alpha\tanh(\beta)\Phi(\alpha, a) = 0. \tag{5.1.21}$$

Similarly, from (5.1.10),

$$\Phi'(\alpha, b^-) - \alpha\tanh(\beta)\Phi(\alpha, b^-) = \Phi'(\alpha, b^+) - \alpha\tanh(\beta)\Phi(\alpha, b^+). \tag{5.1.22}$$

From the boundary conditions (5.1.20)–(5.1.22),

$$B = \lambda A, \tag{5.1.23}$$

$$C = -A\frac{\sinh(\gamma b)}{\sinh[\gamma(a-b)]}e^{\gamma a} \tag{5.1.24}$$

and

$$D = -\lambda A\frac{\sinh(\gamma b)}{\sinh[\gamma(a-b)]}e^{-\gamma a} \tag{5.1.25}$$

where

$$\lambda = \frac{\gamma + \alpha\tanh(\beta)}{\gamma - \alpha\tanh(\beta)}. \tag{5.1.26}$$

By taking the one-sided Fourier transform of (5.1.9) (from $-\infty$ to 0) for the condition $x < 0$, we obtain the two equations:

$$\Phi'_-(\alpha, b^+) - \alpha\tanh(\beta)\Phi_-(\alpha, b^+) = i\tanh(\beta)\phi(0, b) \tag{5.1.27}$$

and

$$\Phi'_-(\alpha, b^-) - \alpha \tanh(\beta)\Phi_-(\alpha, b^-) = i \tanh(\beta)\phi(0, b) \qquad (5.1.28)$$

because $\phi(0, b^-) = \phi(0, b^+) = \phi(0, b)$. Using (5.1.27)–(5.1.28) in conjunction with (5.1.22),

$$\Phi'_+(\alpha, b^-) - \alpha \tanh(\beta)\Phi_+(\alpha, b^-) + i \tanh(\beta)\phi(0, b)$$
$$= \Phi'_+(\alpha, b^+) - \alpha \tanh(\beta)\Phi_+(\alpha, b^+) + i \tanh(\beta)\phi(0, b)$$
$$= P_+(\alpha), \quad \text{say.} \qquad (5.1.29)$$

Finally, the one-sided Fourier transform (from 0 to ∞) of (5.1.11) yields

$$\Phi_+(\alpha, b^-) + \frac{i \exp[-kb \sinh(\beta)]}{\alpha + k \cosh(\beta)} = \Phi_+(\alpha, b^+). \qquad (5.1.30)$$

Then by (5.1.17), (5.1.19), (5.1.27)–(5.1.29) and the definition of B,

$$P_+(\alpha) = 2A[\gamma + \alpha \tanh(\beta)] \sinh(\gamma b). \qquad (5.1.31)$$

Let us now introduce the function $Q_-(\alpha)$ defined by

$$Q_-(\alpha) = \tfrac{1}{2}[\Phi_-(\alpha, b^-) - \Phi(\alpha, b^+)]. \qquad (5.1.32)$$

From (5.1.19), (5.1.22), (5.1.24), (5.1.26), (5.1.30) and $D = \lambda C \, e^{-2\gamma a}$,

$$2Q_-(\alpha) - \frac{i \exp[-kb \sinh(\beta)]}{\alpha + k \cosh(\beta)} = \frac{2A\gamma \sinh(\gamma a)}{[\gamma - \alpha \tanh(\beta)] \sinh[\gamma(a - b)]}. \qquad (5.1.33)$$

Eliminating A between (5.1.31) and (5.1.33), we finally obtain a functional equation of the Wiener-Hopf type:

$$2Q_-(\alpha) - \frac{i \exp[-kb \sinh(\beta)]}{\alpha + k \cosh(\beta)}$$
$$= \frac{P_+(\alpha)}{\gamma^2 - \alpha^2 \tanh(\beta)} \times \frac{\gamma \sinh(\gamma a)}{\sinh(\gamma b) \sinh[\gamma(a - b)]}. \qquad (5.1.34)$$

What makes (5.1.34) a functional equation of the Wiener-Hopf type? Note that $Q_-(\alpha)$ is analytic for $\tau < \tau_0$ while $P_+(\alpha)$ is analytic for $\tau > -\tau_0$. In order for (5.1.34) to be true, we must restrict ourselves to the strip $|\tau| < \tau_0$. Thus, the Wiener-Hopf equation contains complex Fourier transforms which are regular over the common interval of $\tau_- < \tau < \tau_+$ where $Q_-(\alpha)$ is regular for $\tau < \tau_+$ and $\tau > \tau_-$.

A crucial step in solving the Wiener-Hopf equation (5.1.34) is the process of factorization. Factorization is the method of rewriting certain terms in (5.1.34) as a product $M(\alpha) = M_+(\alpha)M_-(\alpha)$ where $M_+(\alpha)$ and $M_-(\alpha)$ are regular and free of zeros in an upper and lower half-plane, respectively. These half-planes share a certain strip of the α-plane in common. Why we want to factor will become apparent in a few paragraphs.

This process of factorization is the most important step in the Wiener-Hopf technique and usually follows two different paths depending upon whether $M(\alpha)$ is a single-valued or multivalued function. Although we have a square root γ present in (5.1.34), this expression is an even function of γ and there are consequently no branch points or cuts. Therefore, $M(\alpha)$ a meromorphic function. In the next section we deal with the situation when $M(\alpha)$ is multivalued.

The common method for factorizing a meromorphic function lies in the following theorem from complex variables:

The Infinite Product Theorem[9]: If $f(z)$ is an entire function of z with simple zeros at $z_1, z_2, \ldots$, then

$$f(z) = f(0) \exp\left[zf'(0)/f(0)\right] \prod_{n=1}^{\infty} \left(1 - \frac{z}{z_n}\right) e^{z/z_n}. \qquad (5.1.35)$$

In our particular problem,

$$M(\alpha) = \frac{\sinh(\gamma b) \sinh[\gamma(a - b)]}{\gamma \sinh(\gamma a)}. \qquad (5.1.36)$$

Applying the infinite product theorem separately to the numerator and denominator of (5.1.36), we immediately find that

$$M_+(\alpha) = M_-(-\alpha)$$

$$= \left\{ \frac{\sin(kb) \sin[k(a - b)]}{k \sin(ka)} \right\}^{1/2}$$

$$\times \exp\left\{ \frac{\alpha i}{\pi}\left[b \ln\left(\frac{a}{b}\right) + (a - b) \ln\left(\frac{a}{a - b}\right) \right] \right\}$$

$$\times \prod_{n=1}^{\infty} (1 + \alpha/\alpha_{nb}) \exp(ib\alpha/n\pi)$$

$$\times \prod_{n=1}^{\infty} [1 + \alpha/\alpha_{n(a-b)}] \exp[i(a - b)\alpha/n\pi]$$

$$\Bigg/ \prod_{n=1}^{\infty} (1 + \alpha/\alpha_{na}) \exp(ia\alpha/n\pi) \qquad (5.1.37)$$

[9] See Titchmarsh, E. C., 1939: *The Theory of Functions.* 2nd Edition. Oxford: Oxford University Press, Section 3.23.

where $\alpha_{nl} = i\sqrt{n^2\pi^2/l^2 - k^2}$ and $l = a$ or b or $(a - b)$. The square root has a positive real part or a negative imaginary part. Note that in this factorization $M_+(\alpha)$ is regular and nonzero in the upper half of the α-plane ($\tau > -k_2$) while $M_-(\alpha)$ is regular and nonzero in the lower half of the α-plane ($\tau < k_2$).

Substituting this factorization into (5.1.34),

$$2[\alpha - k\cosh(\beta)]M_-(\alpha)Q_-(\alpha) - \frac{ie^{-kb\sinh(\beta)}[\alpha - k\cosh(\beta)]M_-(\alpha)}{\alpha + k\cosh(\beta)}$$

$$= \frac{\cosh^2(\beta)}{[\alpha + k\cosh(\beta)]M_+(\alpha)}P_+(\alpha). \qquad (5.1.38)$$

Next, we note that

$$\frac{[\alpha - k\cosh(\beta)]M_-(\alpha)}{\alpha + k\cosh(\beta)}$$

$$= \frac{[\alpha - k\cosh(\beta)]M_-(\alpha) + 2k\cosh(\beta)M_-[-k\cosh(\beta)]}{\alpha + k\cosh(\beta)}$$

$$- \frac{2k\cosh(\beta)M_-[-k\cosh(\beta)]}{\alpha + k\cosh(\beta)}. \qquad (5.1.39)$$

Therefore, (5.1.38) becomes

$$2[\alpha - k\cosh(\beta)]M_-(\alpha)Q_-(\alpha) - i\exp[-kb\sinh(\beta)]$$

$$\times \left\{ \frac{[\alpha - k\cosh(\beta)]M_-(\alpha) + 2k\cosh(\beta)M_-[-k\cosh(\beta)]}{\alpha + k\cosh(\beta)} \right\}$$

$$= \frac{\cosh^2(\beta)}{[\alpha + k\cosh(\beta)]M_+(\alpha)}P_+(\alpha)$$

$$- i\exp[-kb\sinh(\beta)]\frac{2k\cosh(\beta)M_-[-k\cosh(\beta)]}{\alpha + k\cosh(\beta)} \qquad (5.1.40)$$

The primary reason for the factorization and the subsequent algebraic manipulation is the fact that the left side of (5.1.40) is regular in $\tau > -\tau_0$ while the right side is regular in $\tau < \tau_0$. Hence, both sides are regular on the strip $|\tau| < \tau_0$. Then by analytic continuation it follows that (5.1.40) is defined in the entire α-plane and both sides equal to an entire function $p(\alpha)$. To determine $p(\alpha)$ we use a version of Liouville's theorem for polynomials. This theorem is as follows:

Liouville's theorem[10]: If $f(z)$ is analytic for all finite value of z, and as $|z| \to \infty$, $f(z) = O(|z|^k)$, then $f(z)$ is a polynomial of degree $\leq k$.

---

[10] See Titchmarsh, E. C., 1939: *The Theory of Functions*. 2nd Edition. Oxford: Oxford University Press, Section 2.52.

Consequently, upon examining the asymptotic value of (5.1.40) as $|\alpha| \to \infty$ as well as using the edge condition (5.1.12)–(5.1.13), Liouville's theorem shows that $p(\alpha)$ is a constant. Because in the limit of $|\alpha| \to \infty$, $p(\alpha) \to 0$, then $p(\alpha) = 0$. Therefore, from (5.1.40),

$$P_+(\alpha) = \frac{2ikM_+[k\cosh(\beta)]\exp[-kb\sinh(\beta)]}{\cosh(\beta)}M_+(\alpha). \qquad (5.1.41)$$

Knowing $P_+(\alpha)$, we find from (5.1.23), (5.1.24), (5.1.25) and (5.1.31) that

$$A = \frac{EM_+(\alpha)}{[\gamma + \alpha\tanh(\beta)]\sin(\gamma b)}, \qquad (5.1.42)$$

$$B = \frac{EM_+(\alpha)}{[\gamma - \alpha\tanh(\beta)]\sin(\gamma b)}, \qquad (5.1.43)$$

$$C = -\frac{EM_+(\alpha)e^{\gamma a}}{[\gamma + \alpha\tanh(\beta)]\sin[\gamma(a-b)]} \qquad (5.1.44)$$

and

$$D = -\frac{EM_+(\alpha)e^{-\gamma a}}{[\gamma - \alpha\tanh(\beta)]\sin[\gamma(a-b)]} \qquad (5.1.45)$$

where

$$E = \frac{ikM_+[k\cosh(\beta)]\exp[-kb\sinh(\beta)]}{\cosh(\beta)}. \qquad (5.1.46)$$

With these values of A, B C and D we have found $\Phi(\alpha, y)$. Therefore, the final solution requires the inversion of $\Phi(\alpha, y)$. For example, for $-\infty < x < \infty, 0 \le y \le b$,

$$\phi(x,y) = \frac{1}{2\pi}\int_{-\infty}^{\infty}\frac{M_+(\alpha)E}{\sin(\gamma b)}\left[\frac{e^{-\gamma y}}{\gamma + \alpha\tanh(\beta)} + \frac{e^{\gamma y}}{\gamma - \alpha\tanh(\beta)}\right]e^{-i\alpha x}\,d\alpha. \qquad (5.1.47)$$

For $x < 0$ we evaluate (5.1.47) by closing the integration along the real axis with a infinite semicircle in the upper half of the α-plane by Jordan's lemma and using the residue theorem. $M_+(\alpha)$ has simple poles at $\gamma b = n\pi$ where $n = \pm 1, \pm 2, \ldots$ and the zeros of $[\gamma \pm \alpha\tanh(\beta)]$. Upon applying the residue theorem,

$$\phi(x,y) = -\frac{k\sin(\beta)M_+^2[k\cosh(\beta)]\exp[-kb\sinh(\beta)]}{\sinh[kb\sinh(\beta)]}$$
$$\times \exp\{k[-ix\cosh(\beta) + y\sinh(\beta)]\}$$
$$+ \frac{2i\pi E}{b^2}\sum_{n=1}^{\infty}\frac{(-1)^n n M_+(\alpha_n)}{\alpha_{nb}[(n\pi/b)^2 + \alpha_{nb}^2\tanh^2(\beta)]}$$
$$\times \left[\frac{n\pi}{b}\cos\left(\frac{n\pi y}{b}\right) + \alpha_{nb}\tanh(\beta)\sin\left(\frac{n\pi y}{b}\right)\right]e^{-i\alpha_{nb}x}$$

$$\qquad (5.1.48)$$

where $a_{nb} = i\sqrt{n^2\pi^2/b^2 - k^2}$. The first part of the right side of (5.1.48) represents the reflected Kelvin wave traveling in the channel ($0 \le y \le b, x < 0$) to the left. The infinite series represents attenuated, stationary modes.

In a similar manner, we apply the residue theorem to give the solution in the remaining domains. They are

$$
\phi(x,y) = -\frac{\sinh[k(a-b)\sinh(\beta)]\exp[-kb\sinh(\beta)]}{\sinh[ka\sinh(\beta)]}
$$
$$
\times \exp\{k[ix\cosh(\beta) - y\sinh(\beta)]\}
$$
$$
-\frac{2iE}{\alpha}\sum_{n=1}^{\infty}\frac{\sin(\pi nb/2)}{\alpha_{na}M(\alpha_{na})[(n\pi/a)^2 + \alpha_{na}^2\tanh^2(\beta)]}
$$
$$
\times\left[\frac{n\pi}{a}\cos\left(\frac{n\pi y}{a}\right) - \alpha_{na}\tanh(\beta)\sin\left(\frac{n\pi y}{a}\right)\right]e^{i\alpha_{na}x}
$$

$$(5.1.49)$$

for $0 \le y \le b, x > 0$ and

$$
\phi(x,y) = \frac{k\sinh(\beta)M_+^2[k\cosh(\beta)]\exp[-k(a+b)\sinh(\beta)]}{\sinh[kd\sinh(\beta)]}
$$
$$
\times \exp\{k[-ix\cosh(\beta) + y\sinh(\beta)]\}
$$
$$
-\frac{2i\pi E}{d^2}\sum_{n=1}^{\infty}\frac{(-1)^n nM_+(\alpha_{nd})}{[(n\pi/2)^2 + \alpha_{nd}^2\tan^2(\beta)]}
$$
$$
\times\left\{\frac{n\pi}{d}\cos\left[\frac{n\pi(y-a)}{d}\right] + \alpha_{nd}\tanh(\beta)\sin\left[\frac{n\pi(y-a)}{d}\right]\right\}e^{-i\alpha_{nd}x}
$$

$$(5.1.50)$$

for $b \le y \le a, -\infty < x < \infty$ where $d = a - b$.

In the previous example, factorization followed directly from the infinite product theorem. In certain cases, factorization is not so simple. For example[11], consider the Wiener-Hopf equation in ω:

$$
\sqrt{\omega^2 + \lambda^2}U_+(\omega) = \frac{1}{\omega - \kappa} - T_-(\omega)
$$

$$(5.1.51)$$

where $\lambda > 0$, $\kappa < 0$, $U_+(\omega)$ is analytic in the upper complex ω-plane $\text{Im}(\omega) > 0$ and $T_-(\omega)$ is analytic in the half plane $\text{Im}(\omega) < \tau, \lambda > \tau > 0$. We now factor the square root as $\sqrt{\omega + \lambda i}\sqrt{\omega - \lambda i}$. Therefore,

$$
\sqrt{\omega + \lambda i}\, U_+(\omega) = \frac{1}{(\omega - \kappa)\sqrt{\omega - \lambda i}} - \frac{T_-(\omega)}{\sqrt{\omega - \lambda i}}.
$$

$$(5.1.52)$$

---

[11] Taken from Lehner, F. K., V. C. Li and J. R. Rice, 1981: Stress diffusion along rupturing plate boundaries. *J. Geophys. Res.*, **86**, 6155–6169. ©1981 American Geophysical Union.

The left side of (5.1.52) is analytic in the upper half-plane $\text{Im}(\omega) > 0$, while the second term on the right side is analytic in the lower half-plane $\text{Im}(\omega) < \tau$. The problem in factorization lies with the first term on the right side of (5.1.52) where a simple pole lies in the lower half-plane $\text{Im}(\omega) < 0$. However, because this first term is analytic in the strip $0 < \text{Im}(\omega) < \tau$, analytic function theory[12] states that we can express it in two parts which are analytic, respectively, in the overlapping regions $\text{Im}(\omega) > 0$ and $\text{Im}(\omega) < \tau$. We do the splitting by inspection and this yields

$$\frac{1}{(\omega - \kappa)\sqrt{\omega - \lambda i}} = \frac{1}{(\omega - \kappa)\sqrt{\kappa - \lambda i}} + \left[\frac{1}{\sqrt{\omega - \lambda i}} - \frac{1}{\sqrt{\kappa - \lambda i}}\right]\frac{1}{\omega - \kappa}.$$
(5.1.53)

The first term on the right side of (5.1.53) is analytic in the half-plane $\text{Im}(\omega) > 0$ while the second term is certainly analytic in the half-plane $\text{Im}(\omega) < \tau$. Therefore,

$$\sqrt{\omega + \lambda i}\, U_+(\omega) - \frac{1}{(\omega - \kappa)\sqrt{\kappa - \lambda i}}$$

$$= \left[\frac{1}{\sqrt{\omega - \lambda i}} - \frac{1}{\sqrt{\kappa - \lambda i}}\right]\frac{1}{\omega - \kappa} - \frac{T_-(\omega)}{\sqrt{\omega - \lambda i}}.$$
(5.1.54)

is the desired factorization and the analysis may proceed.

Problems

1. Use the Wiener-Hopf technique to solve the partial differential equation[13]

$$\frac{\partial^2 u}{\partial x^2} + \frac{\partial^2 u}{\partial y^2} = \frac{\partial u}{\partial t}, \qquad -\infty < x < \infty, y > 0, t > 0$$

with the boundary conditions

$$u(x, 0, t) = e^{-\epsilon x}, \qquad x > 0,$$

$$\frac{\partial u(x, 0, t)}{\partial y} = -e^{\epsilon x}, \qquad x < 0,$$

[12] See Morse, P. M., and H. Feshbach, 1953: *Methods of Theoretical Physics.* New York: McGraw-Hill Book Co., Inc., p. 987 for a full discussion.

[13] Taken from Huang, S. C., 1985: Unsteady-state heat conduction in semi-infinite regions with mixed-type boundary conditions. *J. Heat Transfer*, **107**, 489–491.

$$\lim_{|x|\to\infty} u(x,y,t) \to 0 \text{ and } \lim_{y\to\infty} u(x,y,t) \to 0$$

and the initial condition $u(x,y,0) = 0$ with $\epsilon \to 0$.

Step 1. Define the Laplace transform

$$\overline{u}(x,y,s) = \int_0^\infty u(x,y,t)e^{-st}dt$$

and the Fourier transforms

$$\overline{U}_+(k,y,s) = \int_0^\infty \overline{u}(x,y,s)e^{-ikx}dx,$$

$$\overline{U}_-(k,y,s) = \int_{-\infty}^0 \overline{u}(x,y,s)e^{-ikx}dx$$

and

$$\overline{U}(k,y,s) = \overline{U}_+(k,y,s) + \overline{U}_-(k,y,s) = \int_{-\infty}^\infty \overline{u}(x,y,s)e^{-ikx}dx.$$

Then show that the partial differential equation and boundary conditions reduce to

$$\frac{d^2\overline{U}}{dy^2} - (k^2 + s)\overline{U} = 0, \qquad 0 < y < \infty$$

with

$$\overline{U}_+(k,0,s) = -\frac{i}{s(k-\epsilon i)}$$

$$\frac{d\overline{U}_-(k,0,s)}{dy} = \frac{i}{s(k+\epsilon i)}$$

and

$$\lim_{y\to\infty} \overline{U}(k,y,s) \to 0.$$

Step 2. Show that the solution to step 1 is

$$\overline{U}(k,s) = A(k,s)\exp(-y\sqrt{k^2+s})$$

with

$$A(k,s) = -\frac{i}{s(k-\epsilon i)} + \overline{U}_-(k,0,s)$$

and

$$-\sqrt{k^2+s}\,A(k,s) = \frac{d\overline{U}_+(k,0,s)}{dy} + \frac{i}{s(k+\epsilon i)}.$$

Step 3. Show that the Wiener-Hopf equation is

$$\frac{d\overline{U}_+(k,0,s)}{dy} + \frac{i}{s(k+\epsilon i)} = -\sqrt{k^2+s}\left[\overline{U}_-(k,0,s) - \frac{i}{s(k-\epsilon i)}\right].$$

Step 4. If we factor $\sqrt{k^2+s} = \sqrt{k-i\sqrt{s}}\sqrt{k+i\sqrt{s}}$, where we take the branch cuts along the imaginary axis in the k-plane from $[-\infty i, -\sqrt{s}i]$ and $[\sqrt{s}i, \infty i]$, show that we may factor the Wiener-Hopf equation in step 3 as

$$\frac{1}{\sqrt{k+i\sqrt{s}}}\frac{d\overline{U}_+(k,0,s)}{dy} + \frac{\sqrt{-i\sqrt{s}}}{s(k-\epsilon i)}$$

$$+ \frac{i}{s(k+\epsilon i)}\left(\frac{1}{\sqrt{k+i\sqrt{s}}} - \frac{1}{\sqrt{i\sqrt{s}}}\right)$$

$$= -\sqrt{k-i\sqrt{s}}\,\overline{U}_-(k,0,s) + \frac{i}{s(k-\epsilon i)}\left(\sqrt{k-i\sqrt{s}}\right.$$

$$\left. -\sqrt{-i\sqrt{s}}\right) - \frac{i}{s(k+\epsilon i)\sqrt{i\sqrt{s}}} = p(k)$$

where $p(k)$ is an entire function, $\overline{U}_+(k,0,s)$ is analytic for $\text{Im}(k) > -\sqrt{s}$ and $\overline{U}_-(k,0,s)$ is analytic for $\text{Im}(k) < \sqrt{s}$.

Step 4. Use Liouville's theorem to show that $p(k) \to 0$ as $|k| \to \infty$ so that

$$A(k,s) = -\frac{i}{s\sqrt{k-i\sqrt{s}}}\left[\frac{\sqrt{-i\sqrt{s}}}{k-\epsilon i} + \frac{1}{(k+\epsilon i)\sqrt{i\sqrt{s}}}\right].$$

Step 5. Use the residue theorem to show that in the limit $\epsilon \to 0$ the inverse of $\overline{U}(k,y,s)$ is

$$\overline{u}(x,y,s) = \frac{1}{s}e^{-y\sqrt{s}} + \overline{u}_1(x,y,s)$$

for $x > 0$ and

$$\overline{u}(x,y,s) = -\frac{1}{s^{3/2}}e^{-y\sqrt{s}} + \overline{u}_1(x,y,s)$$

for $x < 0$ where

$$\overline{u}_1(x,y,s) = \frac{\sqrt{2}}{2\pi s}\int_{-\infty}^{\infty}\frac{\sin[(\theta+i\tau)/2]}{\cos(\theta+i\tau)}\left(1+\frac{1}{\sqrt{s}}\right)\exp[-r\sqrt{s}\cosh(\tau)]\,d\tau,$$

$x = r\cos(\theta)$ and $y = r\sin(\theta)$.

Step 6. Finish the problem by inverting the Laplace transform and showing that

$$u(x, y, t) = \text{erfc}\left(\frac{y}{2\sqrt{t}}\right) + u_1(x, y, t)$$

for $x > 0$ and

$$u(x, y, t) = -\left[2\sqrt{\frac{t}{\pi}}e^{-y^2/(4t)} - y\,\text{erfc}\left(\frac{y}{2\sqrt{t}}\right)\right] + u_1(x, y, t)$$

for $x < 0$ where

$$u_1(x, y, t) = -\frac{\sqrt{2}}{\pi}\int_0^\infty \text{Re}\left\{\frac{\sin[(\theta + i\tau)/2]}{\cos(\theta + i\tau)}\right\}d\tau$$
$$\times\left\{[1 - r\cosh(\tau)]\text{erfc}\left[\frac{r\cosh(\tau)}{2\sqrt{t}}\right]\right.$$
$$\left. + 2\sqrt{\frac{t}{\pi}}\exp\left[-\frac{r^2\cosh^2(\tau)}{4t}\right]\right\}.$$

2. Use the Wiener-Hopf technique to solve the partial differential equation[14]

$$\frac{\partial^2 u}{\partial x^2} + \frac{\partial^2 u}{\partial z^2} - u = 0, \qquad -\infty < x < \infty, 0 < z < 1$$

with the boundary conditions $\lim_{|x|\to\infty} u(x, t) \to 0$,

$$\frac{\partial u(x, 0)}{\partial z} = 0, \qquad -\infty < x < \infty$$

$$u(x, 1) = e^{-x}, \qquad x > 0$$

and

$$\frac{\partial u(x, 1)}{\partial z} = 0, \qquad x < 0.$$

Step 1. Define the following Fourier transforms

$$U(k, z) = \int_{-\infty}^\infty u(x, z)e^{-ikx}dx,$$

---

[14] Taken from Horvay, G., 1961: Temperature distribution in a slab moving from a chamber at one temperature to a chamber at another temperature. *J. Heat Transfer*, **83**, 391–402.

$$U_+(k, z) = \int_0^\infty u(x, z)e^{-ikx}\,dx$$

and

$$U_-(k, z) = \int_{-\infty}^0 u(x, z)e^{-ikx}\,dx$$

so that $U(k, z) = U_+(k, z) + U_-(k, z)$. Then show that we may write the partial differential equation and boundary conditions

$$\frac{d^2U}{dz^2} - m^2 U = 0, \qquad 0 < z < 1$$

with $U'(k, 0) = 0$,

$$U_+(k, 1) = \frac{1}{1 + ik} \qquad \text{and} \qquad U'_-(k, 1) = 0$$

where $m^2 = k^2 + 1$.

Step 2. Show that we can write the solution to step 1

$$U(k, z) = A(k)\frac{\cosh(mz)}{\cosh(m)}, \qquad m^2 = k^2 + 1$$

with

$$U'_+(k, 1) = m\tanh(m)A(k)$$

and

$$U_-(k, 1) = A(k) - \frac{1}{i(k - i)}.$$

Step 3. By eliminating $A(k)$ from the last two equations in step 2, show that we may factor the resulting equation as

$$K_+(k)U'_+(k, 1) = \frac{m^2}{K_-(k)}\left[U_-(k, 1) + \frac{1}{i(k - i)}\right]$$

where $m\coth(m) = K_+(k)K_-(k)$. Note that the left side of the equation is analytic in the upper half-plane, $\text{Im}(k) > -c_1, c_1 > 0$ and the right side of the equation is analytic in the lower half-plane, $\text{Im}(k) < c_2, c_2 > 0$.

Step 4. Use Liouville's theorem to show that each side equals a constant value J. Then show that

$$A(k) = \frac{JK_-(k)}{m^2}.$$

To find J,

$$\frac{J}{2\pi} \int_{-\infty}^{\infty} \frac{K_-(k)}{(k-i)(k+i)} e^{ikx} dk = e^{-x}, \qquad x > 0$$

or $J = 2/K_-(i)$.

Step 5. Use the infinite product theorem to show that

$$K_+(-k) = K_-(k) = \Omega(ik)$$

where

$$\Omega(z) = \prod_{k=1}^{\infty} \frac{\sqrt{1 + \frac{1}{(k-1/2)^2 \pi^2}} + \frac{z}{(k-1/2)\pi}}{\sqrt{1 + \frac{1}{k^2 \pi^2}} + \frac{z}{k\pi}}$$

so that

$$u(x,z) = \frac{1}{\pi \Omega(1)} \int_{-\infty}^{\infty} \frac{\Omega(-ik)}{k^2 + 1} \frac{\cosh(mz)}{\cosh(m)} e^{ikx} dk.$$

Step 6. Use the residue theorem and show that

$$u(x,z) = e^{-x} + \frac{2i}{\Omega(-1)} \sum_{n=0}^{\infty} \frac{\Omega(-i\alpha_n)}{\mu_n \alpha_n} \frac{\cosh(\mu_n z)}{\sinh(\mu_n)} e^{i\alpha_n x}$$

where $x > 0$, $\mu_n = (n+1/2)\pi i$, $\alpha_n = i\sqrt{1 + (2n+1)^2 \pi^2/4}$ and

$$u(x,z) = \frac{e^x}{\Omega(-1)^2} - \frac{2i}{\Omega(-1)} \sum_{n=1}^{\infty} \frac{1}{\alpha_n \Omega(i\alpha_n)} \frac{\cosh(\mu_n z)}{\cosh(\mu_n)} e^{i\alpha_n x}$$

where $x < 0$, $\mu_n = n\pi i$, $\alpha_n = -i\sqrt{1 + n^2 \pi^2}$.

5.2 THE WIENER-HOPF TECHNIQUE WHEN THE FACTORIZATION CONTAINS BRANCH POINTS

In the previous problem we solved a diffraction problem by the Wiener-Hopf technique when the factorization involved a meromorphic function. In this section we solve another diffraction problem where the factorization involves branch points. This problem[15] arises in the diffraction of a plane compression pulse by a rigid, incompressible semi-infinite body in lubricated contact with a homogeneous, isotropic, perfectly elastic half-space. See Fig. 5.2.1. We join the problem mid-way

---

[15] Fredricks, R. W., 1961: Diffraction of an elastic pulse in a loaded half-space. *J. Acoust. Soc. Am.*, **33**, 17–22.

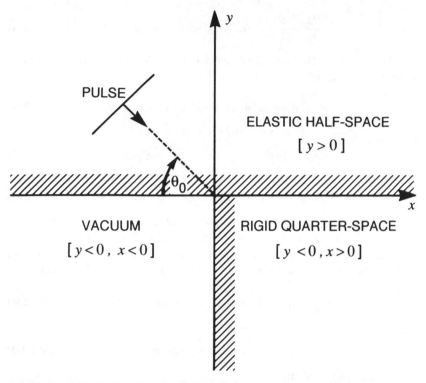

Fig. 5.2.1: Schematic for the diffraction of an elastic pulse by a loaded half-space.

in progress after we have eliminated the temporal dependence through the application of Laplace transforms.

Our problem involves solving the pair of partial differential equations

$$\frac{\partial^2 \phi}{\partial x^2} + \frac{\partial^2 \phi}{\partial y^2} - \frac{s^2}{\alpha^2}\phi = 0 \qquad (5.2.1)$$

and

$$\frac{\partial^2 \psi}{\partial x^2} + \frac{\partial^2 \psi}{\partial y^2} - \frac{s^2}{\beta^2}\psi = 0 \qquad (5.2.2)$$

where ϕ and ψ are the scalar Helmholtz potentials, s is the Laplace transform parameter and α and β are the longitudinal and transverse phase velocities. At $y = 0$, these potentials must satisfy the following boundary conditions:

$$2\frac{\partial^2 \phi}{\partial x \partial y} + \left(\frac{s^2}{\beta^2} - 2\frac{\partial^2}{\partial x^2}\right)\psi = 0, \qquad -\infty < x < \infty, y = 0, \qquad (5.2.3)$$

$$\left(\frac{s^2}{\beta^2} - 2\frac{\partial^2}{\partial x^2}\right)\phi - 2\frac{\partial^2 \phi}{\partial x \partial y} = 0, \qquad -\infty < x < 0, y = 0 \qquad (5.2.4)$$

and

$$\frac{\partial \phi}{\partial y} - \frac{\partial \psi}{\partial x} = 0, \qquad 0 < x < \infty, y = 0. \tag{5.2.5}$$

The method that we shall use to solve (5.2.1)–(5.2.5) consists of writing ϕ and ψ as

$$\phi(x,y) = \phi_i(x,y) + \phi_r(x,y) + \phi_1(x,y) \tag{5.2.6}$$

and

$$\psi(x,y) = \psi_r(x,y) + \psi_1(x,y) \tag{5.2.7}$$

where

$$\phi_i(x,y) = \exp\{-s[x\cos(\theta_0) - y\sin(\theta_0)]/\alpha\} \tag{5.2.8}$$

is the incident, plane compressible pulse which approaches the interface at the angle θ_0 measured in the negative sense from the negative real axis. The functions $\phi_r(x,y)$ and $\psi_r(x,y)$ are the specularly reflected waves from the interface $x < 0$, $y = 0$. We may express these waves by

$$\phi_r = \Lambda_p \exp\{-s[x\cos(\theta_0) + y\sin(\theta_0)]/\alpha\} \tag{5.2.9}$$

and

$$\psi_r = \Lambda_s \exp\{-s[x\cos(\theta_0) + y\sin(\theta_0)]/\beta\} \tag{5.2.10}$$

where $\beta\cos(\theta_0) = \alpha\sin(\phi_0)$,

$$\Lambda_p = \frac{\cos^2(2\phi_0) - (\beta/\alpha)^2 \sin(2\theta_0)\sin(2\phi_0)}{\cos^2(2\phi_0) + (\beta/\alpha)^2 \sin(2\theta_0)\sin(2\phi_0)} \tag{5.2.11}$$

and

$$\Lambda_s = \frac{2(\beta/\alpha)\sin(2\theta_0)\cos(2\phi_0)}{\cos^2(2\phi_0) + (\beta/\alpha)^2 \sin(2\theta_0)\sin(2\phi_0)}. \tag{5.2.12}$$

To evaluate $\phi_1(x,y)$ and $\psi_1(x,y)$, we begin by noting that $\phi_1(x,y)$ and $\psi_1(x,y)$ have the integral representations

$$2\pi i\phi_1(x,y) = \int_{c-\infty i}^{c+\infty i} A(p)\exp[s(px - \eta_1 y)]\,dp \tag{5.2.13}$$

and

$$2\pi i\psi_1(x,y) = \int_{c-\infty i}^{c+\infty i} B(p)\exp[s(px - \eta_2 y)]\,dp \tag{5.2.14}$$

where $\eta_1 = \sqrt{\alpha^{-2} - p^2}$, $\eta_2 = \sqrt{\beta^{-2} - p^2}$ and $-1/\alpha < c < 1/\alpha$. To ensure uniqueness we choose $\text{Re}(\eta_1) \geq 0$ and $\text{Re}(\eta_2) \geq 0$. Equations (5.2.13)–(5.2.14) are Fourier transforms that we have rotated 90° on the complex wavenumber plane.

We next substitute ϕ_i, ϕ_r, ψ_r, ϕ_1 and ψ_1 into the boundary conditions. This leads to

$$\int_{c-\infty i}^{c+\infty i} [2p\eta_1 A(p) - (\beta^{-2} - 2p^2)B(p)]e^{spx}dp = 0, \quad -\infty < x < \infty,$$
(5.2.15)

$$\int_{c-\infty i}^{c+\infty i} [(\beta^{-2} - 2p^2)A(p) + 2p\eta_2 B(p)]e^{spx}dp = 0, \quad -\infty < x < 0$$
(5.2.16)

and

$$\int_{c-\infty i}^{c+\infty i} [\eta_1 A(p) + pB(p)]e^{spx}dp = 2\pi i Q(\theta_0, \phi_0)e^{-sp_0 x}, \quad 0 < x < \infty$$
(5.2.17)

where

$$p_0 = \alpha^{-1}\cos(\theta_0) = \beta^{-1}\cos(\phi_0) \tag{5.2.18}$$

and

$$\alpha Q(\theta_0, \phi_0) = \sin(\theta_0)[1 - \Lambda_p] + \cos(\theta_0)\Lambda_s. \tag{5.2.19}$$

Because (5.2.15) must hold for all x, the bracketed term must vanish identically and

$$B(p) = \frac{2p\eta_1}{\beta^{-2} - 2p^2}A(p). \tag{5.2.20}$$

Using (5.2.20) in (5.2.16) and (5.2.17), we now have the dual integral equations

$$\int_{c-\infty i}^{c+\infty i} R(p)C(p)e^{spx}dp = 0, \quad -\infty < x < 0 \tag{5.2.21}$$

and

$$\int_{c-\infty i}^{c+\infty i} \eta_1 C(p)e^{spx}dp = 2\pi i \beta^2 Q(\theta_0, \phi_0)e^{-sp_0 x}, \quad 0 < x < \infty \tag{5.2.22}$$

where

$$C(p) = \frac{A(p)}{\beta^{-2} - 2p^2} \tag{5.2.23}$$

and

$$R(p) = (\beta^{-2} - 2p^2)^2 + 4p^2\eta_1\eta_2. \tag{5.2.24}$$

We solve these integral equations by introducing an arbitrary function $N_-(p)$ which has all of its singularities to the right of the line $\mathrm{Re}(p) = c$. From Cauchy's integral formula,

$$2\pi i e^{-sp_0 x} = \int_{c-\infty i}^{c+\infty i} \frac{N_-(p)e^{spx}}{N_-(-p_0)(p + p_0)}dp. \tag{5.2.25}$$

Transform Methods for Solving Partial Differential Equations

We have invented $N_-(\ )$ so that we can extend (5.2.22) into the domain $x < 0$ or

$$\int_{c-\infty i}^{c+\infty i} \left[\frac{\eta_1 C(p)}{\beta^2} - \frac{Q(\theta_0, \phi_0)N_-(p)}{N_-(-p_0)(p+p_0)} \right] e^{spx} \, dp = 0, \quad -\infty < x < \infty.$$
(5.2.26)

Because (5.2.26) must hold for any x, the bracketed term inside (5.2.26) must vanish or

$$C(p) = \frac{\beta^2 Q(\theta_0, \phi_0)N_-(p)}{\eta_1 N_-(-p_0)(p+p_0)}.$$
(5.2.27)

Substituting this $C(p)$ into (5.2.21) yields

$$\int_{c-\infty i}^{c+\infty i} \frac{R(p)N_-(p)}{\eta_1(p+p_0)} e^{spx} \, dp = 0, -\infty < x < 0, c > -p_0.$$
(5.2.28)

Let us now introduce the new function

$$K(p) = \frac{2(\alpha^{-2} - \beta^{-2})(\gamma^{-2} - p^2)}{R(p)}$$
(5.2.29)

where $1/\gamma$ is the positive real root of $R(p)$. Note that $K(p)$ tends to unity as $|p| \to \infty$. Substituting (5.2.29) into (5.2.28),

$$\int_{c-\infty i}^{c+\infty i} \frac{(\gamma^{-2} - p^2)N_-(p)}{\eta_1(p+p_0)K(p)} e^{spx} \, dp = 0, -\infty < x < 0.$$
(5.2.30)

At this point we factor $K(p)$ into the product $K_-(p) \times K_+(p)$. However, because $K(p)$ contains branch points we cannot use the infinite product theorem. On the other hand, from Cauchy's integral formula,

$$\log[K(p)] = \log[K_+(p)] + \log[K_-(p)] = \frac{1}{2\pi i} \oint_C \log[K(z)] \frac{dz}{z-p}$$
(5.2.31)

where Fig. 5.2.2 shows the contour C and we use the principal branch of the logarithm. Because $K(z)$ tends to unity as $|z| \to \infty$, we may deform C into the path B_+ and B_- shown in Fig. 5.2.2 because $K(z)$ has the branch points $z = \pm 1/\alpha$ and $z = \pm 1/\beta$ and no other singularities. Assuming that $z = p$ does not lie on either branch cut,

$$\log[K_\pm(p)] = \frac{1}{2\pi i} \int_{B_\pm} \log[K(z)] \frac{dz}{z-p}$$
(5.2.32)

or

$$K_\pm(p) = \exp\left\{ \frac{1}{\pi} \int_{\alpha^{-1}}^{\beta^{-1}} \tan^{-1}\left[\frac{4\zeta^2\sqrt{\beta^{-2} - \zeta^2}\sqrt{\zeta^2 - \alpha^{-2}}}{(2\zeta^2 - \beta^{-2})^2} \right] \frac{d\zeta}{\zeta \pm p} \right\}.$$
(5.2.33)

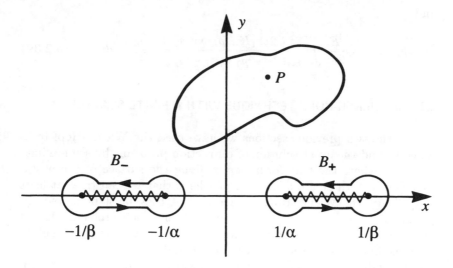

Fig. 5.2.2: Paths of integration for decomposition of the function $K(p)$.

Note that $K_-(-p) = K_+(p)$.

Finally, we introduce the new functions

$$L_\pm(p) = \frac{K_\pm(p)\sqrt{\alpha^{-1} \pm p}}{\gamma^{-1} \pm p} \qquad (5.2.34)$$

so that a general solution to (5.2.30) is

$$N_-(p) = L_-(p)P(p) \qquad (5.2.35)$$

where $P(\alpha)$ is a polynomial in p such that $P(-p_0) \neq 0$. Therefore,

$$C(p) = E\frac{K_-(p)}{(p + p_0)(\gamma^{-1} - p)\sqrt{p + \alpha^{-1}}}\frac{P(p)}{P(-p_0)} \qquad (5.2.36)$$

where

$$E = \frac{\beta^2 Q(\theta_0, \phi_0)(\gamma^{-1} + p_0)}{K_+(p_0)\sqrt{p_0 + \alpha^{-1}}}. \qquad (5.2.37)$$

From Liouville's theorem in the limit of $|p| \to \infty$, $P(p)$ must equal a constant. An asymptotic analysis shows that this constant is $P(-p_0)$. Therefore, the final solution, as far as we are concerned, is given by the integrals

$$\phi_1(x, y) = \frac{E}{2\pi i}\int_{c-\infty i}^{c+\infty i}\frac{(\beta^2 - 2p^2)K_-(p)\exp[s(px - \eta_1 y)]}{(p + p_0)(\gamma^{-1} - p)\sqrt{\alpha^{-1} + p}}dp \qquad (5.2.38)$$

and

$$\psi_1(x, y) = \frac{E}{2\pi i} \int_{c-\infty i}^{c+\infty i} \frac{2p\eta_1 K_-(p) \exp[s(px - \eta_2 y)]}{(p + p_0)(\gamma^{-1} - p)\sqrt{\alpha^{-1} + p}} dp. \qquad (5.2.39)$$

5.3 THE WIENER-HOPF TECHNIQUE WITH A FINITE SCATTERER

In the two previous sections we have used the Wiener-Hopf technique to find the exact solution of diffraction problems by semi-infinite obstacles where the transform axis is along the surface of the obstacle. However, there are additional problems that we can treat at least formally by the Wiener-Hopf techniques but they must be solved approximately. In this section we shall work one such problem, that of diffraction by a strip of finite width. As before, we assume a slightly lossy medium so that there are complex k and ω.

Consider a dock[16] of finite width $2a$ which occupies a part ($|x| < a$, $-\infty < y < \infty$) of an infinite ocean with depth h. Our coordinate system has its origin at the bottom of the ocean. Assuming that all of the disturbances are of the form $e^{i(ky-\omega t)}$ with $k \neq 0$, we find the reflected and transmitted waves that result from an incident wave of the form

$$\phi_i(x, z) = \cosh(c_0 z)e^{-i\kappa x} \qquad (5.3.1)$$

where ic_0 is the root of the equation $\beta \cos(c_0 h) + \sin(c_0 h) = 0$, $\beta = \omega^2/g$ and $\kappa^2 = c_0^2 - k^2$.

Let us denote the reflected and transmitted velocity potential by $\phi(x, z)$. Because $\phi_i + \phi$ gives the complete wave field, ϕ must satisfy the governing equation:

$$\frac{\partial^2 \phi}{\partial x^2} + \frac{\partial^2 \phi}{\partial z^2} - k^2\phi = 0, \quad -\infty < x < \infty, \quad 0 \leq z \leq h. \qquad (5.3.2)$$

Furthermore, ϕ must satisfy certain boundary conditions. At the free surface

$$\frac{\partial \phi}{\partial z} - \beta\phi = 0, \quad |x| > a, \quad z = h \qquad (5.3.3)$$

while at the bottom of the dock there is no vertical velocity so that

$$\frac{\partial \phi}{\partial z} = -c_0 \sinh(c_0 h)e^{-i\kappa x}, \quad |x| < a, \quad z = h. \qquad (5.3.4)$$

---

[16] Reprinted from *J. Appl. Math. Mech.*, **34**, V. F. Vitiuk, Diffraction of surface waves at a dock of finite width, 27–35, ©1970, with kind permission from Pergamon Press Ltd., Headington Hill Hall, Oxford OX3 0BX, UK.

At the bottom,

$$\frac{\partial \phi}{\partial z} = 0, \quad -\infty < x < \infty, \quad z = 0. \tag{5.3.5}$$

Finally, $|\phi(x, z)|$ must remain finite as $r = \sqrt{(x \pm a)^2 + (z - h)^2} \to 0$.
Let us define the following Fourier transforms:

$$\Phi_+(\alpha, z) = \int_a^\infty \phi(x, z) e^{i\alpha(x-a)} dx, \quad \tau > \tau_- \tag{5.3.6}$$

$$\Phi_1(\alpha, z) = \int_{-a}^a \phi(x, z) e^{i\alpha x} dx \tag{5.3.7}$$

$$\Phi_-(\alpha, z) = \int_{-\infty}^{-a} \phi(x, z) e^{i\alpha(x+a)} dx, \quad \tau < \tau_+ \tag{5.3.8}$$

where $\tau_- < 0$ and $\tau_+ > 0$. From these relationships,

$$\Phi(\alpha, z) = \int_{-\infty}^\infty \phi(x, z) e^{i\alpha x} dx \tag{5.3.9}$$

$$= e^{i\alpha a} \Phi_+(\alpha, z) + \Phi_1(\alpha, z) + e^{-i\alpha a} \Phi_-(\alpha, z). \tag{5.3.10}$$

Taking the two-sided Fourier transform of (5.3.2) and (5.3.5), the solution to these equations is

$$\Phi(\alpha, z) = A(\alpha) \cosh(\gamma z) \tag{5.3.11}$$

where $\gamma = \sqrt{\alpha^2 + k^2}$. Note that because $\Phi(\alpha, z)$ is an even function of γ, there are no branch points and cuts. Taking the one-sided Fourier transform from $-\infty$ to $-a$ of (5.3.3),

$$\Phi'_-(\alpha, h) - \beta \Phi_-(\alpha, h) = 0 \tag{5.3.12}$$

while the one-sided Fourier transform from a to ∞ of (5.3.3) yields

$$\Phi'_+(\alpha, h) - \beta \Phi_+(\alpha, h) = 0. \tag{5.3.13}$$

Finally, from (5.3.4),

$$\Phi'_1(\alpha, h) = -2c_0 \sinh(c_0 h) \frac{\sin[(\alpha - \kappa)a]}{\alpha - \kappa}. \tag{5.3.14}$$

From (5.3.10) and (5.3.11),

$$\Phi'(\alpha, h) = e^{i\alpha a} \Phi'_+(\alpha, h) + \Phi'_1(\alpha, h) + e^{-i\alpha a} \Phi'_-(\alpha, h) = A(\alpha) \gamma \sinh(\gamma h) \tag{5.3.15}$$

and

$$\Phi(\alpha, h) = e^{i\alpha a}\Phi_+(\alpha, h) + \Phi_1(\alpha, h) + e^{-i\alpha a}\Phi_-(\alpha, h) = A(\alpha)\cosh(\gamma h).$$
$$(5.3.16)$$

Using (5.3.12)–(5.3.14) to eliminate $\Phi'_+(\alpha, h)$, $\Phi'_-(\alpha, h)$ and $\Phi'_1(\alpha, h)$ and then combining (5.3.15) and (5.3.16),

$$e^{i\alpha a}\Phi_+(\alpha, h) + K(\alpha)\Phi_1(\alpha, h) + e^{-i\alpha a}\Phi_-(\alpha, h)$$
$$= -\frac{2c_0\sinh(c_0 h)\sin[(\alpha - \kappa)a]\cosh(\gamma h)}{[\gamma\sinh(\gamma h) - \beta\cosh(\gamma h)](\alpha - \kappa)} \qquad (5.3.17)$$

where

$$K(\alpha) = \frac{\gamma\sinh(\gamma h)}{\gamma\sinh(\gamma h) - \beta\cosh(\gamma h)}. \qquad (5.3.18)$$

We may also use (5.3.16) to find $A(\alpha)$ so that

$$\Phi(\alpha, z) = \left\{ \frac{\beta}{\gamma\sinh(\gamma h)} \left[e^{i\alpha a}\Phi_+(\alpha, h) + e^{-i\alpha a}\Phi_-(\alpha, h) \right] \right.$$
$$\left. - \frac{2c_0\sinh(c_0 h)\sin[(\alpha - \kappa)a]}{\gamma(\alpha - \kappa)\sinh(\gamma h)} \right\}\cosh(\gamma z). \qquad (5.3.19)$$

Our first step in solving (5.3.17) is the factorization of $K(\alpha)$. From the infinite product theorem (5.1.35),

$$K_+(\alpha) = K_-(-\alpha) \qquad (5.3.20)$$
$$= i\left(\frac{h}{\beta}\right)^{1/2}\left(\frac{k - i\alpha}{\delta + \alpha}\right)\frac{\rho_0}{h}\prod_{n=1}^{\infty}\frac{\sqrt{1 + k^2 h^2/n^2\pi^2} - i\alpha h/n\pi}{\sqrt{1 + k^2 h^2/\rho_n^2} - i\alpha h/\rho_n}$$
$$(5.3.21)$$

where $K(\alpha) = K_+(\alpha)K_-(\alpha)$, $\rho_0^2 = h^2(\delta^2 + k^2)$ and $\pm\rho_0/h$ and $\pm i\rho_n/h$ are the roots of $\rho\sinh(\rho h) - \beta\cosh(\rho h) = 0$.

Let us now multiply (5.3.17) by $e^{-i\alpha a}/K_+(\alpha)$. Then

$$\frac{\Phi_+(\alpha, h)}{K_+(\alpha)} + \frac{2c_0\sinh(c_0 h)e^{-i\alpha a}\sin[(\alpha - \kappa)a]\cosh(\gamma h)}{K_+(\alpha)[\gamma\sinh(\gamma h) - \beta\cosh(\gamma h)](\alpha - \kappa)}$$
$$= -S_-(\alpha, h)K_-(\alpha) - e^{-2i\alpha a}\frac{\Phi_-(\alpha, h)}{K_+(\alpha)} \qquad (5.3.22)$$

where $S_-(\alpha, h) = \Phi_1(\alpha, h)e^{-i\alpha a}$ and $\tau_- < \tau < 0$. Now, the first term on the left side of (5.3.22) is regular and bounded for $\tau > \tau_-$ while the first term on the right side of (5.3.22) is regular and bounded for $\tau < \tau_+$. The remaining terms are regular and bounded on $\tau_- < \tau < \tau_+$. Thus, we wish to break up these latter terms into a part that is regular on $\tau > \tau_-$ and another part that is regular on $\tau < \tau_+$.

The method used in this separation is essentially Cauchy's integral formula. By selecting appropriate infinite semi-circular contours, we can rewrite (5.3.22) as

$$\frac{1}{2\pi i}\int_{-\infty+f_--i}^{\infty+f_--i}\frac{2c_0\sinh(c_0h)e^{-i\xi a}\sin[(\xi-\kappa)a]\cosh(\eta h)}{K_+(\alpha)[\eta\sinh(\eta h)-\beta\cosh(\eta h)](\xi-\kappa)(\xi-\alpha)}d\xi$$

$$+\frac{\Phi_+(\alpha,h)}{K_+(\alpha)}+\frac{1}{2\pi i}\int_{-\infty+c_--i}^{\infty+c_--i}\frac{e^{-2i\xi a}\Phi_-(\xi,h)}{K_+(\xi)(\xi-\alpha)}d\xi=-S_-(\alpha,h)K_-(\alpha)$$

$$+\frac{1}{2\pi i}\int_{-\infty+f_++i}^{\infty+f_++i}\frac{2c_0\sinh(c_0h)e^{-i\xi a}\sin[(\xi-\kappa)a]\cosh(\eta h)}{K_+(\alpha)[\eta\sinh(\eta h)-\beta\cosh(\eta h)](\xi-\kappa)(\xi-\alpha)}d\xi$$

$$-\frac{1}{2\pi i}\int_{-\infty+c_++i}^{\infty+c_++i}\frac{e^{-2i\xi a}\Phi_-(\xi,h)}{K_+(\xi)(\xi-\alpha)}d\xi \tag{5.3.23}$$

where $\tau_- < c_- < \tau < c_+ < \tau_+$, $\tau_- < f_- < \tau < f_+ < \tau_+$ and $\eta^2 = \xi^2 + k^2$. Once again, the left side of (5.3.23) is regular for $\tau > \tau_-$, the right side of (5.3.23) is regular for $\tau < \tau_+$ and both sides are regular in the common strip $\tau_- < \tau < \tau_+$. From analytic continuation, both sides of (5.3.23) must equal an entire function, say $P(\alpha)$. Finally, from the asymptotic behavior of $K_+(\alpha)$, $K_-(\alpha)$ and the integrals as $|\alpha| \to \infty$, Liouville's theorem gives $P(\alpha) = 0$. This leads to two equations as we set each side to zero, one of which is

$$\frac{\Phi_+(\alpha,h)}{K_+(\alpha)}+\frac{1}{2\pi i}\int_{-\infty+c_--i}^{\infty+c_--i}\frac{e^{-2i\xi a}\Phi_-(\xi,h)}{K_+(\xi)(\xi-\alpha)}d\xi$$

$$=-\frac{c_0\sinh(c_0h)}{\pi i}\int_{-\infty+f_--i}^{\infty+f_--i}\frac{e^{-i\xi a}}{K_+(\xi)(\xi-\kappa)(\xi-\alpha)}$$

$$\times\frac{\sin[(\xi-\kappa)a]\cosh(\eta h)}{\eta\sinh(\eta h)-\beta\cosh(\eta h)}d\xi. \tag{5.3.24}$$

In a similar manner we can multiple (5.3.17) by $e^{i\alpha a}/K_-(\alpha)$. In this case,

$$\frac{\Phi_-(\alpha,h)}{K_-(\alpha)}+\frac{2c_0\sinh(c_0h)e^{i\alpha a}\sin[(\alpha-\kappa)a]\cosh(\gamma h)}{K_-(\alpha)[\gamma\sinh(\gamma h)-\beta\cosh(\gamma h)](\alpha-\kappa)}$$

$$=-S_+(\alpha,h)K_+(\alpha)-e^{2i\alpha a}\frac{\Phi_+(\alpha,h)}{K_-(\alpha)} \tag{5.3.25}$$

where $S_+(\alpha,h) = \Phi_1(\alpha,h)e^{i\alpha a}$ and $0 < \tau < \tau_+$. After factorization by splitting,

$$\frac{\Phi_-(\alpha,h)}{K_-(\alpha)}-\frac{1}{2\pi i}\int_{-\infty+c_++i}^{\infty+c_++i}\frac{e^{2i\xi a}\Phi_+(\xi,h)}{K_-(\xi)(\xi-\alpha)}d\xi$$

$$= \frac{c_0 \sinh(c_0 h)}{\pi i} \int_{-\infty+f_+i}^{\infty+f_+i} \frac{e^{i\xi a}}{K_-(\xi)(\xi - \kappa)(\xi - \alpha)}$$

$$\times \frac{\sin[(\xi - \kappa)a] \cosh(\eta h)}{\eta \sinh(\eta h) - \beta \cosh(\eta h)} d\xi. \qquad (5.3.26)$$

Let us introduce two positive, real constants l and m. We use them to simplify (5.3.24) and (5.3.26) by setting $c_+ = -c_- = l$ and $f_+ = -f_- = m$. Then we can write (5.3.24) and (5.3.26) as

$$\frac{\Phi_+(\alpha, h)}{K_+(\alpha)} - \frac{1}{2\pi i} \int_{-\infty+li}^{\infty+li} \frac{e^{2i\xi a} \Phi_-(-\xi, h)}{K_-(\xi)(\xi + \alpha)} d\xi$$

$$= \frac{c_0 \sinh(c_0 h)}{\pi i} \int_{-\infty+mi}^{\infty+mi} \frac{e^{i\xi a}}{K_-(\xi)(\xi + \kappa)(\xi + \alpha)}$$

$$\times \frac{\sin[(\xi + \kappa)a] \cosh(\eta h)}{\eta \sinh(\eta h) - \beta \cosh(\eta h)} d\xi \qquad (5.3.27)$$

and

$$\frac{\Phi_-(-\alpha, h)}{K_+(\alpha)} - \frac{1}{2\pi i} \int_{-\infty+li}^{\infty+li} \frac{e^{2i\xi a} \Phi_+(\xi, h)}{K_-(\xi)(\xi + \alpha)} d\xi$$

$$= \frac{c_0 \sinh(c_0 h)}{\pi i} \int_{-\infty+mi}^{\infty+mi} \frac{e^{i\xi a}}{K_-(\xi)(\xi - \kappa)(\xi + \alpha)}$$

$$\times \frac{\sin[(\xi - \kappa)a] \cosh(\eta h)}{\eta \sinh(\eta h) - \beta \cosh(\eta h)} d\xi \qquad (5.3.28)$$

or

$$\frac{G_+^+(\alpha)}{K_+(\alpha)} + \frac{1}{2\pi i} \int_{-\infty+li}^{\infty+li} \frac{e^{2i\xi a} G_+^+(\xi)}{K_-(\xi)(\xi + \alpha)} d\xi$$

$$= \frac{c_0 \sinh(c_0 h)}{\pi i} \int_{-\infty+mi}^{\infty+mi} \frac{e^{i\xi a} \cosh(\eta h)}{K_-(\xi)[\eta \sinh(\eta h) - \beta \cosh(\eta h)]}$$

$$\times \left\{ \frac{\sin[(\xi + \kappa)a]}{\xi + \kappa} - \frac{\sin[(\xi - \kappa)a]}{\xi - \kappa} \right\} \frac{d\xi}{\xi + \alpha} \qquad (5.3.29)$$

and

$$\frac{G_+^-(\alpha)}{K_+(\alpha)} - \frac{1}{2\pi i} \int_{-\infty+li}^{\infty+li} \frac{e^{2i\xi a} G_+^-(\xi)}{K_-(\xi)(\xi + \alpha)} d\xi$$

$$= \frac{c_0 \sinh(c_0 h)}{\pi i} \int_{-\infty+mi}^{\infty+mi} \frac{e^{i\xi a} \cosh(\eta h)}{K_-(\xi)[\eta \sinh(\eta h) - \beta \cosh(\eta h)]}$$

$$\times \left\{ \frac{\sin[(\xi + \kappa)a]}{\xi + \kappa} + \frac{\sin[(\xi - \kappa)a]}{\xi - \kappa} \right\} \frac{d\xi}{\xi + \alpha} \qquad (5.3.30)$$

where $G_+^+(\alpha) = \Phi_+(\alpha, h) - \Phi_-(-\alpha, h)$ and $G_+^-(\alpha) = \Phi_+(\alpha, h) + \Phi_-(-\alpha, h)$. Recalling that

$$K_+(\alpha)K_-(\alpha) = \frac{\gamma \sinh(\gamma h)}{\gamma \sinh(\gamma h) - \beta \cosh(\beta h)}, \tag{5.3.31}$$

we can rewrite (5.3.29)–(5.3.30) as

$$\frac{G_+^+(\alpha)}{K_+(\alpha)} + \frac{1}{2\pi i} \int_{-\infty+li}^{\infty+li} \frac{e^{2i\xi a} G_+^+(\xi)[\eta \sinh(\eta h) - \beta \cosh(\eta h)] K_+(\xi)}{\eta \sinh(\eta h)(\xi + \alpha)} d\xi$$

$$= \frac{c_0 \sinh(c_0 h)}{\pi i} \int_{-\infty+mi}^{\infty+mi} \frac{e^{i\xi a} \cosh(\eta h) K_+(\xi)}{\eta \sinh(\eta h)}$$

$$\times \left\{ \frac{\sin[(\xi + \kappa)a]}{\xi + \kappa} - \frac{\sin[(\xi - \kappa)a]}{\xi - \kappa} \right\} \frac{d\xi}{\xi + \alpha} \tag{5.3.32}$$

and

$$\frac{G_+^-(\alpha)}{K_+(\alpha)} - \frac{1}{2\pi i} \int_{-\infty+li}^{\infty+li} \frac{e^{2i\xi a} G_+^-(\xi)[\eta \sinh(\eta h) - \beta \cosh(\eta h)] K_+(\xi)}{\eta \sinh(\eta h)(\xi + \alpha)} d\xi$$

$$= \frac{c_0 \sinh(c_0 h)}{\pi i} \int_{-\infty+mi}^{\infty+mi} \frac{e^{i\xi a} \cosh(\eta h) K_+(\xi)}{\eta \sinh(\eta h)}$$

$$\times \left\{ \frac{\sin[(\xi + \kappa)a]}{\xi + \kappa} + \frac{\sin[(\xi - \kappa)a]}{\xi - \kappa} \right\} \frac{d\xi}{\xi + \alpha}. \tag{5.3.33}$$

We now evaluate the integrals in (5.3.32)–(5.3.33) by the residue theorem after closing the contour with a semicircle in the upper half of the ξ-plane. The enclosed singularities are at $\xi = ki$ and $\mu_n i$ where $\mu_n = \sqrt{k^2 + n^2\pi^2/h^2}$ with $n = 1, 2, 3, \ldots$ Applying the residue theorem,

$$\frac{G_+^+(\alpha)}{K_+(\alpha)} + \frac{i\beta}{2kh} \frac{G_+^+(ik)e^{-2ka}K_+(ik)}{ik + \alpha}$$

$$+ \frac{i\beta}{h} \sum_{n=1}^{\infty} \frac{G_+^+(i\mu_n)e^{-2\mu_n a}K_+(i\mu_n)}{\mu_n(i\mu_n + \alpha)}$$

$$= -\frac{2ic_0 \sinh(c_0 h)}{h} \left(\frac{1}{2k} \frac{K_+(ik)}{ik + \alpha} \left\{ \frac{\sin[(ik + \kappa)a]}{ik + \kappa} \right. \right.$$

$$\left. - \frac{\sin[(ik - \kappa)a]}{ik - \kappa} \right\} e^{-ka} + \sum_{n=1}^{\infty} \frac{K_+(i\mu_n)e^{-\mu_n a}}{\mu_n(i\mu_n + \alpha)}$$

$$\left. \left\{ \frac{\sin[(i\mu_n + \kappa)a]}{i\mu_n + \kappa} - \frac{\sin[(i\mu_n - \kappa)a]}{i\mu_n - \kappa} \right\} \right) \tag{5.3.34}$$

and

$$\frac{G_+^-(\alpha)}{K_+(\alpha)} - \frac{i\beta}{2kh} \frac{G_+^-(ik)e^{-2ka}K_+(ik)}{ik+\alpha}$$

$$- \frac{i\beta}{h} \sum_{n=1}^{\infty} \frac{G_+^-(i\mu_n)e^{-2\mu_n a}K_+(i\mu_n)}{\mu_n(i\mu_n+\alpha)}$$

$$= -\frac{2ic_0 \sinh(c_0 h)}{h} \left(\frac{1}{2k} \frac{K_+(ik)}{ik+\alpha} \left\{ \frac{\sin[(ik+\kappa)a]}{ik+\kappa} \right. \right.$$

$$+ \left. \frac{\sin[(ik-\kappa)a]}{ik-\kappa} \right\} e^{-ka} + \sum_{n=1}^{\infty} \frac{K_+(i\mu_n)e^{-\mu_n a}}{\mu_n(i\mu_n+\alpha)}$$

$$\left. \left\{ \frac{\sin[(i\mu_n+\kappa)a]}{i\mu_n+\kappa} + \frac{\sin[(i\mu_n-\kappa)a]}{i\mu_n-\kappa} \right\} \right). \quad (5.3.35)$$

Upon setting $\alpha = ki$ or $\mu_j i$ where $j = 1, 2, 3, \ldots$, we obtain the following system of equations:

$$G_+^+(ik) = \left[\frac{1}{K_+(ik)} + \frac{\beta K_+(ik)}{4k^2 h} e^{-2ka} \right]^{-1}$$

$$\times \left[-\frac{\beta}{h} \sum_{n=1}^{\infty} \frac{e^{-2\mu_n a}G_+^+(i\mu_n)K(i\mu_n)}{\mu_n(\mu_n+k)} - \frac{2c_0 \sinh(c_0 h)}{h} \right.$$

$$\times \left(\frac{1}{2k} \frac{K_+(ik)}{2k} \left\{ \frac{\sin[(ik+\kappa)a]}{ik+\kappa} - \frac{\sin[(ik-\kappa)a]}{ik-\kappa} \right\} e^{-ka} \right.$$

$$\left. \left. + \sum_{n=1}^{\infty} \frac{K_+(i\mu_n)e^{-\mu_n a}}{\mu_n(\mu_n+k)} \left\{ \frac{\sin[(i\mu_n+\kappa)a]}{i\mu_n+\kappa} - \frac{\sin[(i\mu_n-\kappa)a]}{i\mu_n-\kappa} \right\} \right) \right],$$

$$(5.3.36)$$

$$G_+^-(ik) = \left[\frac{1}{K_+(ik)} - \frac{\beta K_+(ik)}{4k^2 h} e^{-2ka} \right]^{-1}$$

$$\times \left[\frac{\beta}{h} \sum_{n=1}^{\infty} \frac{e^{-2\mu_n a}G_+^-(i\mu_n)K(i\mu_n)}{\mu_n(\mu_n+k)} - \frac{2c_0 \sinh(c_0 h)}{h} \right.$$

$$\times \left(\frac{1}{2k} \frac{K_+(ik)}{2k} \left\{ \frac{\sin[(ik+\kappa)a]}{ik+\kappa} + \frac{\sin[(ik-\kappa)a]}{ik-\kappa} \right\} e^{-ka} \right.$$

$$\left. \left. + \sum_{n=1}^{\infty} \frac{K_+(i\mu_n)e^{-\mu_n a}}{\mu_n(\mu_n+k)} \left\{ \frac{\sin[(i\mu_n+\kappa)a]}{i\mu_n+\kappa} + \frac{\sin[(i\mu_n-\kappa)a]}{i\mu_n-\kappa} \right\} \right) \right],$$

$$(5.3.37)$$

$$\frac{G_+^+(i\mu_j)}{K_+(i\mu_j)} + \frac{\beta K_+(ik)}{2kh} \frac{G_+^+(ik)e^{-2ka}}{k+\mu_j}$$

$$+ \frac{\beta}{h} \sum_{n=1}^{\infty} \frac{e^{-2\mu_n a} G_+^+(i\mu_n) K(i\mu_n)}{\mu_n(\mu_n + \mu_j)}$$

$$= -\frac{2c_0 \sinh(c_0 h)}{h} \left(\frac{1}{2k} \frac{K_+(ik)}{k + \mu_j} \left\{ \frac{\sin[(ik + \kappa)a]}{ik + \kappa} - \frac{\sin[(ik - \kappa)a]}{ik - \kappa} \right\} e^{-ka} \right.$$

$$\left. + \sum_{n=1}^{\infty} \frac{K_+(i\mu_n) e^{-\mu_n a}}{\mu_n(\mu_n + \mu_j)} \left\{ \frac{\sin[(i\mu_n + \kappa)a]}{i\mu_n + \kappa} - \frac{\sin[(i\mu_n - \kappa)a]}{i\mu_n - \kappa} \right\} \right)$$

$$(5.3.38)$$

and

$$\frac{G_+^-(i\mu_j)}{K_+(i\mu_j)} - \frac{\beta K_+(ik)}{2kh} \frac{G_+^-(ik) e^{-2ka}}{k + \mu_j}$$

$$- \frac{\beta}{h} \sum_{n=1}^{\infty} \frac{e^{-2\mu_n a} G_+^-(i\mu_n) K(i\mu_n)}{\mu_n(\mu_n + \mu_j)}$$

$$= -\frac{2c_0 \sinh(c_0 h)}{h} \left(\frac{1}{2k} \frac{K_+(ik)}{k + \mu_j} \left\{ \frac{\sin[(ik + \kappa)a]}{ik + \kappa} + \frac{\sin[(ik - \kappa)a]}{ik - \kappa} \right\} e^{-ka} \right.$$

$$\left. + \sum_{n=1}^{\infty} \frac{K_+(i\mu_n) e^{-\mu_n a}}{\mu_n(\mu_n + \mu_j)} \left\{ \frac{\sin[(i\mu_n + \kappa)a]}{i\mu_n + \kappa} + \frac{\sin[(i\mu_n - \kappa)a]}{i\mu_n - \kappa} \right\} \right).$$

$$(5.3.39)$$

Eqs. (5.3.36)–(5.3.39) are completely regular if

$$\sum_{n=1}^{\infty} \left| \frac{\beta}{h} \frac{K_+(i\mu_n)}{\mu_n} e^{-2\mu_n a} K_+(i\mu_j) \left\{ \frac{1}{\mu_n + \mu_j} \right. \right.$$

$$\left. \left. - \frac{2k\beta K_+^2(ik)}{[4k^2 h e^{2ka} + \beta K_+^2(ik)](k + \mu_n)(k + \mu_j)} \right\} \right| < 1$$

$$(5.3.40)$$

and

$$\sum_{n=1}^{\infty} \left| \frac{\beta}{h} \frac{K_+(i\mu_n)}{\mu_n} e^{-2\mu_n a} K_+(i\mu_j) \left\{ -\frac{1}{\mu_n + \mu_j} \right. \right.$$

$$\left. \left. - \frac{2k\beta K_+^2(ik)}{[4k^2 h e^{2ka} - \beta K_+^2(ik)](k + \mu_n)(k + \mu_j)} \right\} \right| < 1.$$

$$(5.3.41)$$

Eqs. (5.3.40)–(5.3.41) give those values of a/h for which (5.3.36)–(5.3.39) are regular. Under these conditions we could, in principle, solve for $G_+^+(\alpha)$ and $G_+^-(\alpha)$ and then $\Phi_+(\alpha, h)$ and $\Phi_-(\alpha, h)$. However, in most cases, we would have to solve this system numerically, or at least by successive correction. However, in the limit of a wide dock compared to the depth of the water, $2a/h >> 1$, we can simplify $G_+(\alpha)$ and $G_-(\alpha)$ to

$$
\begin{aligned}
G_+^+(\alpha) = -iK_+(\alpha)\bigg[& \frac{e^{-ka}}{kh}\frac{K_+(ik)}{ik+\alpha}\bigg(\frac{\beta}{2}G_+^+(ik)e^{-ka} \\
& + c_0\sinh(c_0h)\bigg\{\frac{\sin[(ik+\kappa)a]}{ik+\kappa} - \frac{\sin[(ik-\kappa)a]}{ik-\kappa}\bigg\}\bigg) \\
& + \frac{2ic_0\sinh(c_0h)}{h}\sum_{n=1}^{\infty}\frac{K_+(i\mu_n)}{\mu_n(i\mu_n+\alpha)}\bigg(\frac{e^{-i\kappa x}}{i\mu_n+\kappa} - \frac{e^{i\kappa x}}{i\mu_n-\kappa}\bigg)\bigg]
\end{aligned}
$$

$$(5.3.42)$$

and

$$
\begin{aligned}
G_+^-(\alpha) = -iK_+(\alpha)\bigg[& \frac{e^{-ka}}{kh}\frac{K_+(ik)}{ik+\alpha}\bigg(-\frac{\beta}{2}G_+^-(ik)e^{-ka} \\
& + c_0\sinh(c_0h)\bigg\{\frac{\sin[(ik+\kappa)a]}{ik+\kappa} + \frac{\sin[(ik-\kappa)a]}{ik-\kappa}\bigg\}\bigg) \\
& + \frac{2ic_0\sinh(c_0h)}{h}\sum_{n=1}^{\infty}\frac{K_+(i\mu_n)}{\mu_n(i\mu_n+\alpha)}\bigg(\frac{e^{-i\kappa x}}{i\mu_n+\kappa} + \frac{e^{i\kappa x}}{i\mu_n-\kappa}\bigg)\bigg].
\end{aligned}
$$

$$(5.3.43)$$

In this case, (5.3.36)–(5.3.39) gives $G_+^+(ik)$ and $G_+^-(ik)$ as

$$
\begin{aligned}
G_+^+(ik) = -\frac{2c_0\sinh(c_0h)}{h}&\bigg[\frac{1}{K_+(ik)} + \frac{\beta e^{-2ka}}{4k^2h}K_+(ik)\bigg]^{-1} \\
&\times\bigg(\frac{K_+(ik)}{k^2}\bigg\{\frac{\sin[(ik+\kappa)a]}{ik+\kappa} - \frac{\sin[(ik-\kappa)a]}{ik-\kappa}\bigg\}e^{-ka} \\
&+ \sum_{n=1}^{\infty}\frac{K_+(i\mu_n)}{\mu_n(\mu_n+k)}\bigg[\frac{e^{-i\kappa a}}{i\mu_n+\kappa} - \frac{e^{i\kappa a}}{i\mu_n-\kappa}\bigg]\bigg)
\end{aligned}
$$

$$(5.3.44)$$

and

$$
\begin{aligned}
G_+^-(ik) = -\frac{2c_0\sinh(c_0h)}{h}&\bigg[\frac{1}{K_+(ik)} - \frac{\beta e^{-2ka}}{4k^2h}K_+(ik)\bigg]^{-1} \\
&\times\bigg(\frac{K_+(ik)}{k^2}\bigg\{\frac{\sin[(ik+\kappa)a]}{ik+\kappa} + \frac{\sin[(ik-\kappa)a]}{ik-\kappa}\bigg\}e^{-ka} \\
&+ \sum_{n=1}^{\infty}\frac{K_+(i\mu_n)}{\mu_n(\mu_n+k)}\bigg[\frac{e^{-i\kappa a}}{i\mu_n+\kappa} + \frac{e^{i\kappa a}}{i\mu_n-\kappa}\bigg]\bigg).
\end{aligned}
$$

$$(5.3.45)$$

Using the definition of $G_+^+(\alpha)$ and $G_+^-(\alpha)$, we obtain for $\Phi_+(\alpha, h)$ and $\Phi_-(\alpha, h)$ that

$$
\begin{aligned}
\Phi_+(\alpha, h) = -iK_+(\alpha)\Bigg(\frac{e^{-ka}}{kh}\frac{K_+(ik)}{ik+\alpha}\bigg\{\frac{\beta}{2}\bigg[\frac{G_+^+(ik)-G_+^-(ik)}{2}\bigg]e^{-ka} \\
+ c_0\sinh(c_0h)\frac{\sin[(ik-\kappa)a]}{ik-\kappa}\bigg\} \\
+ \frac{ic_0\sinh(c_0h)}{h}\sum_{n=1}^{\infty}\frac{K_+(i\mu_n)}{\mu_n(i\mu_n+\alpha)}\frac{e^{-i\kappa a}}{i\mu_n+\kappa}\Bigg)
\end{aligned}
$$

$$(5.3.46)$$

and

$$
\begin{aligned}
\Phi_-(\alpha, h) = -iK_-(\alpha)\Bigg(\frac{e^{-ka}}{kh}\frac{K_+(ik)}{ik-\alpha}\bigg\{-\frac{\beta}{2}\bigg[\frac{G_+^+(ik)+G_+^-(ik)}{2}\bigg]e^{-ka} \\
+ c_0\sinh(c_0h)\frac{\sin[(ik-\kappa)a]}{ik-\kappa}\bigg\} \\
+ \frac{ic_0\sinh(c_0h)}{h}\sum_{n=1}^{\infty}\frac{K_+(i\mu_n)}{\mu_n(i\mu_n-\alpha)}\frac{e^{i\kappa a}}{i\mu_n-\kappa}\Bigg).
\end{aligned}
$$

$$(5.3.47)$$

Upon substituting (5.3.46)–(5.3.47) into (5.3.9) and applying the inverse Fourier transform, we finally obtain the velocity potential for the various domains. For $x < -a$,

$$
\begin{aligned}
\phi(x, z) = (a_0 + ib_0)\cosh(c_0 z)e^{-i\delta x}\Bigg(\frac{e^{-ka}}{kh}\frac{K_+(ik)}{ik-\delta} \\
\times\bigg\{\frac{\beta}{2}\bigg[\frac{G_+^+(ik)+G_+^-(ik)}{2}\bigg]-c_0\sinh(c_0h)\frac{\sin[(ik-\kappa)]}{ik-\kappa}\bigg\} \\
-\frac{ic_0\sinh(c_0h)}{h}\sum_{n=1}^{\infty}\frac{K_+(i\mu_n)}{\mu_n(i\mu_n-\delta)}\frac{e^{i\kappa a}}{i\mu_n-\kappa}\Bigg) \\
-\sum_{p=1}^{\infty}(a_p + ib_p)e^{\nu_p x}\cos(\rho_p z/h)\Bigg(\frac{e^{-ka}}{kh}\frac{K_+(ik)}{ik-\nu_p} \\
\times\bigg\{\frac{\beta}{2}\bigg[\frac{G_+^+(ik)+G_+^-(ik)}{2}\bigg]-c_0\sinh(c_0h)\frac{\sin[(ik-\kappa)]}{ik-\kappa}\bigg\} \\
-\frac{ic_0\sinh(c_0h)}{h}\sum_{n=1}^{\infty}\frac{K_+(i\mu_n)}{\mu_n(\mu_n-\nu_p)}\frac{e^{i\kappa a}}{i\mu_n-\kappa}\Bigg).
\end{aligned}
$$

$$(5.3.48)$$

For $|x| < a$,

$$\phi(x,z) = 2c_0 h \sinh(c_0 h) \sum_{n=1}^{\infty} \frac{(-1)^n e^{-\mu_n a}}{\mu_n(n^2\pi^2 + c_0^2 h^2)} \left[\mu_n \cos(i\mu_n x + \kappa a) \right.$$
$$\left. - \kappa \sin(i\mu_n x + \kappa a) \right] \cos(n\pi z/h)$$
$$+ \frac{e^{-ka}}{kh} \left\{ \frac{\beta}{2} \left[G_+^+(ik)\sinh(kx) + G_+^-(ik)\cosh(kx) \right] \right.$$
$$+ \frac{\sinh(c_0 h)}{c_0} \left[k\cos(ikx + \kappa a) - \kappa \sin(ikx + \kappa a) \right] \Big\}$$
$$- e^{-i\kappa x} \cosh(c_0 z).$$

$$(5.3.49)$$

For $x > a$,

$$\phi(x,z) = (a_0 + ib_0)\cosh(c_0 z)e^{i\delta x}\left(\frac{e^{-ka}}{kh} \right.$$
$$\times \left\{ \frac{\beta}{2}\left[\frac{G_+^-(ik) - G_+^+(ik)}{2} \right] - c_0 \sinh(c_0 h)\frac{\sin[(ik+\kappa)a]}{ik+\kappa} \right\}$$
$$\left. - \frac{ic_0 \sinh(c_0 h)}{h}\sum_{n=1}^{\infty} \frac{K_+(i\mu_n)}{\mu_n(i\mu_n + k)}\frac{e^{-i\kappa a}}{i\mu_n + \delta} \right)$$
$$- \sum_{p=1}^{\infty} (a_p + ib_p)e^{-\nu_p x}\cos(\rho_p z/h)\left(\frac{e^{-ka}}{kh}\frac{K_+(ik)}{k - \nu_p} \right.$$
$$\times \left\{ \frac{\beta}{2}\left[\frac{G_+^-(ik) - G_+^+(ik)}{2} \right] - c_0 \sinh(c_0 h)\frac{\sin[(ik+\kappa)a]}{ik+\kappa} \right\}$$
$$\left. - \frac{ic_0 \sinh(c_0 h)}{h}\sum_{n=1}^{\infty} \frac{K_+(i\mu_n)}{\mu_n(\mu_n - \nu_p)}\frac{e^{-i\kappa a}}{i\mu_n + \kappa} \right)$$

$$(5.3.50)$$

where

$$\nu_p = \sqrt{k^2 + \rho_p^2/h^2},$$

$$(5.3.51)$$

$$a_0 + b_0 i = \frac{\beta\rho_0^2 e^{-i\delta a}}{K_+(\delta)\delta h[\beta^2 h^2 - \beta h - \rho_0^2]\cosh(\rho_0)}$$

$$(5.3.52)$$

and

$$a_p + b_p i = \frac{\beta\rho_p^2 e^{\nu_p a}}{K_+(i\nu_p)\nu_p h[\beta h - \beta^2 h^2 - \rho_p^2]\cos(\rho_p)}.$$

$$(5.3.53)$$

Papers Using the Wiener-Hopf Technique

Abrahams, I. D., 1982: Scattering of sound by large finite geometries. *IMA J. Appl. Math.*, **29**, 79–97.

Abrahams, I. D., 1986: Scattering of sound by a semi-infinite elastic plate with a soft backing; a matrix Wiener-Hopf problem. *IMA J. Appl. Math.*, **37**, 227–245.

Abrahams, I. D., 1987: Scattering of sound by two parallel semi-infinite screens. *Wave Motion*, **9**, 289–300.

Abrahams, I. D., and G. R. Wickham, 1988: On the scattering of sound by two semi-infinite parallel staggered plates. I. Explicit matrix Wiener-Hopf factorization. *Proc. R. Soc. Lond.*, **A420**, 131–156.

Abrahams, I. D., and G. R. Wickham, 1990: Acoustic scattering by two parallel slightly staggered rigid plates. *Wave Motion*, **12**, 281–297.

Abrahams, I. D., and G. R. Wickham, 1990: General Wiener-Hopf factorization of matrix kernels with exponential phase factors. *SIAM J. Appl. Math.*, **50**, 819–838.

Abrahams, I. D., and G. R. Wickham, 1991: The scattering of water waves by two semi-infinite opposed vertical walls. *Wave Motion*, **14**, 145–168.

Achenbach, J. D., and A. K. Gautesen, 1977: Elastodynamic stress-intensity factors for a semi-infinite crack under 3-D loading. *J. Appl. Mech.*, **44**, 243–249.

Achenbach, J. D., and A. K. Gautesen, 1977: Geometrical theory of diffraction for three-D elastodynamics. *J. Acoust. Soc. Am.*, **61**, 413–421.

Adamczyk, J. J., and M. E. Goldstein, 1978: Unsteady flow in a supersonic cascade with subsonic leading-edge locus. *AIAA J.*, **16**, 1248–1254.

Aivazyan, Yu. M., and D. M. Sedrakyan, 1965: Radiation from the open end of a planar semiinfinite waveguide. *Sov. Phys. Tech. Phys.*, **10**, 358–361.

Aizin, L. B., 1992: Sound generation by a Tollmien-Schlichting wave at the end of a plate in a flow. *J. Appl. Mech. Tech. Phys.*, **33**, 355–362.

Albertsen, N. Chr., and P. Skov-Madsen, 1983: A compact septum polarizer. *IEEE Trans. Microwave Theory Tech.*, **MTT-31**, 654–660.

Alblas, J. B., 1957: On the diffraction of sound waves in a viscous medium. *Appl. Sci. Res.*, **A6**, 237–262.

Alblas, J. B., 1958: On the generation of water waves by a vibrating strip. *Appl. Sci. Res.*, **A7**, 224–236.

Alblas, J. B., and M. Kuipers, 1969: Contact problems of a rectangular block on an elastic layer of finite thickness. Part I: The thin layer. *Acta Mech.*, **8**, 133–145.

Alblas, J. B., and M. Kuipers, 1970: On the two dimensional problem of a cylindrical stamp pressed into a thin elastic layer. *Acta Mech.*,

9, 292–311.

Amatuni, A. Ts., 1965: Solution of the problem of transitional radiation in a plasma-like medium. *Sov. Phys. Tech. Phys.*, **9**, 1049–1056.

Anderson, I., 1979: Plane wave diffraction by a thin dielectric half-plane. *IEEE Trans. Antennas Propagat.*, **AP-27**, 584–589.

Ando, Y., 1969/70: On the sound radiation from semi-infinite circular pipe of certain wall thickness. *Acustica*, **22**, 219–225.

Ando, Y., and T. Koizumi, 1976: Sound radiation from a semi-infinite circular pipe having an arbitrary profile of orifice. *J. Acoust. Soc. Am.*, **59**, 1033–1039.

Angel, Y. C., 1990: Singular integral equation for antiplane-wave scattering by a semi-infinite crack. *J. Elasticity*, **23**, 53–67.

Angulo, C. M., and W. S. C. Chang, 1959: The launching of surface waves by a parallel plate waveguide. *IEEE Trans. Antennas Propagat.*, **AP-7**, 359–368.

Aoki, K., and J. Osteryoung, 1981: Diffusion-controlled current at the stationary finite disk electrode. *J. Electroanal. Chem.*, **122**, 19–35.

Aoki, K., T. Miyazaki, K. Uchida and Y. Shimada, 1982: On junction of two semi-infinite dielectric guides. *Radio Sci.*, **17**, 11–19.

Aoki, K., K. Tokuda and H. Matsuda, 1987: Hydrodynamic voltammetry at channel electrodes. Part IX. Edge effects at rectangular channel flow microelectrodes. *J. Electroanal. Chem.*, **217**, 33–47.

Aoki, K., K. Tokuda and H. Matsuda, 1987: Theory of chrono-amperometric curves at microband electrodes. *J. Electroanal. Chem.*, **225**, 19–32.

Aoki, K., K. Tokuda and H. Matsuda, 1987: Theory of stationary current-potential curves at microdisk electrodes for quasi-reversible and totally irreversible electrode reactions. *J. Electroanal. Chem.*, **235**, 87–96.

Arora, R. K., 1965: Bifurcation of a parallel-plate waveguide by a unidirectionally conducting screen. *Proc. R. Soc. Edin.*, **A67**, 50–68.

Arora, R. K., and S. Vijayaraghavan, 1970: Scattering of a shielded surface wave by a wall-impedance discontinuity. *IEEE Trans. Microwave Theory Tech.*, **MTT-18**, 734–736.

Asghar, S., and G. H. Zahid, 1985: Diffraction of Kelvin waves by a finite barrier. *Zeit. angew. Math. Phys.*, **36**, 712–722.

Asghar, S., and G. H. Zahid, 1986: Field in an open-ended waveguide satisfying impedance boundary conditions. *Zeit. angew. Math. Phys.*, **37**, 194–205.

Asghar, S., and F. D. Zaman, 1986: Diffraction of Love waves by a finite rigid barrier. *Bull. Seism. Soc. Am.*, **76**, 241–257.

Asghar, S., 1988: Acoustic diffraction by an absorbing finite strip in a moving liquid. *J. Acoust. Soc. Am.*, **83**, 812–816.

Asghar, S., B. Ahmad and M. Ayub, 1991: Point-source diffraction by an absorbing half-plane. *IMA J. Appl. Math.*, **46**, 217–224.

Atkinson, C., 1975: On the stress intensity factors associated with cracks interacting with an interface between two elastic media. *Int. J. Engng. Sci.*, **13**, 489–504.

Atkinson, C., and R. D. List, 1978: Steady state crack propagation into media with spatially varying elastic properties. *Int. J. Engng. Sci.*, **16**, 717–730.

Atkinson, C., 1979: A note on some dynamic crack problems in linear viscoelasticity. *Arch. Mech.*, **31**, 829–849.

Atkinson, C., 1981: The growth kinetics of individual ledges during solid-solid phase transformation. *Proc. R. Soc. Lond.*, **A378**, 351–368.

Atkinson, C., 1982: Diffusion controlled ledge growth in a medium of finite extent. *J. Appl. Phys.*, **53**, 5689–5696.

Atkinson, C., and C. R. Champion, 1984: Some boundary-value problems for the equation $\nabla \bullet (|\nabla\phi|^N \nabla\phi) = 0$. *Quart. J. Mech. Appl. Math.*, **37**, 401–419.

Austin, D. M., and R. D. Gregory, 1988: On the bending and flexure of a plate restrained by smooth-rigid clamps. *Quart. J. Mech. Appl. Math.*, **41**, 549–562.

Avdeyev, Ye. V., and G. V. Voskresenskiy, 1967: The radiation accompanying the uniform motion of a charged filament in the vicinity of a comb structure. General solution. *Radio Engng. Electron. Phys.*, **12**, 432–440.

Bailin, L. L., 1951: An analysis of the effect of the discontinuity in a bifurcated circular guide upon plane longitudinal waves. *J. Res. NBS*, **47**, 315–335.

Baker, B. R., 1966: Ductile yielding and brittle fracture at the ends of parallel cracks in a stretched orthotropic sheet. *Int. J. Fract. Mech.*, **2**, 576–596.

Baksht, F. G., 1970: Electron energy distribution in the electrode sheath in a weakly ionized plasma. *Sov. Phys. Tech. Phys.*, **14**, 1196–1204.

Baksht, F. G., V. G. Ivanov and B. Ya. Moizhes, 1972: Calculation of the electron emission probability from dielectrics in the case of secondary electron emission and the external photoeffect. *Sov. Phys. Solid State*, **13**, 2436–2443.

Baldwin, G. L., and A. E. Heins, 1954: On the diffraction of a plane wave by an infinite plane grating. *Math. Scand.*, **2**, 103–118.

Banks-Sills, L., and Y. Benveniste, 1983: Steady interface crack propagation between two viscoelastic standard solids. *Int. J. Fract.*, **21**, 243–260.

Bates, C. P., and R. Mittra, 1968: Waveguide excitation of dielectric and plasma slabs. *Radio Sci.*, **3**, 251–266.

Bates, C. P., and R. Mittra, 1969: A factorization procedure for Wiener-Hopf kernels. *IEEE Trans. Antennas Propagat.*, **AP-17**, 102–103.

Bates, C. P., 1970: Internal reflections at the open end of a semi-infinite waveguide. *IEEE Trans. Antennas Propagat.*, **AP-18**, 230–235.

Beilis, A., J. W. Dash and A. Farrow, 1987: Analytic "α-pole" approach to an electromagnetic scattering problem. *IEEE Trans. Electromagn. Compat.*, **EMC-29**, 175–185.

Bera, R., and B. Patra, 1979: On the shear of an elastic wedge as a mixed boundary value problem. *Acta Mech.*, **33**, 307–316.

Bierman, G. J., 1971: A particular class of singular integral equations. *SIAM J. Appl. Math.*, **20**, 99–109.

Blatter, G., 1985: Scattering of atomic beams off a single surface step: An exact solution. *Ann. Physics*, **162**, 100–131.

Block, L. M., 1982: Shear and normal impact loadings on one face of a narrow slit. *Int. J. Solid Structures*, **18**, 467–477.

Block, L. M., 1992: Transient thermal effects in edge dislocation generation near a crack edge. *Int. J. Solids Structures*, **29**, 2217–2234.

Block, L. M., and J. P. Thomas, 1992: Thermal effects in rudimentary crack edge inelastic zone growth under stress wave loading. *Acta Mech.*, **93**, 223–239.

Boersma, J., J. J. E. Indenkleef and H. K. Kuiken, 1984: A diffusion problem in semiconductor technology. *J. Engng. Math.*, **18**, 315–333.

Bolotovskiĭ, B. M., and G. V. Voskresenskii, 1966: Diffraction radiation. *Sov. Phys. Uspekhi*, **9**, 73–96.

Bolotovskiĭ, B. M., and G. V. Voskresenskii, 1968: Emission from charged particles in periodic structures. *Sov. Phys. Uspekhi*, **11**, 143–162.

Bose, S. K., 1968: On the sudden subsidence of half the surface of an elastic half-space. *Bull. Cal. Math. Soc.*, **60**, 117–128.

Boyd, W. G. C., 1974: High-frequency scattering in a certain stratified medium. The two-part problem. *Proc. R. Soc. Edin.*, **A72**, 149–178.

Boyd, W. G. C., 1977: High-frequency scattering in a certain stratified medium: The three-part problem. *Proc. R. Soc. Lond.*, **A356**, 315–343.

Breithaupt, R. W., 1963: Diffraction of a cylindrical surface wave by a discontinuity in surface reactance. *Proc. IEEE*, **51**, 1455–1463.

Breithaupt, R. W., 1966: Diffraction of a general surface wave mode by a surface reactance discontinuity. *IEEE Trans. Antennas Propagat.*, **AP-14**, 290–297.

Broome, N. L., 1979: Improvements to nonnumerical methods for calculating the transient behavior of linear and aperture antennas. *IEEE Trans. Antennas Propagat.*, **AP-27**, 51–62.

Brykina, I. G., 1988: An analytical solution of the problem of convective diffusion in the neighborhood of a discontinuity of the catalytic properties of a surface. *J. Appl. Math. Mech.*, **52**, 191–197.

Buchwald, V. T., and H. E. Doran, 1965: Eigenfunctions of plane elastostatics. II. A mixed boundary value problem of the strip. *Proc. R. Soc. Lond.*, **A284**, 69–82.

Buchwald, V. T., 1968: The diffraction of Kelvin waves at a corner. *J. Fluid Mech.*, **31**, 193–205.

Budden, P. J., and J. Norbury, 1984: Stability of a subcritical flow under a sluice gate. *Quart. J. Mech. Appl. Math.*, **37**, 293–310.

Burke, J. E., 1964: Scattering of surface waves on an infinitely deep fluid. *J. Math. Phys.*, **5**, 805–819.

Büyükaksoy, A., 1985: Diffraction coefficients related to cylindrically curved soft-hard surfaces. *Ann. Télécommun.*, **40**, 402–410.

Büyükaksoy, A., G. Uzgören and A. H. Serbest, 1989: Diffraction of an obliquely incident plane wave by the discontinuity of a two part thin dielectric plane. *Int. J. Engng. Sci.*, **27**, 701–710.

Candel, S. M., 1973: Diffraction of a plane wave by a half plane in a subsonic and supersonic medium. *J. Acoust. Soc. Am.*, **54**, 1008–1016.

Candel, S. M., and C. Crance, 1981: Direct Fourier synthesis of waves: Application to acoustic source radiation. *AIAA J.*, **19**, 290–295.

Cannell, P. A., 1975: Edge scattering of aerodynamic sound by a lightly loaded elastic half-plane. *Proc. R. Soc. Lond.*, **A347**, 213–238.

Cannell, P. A., 1976: Acoustic edge scattering by a heavily loaded elastic half-plane. *Proc. R. Soc. Lond.*, **A350**, 71–89.

Cargill, A. M., 1982: Low frequency sound radiation and generation due to the interaction of unsteady flow with a jet pipe. *J. Fluid Mech.*, **121**, 59–105.

Carlson, J. F., and A. E. Heins, 1946: The reflection of an electromagnetic plane wave by an infinite set of plates. I. *Quart. Appl. Math.*, **4**, 313–329.

Case, K. M., 1960: Edge effects and the stability of plane Couette flow. *Phys. Fluids*, **3**, 432–435.

Chakrabarti, A., 1977: Diffraction by a uni-directionally conducting strip. *Indian J. Pure Appl. Math.*, **8**, 702–717.

Chakrabarti, A., 1979: Diffraction by a strip under mixed boundary conditions. *J. Indian Inst. Sci.*, **B61**, 163–176.

Chakrabarti, A., and S. Dowerah, 1984: Diffraction by a periodically corrugated strip. *J. Tech. Phys.*, **25**, 113–126.

Chakrabarti, A., and S. Dowerah, 1984: Traveling waves in a parallel plate waveguide with periodic wall perturbations. *Can. J. Phys.*, **62**, 271–284.

Chakrabarti, A., 1986: Diffraction by a dielectric half-plane. *IEEE Trans. Antennas Propagat.*, **AP-34**, 830–833.

Champion, C. R., 1988: The stress intensity factor history for an advancing crack under three-dimensional loading. *Int. J. Solids Structures*, **24**, 285–300.

Chang, D. C., 1969: VLF wave propagation along a mixed path in the curved earth-ionosphere waveguide. *Radio Sci.*, **4**, 335–345.

Chang, D. C., 1973: Radiation of a buried dipole in the presence of a semi-infinite metallic tube. *Radio Sci.*, **8**, 147–154.

Chang, S.-J., 1971: Diffraction of plane dilatational waves by a finite crack. *Quart. J. Mech. Appl. Math.*, **24**, 423–443.

Chang, S.-J., and S. M. Ohr, 1984: Inclined pileup of screw dislocations at the crack tip without a dislocation-free zone. *J. Appl. Phys.*, **55**, 3505–3513.

Chao, S.-Y., L. J. Pietrafesa and G. S. Janowitz, 1979: The scattering of continental shelf waves by an isolated topographic irregularity. *J. Phys. Oceanogr.*, **9**, 687–695.

Chattopadhyay, A., and U. Bandyopadhyay, 1988: Propagation of a crack due to shear waves in a medium of monoclinic type. *Acta Mech.*, **71**, 145–156.

Cherepanov, G. P., and V. D. Kuliev, 1975: On crack twinning. *Int. J. Fract.*, **11**, 29–38.

Cherepanov, G. P., 1976: Equilibrium of a slope with a tectonic crack. *J. Appl. Math. Mech.*, **40**, 119–133.

Cherepanov, G. P., 1976: Plastic rupture lines at the tip of a crack. *J. Appl. Math. Mech.*, **40**, 666–674.

Chester, W., 1950: The propagation of sound waves in an open-ended channel. *Phil. Mag.*, *Ser. 7*, **41**, 11–33.

Chew, W. C., and J. A. Kong, 1981: Asymptotic formula for the capacitance of two oppositely charged discs. *Math. Proc. Camb. Phil. Soc.*, **89**, 373–384.

Chew, W. C., and J. A. Kong, 1982: Microstrip capacitance for a circular disk through matched asymptotic expansions. *SIAM J. Appl. Math.*, **42**, 302–317.

Choi, S. R., and Y. Y. Earmme, 1991: Green's function of semi-infinite kinked crack under anti-plane shear. *Int. J. Fract.*, **51**, R3–R11.

Chuang, C. A., C. S. Liang and S.-W. Lee, 1975: High frequency scattering from an open-ended semi-infinite cylinder. *IEEE Trans. Antennas Propagat.*, **AP-23**, 770–776.

Clarke, J. F., 1967: The laminar diffusion flame in Oseen flow: The stoichiometric Burke-Schumann flame and frozen flow. *Proc. R. Soc. Lond.*, **A296**, 519–545.

Clausert, H., 1968: A parallel-plate waveguide, asymmetrically bifurcated by a unidirectionally conducting half screen. *Proc. Camb. Phil. Soc.*, **64**, 559–563.

Clemmow, P. C., 1951: A method for the exact solution of a class of two-dimensional diffraction problems. *Proc. R. Soc. Lond.*, **A205**, 286–308.

Clemmow, P. C., 1953: Radio propagation over a flat earth across a boundary separating two different media. *Phil. Trans. R. Soc. Lond.*, **A246**, 1–55.

Clovis, L. F., and H. G. Pinsent, 1983: The incidence of internal waves onto a thin submerged barrier. *Geophys. Astrophys. Fluid Dynamics*, **25**, 191–212.

Comstock, C., 1966: Transmission and reflection of electromagnetic waves normally incident on a warm plasma. *Phys. Fluids*, **9**, 1514–1521.

Craster, R. V., and C. Atkinson, 1992: Interfacial fracture in elastic diffusive media. *Int. J. Solids Structures*, **29**, 1463–1498.

Crease, J., 1956: Long waves on a rotating earth in the presence of a semi-infinite barrier. *J. Fluid Mech.*, **1**, 86–96.

Crighton, D. G., and F. G. Leppington, 1974: Radiative properties of the semi-infinite vortex sheet: The initial-value problem. *J. Fluid Mech.*, **64**, 393–414.

Crighton, D. G., and D. Innes, 1984: The modes, resonances and forced response of elastic structures under heavy fluid loading. *Phil. Trans. R. Soc. Lond.*, **A312**, 295–341.

Dahl, P. H., and G. V. Frisk, 1991: Diffraction from the juncture of pressure release and locally reacting half-planes. *J. Acoust. Soc. Am.*, **90**, 1093–1100.

Daniele, V. G., I. Montrosset and R. S. Zich, 1979: Wiener-Hopf solution for the junction between a smooth and a corrugated cylindrical waveguide. *Radio Sci.*, **14**, 943–955.

Das, S. C., and S. C. Prasad, 1972: Thermal stresses in an elastic cone due to a mixed thermal boundary condition. *Indian J. Pure Appl. Math.*, **3**, 248–262.

Das Gupta, S. P., 1970: Diffraction by a corrugated half-plane. *Proc. Vibr. Prob.*, **11**, 413–424.

Davidson, R. F., 1988: Waves below first cutoff in a duct. *J. Austral. Math. Soc.*, **B29**, 448–460.

Davies, B., 1978: *Integral Transforms and Their Applications*. New York: Springer-Verlag, Chapters 18–19.

Davies, H. G., 1974: Natural motion of a fluid-loaded semi-infinite membrane. *J. Acoust. Soc. Am.*, **55**, 213–219.

Davies, H. G., 1975: Edge-mode radiation from fluid-loaded semi-infinite membranes. *J. Sound Vib.*, **40**, 179–189.

Davis, A. M. J., 1987: Continential shelf wave scattering by a semi-infinite coastline. *Geophys. Astrophys. Fluid Dynamics*, **39**, 25–55.

Davis, A. M. J., 1990: Continential shelf wave scattering by a semi-infinite coastline; partial removal of the "rigid lid". *Geophys. Astro-*

phys. Fluid Dynamics, **50**, 175–194.

Dawson, T. W., and J. T. Weaver, 1979: H-polarization induction in two thin half-sheets. *Geophys. J. R. Astr. Soc.*, **56**, 419–438.

Dawson, T. W., J. W. Weaver and U. Raval, 1982: B-polarization induction in two generalized thin sheets at the surface of a conducting half-space. *Geophys. J. R. Astr. Soc.*, **69**, 209–234.

Dawson, T. W., 1983: E-polarization induction in two thin half-sheets. *Geophys. J. R. Astr. Soc.*, **73**, 83–107.

DeBruijn, A., 1973: A mathematical analysis concerning the edge effect of sound absorbing materials. *Acustica*, **28**, 33–44.

Delogne, P., 1969: Problèmes de diffraction sur guide unifilaire. *Ann. Télécommunic.*, **24**, 405–426.

DeSanto, J. A., 1971: Scattering from a periodic corrugated structure: Thin comb with soft boundaries. *J. Math. Phys.*, **12**, 1913–1923.

DeSanto, J. A., 1973: Scattering from a periodic corrugated surface: Semi-infinite alternately filled plates. *J. Acoust. Soc. Am.*, **53**, 719–734.

Deshwal, P. S., 1971: Diffraction of a compressional wave by a rigid barrier in a liquid half-space. *PAGEOPH*, **85**, 107–124.

Deshwal, P. S., 1971: Diffraction of a compressional wave by a rigid barrier in a liquid layer - Part I. *PAGEOPH*, **88**, 12–28.

Deshwal, P. S., 1971: Diffraction of a compressional wave by a rigid barrier in a liquid layer - Part II. *PAGEOPH*, **91**, 14–33.

Deshwal, P. S., 1981: Diffraction of compressional waves by a strip in a liquid halfspace. *J. Math. Phys. Sci.*, **15**, 263–281.

Deshwal, P. S., 1987: Rayleigh wave scattering due to a surface impedance in a liquid layer. *J. Math. Phys. Sci.*, **21**, 399–421.

Deshwal, P. S., and K. K. Mann, 1988: Rayleigh wave scattering at coastal regions. *J. Acoust. Soc. Am.*, **84**, 286–291.

Deshwal, P. S., and N. Mohan, 1988: P-wave scattering at a coastal region in a shallow ocean. *Indian J. Pure Appl. Mech.*, **19**, 1020–1030.

Diprima, R. C., 1957: On the diffusion of tides into permeable rock of finite depth. *Quart. Appl. Math.*, **15**, 329–339.

Dobrott, D. R., 1966: Propagation of VLF waves past a coastline. *Radio Sci.*, **1**, 1411–1424.

Dorfman, L. G., and V. V. Filatov, 1966: Scattering coefficients caused by abrupt change in electrical properties of the narrow wall of a rectangular waveguide. *Radio Engng. Electron. Phys.*, **11**, 170–177.

Dowerah, S., and A. Chakrabarti, 1985: Diffraction by two staggered half-planes with periodic wall perturbations. *Indian J. Pure Appl. Math.*, **16**, 411–447.

Dowerah, S., and A. Chakrabarti, 1987: Double knife-edge diffraction with mixed boundary conditions on one of the scatterers. *J. Sound*

Vib., **116**, 49–70.

Dowerah, S., and A. Chakrabarti, 1988: Extinction cross section of a dielectric strip. *IEEE Trans. Antennas Propagat.*, **AP-36**, 696–706.

Dua, S. S., and C. L. Tien, 1976: Two-dimensional analysis of conduction-controlled rewetting with precursory cooling. *J. Heat Transfer*, **98**, 407–413.

Du Cloux, R., 1984: Pulsed electromagnetic radiation from a line source in the presence of a semi-infinite screen in the plane interface of two different media. *Wave Motion*, **6**, 459–476.

Durbin, P. A., 1989: Stokes flow near a moving contact line with yield-stress boundary condition. *Quart. J. Mech. Appl. Math.*, **42**, 99–113.

Eaves, R. E., and D. M. Bolle, 1970: Modes of shielded slot lines. *Arch. der elektrischen Übertragung*, **24**, 389–394.

Eid, Y., S. Morsy, A. Abboud, S. El-Konsol, A. Hussein and I. Hamouda, 1976: Investigation of the space-dependent energy and angular neutron spectrum in two adjacent moderating media. *Atomkernenergie*, **28**, 259–264.

Elliott, J. P., 1955: Milne's problem with a point source. *Proc. R. Soc. Lond.*, **A228**, 424–433.

Elmoazzen, Y. E., and L. Shafai, 1973: Mutual coupling between parallel-plate waveguides. *IEEE Trans. Microwave Theory Tech.*, **MTT-21**, 825–833.

Elmoazzen, Y., and L. Shafai, 1974: Mutual coupling between two circular waveguides. *IEEE Trans. Antennas Propagat.*, **AP-22**, 751–760.

Englert, G. W., 1984: Interaction of upstream flow distortions with high-Mach-number cascades. *J. Engng. Gas Turbines Power*, **106**, 260–270.

Evans, D. V., 1972: The application of a new source potential to the problem of the transmission of water waves over a shelf of arbitrary profile. *Proc. Camb. Phil. Soc.*, **71**, 391–410.

Evans, D. V., 1984: A note on the cooling of a cylinder entering a fluid. *IMA J. Appl. Math.*, **33**, 49–54.

Evans, D. V., 1985: The solution of a class of boundary-value problems with smoothly varying boundary conditions. *Quart. J. Mech. Appl. Math.*, **38**, 521–536.

Fainstein-Pedraza, D., and G. F. Bolling, 1975: Superdendritic growth. II. A mathematical analysis. *J. Crystal Growth*, **28**, 319–333.

Faulkner, T. R., 1966: The diffraction of an obliquely incident surface wave by a vertical barrier of finite depth. *Proc. Camb. Phil. Soc.*, **62**, 829–838.

Fernandes, C. A., and A. M. Barbosa, 1990: Radiation from a sheath helix excited by a circular waveguide: a Wiener-Hopf analysis. *IEE Proc., Part H*, **137**, 269–275.

Fong, T. T., 1972: Radiation from an open-ended waveguide with extended dielectric loading. *Radio Sci.*, **7**, 965–972.

Fonseca, J. G., J. D. Eshelby and C. Atkinson, 1971: The fracture mechanics of flint-knapping and allied processes. *Int. J. Fract.*, **7**, 421–433.

Foster, M. R., 1980: A flat plate in a rotating, stratified flow. *J. Engng. Math.*, **14**, 117–132.

Fouad, A. A., 1966: End losses in a magnetohydrodynamic channel: DC channel with fluid having large magnetic Reynolds number. *IEEE Trans. Electron Devices*, **ED-13**, 554–561.

Fredricks, R. W., and L. Knopoff, 1960: The reflection of Rayleigh waves by a high impedance obstacle on a half-space. *Geophysics*, **25**, 1195–1202.

Fredricks, R. W., 1961: Diffraction of an elastic pulse in a loaded half-space. *J. Acoust. Soc. Am.*, **33**, 17–22.

Freund, L. B., and J. D. Achenbach, 1967: Diffraction of a plane pulse by a semi-infinite barrier at a fluid-solid interface. *J. Appl. Mech.*, **34**, 571–578.

Freund, L. B., and J. D. Achenbach, 1968: Diffraction of a plane pulse by a closed crack at the interface of elastic solids. *Zeit. angew. Math. Mech.*, **48**, 173–185.

Freund, L. B., and J. D. Achenbach, 1968: Waves in a semi-infinite plate in smooth contact with a harmonically disturbed half-space. *Int. J. Solids Structures*, **4**, 605–621.

Friedrich, N., 1980: Zur Faktorisierung einiger Kerne, die bei Ausstrahlungsproblemen der Elektrodynamik und Aeroelastizität auftreten. *J. reine angew. Math.*, **316**, 15–30.

Gautesen, A. K., 1971: Symmetric Oseen flow past a semi-infinite plate and two force singularities. *Zeit. angew. Math. Phys.*, **22**, 144–155.

Gautesen, A. K., 1971: Oseen flow past a semi-infinite flat plate and a force singularity. *Zeit. angew. Math. Phys.*, **22**, 247–257.

Gazarian, Iu. L., 1957: Waveduct propagation of sound for one particular class of laminarly-inhomogeneous media. *Sov. Phys. Acoust.*, **3**, 135–149.

Georgiadis, H. G., 1987: Moving punch on a highly orthotropic elastic layer. *Acta Mech.*, **68**, 193–202.

Georgiadis, H. G., J. R. Barber and F. Ben Ammar, 1991: An asymptotic solution for short-time transient heat conduction between two dissimilar bodies in contact. *Quart. J. Mech. Appl. Math.*, **44**, 303–322.

Ghosh, M. L., 1962: Reflection of Love wave from a rigid obstacle. *Zeit. Geophys.*, **28**, 223–230.

Ghosh, M. L., 1973: On reflection and diffraction of pressure waves from floating ice. *PAGEOPH*, **111**, 2163–2176.

Ghosh, M. L., 1974: On the propagation of Love's waves in an elastic layer in the presence of a vertical crack. *Proc. Vibr. Prob.*, **15**, 147–165.

Ghosh, S. C., 1968: Diffraction of SH-waves originating from a moving point source by a rigid quarter space. *PAGEOPH*, **70**, 22–27.

Gol'dshtein, R. V., and A. V. Marchenko, 1989: The diffraction of plane gravitational waves by the edge of an ice cover. *J. Appl. Math. Mech.*, **53**, 731–736.

Goldstein, M. E., W. Braun and J. J. Adamczyk, 1977: Unsteady flow in a supersonic cascade with strong in-passage shocks. *J. Fluid Mech.*, **83**, 569–604.

Goldstein, M. E., 1978: Characteristics of the unsteady motion on transversely sheared mean flow. *J. Fluid Mech.*, **84**, 305–329.

Goldstein, M. E., 1979: Scattering and distortion of the unsteady motion on transversely sheared mean flows. *J. Fluid Mech.*, **91**, 601–632.

Goldstein, M. E., 1981: The coupling between flow instabilities and incident disturbances at a leading edge. *J. Fluid Mech.*, **104**, 217–246.

Göksu, Ç., and F. Güneş, 1991: High frequency surface currents induced on a soft-hard cylindrical strip. *IEE Proc., Pt. A*, **138**, 113–118.

Graebel, W. P., 1965: Slow viscous shear flow past a plate in a channel. *Phys. Fluids*, **8**, 1929–1935.

Greene, T. R., and A. E. Heins, 1953: Wave waves over a channel of infinite depth. *Quart. Appl. Math.*, **11**, 201–214.

Greenspan, H. P., and L. A. Peletier, 1962: Some exact solutions of magneto-hydrodynamic viscous flow problems. *J. Phys. and Math.*, **41**, 116–131.

Gregory, R. D., 1966: The attenuation of a Rayleigh wave in a half-space by a surface impedance. *Proc. Camb. Phil. Soc.*, **62**, 811–827.

Gregory, R. D., 1977: A circular disc containing a radial edge crack opened by a constant internal pressure. *Math. Proc. Camb. Phil. Soc.*, **81**, 497–521.

Gregory, R. D., 1979: The edge-cracked circular disc under symmetric pin-loading. *Math. Proc. Camb. Phil. Soc.*, **85**, 523–538.

Gregory, R. D., 1989: The spinning circular disc with a radial edge crack; an exact solution. *Int. J. Fract.*, **41**, 39–50.

Haines, C. R., 1981: A Wiener-Hopf approach to Kelvin wave generation by a semi-infinite barrier and a depth discontinuity. *Quart. J. Mech. Appl. Math.*, **34**, 139–151.

Hallén, E., 1956: Exact treatment of antenna current wave reflection at the end of a tube-shaped cylindrical antenna. *IEEE Antennas Propagat.*, **AP-3**, 479–491.

Hamblin, P. F., 1980: An analysis of advective diffusion in branching channels. *J. Fluid Mech.*, **99**, 101–110.

Hanzawa, H., M. Kishida and M. Asano, 1981: Dynamic interference between a crack and a plane boundary: Dynamic stress intensity factor induced by a plane harmonic SH-wave. *Bull. JSME*, **24**, 895–901.

Hazeltine, R. D., M. N. Rosenbluth and A. M. Sessler, 1971: Diffraction radiation by a line charge moving past a comb: A model of radiation losses in an electron ring accelerator. *J. Math. Phys.*, **12**, 502–514.

Hebenstreit, H., 1976: Oberflächenwellen bei verschiedenen Anregungsformen. *Acta Phys. Austr.*, **44**, 259–277.

Hector, D. L., 1967: On the linearized M. H. D. flow past a semi-infinite flat plate in the presence of a transverse magnetic field. *J. Phys. and Math.*, **46**, 408–424.

Heins, A. E., and J. F. Carlson, 1947: The reflection of an electromagnetic plane wave by an infinite set of plates. II. *Quart. Appl. Math.*, **5**, 82–88.

Heins, A. E., and H. Feshbach, 1947: The coupling of two acoustical ducts. *J. Math. and Phys.*, **26**, 143–155.

Heins, A. E., 1948: The radiation and transmission properties of a pair of semi-infinite parallel plates. *Quart. Appl. Math.*, **6**, 157–166, 215–220.

Heins, A. E., 1948: Water waves over a channel of finite depth with a dock. *Am. J. Math.*, **70**, 730–748.

Heins, A. E., 1950: Water waves over a channel of finite depth with a submerged plane barrier. *Can. J. Math.*, **2**, 210–222.

Heins, A. E., 1950: The reflection of an electromagnetic plane wave by an infinite set of plates. III. *Quart. Appl. Math.*, **8**, 281–291.

Heins, A. E., 1951: Some remarks on the coupling of two ducts. *J. Math. and Phys.*, **30**, 164–169.

Heins, A. E., 1956: The scope and limitations of the method of Wiener and Hopf. *Commun. Pure Appl. Math.*, **9**, 447–466.

Heins, A. E., 1957: The Green's function for periodic structures in diffraction theory with an application to parallel plate media. I. *J. Math. Mech.*, **6**, 401–426.

Heitman, W. G., and P. M. van den Berg, 1975: Diffraction of electromagnetic waves by a semi-infinite screen in a layered medium. *Can. J. Phys.*, **53**, 1305–1317.

Hemp, J., 1987: The virtual current in electromagnetic flowmeters with end effects. *Quart. J. Mech. Appl. Math.*, **40**, 507–525.

Herrmann, J. M., and L. Schovanec, 1990: Quasi-static mode III fracture in a nonhomogeneous viscoelastic body. *Acta Mech.*, **85**, 235–249.

Holford, R. L., 1964: Short surface waves in the presence of a finite dock. *Proc. Camb. Phil. Soc.*, **60**, 957–983, 985–1011.

Horvay, G., and M. DaCosta, 1964: Temperature distribution in a cylindrical rod moving from a chamber at one temperature to a chamber at another temperature. *J. Heat Transfer*, **86**, 265–266.

Hsieh, W. W., and V. T. Buchwald, 1984: The scattering of a continental shelf wave by a semi-infinite barrier located along the outer edge of a step shelf. *Geophys. Astrophys. Fluid Dynamics*, **28**, 257–276.

Huang, S. C., 1985: Unsteady-state heat conduction in semi-infinite regions with mixed-type boundary conditions. *J. Heat Transfer*, **107**, 489–491.

Hurd, R. A., 1960: Diffraction by a unidirectionally conducting half-plane. *Can. J. Phys.*, **38**, 168–175.

Hurd, R. A., 1965: The admittance of a linear antenna in a uniaxial medium. *Can. J. Phys.*, **43**, 2276–2309.

Hurd, R. A., 1966: Admittance of a long linear antenna. *Can. J. Phys.*, **44**, 1723–1744.

Hurd, R. A., 1972: Mutual admittance of two collinear antennas. *J. Appl. Phys.*, **43**, 3701–3707.

Hutter, K., and V. O. S. Olunloyo, 1980: On the distribution of stress and velocity in an ice strip, which is partly sliding over and partly adhering to its bed, by using a Newtonian viscous approximation. *Proc. R. Soc. Lond.*, **A373**, 385–403.

Idemen, M., 1975: On the scalar scattering by a strip in a dissipative medium. *J. Engng. Math.*, **9**, 93–102.

Igarashi, A., 1964: Simultaneous Wiener-Hopf equations and their application to diffraction problems in electromagnetic theory. I. *J. Phys. Soc. Japan*, **19**, 1213–1221.

Igarashi, A., 1968: Simultaneous Wiener-Hopf equations and their application to diffraction problems in electromagnetic theory. II. *J. Phys. Soc. Japan*, **25**, 260–271.

Igarashi, A., 1969: Simultaneous Wiener-Hopf equations and their application to diffraction problems in electromagnetic theory. IV. *J. Phys. Soc. Japan*, **26**, 549–560.

Igushkin, L. P., and É. I. Urazakov, 1967: Motion of symmetric current-carrying plasmoids along the axis of a semiinfinite cylindrical waveguide. *Sov. Phys. Tech. Phys.*, **12**, 27–32.

Ittipiboon, A., and M. Hamid, 1981: Application of the Wiener-Hopf technique to dielectric slab waveguide discontinuities. *IEE Proc., Part H*, **128**, 188–196.

Jeong, J.-T., and M.-U. Kim, 1983: Slow viscous flow around an inclined fence on a plane. *J. Phys. Soc. Japan*, **52**, 2356–2363.

Jeong, J.-T., and M.-U. Kim, 1985: Slow viscous flow due to sliding of a semi-infinite plate over a plane. *J. Phys. Soc. Japan*, **54**, 1789–1799.

Jeong, J.-T., and M.-U. Kim, 1988: Two-dimensional slow viscous flow from a converging nozzle. *J. Phys. Soc. Japan*, **57**, 856–865.

Jeong, J.-T., and M.-U. Kim, 1990: Two-dimensional slow viscous flow due to the rotation of a body with cusped edges. *J. Phys. Soc. Japan*, **59**, 3194–3202.

Johansen, E. L., 1962: Scattering coefficients for wall impedance changes in waveguides. *IEEE Trans. Microwave Theory Tech.*, **10**, 26–29.

Johansen, E. L., 1965: The radiation properties of a parallel-plane wave-guide in a transversely magnetized, homogeneous plasma. *IEEE Trans. Microwave Theory Tech.*, **MTT-13**, 77–83.

Johansen, E. L., 1967: Surface wave scattering by a step. *IEEE Trans. Antennas Propagat.*, **AP-15**, 442–448.

Johansen, E. L., 1968: Surface wave radiation from a thick, semi-infinite plane with a reactive surface. *IEEE Trans. Antennas Propagat.*, **AP-16**, 391–398.

Johnson, G. W., and K. Ogimoto, 1980: Sound radiation from a finite length unflanged circular duct with uniform axial flow. I. Theoretical analysis. *J. Acoust. Soc. Am.*, **68**, 1858–1870.

Jones, D. S., 1952: Diffraction by a wave-guide of finite length. *Proc. Camb. Phil. Soc.*, **48**, 118–134.

Jones, D. S., 1952: A simplifying technique in the solution of a class of diffraction problems. *Quart. J. Math.*, *Ser. 2*, **3**, 189–196.

Jones, D. S., 1953: Diffraction by a thick semi-infinite plate. *Proc. R. Soc. Lond.*, **A217**, 153–175.

Jones, D. S., 1955: The scattering of a scalar wave by a semi-infinite rod of circular cross section. *Phil Trans. R. Soc. Lond.*, **A247**, 499–528.

Jones, D. S., 1973: Double knife-edge diffraction and ray theory. *Quart. J. Mech. Appl. Math.*, **26**, 1–18.

Jull, E. V., 1967: Aperture fields in an anisotropic medium. *Radio Sci.*, **2**, 837–852.

Jull, E. V., 1968: Diffraction by a wide unidirectionally conducting strip. *Can. J. Phys.*, **46**, 2107–2117.

Kane, J., 1960: The efficiency of lauching surface waves on a reactive half plane by an arbitrary antenna. *IEEE Trans. Antennas Propagat.*, **AP-8**, 500–507.

Kane, J., 1962: The propagation of Rayleigh waves past a fluid-loaded boundary. *J. Phys. and Math.*, **41**, 179–190.

Kapila, A. K., and G. S. S. Ludford, 1977: MHD with inertia: Flow over blunt obstacles in channels. *Int. J. Engng. Sci.*, **15**, 465–480.

Kapoulitsas, G. M., 1977: Diffraction of long waves by a semi-infinite vertical barrier on a rotating earth. *Int. J. Theoret. Phys.*, **16**, 763–773.

Kapoulitsas, G. M., 1979: Diffraction of Kelvin waves from a rotating channel with an infinite and a semi-infinite barrier. *J. Phys. A*, **12**, 733–742.

Kapoulitsas, G. M., 1980: Scattering of long waves in a rotating, bifurcated channel. *Int. J. Theoret. Phys.*, **19**, 773–788.

Kapoulitsas, G. M., 1984: Diffraction of long waves by a step. *Acta Mech.*, **52**, 77–91.

Kapoulitsas, G. M., 1984: Propagation of long waves into a set of parallel vertical barriers on a rotating earth. *Wave Motion*, **6**, 1–14.

Kashyap, S. C., and M. A. K. Hamid, 1971: Diffraction characteristics of a slit in a thick conducting screen. *IEEE Trans. Antennas Propagat.*, **AP-19**, 499–507.

Kashyap, S. C., 1974: Diffraction characteristics of a slit formed by two staggered parallel planes. *J. Math. Phys.*, **15**, 1944–1949.

Kay, A. F., 1959: Scattering of a surface wave by a discontinuity in reactance. *IEEE Trans. Antennas Propagat.*, **AP-7**, 22–31.

Kazi, M. H., 1975: Diffraction of Love waves by perfectly rigid and perfectly weak half-plane. *Bull. Seism. Soc. Am.*, **65**, 1461–1479.

Kempton, A. J., 1976: Heat diffusion as a source of aerodynamic sound. *J. Fluid Mech.*, **78**, 1–31.

Keogh, P. S., 1985: High-frequency scattering by a Griffith crack. I: A crack Green's function. *Quart. J. Mech. Appl. Math.*, **38**, 185–204.

Keogh, P. S., 1985: High-frequency scattering by a Griffith crack. II: Incident plane and cylindrical waves. *Quart. J. Mech. Appl. Math.*, **38**, 205–232.

Keogh, P. S., 1986: High-frequency scattering of a normally incident plane compressional wave by a penny-shaped crack. *Quart J. Mech. Appl. Math.*, **39**, 535–566.

Kim, D. H., and L. L. Koss, 1990: Sound radiation from a circular duct with axial temperature gradients. *J. Sound Vib.*, **141**, 1–16.

Kim, M.-U., and M. K. Chung, 1984: Two-dimensional slow viscous flow past a plate midway between an infinite channel. *J. Phys. Soc. Japan*, **53**, 156–166.

Kim, M.-U., J.-T. Jeong and T.-H. Lee, 1986: Slow viscous flow due to the rotation of two finite hinged plates. *J. Phys. Soc. Japan*, **55**, 115–127.

Kobayashi, K., 1984: On the factorization of certain kernels arising in functional equations of the Wiener-Hopf type. *J. Phys. Soc. Japan*, **53**, 2885–2898.

Kobayashi, K., and T. Inoue, 1988: Diffraction of a plane wave by an inclined parallel plate grating. *IEEE Trans. Antennas Propagat.*, **AP-36**, 1424–1434.

Kobayashi, K., and K. Miura, 1989: Diffraction of a plane wave by a thick strip grating. *IEEE Trans. Antennas Propagat.*, **AP-37**, 459–470.

Koch, W., and G. S. S. Ludford, 1970: Diffusion in shear flow past a semi-infinite flat plate. *Acta Mech.*, **10**, 229–250.

Koch, W., 1970: On the heat transfer from a finite plate in channel flow for vanishing Prandtl number. *Zeit. angew. Math. Phys.*, **21**, 910–918.

Koch, W., G. S. S. Ludford and A. R. Seebass, 1971: Diffusion in shear flow past a semi-infinite flat plate. Part II: Viscous effects. *Acta*

Mech., **12**, 99–120.

Koch, W., 1971: On the transmission of sound waves through a blade row. *J. Sound Vib.*, **18**, 111–128.

Koch, W., 1974: Radiation - convection interaction for a laterally bounded flow past a hot plate. *Int. J. Heat Mass Transfer*, **17**, 915–931.

Koch, W., 1975: Heat transfer from two hot plates in laterally bounded uniform flow as an example of an n-part Wiener-Hopf-problem. *Zeit. angew. Math. Phys.*, **26**, 187–198.

Koch, W., 1977: Attenuation of sound in multi-element acoustically lined rectangular ducts in the absence of mean flow. *J. Sound Vib.*, **52**, 459–496.

Koch, W., 1977: Radiation of sound from a two-dimensional acoustically lined duct. *J. Sound Vib.*, **55**, 255–274.

Koch, W., and W. Möhring, 1983: Eigensolutions for liners in uniform mean flow ducts. *AIAA J.*, **21**, 200–213.

Konovalyuk, I. P., 1969: Diffraction of a plane sound wave by a plate reinforced with stiffness members. *Sov. Phys. Acoust.*, **14**, 465–469.

Korovkin, A. N., and D. D. Plakhov, 1973: Sound diffraction by a baffle coupled with an elastic plate. *Sov. Phys. Acoust.*, **19**, 459–462.

Kostelnicek, R. J., and R. Mittra, 1971: Radiation from a parallel-plate waveguide into a dielectric or plasma layer. *Radio Sci.*, **6**, 981–990.

Kouzov, D. P., 1969: Diffraction of a cylindrical hydroacoustic wave at the joint of two semi-infinite plates. *J. Appl. Math. Mech.*, **33**, 225–234.

Kovalenko, G. P., 1974: Pressure pulse on the boundary of an elastic homogeneous half-plane. *J. Appl. Mech. Tech. Phys.*, **15**, 832–836.

Kraut, E. A., 1968: Diffraction of elastic waves by a rigid 90° wedge. Part I. *Bull. Seism. Soc. Am.*, **58**, 1083–1096.

Kudriavtsev, B. A., V. Z. Parton and B. D. Rubinskii, 1980: Magnetothermoelastic field in a body with a semi-infinite cut. *J. Appl. Math. Mech.*, **44**, 646–650.

Kuhn, G., and M. Matczyński, 1974: Beitrag zum gemischten Randwertproblem am Streifen. *Zeit. angew. Math. Mech.*, **54**, T88–T91.

Kuhn, G., and M. Matczyński, 1975: Elastic strip with a crack under periodic loading. *Arch. Mech.*, **27**, 459–472.

Kuhn, G., and M. Matczyński, 1975: Analytische Ermittlung des Spannungsintensitätsfaktors eines ebenen Rißproblems unter periodischer Belastung. *Zeit. angew. Math. Mech.*, **55**, T99–T102.

Kuiken, H. K., 1984: Etching: A two-dimensional mathematical approach. *Proc. R. Soc. Lond.*, **A392**, 199–225.

Kuiken, H. K., 1985: Edge effects in crystal growth under intermediate diffusive-kinetic control. *IMA J. Appl. Math.*, **35**, 117–129.

Kuiken, H. K., 1986: On the influence of longitudinal diffusion in time-dependent convective-diffusive systems. *J. Fluid Mech.*, **165**, 147–

162.

Kuo, M. K., and S. H. Cheng, 1991: Elastodynamic responses due to anti-plane point impact loadings on the faces of an interface crack along dissimilar anisotropic materials. *Int. J. Solids Structures*, **28**, 751–768.

Kurushin, E. P., and E. I. Nefedov, 1969: Diffraction of H-waves on a wide slit in a plane multiwave waveguide. *Radio Engng. Electron. Phys.*, **14**, 176–182.

Lam, J., 1967: Radiation of a point charge moving uniformly over an infinite array of conducting half-planes. *J. Math. Phys.*, **8**, 1053–1060.

Lamb, G. L., 1959: Diffraction of a plane sound wave by a semi-infinite thin elastic plate. *J. Acoust. Soc. Am.*, **31**, 929–935.

Latham, R. W., and K. S. H. Lee, 1968: Magnetic field leakage into a semi-infinite pipe. *Can. J. Phys.*, **46**, 1455–1462.

Lawrie, J. B., 1986: Vibrations of a heavily loaded, semi-infinite cylindrical elastic shell. I. *Proc. R. Soc. Lond.*, **A408**, 103–128.

Lawrie, J. B., 1988: Axisymmetric radiation from a finite gap in an infinite, rigid, circular duct. *IMA J. Appl. Math.*, **40**, 113–128.

Lawrie, J. B., 1988: Scattering of sound by a heavily loaded finite cylindrical elastic shell. *Quart. J. Mech. Appl. Math.*, **41**, 445–467.

Laurmann, J. A., 1961: Linearized slip flow past a semi-infinite flat plate. *J. Fluid Mech.*, **11**, 82–96.

Lee, S. W., 1967: Radiation from an infinite array of parallel-plate waveguides with thick walls. *IEEE Trans. Microwave Theory Tech.*, **MTT-15**, 364–371.

Lee, S. W., 1969: Cylindrical antenna in uniaxial resonant plasmas. *Radio Sci.*, **4**, 179–189.

Lee, S. W., and R. Mittra, 1969: Admittance of a solid cylindrical antenna. *Can. J. Phys.*, **47**, 1959–1970.

Lee, S. W., 1970: Ray theory of diffraction by open-ended waveguides. I. Field in waveguides. *J. Math. Phys.*, **11**, 2830–2850.

Lee, S. W., 1972: Ray theory of diffraction by open-ended waveguides. II. Applications. *J. Math. Phys.*, **13**, 656–664.

Lee, S.-W., V. Jamnejad and R. Mittra, 1973: Near field of scattering by a hollow semi-infinite cylinder and its application to sensor booms. *IEEE Trans. Antennas Propagat.*, **AP-21**, 182–188.

Lee, S. W., and J. Boersma, 1975: Ray-optical analysis of fields on shadow boundaries of two parallel plates. *J. Math. Phys.*, **16**, 1746–1764.

Lehman, G. W., 1970: Diffraction of electromagnetic waves by planar dielectric structures. I. Transverse electric excitation. *J. Math. Phys.*, **11**, 1522–1535.

Lehner, F. K., V. C. Li and J. R. Rice, 1981: Stress diffusion along rupturing plate boundaries. *J. Geophys. Res.*, **86**, 6155–6169.

Leib, S. J., and M. E. Goldstein, 1986: The generation of capillary instabilities on a liquid jet. *J. Fluid Mech.*, **168**, 479–500.

Lenau, C. W., 1980: Confined flow in aquifer containing thin layer. *J. Eng. Mech. Div. ASCE*, **106**, 719–737.

Lennox, S. C., and D. C. Pack, 1963: The flow in compound jets. *J. Fluid Mech.*, **15**, 513–526.

Leppington, F. G., 1983: Travelling waves in a dielectric slab with an abrupt change in thickness. *Proc. R. Soc. Lond.*, **A386**, 443–460.

Levine, H., and J. Schwinger, 1948: On the radiation of sound from an unflanged circular pipe. *Phys. Rev., Series 2*, **73**, 383–406.

Levine, H., 1954: On the theory of sound reflection in an open-ended cylindrical tube. *J. Acoust. Soc. Am.*, **26**, 200–211.

Levine, H., 1982: On a mixed boundary value problem of diffusion type. *Appl. Sci. Res.*, **39**, 261–276.

Levinson, Y. B., and E. V. Sukhorukov, 1991: Bending of electron edge states in a magnetic field. *J. Phys., Condens. Matter*, **3**, 7291–7306.

Linton, C. M., and D. V. Evans, 1991: Trapped modes above a submerged horizontal plate. *Quart. J. Mech. Appl. Math.*, **44**, 487–506.

Lovis, D., 1971: The relationship between the electric and magnetic fields with an arbitrary field distribution on an open-ended parallel-plate waveguide and its radiation characteristic. *Arch. für Electronik und Übertragungstechnik*, **25**, 502–508.

Low, R. D., and H. J. Weiss, 1962: On a mixed boundary value problem for an infinite elastic cone. *Zeit. angew. Math. Phys.*, **13**, 232–242.

Ludford, G. S. S., and V. O. S. Olunloyo, 1972: Further results concerning the forces on a flat plate in a Couette flow. *Zeit. angew. Math. Phys.*, **23**, 729–744.

Lyamshev, L. M., 1967: Sound diffraction by a semiinfinite elastic plate in a moving media. *Sov. Phys. Acoust.*, **12**, 291–294.

Lyamshev, L. M., 1967: Scattering of sound by a semiinfinite elastic tube in a moving medium. *Sov. Phys. Acoust.*, **13**, 71–77.

Ma, C. C., and P. Burgers, 1986: Mode-III crack kinking with delay time: An analytical approximation. *Int. J. Solids Structures*, **22**, 883–899.

Ma, C.-C., 1990: A dynamic crack model with viscous resistance to crack opening: Antiplane shear mode. *Engng. Fracture Mech.*, **37**, 127–144.

Mandal, S. C., and M. L. Ghosh, 1990: Moving punch on a viscoelastic semi-infinite medium. *Indian J. Pure Appl. Math.*, **21**, 847–864.

Mani, R., 1973: Refraction of acoustic duct waveguide modes by exhaust jets. *Quart. Appl. Math.*, **30**, 501–520.

Mann, K. K., and P. S. Deshwal, 1986: Rayleigh wave scattering by a plane barrier in a shallow ocean. *Indian J. Pure Appl. Math.*, **17**, 1056–1066.

Marin, L., 1969: On the coupling between two adjacent rectangular waveguides through a slot in the common wall. *Acta Polytech. Scandinavica, *Electrical Engineering Ser. No. 20*, 39 pp.

Marin, L., 1971: Analysis of a slotted circular waveguide. *Acta Polytech. Scandinavica, Electrical Engineering Ser. No. 28*, 23 pp.

Marshak, R. E., 1947: The Milne problem for a large plane slab with constant source and anisotropic scattering. *Phys. Rev., Ser. 2*, **72**, 47–50.

Martinez, R., and S. E. Widnall, 1980: Unified aerodynamic-acoustic theory for a thin rectangular wing encountering a gust. *AIAA J.*, **18**, 636–645.

Mason, R. J., 1968: Electric-field penetration into a plasma with a fractionally accommodating boundary. *J. Math. Phys.*, **9**, 868–874.

Matczyński, M., 1965: Axi-symmetric problem for a partly clamped elastic rod. *Arch. Mech. Stos.*, **17**, 43–64.

Matczyński, M., 1973: Quasi-static problem of a crack in an elastic strip subject to antiplane state of strain. *Arch. Mech.*, **25**, 851–860.

Matczyński, M., 1974: Quasistatic problem of a non-homogeneous elastic layer containing a crack. *Acta Mech.*, **19**, 153–168.

Matsui, E., 1971: Free-field correction for laboratory standard microphones mounted on a semiinfinite rod. *J. Acoust. Soc. Am.*, **49**, 1475–1483.

Maue, A.-W., 1953: Die Beugung elastischer Wellen an der Halbebene. *Zeit. ange. Math. Mech.*, **33**, 1–10.

Meijers, P., 1968: The contact problem of a rigid cylinder on an elastic layer. *Appl. Sci. Res.*, **18**, 353–383.

Meister, E., 1962: Zum Dirichlet-Problem der Helmholtzschen Schwingungsgleichung für ein gestaffeltes Streckengitter. *Arch. Rat. Mech. Anal.*, **10**, 67–100.

Meister, E., 1962: Zum Neumann-Problem der Helmholtzschen Schwingungsgleichung für ein gestaffeltes Streckengitter. *Arch. Rat. Mech. Anal.*, **10**, 127–148.

Meister, E., 1965: Zur Theorie der ebenen, instationären Unterschallströmung um ein schwingendes Profil im Kanal. *Zeit. angew. Math. Phys.*, **16**, 770-780.

Meister, E., 1969: Theorie instationärer Unterschallströmungen durch ein schwingendes Gitter im Windkanal. *Zeit. angew. Math. Mech.*, **49**, 481–494.

Merkin, J. H., and D. J. Needham, 1987: The natural convection flow above a heated wall in a saturated porous medium. *Quart. J. Mech. Appl. Math.*, **40**, 559–574.

Miles, J. W., 1951: The oscillating rectangular airfoil at supersonic speeds. *Quart. Appl. Math.*, **9**, 47–65.

Millar, R. F., 1964: Plane wave spectra in grating theory. III. Scattering by a semiinfinite grating of identical cylinders. *Can. J. Phys.*, **42**, 1149–1184.

Millar, R. F., 1966: Plane wave spectra in grating theory. V. Scattering by a semi-infinite grating of isotropic scatterers. *Can. J. Phys.*, **44**, 2839–2874.

Millar, R. F., 1967: Propagation of electromagnetic waves near a coast-line on a flat earth. *Radio Sci.*, **2**, 261–286.

Misiakos, K., C. H. Wang, A. Neugroschel and F. A. Lindholm, 1990: Simultaneous extraction of minority-carrier transport parameters in crystalline semiconductors by lateral photocurrent. *J. Appl. Phys.*, **67**, 321–333.

Mitsioulis, G., 1990: A Wiener-Hopf theory for a semi-infinite dielectric slab. *Can. J. Phys.*, **68**, 1348–1351.

Mitsioulis, G., 1991: Renormalization of the energies stored around a Wiener-Hopf structure: I. *Can. J. Phys.*, **69**, 875–890.

Mittra, R., and S. W. Lee, 1970: On the solution of a generalized Wiener-Hopf equation. *J. Math. Phys.*, **11**, 775–783.

Mohan, N., and P. S. Deshwal, 1989: Scattering of a compressional wave at the corner of a quarter space. *Indian J. Pure Appl. Math.*, **20**, 386–394.

Morioka, S., 1980: Steady two-dimensional jet-flow of a dispersive compressible fluid. *J. Phys. Soc. Japan*, **48**, 1009–1017.

Morsy, S., and I. Kamha, 1976: Flux distortion in three-dimensional reactors due to plane absorbers. *Atomkernenergie*, **28**, 265–270.

Mukhopadhyay, D., 1977: Stresses in semi-infinite strip under symmetric load on the lateral sides. *Indian J. Pure Appl. Math.*, **8**, 663–671.

Munt, R. M., 1977: The interaction of sound with a subsonic jet issuing from a semi-infinite cylindrical pipe. *J. Fluid Mech.*, **83**, 609–640.

Munt, R. M., 1990: Acoustic transmission properties of a jet pipe with subsonic jet flow: I. The cold jet reflection coefficient. *J. Sound Vib.*, **142**, 413–436.

Murphy, D. G., and A. J. Willmott, 1991: Rossby wave scattering by a meridional line barrier in an infinitely long zonal channel. *J. Phys. Oceanogr.*, **21**, 621–634.

Nasalski, W., 1983: Leaky and surface wave diffraction by an asymmetric impedance half plane. *Can. J. Phys.*, **61**, 906–918.

Nicoll, M. A., and J. T. Weaver, 1977: H-polarization induction over an ocean edge coupled to the mantle by a conducting crust. *Geophys. J. R. Astr. Soc.*, **49**, 427–441.

Noble, B., 1958: *Methods Based on the Wiener-Hopf Technique for the Solution of Partial Differential Equations*. London: Pergamon Press, 246 pp.

Norris, A. N., and J. D. Achenbach, 1984: Elastic wave diffraction by a semi-infinite crack in a transversely isotropic material. *Quart. J. Mech. Appl. Math.*, **37**, 565–580.

Nuismer, R. J., and J. D. Achenbach, 1972: Dynamically induced fracture. *J. Mech. Phys. Solids*, **20**, 203–222.

Olek, S., 1989: Rewetting of a solid cylinder with precursory cooling. *Appl. Sci. Res.*, **46**, 347–364.

Olmstead, W. E., 1966: An exact solution for Oseen flow past a half plane and a horizontal force singularity. *J. Math. and Phys.*, **45**, 156–161.

Orszag, S. A., and S. C. Crow, 1970: Instability of a vortex sheet leaving a semi-infinite plate. *Stud. Appl. Math.*, **49**, 167–181.

Packham, B. A., 1969: Reflexion of Kelvin waves at the open end of a rotating semi-infinite channel. *J. Fluid Mech.*, **39**, 321–328.

Pal, A., and L. W. Pearson, 1986: A new approach to the electromagnetic diffraction problem of a perfectly conducting half-plane screen. *IEEE Trans. Antennas Propagat.*, **AP-34**, 1281–1287.

Pal, S. C., and M. L. Ghosh, 1990: High frequency scattering of antiplane shear waves by an interface crack. *Indian J. Pure Appl. Math.*, **21**, 1107–1124.

Papadopoulos, V. M., 1956: Scattering by a semi-infinite resistive strip of dominant-mode propagation in an infinite rectangular wave-guide. *Proc. Camb. Phil. Soc.*, **52**, 553–563.

Papadopoulos, V. M., 1957: The scattering effect of a junction between two circular waveguides. *Quart. J. Mech. Appl. Math.*, **10**, 191–209.

Pathak, P. H., and R. G. Kouyoumjian, 1979: Surface wave diffraction by a truncated dielectric slab recessed in a perfectly conducting surface. *Radio Sci.*, **14**, 405–417.

Pearson, J. D., 1953: The diffraction of electro-magnetic waves by a semi-infinite circular wave guide. *Proc. Camb. Phil. Soc.*, **49**, 659–667.

Pierucci, M., 1978: Acoustic radiation due to a fluid loading discontinuity on an infinite membrane. *J. Acoust. Soc. Am.*, **64**, 223–231.

Pinsent, H. G., 1971: The effect of a depth discontinuity on Kelvin wave diffraction. *J. Fluid Mech.*, **45**, 747–758.

Plis, A. I., and V. I. Plis, 1980: Diffraction of Kelvin waves at the open end of a plane-parallel channel. *J. Appl. Math. Mech.*, **44**, 45–50.

Plis, V. I., 1981: Propagation of Kelvin waves from a channel into a semibounded tank. *J. Appl. Math. Mech.*, **45**, 785–790.

Plis, V. I., 1986: Diffraction of Kelvin waves in a rotating semibounded basin containing a semi-infinite wall. *J. Appl. Math. Mech.*, **50**, 286–290.

Poddar, M., 1982: Very low-frequency electromagnetic response of a perfectly conducting half-plane in a layered half-space. *Geophysics*, **47**, 1059–1067.

Popelar, C. H., and C. Atkinson, 1980: Dynamic crack propagation in a viscoelastic strip. *J. Mech. Phys. Solids*, **28**, 79–93.

Pridmore-Brown, D. C., 1968: A Wiener-Hopf solution of a radiation problem in conical geometry. *J. Phys. and Math.*, **47**, 79–94.

Przeździecki, S., and R. A. Hurd, 1977: Diffraction by a half-plane perpendicular to the distinguished axis of a general gyrotropic medium. *Can. J. Phys.*, **55**, 305–324.

Przeździecki, S., and R. A. Hurd, 1981: Diffraction by a half plane perpendicular to the distinguished axis of a gyrotropic medium (oblique incidence). *Can. J. Phys.*, **59**, 403–424.

Radlow, J., and W. B. Ericson, 1962: Transverse magnetohydrodynamic flow past a semi-infinite plate. *Phys. Fluids*, **5**, 1428–1434.

Ramakrishnan, R., 1982: A note on the calculation of Wiener-Hopf split functions. *J. Sound Vib.*, **81**, 592–595.

Ramirez, J.-C., 1987: The three-dimensional stress intensity factor due to the motion of a load on the faces of a crack. *Quart. Appl. Math.*, **45**, 361–375.

Ranta, M. A., 1964: An application of integral transformations and the Wiener-Hopf techniques to the supersonic flow past an oscillating, nearly circular, slender body and past an elastic, two-dimensional, thin wing. *Acta Polytech. Scandinavica, Mechanical Engineering Ser. No. 18*, 69 pp.

Ranta, M. A., 1969: On the application of perturbation techniques to the study of missile flight with special references to internal mass flow, thrust malalignment, and the effect of gravity. *Acta Polytech. Scandinavica, Mechanical Engineering Ser. No. 43*, 41 pp.

Rao, T. C. K., and M. A. K. Hamid, 1980: On the coaxial excitation of the modified Goubau line. *Radio Sci.*, **15**, 17–24.

Raval, U., J. T. Weaver and T. W. Dawson, 1981: The ocean-coast effect re-examined. *Geophys. J. R. Astr. Soc.*, **67**, 115–123.

Rawlins, A. D., 1975: Acoustic diffraction by an absorbing semi-infinite half plane in a moving fluid. *Proc. R. Soc. Edin.*, **A72**, 337–357.

Rawlins, A. D., 1975: The solution of a mixed boundary value problem in the theory of diffraction by a semi-infinite plane. *Proc. R. Soc. Lond.*, **A346**, 469–484.

Rawlins, A. D., 1976: Diffraction of sound by a rigid screen with a soft or perfectly adsorbing edge. *J. Sound Vib.*, **45**, 53–67.

Rawlins, A. D., 1976: Diffraction of sound by a rigid screen with absorbent edge. *J. Sound Vib.*, **47**, 523–541.

Rawlins, A. D., 1977: The engine-over-the-wing noise problem. *J. Sound Vib.*, **50**, 553–569.

Rawlins, A. D., 1977: Diffraction by an acoustically penetrable or an electromagnetically dielectric half plane. *Int. J. Engng. Sci.*, **15**, 569–578.

Rawlins, A. D., 1978: Radiation of sound from an unflanged rigid cylindrical duct with an acoustically absorbing internal surface. *Proc. R.*

Soc. Lond., **A361**, 65–91.

Rawlins, A. D., 1980: Simultaneous Wiener-Hopf equations. *Can. J. Phys.*, **58**, 420–428.

Rawlins, A. D., 1984: The solution of a mixed boundary value problem in the theory of diffraction. *J. Engng. Math.*, **18**, 37–62.

Rawlins, A. D., E. Meister and F.-O. Speck, 1991: Diffraction by an acoustically transmissive or an electromagnetic dielectric half-plane. *Math. Meth. App. Sci.*, **14**, 387–402.

Reuter, G. E. H., and E. H. Sondheimer, 1948: The theory of the anomalous skin effect in metals. *Proc. R. Soc. Lond.*, **A195**, 336–364.

Rice, J. R., and D. A. Simons, 1976: The stabilization of spreading shear faults by coupled deformation-diffusion effects in fluid-infiltrated porous materials. *J. Geophys. Res.*, **81**, 5322–5334.

Richardson, S., 1970: A 'stick-slip' problem related to the motion of a free jet at low Reynolds numbers. *Proc. Camb. Phil. Soc.*, **67**, 477–489.

Rienstra, S. W., 1983: A small Strouhal number analysis for acoustic wave-jet flow-pipe interactions. *J. Sound Vib.*, **86**, 539–556.

Rienstra, S. W., 1984: Acoustic radiation from a semi-infinite annular duct in a uniform subsonic mean flow. *J. Sound Vib.*, **94**, 267–288.

Rojas, R. G., H. C. Ly and P. H. Pathak, 1991: Electromagnetic plane wave diffraction by a planar junction of two thin dielectric/ferrite half planes. *Radio Sci.*, **26**, 641–660.

Rokhlin, S., 1980: Diffraction of Lamb waves by a finite crack in an elastic layer. *J. Acoust. Soc. Am.*, **67**, 1157–1165.

Rose, L. R. F., 1976: An approximate (Wiener-Hopf) kernel for dynamic crack problems in linear elasticity and viscoelasticity. *Proc. R. Soc. Lond.*, **A349**, 497–521.

Rosenbaum, S., 1967: Edge diffraction in an arbitrary anisotropic medium. I. *Can. J. Phys.*, **45**, 3479–3502.

Roy, A., and N. Visalakshi, 1982: Diffraction of shear waves by an edge crack in an elastic wedge. *Int. J. Engng. Sci.*, **20**, 553–563.

Rulf, B., 1967: Diffraction by a halfplane in a uniaxial medium. *SIAM J. Appl. Math.*, **15**, 120–127.

Rulf, B., and R. A. Hurd, 1978: Radiation from an open waveguide with reactive walls. *IEEE Trans. Antennas Propagat.*, **AP-26**, 668–673.

Ryvkin, M., and L. Banks-Sills, 1993: Steady-state mode III propagation of an interface crack in an inhomogeneous viscoelastic strip. *Int. J. Solids Structures*, **30**, 483–498.

Sander, L. M., 1985: Exact solution for the peripheral photoresponse of a p-n junction. *J. Appl. Phys.*, **57**, 2057–2059.

Savkar, S. D., 1975: Radiation of cylindrical duct acoustic modes with flow mismatch. *J. Sound Vib.*, **42**, 363–386.

Schmidt, G. H., 1981: Linearized stern flow of a two-dimensional shallow-draft ship. *J. Ship Res.*, **25**, 236–242.

Schorr, B., 1967: Das Anfangs-Rantwertproblem eines dünnen Profils in kompressibler Unterschallströmung. *Zeit. angew. Math. Phys.*, **18**, 149–164.

Seebass, R., K. Tamada and T. Miyagi, 1966: Oseen flow past a finite flat plate. *Phys. Fluids*, **9**, 1697–1703.

Senior, T. B. A., 1952: Diffraction by a semi-infinite metallic sheet. *Proc. R. Soc. Lond.*, **A213**, 436–458.

Senior, T. B. A., 1960: Diffraction by an imperfectly conducting half-plane at oblique incidence. *Appl. Sci. Res.*, **B8**, 35–61.

Serbest, A. H., G. Uzgören and A. Büyükaksoy, 1991: Diffraction of plane waves by a resistive strip residing between two impedance half-planes. *Ann. Télécommunic.*, **46**, 359–366.

Seshadri, S. R., and T. T. Wu, 1960: High-frequency diffraction of plane waves by an infinite slit for grazing incidence. *IEEE Trans. Antennas Propagat.*, **AP-8**, 37–42.

Seshadri, S. R., 1961: Diffraction of a plane wave by an infinite slit in a unidirectionally conducting screen. *IEEE Trans. Antennas Propagat.*, **AP-9**, 199–207.

Seshadri, S. R., and T. T. Wu, 1963: Diffraction by a circular aperture in a unidirectionally conducting screen. *IEEE Trans. Antennas Propagat.*, **AP-11**, 56–67.

Sharma, U., K. Viswanathan, K. Singh and M. L. Gogna, 1991: Diffraction of obliquely incident surface waves by a soft vertical plane barrier of finite depth. *Indian J. Pure Appl. Math.*, **22**, 337–353.

Shaw, D. C., 1985: The asymptotic behavior of a curved line source plume within an enclosure. *IMA J. Appl. Math.*, **35**, 71–89.

Shaw, D. C., 1985: Some aspects of the French flexible bag wave-energy device. *Quart. Appl. Math.*, **43**, 337–358.

Shen, H.-M., and T. T. Wu, 1989: The universal current distribution near the end of a tubular antenna. *J. Math. Phys.*, **30**, 2721–2729.

Shindo, Y., 1983: Dynamic singular stresses for a Griffith crack in a soft ferromagnetic elastic solid subjected to a uniform magnetic field. *J. Appl. Mech.*, **50**, 50–56.

Shindo, Y., 1984: Diffraction of waves and singular stresses in a soft ferromagnetic elastic solid with two coplanar Griffith cracks. *J. Acoust. Soc. Am.*, **75**, 50–57.

Shirai, H., and L. B. Felsen, 1986: Spectral method for multiple edge diffraction by a flat strip. *Wave Motion*, **8**, 499–524.

Sills, L. B., and Y. Benveniste, 1981: Steady state propagation of a mode III interface crack between dissimilar viscoelastic media. *Int. J. Engng. Sci.*, **19**, 1255–1268.

Simons, D. A., 1975: Scattering of a Love wave by the edge of a thin surface layer. *J. Appl. Mech.*, **42**, 842–846.

Simons, D. A., 1976: Scattering of a Rayleigh wave by the edge of a thin surface layer of negligible inertia. *J. Acoust. Soc. Am.*, **59**, 12–18.

Sinha, T. K., 1980: Propagation of Love waves in elastic layer of variable thickness resting on a rigid half space. *J. Tech. Phys.*, **21**, 275–289.

Skelton, E. A., 1991: Acoustic scattering by a rigid barrier between two fluid-loaded parallel elastic plates. *Proc. R. Soc. Lond.*, **A435**, 217–232.

Sparenberg, J. A., 1958: On a shrink-fit problem. *Appl. Sci. Res.*, **A7**, 109–120.

Sparenberg, J. A., 1974: On the linear theory of an actuator disk in a viscous fluid. *J. Ship Res.*, **18**, 16–21.

Springer, S. G., and T. J. Pedley, 1973: The solution of heat transfer problems by the Wiener-Hopf technique. *Proc. R. Soc. Lond.*, **A333**, 347–362.

Stakhanov, I. P., 1968: Nonequilibrium ionization in a low-voltage discharge. II. *Sov. Phys. Tech. Phys.*, **12**, 935–942.

Stakhanov, I. P., 1968: Nonequilibrium in a low-voltage arc for large inelastic scattering cross sections. *Sov. Phys. Tech. Phys.*, **12**, 1522–1529.

Surampudi, S. P., and J. J. Adamczyk, 1986: Unsteady transonic flow over cascade blades. *AIAA J.*, **24**, 293–302.

Takahashi, K., 1967: The electromagnetic field of a coaxial antenna. *Elect. Engng. Japan*, **87(12)**, 35–44.

Takahashi, K., and S. Kaji, 1991: Analytical study on plate edge noise (Trailing edge noise caused by vorticity waves). *JSME Int. J., Ser. II*, **34**, 431–438.

Thau, S. A., and Y.-H. Lu, 1970: Diffraction of transient horizontal shear waves by a finite crack and a finite rigid ribbon. *Int. J. Engng. Sci.*, **8**, 857–874.

Thomas, R. M., 1988: Methods for calculating the conduction-controlled rewetting of a cladded rod. *Nucl. Engng. Des.*, **110**, 1–16.

Tien, C. L., and L. S. Yao, 1975: Analysis of conduction-controlled rewetting of a vertical surface. *J. Heat Transfer*, **97**, 161–165.

Trogden, S. A., and D. D. Joseph, 1980: The stick-slip problem for a round jet. I. Large surface tension. *Rheol. Acta*, **19**, 404–420.

Uchida, K., and K. Aoki, 1984: Scattering of surface waves on transverse discontinuities in symmetrical three-layer dielectric waveguides. *IEEE Trans. Microwave Theory Tech.*, **MTT-32**, 11–19.

Uzgören, G., A. Büyükaksoy and A. H. Serbest, 1989: Diffraction coefficient related to a discontinuity formed by impedance and resistive half-planes. *IEE Proc., Part H*, **136**, 19–23.

Uzgören, G., A. Büyükaksoy and A. H. Serbert, 1990: Plane wave diffraction by the discontinuity formed by resistive and impedance half-planes: Oblique incident case. *Ann. Télécommunic.*, **45**, 410–418.

Vaisleib, Yu. V., 1971: Sound scattering by a finite cone. *Sov. Phys. Acoust.*, **17**, 26–33.

Vaisleib, Yu. V., 1971: Perfectly conducting finite cone in the field of a point charge and related electrostatic problems. *Sov. Phys. Tech. Phys.*, **15**, 1395–1404.

Vajnshtejn, L. A., 1948: Rigorous solution of the problem of an open-ended parallel-plate waveguide (in Russian). *Ivz. Akad. Nauk. USSR, Ser. Fiz.*, **12**, 144–165.

Vajnshtejn, L. A., 1948: On the theory of diffraction by two parallel half-planes (in Russian). *Ivz. Akad. Nauk. USSR, Ser. Fiz.*, **12**, 166–180.

Vajnshtejn, L. A., 1948: Theory of symmetric waves in a cylindrical waveguide with an open end (in Russian). *Zh. Tekh. Fiz.*, **18**, 1543–1564.

Vajnshtejn, L. A., 1949: The theory of sound waves in open tubes (in Russian). *Zh. Tekh. Fiz.*, **19**, 911–930.

Vanblaricum, G. F., and R. Mittra, 1969: A modified residue-calculus technique for solving a class of boundary value problems. Part I: Waveguide discontinuities. *IEEE Trans. Microwave Theory Tech.*, **MTT-17**, 302–309.

Van der Pauw, L. J., 1973: Diffraction of a Bleustein-Gulyaev wave by a conductive semi-infinite surface layer. *J. Acoust. Soc. Am.*, **53**, 1107–1115.

Venkataratnam, K., and T. P. Rao, 1982: Analysis of end effects in a linear induction motor by the Wiener-Hopf technique. *IEE Proc., Part B*, **129**, 364–372.

Vijayaraghavan, S., and R. K. Arora, 1971: Scattering of a shielded surface wave in a coaxial waveguide by a wall impedance discontinuity. *IEEE Trans. Microwave Theory Tech.*, **MTT-19**, 736–739.

Vitiuk, V. F., 1970: Diffraction of surface waves at a dock of finite width. *J. Appl. Math. Mech.*, **34**, 27–35.

Volakis, J. L., 1987: Scattering by a thick impedance half plane. *Radio Sci.*, **22**, 13–25.

Volakis, J. L., and M. A. Ricoy, 1987: Diffraction by a thick perfectly conducting half-plane. *IEEE Trans. Antennas Propagat.*, **AP-35**, 62–72.

Volakis, J. L., and M. A. Ricoy, 1989: H-polarization diffraction by a thick metal-dielectric join. *IEEE Trans. Antennas Propagat.*, **AP-37**, 1453–1462.

Volakis, J. L., and J. D. Collins, 1990: Electromagnetic scattering from a resistive half plane on a dielectric interface. *Wave Motion*, **12**, 81–96.

Von Roos, O., 1978: Analysis of the interaction of an electron beam with a solar cell. II. *Solid-State Electronics*, **21**, 1101–1108.

Vorotnytsev, M. A., 1981: Distribution of the potential in the electric double layer at the contact between two different semiinfinite planar electrodes. *Sov. Electrochem.*, **17**, 472–479.

Waechter, R. T., 1968: Steady electrically driven flows. *Proc. Camb. Phil. Soc.*, **64**, 871–894.

Wait, J. R., 1970: On launching an azimuthal surface wave on a cylindrical impedance boundary. *Acta Phys. Austr*, **32**, 122–130.

Wait, J. R., 1970: Propagation of electromagnetic waves over a smooth multisection curved earth – an exact theory. *J. Math. Phys.*, **11**, 2851–2860.

Wait, J. R., 1970: Factorization method applied to electromagnetic wave propagation in a curved waveguide with nonuniform walls. *Radio Sci.*, **5**, 1059–1068.

Weidelt, P., 1971: The electromagnetic induction in two thin half-sheets. *Zeit. Geophys.*, **37**, 649–665.

Weidelt, P., 1983: The harmonic and transient electromagnetic response of a thin dipping dike. *Geophysics*, **48**, 934–952.

Weitz, M., and J. B. Keller, 1950: Reflection of water waves from floating ice in water of finite depth. *Commun. Pure Appl. Math.*, **3**, 305–318.

Wenger, N. C., 1965: The launching of surface waves on an axial-cylindrical reactive surface. *IEEE Trans. Antennas Propagat.*, **AP-13**, 126–134.

Wenger, N. C., 1967: Resonant frequency of open-ended cylindrical cavity. *IEEE Trans. Microwave Theory Tech.*, **MTT-15**, 334–340.

Wesenberg, D. L., and L. M. Murphy, 1974: Steady-state response of an infinite plate to an exponential edge load along a fusing crack – A model for the dynamic edge fusion of two semi-infinite plates. *Int. J. Mech. Sci.*, **16**, 91–103.

Wichman, I. S., 1983: Flame spread in an opposed flow with a linear velocity gradient. *Combust. Flame*, **50**, 287–304.

Wickham, G. R., 1980: Short-wave radiation from a rigid strip in smooth contact with a semi-infinite elastic solid. *Quart. J. Mech. Appl. Math.*, **33**, 409–433.

Wijngaarden, L. van, 1966: Asymptotic solution of a diffusion problem with mixed boundary conditions. *Koninkl. Ned. Akad. Weterschap.*, **B69**, 263–276.

Williams, M. M. R., 1965: Neutron flux perturbations due to infinite plane absorbers. II: Exponential flux. *Brit. J. Appl. Phys.*, **16**, 1841–1852.

Williams, M. M. R., 1966: An exact solution of the two group Milne problem by the method of Wiener and Hopf. *J. Math. and Phys.*, **45**, 64–76.

Williams, M. M. R., 1969: The temperature distribution in a radiating fluid flowing over a flat plate. *Quart. J. Mech. Appl. Math.*, **22**, 487–500.

Williams, M. M. R., 1976: The energy spectrum of sputtered atoms. *Phil. Mag., Ser. 8*, **34**, 669–683.

Williams, M. M. R., 1979: The spatial dependence of the energy spectrum of slowing down particles – I. Applications to reactor physics and atomic sputtering. *Ann. Nucl. Energy*, **6**, 145–173.

Williams, W. E., 1954: Diffraction by two parallel planes of finite length. *Proc. Camb. Phil. Soc.*, **50**, 309–318.

Williams, W. E., 1956: Diffraction by a cylinder of finite length. *Proc. Camb. Phil. Soc.*, **52**, 322–335.

Williams, W. E., 1957: Step discontinuities in waveguides. *IEEE Antennas Propagat.*, **AP-5**, 191–198.

Williams, W. E., and A. Chakrabarti, 1982: Reflection at a discontinuity in a transmission line. *IMA J. Appl. Math.*, **28**, 185–195.

Willis, J. R., 1967: Crack propagation in viscoelastic media. *J. Mech. Phys. Solids*, **15**, 229–240.

Wolfersdorf, L. von, 1964: Eine magnetohydrodynamische Kanalströmung bei inhomogenem elecktrischem Verhalten der Kanalwände. *Rev. Roum. Sci. Tech. - Méc. Appl.*, **9**, 963–976.

Wu, C. P., 1967: Diffraction of a plane electromagnetic wave by an infinite set of parallel metallic plates in an anisotropic plasma. *Can. J. Phys.*, **45**, 1911–1923.

Wu, T. T., 1961: Theory of the dipole antenna and the two-wire transmission line. *J. Math. Phys.*, **2**, 550–574.

Yamada, T., and R. Sato, 1976: SH wave propagation in a medium having step-shaped discontinuity. *J. Phys. Earth*, **24**, 105–130.

Yoneyama, T., and S. Nishida, 1976: Reflection and transmission of Rayleigh waves by the edge of a deposited thin film. *J. Acoust. Soc. Am.*, **55**, 738–743.

Yoneyama, T., and S. Nishida, 1974: Diffraction of elastic surface wave at the edge of deposited thin films. *Electron. Communic. Japan*, **A57(6)**, 35–44.

Yoneyama, T., and S. Nishida, 1974: Transmission and reflection of Rayleigh waves by a high impedance obstacle of finite length. *J. Acoust. Soc. Am.*, **60**, 90–94.

Yoshidomi, K., and K. Aoki, 1988: Scattering of an E-polarized plane wave by two parallel rectangular impedance cylinders. *Radio Sci.*, **23**, 471–480.

Zaitseva, I. A., and A. A. Zolotarev, 1990: Propagation and attenuation of waves generated in a layer by a plane source. *J. Appl. Mech. Tech. Phys.*, **31**, 696–701.

Zaman, F. D., S. Asghar and M. Ahmad, 1987: Diffraction of SH-waves in a layered plate. *J. Tech. Phys.*, **28**, 143–152.

Zhong, Z., W. Yang and S.-W. Yu, 1990: Further result on symmetric bending of cracked Reissner plate. *Theoretical Appl. Fract. Mech.*, **12**, 231–240.

Zyryanov, V. N., 1974: On wind-driven currents in a strait. *Acad. Sci. USSR, Oceanology*, **14**, 386–391.

Worked Solutions
To Some of the Problems

Section 1.1

1.

$$\mathcal{L}(e^{at}) = \int_0^\infty e^{-(s-a)t}dt = -\frac{e^{-(s-a)t}}{s-a}\bigg|_0^\infty = \frac{1}{s-a}$$

$$\mathcal{L}(t^n) = \int_0^\infty t^n e^{-st}dt = n!e^{-st}\sum_{m=0}^n \frac{t^{n-m}}{(n-m)!s^{m+1}}\bigg|_0^\infty = \frac{n!}{s^{n+1}}$$

$$\mathcal{L}[\sin(at)] = \int_0^\infty \sin(at)e^{-st}dt$$

$$= \frac{e^{-st}}{s^2+a^2}\left[-s\sin(at) - a\cos(at)\right]\bigg|_0^\infty = \frac{a}{s^2+a^2}$$

$$\mathcal{L}[\cos(at)] = \int_0^\infty \cos(at)e^{-st}dt$$

$$= \frac{e^{-st}}{s^2+a^2}\left[-s\cos(at) + a\sin(at)\right]\bigg|_0^\infty = \frac{s}{s^2+a^2}$$

2.

$$t^2 * \sin(at) = \int_0^t (t - x)^2 \sin(ax)\, dx$$

$$= -\frac{1}{a}(t - x)^2 \cos(ax)\Big|_0^t - \frac{2}{a}\int_0^t (t - x)\cos(ax)\, dx$$

$$= \frac{t^2}{a} - \frac{2}{a^2}(t - x)\sin(ax)\Big|_0^t - \frac{2}{a^2}\int_0^t \sin(ax)\, dx$$

$$= \frac{t^2}{a} + \frac{2}{a^3}\cos(ax)\Big|_0^t = \frac{t^2}{a} - \frac{2}{a^3}[1 - \cos(at)]$$

$$= \frac{t^2}{a} - \frac{4}{a^3}\sin^2(at/2)$$

$$\mathcal{L}\left[\frac{t^2}{a} - \frac{4}{a^3}\sin^2(at/2)\right] = \frac{2}{s^3}\left(\frac{a}{s^2 + a^2}\right) = \mathcal{L}(t^2)\mathcal{L}[\sin(at)]$$

3. Let $b \geq a$. The case $b < a$ follows by analog. Then

$$H(t - b) * H(t - a) = \int_0^t H(t - b - x)H(x - a)\, dx.$$

Therefore, if $t < a$, the convolution equals zero. If $t > a$

$$H(t - b) * H(t - a) = \int_a^t H(t - b - x)\, dx = -\int_{t-b-a}^{-b} H(\eta)\, d\eta$$

if $\eta = t - b - x$. If $t < b + a$, the convolution equals zero because the Heaviside always equals zero. On the other hand, if $t > b + a$,

$$H(t - b) * H(t - a) = -\int_{t-b-a}^0 d\eta = t - a - b$$

or

$$H(t - b) * H(t - a) = (t - a - b)H(t - a - b).$$

Then

$$\mathcal{L}[H(t - b) * H(t - a)] = \frac{e^{-as-bs}}{s^2} = \left(\frac{e^{-as}}{s}\right)\left(\frac{e^{-bs}}{s}\right)$$

$$= \mathcal{L}[H(t - b)]\mathcal{L}[H(t - a)].$$

4.

$$t * [H(t) - H(t - 2)] = \int_0^t (t - x)[H(x) - H(x - 2)]\, dx.$$

Worked Solutions to Some of the Problems

If $0 < t < 2$,

$$t * [H(t) - H(t-2)] = \int_0^t (t-x)\,dx = t^2/2.$$

If $t > 2$,

$$t * [H(t) - H(t-2)] = \int_0^2 (t-x)\,dx = 2t - 2.$$

Using Heaviside's step function,

$$t * [H(t) - H(t-2)] = \frac{t^2}{2} - \frac{(t-2)^2}{2} H(t-2).$$

From the second shifting theorem,

$$\mathcal{L}\{t * [H(t) - H(t-2)]\} = \frac{1}{s^3} - \frac{1}{s^3}e^{-2s} = \frac{1}{s^2}\left(\frac{1}{s} - \frac{e^{-2s}}{s}\right)$$

$$= \mathcal{L}(t)\mathcal{L}[H(t) - H(t-2)].$$

Section 1.2

1.

$$F(\omega) = \int_{-\infty}^0 e^{(a-\omega i)t}\,dt + \int_0^\infty e^{-(a+\omega i)t}\,dt$$

$$= \frac{1}{a - \omega i} + \frac{1}{a + \omega i} = \frac{2a}{\omega^2 + a^2}.$$

2.

$$F(\omega) = \int_{-\infty}^0 te^{(a-\omega i)t}\,dt + \int_0^\infty te^{-(a+\omega i)t}\,dt$$

$$= -\frac{1}{(a - \omega i)^2} + \frac{1}{(a + \omega i)^2} = -\frac{4a\omega i}{(\omega^2 + a^2)^2}.$$

3.

$$F(\omega) = \int_0^\infty e^{-(1+i+i\omega)t}\,dt - \int_{-\infty}^0 e^{(1-i-i\omega)t}\,dt$$

$$= \frac{1}{1 + (\omega + 1)i} - \frac{1}{1 - (\omega + 1)i} = -\frac{2i(\omega + 1)}{(\omega + 1)^2 + 1}.$$

4.

$$F(\omega) = \frac{1}{2i} \int_0^1 \left[e^{i(1-\omega)t} - e^{-i(1+\omega)t} \right] dt$$

$$= -\frac{1}{2} \left[\frac{1 - \cos(\omega - 1) + i\sin(\omega - 1)}{\omega - 1} \right.$$

$$\left. + \frac{\cos(\omega + 1) - i\sin(\omega + 1) - 1}{\omega + 1} \right]$$

$$= -\frac{1}{2} \left[\frac{1 - \cos(\omega - 1)}{\omega - 1} + \frac{\cos(\omega + 1) - 1}{\omega + 1} \right]$$

$$- \frac{i}{2} \left[\frac{\sin(\omega - 1)}{\omega - 1} - \frac{\sin(\omega + 1)}{\omega + 1} \right]$$

5.

$$e^t H(-t) * [H(t) - H(t - 2)] = \int_{-\infty}^{\infty} e^{t-x} H(x - t)[H(x) - H(x - 2)]\, dx$$

$$= \int_0^2 e^{t-x} H(x - t)\, dx.$$

If $t > 2$, the integrand is always zero and the convolution equals zero. If $t < 0$,

$$e^t H(-t) * [H(t) - H(t - 2)] = \int_0^2 e^{t-x} dx = e^t - e^{t-2}.$$

Finally, if $0 < t < 2$,

$$e^t H(-t) * [H(t) - H(t - 2)] = \int_t^2 e^{t-x} dx = 1 - e^{t-2}.$$

6.

$$[H(t) - H(t - 2)] * [H(t) - H(t - 2)]$$

$$= \int_{-\infty}^{\infty} [H(t - x) - H(t - x - 2)]$$

$$\times [H(x) - H(x - 2)]\, dx$$

$$= \int_0^2 [H(t - x) - H(t - x - 2)]\, dx.$$

If $t < 0$ or $t > 4$, the integrand is always zero and the convolution equals zero. For $0 < t < 2$, $H(t - x - 2)$ equals zero for $0 < x < 2$. Therefore,

$$[H(t) - H(t - 2)] * [H(t) - H(t - 2)] = \int_0^t dx = t.$$

For $2 < t < 4$, $H(t - x)$ equals one where $0 < x < 2$ while $H(t - x - 2)$ equals one for $0 < x < t - 2$. Therefore,

$$[H(t) - H(t - 2)] * [H(t) - H(t - 2)] = \int_0^2 dx - \int_0^{t-2} dx = 4 - t.$$

Section 1.3

1. $y(x) = c_1 \cos(2x) + c_2 \sin(2x) + 2x^3 - 5x^2 + x - 2$

2. $y(x) = c_1 + c_2 e^{-4x} - \frac{1}{10} \cos(2x) - \frac{1}{20} \sin(2x) + \frac{3}{25} \cos(3x) - \frac{4}{25} \sin(3x)$

3. $y(x) = c_1 e^{2x} + c_2 e^{4x} + e^x$

4. $y(x) = c_1 e^{-2x} + c_2 x e^{-2x} + x e^{-x} - 2e^{-x}$

5. $y(x) = c_1 \cos(x) + c_2 \sin(x) - \frac{2}{5} e^x \cos(x) + \frac{1}{5} e^x \sin(x)$

6. $y(x) = c_1 \cos(x) + c_2 \sin(x) + \frac{1}{2} x \sin(x) - \sin(2x)$

7. A quick check shows that the general solution is

$$y(x) = A \sinh[(1 - x)\sqrt{a^2 + s}] + B \cosh[(1 - x)\sqrt{a^2 + s}].$$

We have written the solution in this form because $y(1) = B = 0$. Therefore,

$$y(-1) = A \sinh(2\sqrt{a^2 + s}) = \frac{1}{s}$$

and

$$y(x) = \frac{\sinh[(1 - x)\sqrt{a^2 + s}]}{s \sinh(2\sqrt{a^2 + s})}.$$

8. The general solution is

$$y(r) = A I_0(r\sqrt{s}) + B K_0(r\sqrt{s}) + \frac{1}{s}.$$

Because $K_0(z) \to \infty$ as $z \to 0$, $B = 0$. Using the condition at $r = a$,

$$A \sqrt{s} I_0'(a\sqrt{s}) = -\frac{1}{s}.$$

However, because $I_0'(z) = I_1(z)$,

$$y(r) = \frac{1}{s} - \frac{I_0(r\sqrt{s})}{s^{3/2} I_1(a\sqrt{s})}.$$

Section 1.4

1.

$$\int_{-\infty}^{\infty} \frac{\sin(x)}{x^2 + 4x + 5}\, dx = \text{Im}\left(\oint_C \frac{e^{iz}}{z^2 + 4z + 5}\, dz\right)$$

where C is a contour that contains the real axis plus an infinitely large semi-circle in the upper half-plane. There is only one pole inside the contour at $z = -2 + i$. Therefore, the residue is

$$\text{Res}(z = -2 + i) = \lim_{z \to -2+i} \frac{(z + 2 - i)e^{iz}}{z^2 + 4z + 5} = \lim_{z \to -2+i} \frac{e^{iz}}{2z + 4} = \frac{e^{-2i}}{2ei}.$$

From the residue theorem,

$$\oint_C \frac{e^{iz}}{z^2 + 4z + 5}\, dz = \frac{\pi}{e}[\cos(2) - i\sin(2)]$$

and

$$\int_{-\infty}^{\infty} \frac{\sin(x)}{x^2 + 4x + 5}\, dx = -\frac{\pi}{e}\sin(2).$$

2.

$$\int_0^{\infty} \frac{\cos(x)}{(x^2 + 1)^2}\, dx = \frac{1}{2}\int_{-\infty}^{\infty} \frac{\cos(x)}{(x^2 + 1)^2}\, dx = \frac{1}{2}\text{Re}\left[\oint_C \frac{e^{iz}}{(z^2 + 1)^2}\, dz\right]$$

where C is a contour that contains the real axis plus an infinitely large semi-circle in the upper half-plane. There is only one pole inside the contour at $z = i$ and it is second order. Therefore, the residue is

$$\text{Res}(z = i) = \lim_{z \to i} \frac{d}{dz}\left[\frac{(z - i)^2 e^{iz}}{(z - i)^2(z + i)^2}\right]$$

$$= \lim_{z \to i}\left[\frac{ie^{iz}}{(z + i)^2} - \frac{2e^{iz}}{(z + i)^3}\right] = \frac{i}{2e}.$$

From the residue theorem,

$$\oint_C \frac{e^{iz}}{(z^2 + 1)^2}\, dz = \frac{\pi}{e} \quad \text{and} \quad \int_0^{\infty} \frac{\cos(x)}{(x^2 + 1)^2}\, dx = \frac{\pi}{2e}.$$

3.

$$\int_{-\infty}^{\infty} \frac{x\sin(ax)}{x^2 + 4}\, dx = \text{Im}\left(\oint_C \frac{ze^{iaz}}{z^2 + 4}\, dz\right)$$

where C is a contour that contains the real axis plus an infinitely large semi-circle in the upper half-plane. There is only one pole inside the contour at $z = 2i$. Therefore, the residue is

$$\text{Res}(z = 2i) = \lim_{z \to 2i} \frac{(z - 2i)ze^{iaz}}{(z - 2i)(z + 2i)} = \lim_{z \to 2i} \frac{ze^{iaz}}{z + 2i} = \frac{e^{-2a}}{2}.$$

Therefore, from the residue theorem,

$$\oint_C \frac{ze^{iaz}}{z^2 + 4}\, dz = \pi i e^{-2a} \quad \text{and} \quad \int_{-\infty}^{\infty} \frac{x \sin(ax)}{x^2 + 4}\, dx = \pi e^{-2a}.$$

4.

$$\int_{-\infty}^{\infty} \frac{x \cos(\pi x)}{x^2 + 2x + 5}\, dx = \text{Re}\left(\oint_C \frac{ze^{i\pi z}}{z^2 + 2z + 5}\, dz \right)$$

where C is a contour that contains the real axis plus an infinitely large semi-circle in the upper half-plane. There is only one pole inside the contour at $z = -1 + 2i$. Therefore, the residue is

$$\text{Res}(z = -1 + 2i) = \lim_{z \to -1+2i} \frac{(z + 1 - 2i)ze^{i\pi z}}{z^2 + 2z + 5}$$

$$= \lim_{z \to -1+2i} \frac{ze^{i\pi z}}{2z + 2} = \frac{e^{-2\pi}}{4i}(1 - 2i).$$

Therefore, from the residue theorem,

$$\oint_C \frac{ze^{i\pi z}}{z^2 + 2z + 5}\, dz = \frac{\pi e^{-2\pi}}{2}(1 - 2i)$$

and

$$\int_{-\infty}^{\infty} \frac{x \cos(\pi x)}{x^2 + 2x + 5}\, dx = \frac{\pi e^{-2\pi}}{2}.$$

5.

$$\int_{-\infty}^{\infty} \frac{\cos(ax)}{(x^2 + h^2)^2}\, dx = \text{Re}\left[\oint_C \frac{e^{iaz}}{(z^2 + h^2)^2}\, dz \right]$$

where C is a contour that contains the real axis plus an infinitely large semi-circle in the upper half-plane. There is only one pole inside the contour at $z = hi$ and it is second order. Therefore, the residue is

$$\text{Res}(z = hi) = \lim_{z \to hi} \frac{d}{dz}\left[(z - hi)^2 \frac{e^{iaz}}{(z - hi)^2(z + hi)^2} \right]$$

$$= \lim_{z \to hi}\left[\frac{iae^{iaz}}{(z + hi)^2} - \frac{2e^{iaz}}{(z + hi)^3} \right] = -\frac{i(1 + ah)e^{-ah}}{4h^3}.$$

Therefore, from the residue theorem,

$$\oint_C \frac{e^{iaz}}{(z^2 + h^2)^2} \, dz = \frac{\pi(1 + ah)e^{-ah}}{2h^3}$$

and

$$\int_{-\infty}^{\infty} \frac{\cos(ax)}{(x^2 + h^2)^2} \, dx = \frac{\pi(1 + ah)e^{-ah}}{2h^3}.$$

6.

$$\int_{-\infty}^{\infty} \frac{\cos(ax)}{(x - a)^2 + h^2} \, dx = \mathrm{Re}\left[\oint_C \frac{e^{iaz}}{(z - a)^2 + h^2} \, dz\right]$$

where C is a contour that contains the real axis plus an infinitely large semi-circle in the upper half-plane. There is only one simple pole inside the contour at $z = a + hi$. Therefore, the residue is

$$\mathrm{Res}(z = a + hi) = \lim_{z \to a + hi}\left[(z - a - hi)\frac{e^{iaz}}{(z - a)^2 + h^2}\right] = \frac{e^{ia^2 - ah}}{2hi}.$$

Therefore, from the residue theorem,

$$\oint_C \frac{e^{iaz}}{(z - a)^2 + h^2} \, dz = \frac{\pi e^{-ah}[\cos(a^2) + i\sin(a^2)]}{h}$$

and

$$\int_{-\infty}^{\infty} \frac{\cos(ax)}{(x - a)^2 + h^2} \, dx = \frac{\pi e^{-ah}\cos(a^2)}{h}.$$

7.

$$\int_{-\infty}^{\infty} \frac{\sin(ax)}{(x - a)^2 + h^2} \, dx = \mathrm{Im}\left[\oint_C \frac{e^{iaz}}{(z - a)^2 + h^2} \, dz\right]$$

where C is a contour that contains the real axis plus an infinitely large semi-circle in the upper half-plane. Using the results from problem 6,

$$\int_{-\infty}^{\infty} \frac{\sin(ax)}{(x - a)^2 + h^2} \, dx = \frac{\pi e^{-ah}\sin(a^2)}{h}.$$

8.

$$\int_0^{\infty} \frac{\cos(ax)}{x^4 + 1} \, dx = \frac{1}{2}\int_{-\infty}^{\infty} \frac{\cos(ax)}{x^4 + 1} \, dx = \frac{1}{2}\mathrm{Re}\left(\oint_C \frac{e^{iaz}}{z^4 + 1} \, dz\right)$$

where C is a contour that contains the real axis plus an infinitely large semi-circle in the upper half-plane. There are two simple poles inside the contour at $z = e^{\pi i/4}$ and $z = e^{3\pi i/4}$. Therefore, the residue is

$$\text{Res}(z = e^{\pi i/4}) = \lim_{z \to e^{\pi i/4}} \left[(z - e^{\pi i/4}) \frac{e^{iaz}}{z^4 + 1} \right] = \frac{\exp(ai/\sqrt{2} - a/\sqrt{2})}{4e^{3\pi i/4}}$$

$$= -\frac{1+i}{4\sqrt{2}} e^{-a/\sqrt{2}} \left[\cos\left(\frac{a}{\sqrt{2}}\right) + i\sin\left(\frac{a}{\sqrt{2}}\right) \right]$$

and

$$\text{Res}(z = e^{3\pi i/4}) = \lim_{z \to e^{3\pi i/4}} \left[\frac{(z - e^{3\pi i/4})e^{iaz}}{z^4 + 1} \right] = \frac{\exp(-ai/\sqrt{2} - a/\sqrt{2})}{4e^{9\pi i/4}}$$

$$= \frac{1-i}{4\sqrt{2}} e^{-a/\sqrt{2}} \left[\cos\left(\frac{a}{\sqrt{2}}\right) - i\sin\left(\frac{a}{\sqrt{2}}\right) \right].$$

Therefore, from the residue theorem,

$$\oint_C \frac{e^{iaz}}{z^4 + 1} dz = \frac{\pi}{2\sqrt{2}} e^{-a/\sqrt{2}} \left[\sin\left(\frac{a}{\sqrt{2}}\right) + \cos\left(\frac{a}{\sqrt{2}}\right) \right]$$

and

$$\int_0^\infty \frac{\cos(ax)}{x^4 + 1} dx = \frac{\pi}{4\sqrt{2}} e^{-a/\sqrt{2}} \left[\sin\left(\frac{a}{\sqrt{2}}\right) + \cos\left(\frac{a}{\sqrt{2}}\right) \right].$$

9.

$$\int_0^\infty \frac{x\sin(ax)}{x^4 + 1} dx = \frac{1}{2} \int_{-\infty}^\infty \frac{x\sin(ax)}{x^4 + 1} dx = \frac{1}{2}\text{Im}\left(\oint_C \frac{ze^{iaz}}{z^4 + 1} dz \right)$$

where C is a contour that contains the real axis plus an infinitely large semi-circle in the upper half-plane. There are two simple poles inside the contour at $z = e^{\pi i/4}$ and $z = e^{3\pi i/4}$. Therefore, the residues are

$$\text{Res}(z = e^{\pi i/4}) = \lim_{z \to e^{\pi i/4}} \left[(z - e^{\pi i/4}) \frac{ze^{iaz}}{z^4 + 1} \right]$$

$$= \frac{e^{\pi i/4} \exp(ai/\sqrt{2} - a/\sqrt{2})}{4e^{3\pi i/4}}$$

$$= -\frac{i}{4} e^{-a/\sqrt{2}} \left[\cos\left(\frac{a}{\sqrt{2}}\right) + i\sin\left(\frac{a}{\sqrt{2}}\right) \right]$$

and

$$\text{Res}(z = e^{3\pi i/4}) = \lim_{z \to e^{3\pi i/4}} \left[(z - e^{3\pi i/4}) \frac{z e^{iaz}}{z^4 + 1} \right]$$

$$= \frac{e^{3\pi i/4} \exp(-ai/\sqrt{2} - a/\sqrt{2})}{4 e^{9\pi i/4}}$$

$$= \frac{i}{4} e^{-a/\sqrt{2}} \left[\cos\left(\frac{a}{\sqrt{2}}\right) - i \sin\left(\frac{a}{\sqrt{2}}\right) \right].$$

Therefore, from the residue theorem,

$$\oint_C \frac{z e^{iaz}}{z^4 + 1} \, dz = \pi i e^{-a/\sqrt{2}} \sin\left(\frac{a}{\sqrt{2}}\right)$$

and

$$\int_0^\infty \frac{x \sin(ax)}{x^4 + 1} \, dx = \frac{\pi}{2} e^{-a/\sqrt{2}} \sin\left(\frac{a}{\sqrt{2}}\right).$$

10.

$$\int_0^\infty \frac{\cos(ax)}{(x^2 + h^2)(x^2 + 1)} \, dx = \frac{1}{2} \int_{-\infty}^\infty \frac{\cos(ax)}{(x^2 + h^2)(x^2 + 1)} \, dx$$

$$= \frac{1}{2} \text{Re} \left[\oint_C \frac{e^{iaz}}{(z^2 + h^2)(z^2 + 1)} \, dz \right]$$

where C is a contour that contains the real axis plus an infinitely large semi-circle in the upper half-plane. There are two simple poles inside the contour at $z = hi$ and $z = i$. Therefore, the residues are

$$\text{Res}(z = hi) = \lim_{z \to hi} \left[(z - hi) \frac{e^{iaz}}{(z - hi)(z + hi)(z^2 + 1)} \right] = \frac{e^{-ah}}{(2hi)(1 - h^2)}$$

and

$$\text{Res}(z = i) = \lim_{z \to i} \left[(z - i) \frac{e^{iaz}}{(z^2 + h^2)(z - i)(z + i)} \right] = \frac{e^{-a}}{(2i)(h^2 - 1)}.$$

Therefore, from the residue theorem,

$$\oint_C \frac{e^{iaz}}{(z^2 + h^2)(z^2 + 1)} \, dz = 2\pi i \left[\frac{e^{-ah}}{(2hi)(1 - h^2)} + \frac{e^{-a}}{(2i)(h^2 - 1)} \right]$$

and

$$\int_0^\infty \frac{\cos(ax)}{(x^2 + h^2)(x^2 + 1)} \, dx = \frac{\pi}{2} \left[\frac{e^{-ha}}{h(1 - h^2)} + \frac{e^{-a}}{h^2 - 1} \right].$$

11.

$$\int_0^\infty \frac{x\sin(ax)}{(x^2+h^2)(x^2+1)}\,dx = \frac{1}{2}\int_{-\infty}^\infty \frac{x\sin(ax)}{(x^2+h^2)(x^2+1)}\,dx$$

$$= \frac{1}{2}\text{Im}\left[\oint_C \frac{ze^{iaz}}{(z^2+h^2)(z^2+1)}\,dz\right]$$

where C is a contour that contains the real axis plus an infinitely large semi-circle in the upper half-plane. There are two simple poles inside the contour at $z = hi$ and $z = i$. Therefore, the residues are

$$\text{Res}(z=hi) = \lim_{z\to hi}\left[(z-hi)\frac{ze^{iaz}}{(z-hi)(z+hi)(z^2+1)}\right] = -\frac{e^{-ah}}{2(h^2-1)}$$

and

$$\text{Res}(z=i) = \lim_{z\to i}\left[(z-i)\frac{ze^{iaz}}{(z^2+h^2)(z-i)(z+i)}\right] = \frac{e^{-a}}{2(h^2-1)}.$$

Therefore, from the residue theorem,

$$\oint_C \frac{ze^{iaz}}{(z^2+h^2)(z^2+1)}\,dz = 2\pi i\left[\frac{e^{-a}}{2(h^2-1)} - \frac{e^{-ah}}{2(h^2-1)}\right]$$

and

$$\int_0^\infty \frac{x\sin(ax)}{(x^2+h^2)(x^2+1)}\,dx = \frac{\pi}{2}\left[\frac{e^{-a}}{h^2-1} - \frac{e^{-ah}}{h^2-1}\right].$$

12.

$$\int_0^\infty \frac{8x^2}{(x^2+1)^2[(1+a/h)x^2+(a/h-1)]}\,dx$$

$$= \int_{-\infty}^\infty \frac{4x^2}{(x^2+1)^2[(1+a/h)x^2+(a/h-1)]}\,dx$$

$$= \frac{1}{1+a/h}\oint_C \frac{4z^2}{(z^2+1)^2[z^2+(a-h)/(a+h)]}\,dz$$

where C is an infinite semicircle in the upper half of the z-plane. There are two singularities: a second-order pole $z = i$ and a simple pole $z = i\sqrt{(a-h)/(a+h)}$. Now

$$\text{Res}(z=i) = \lim_{z\to i}\frac{d}{dz}\left\{\frac{4z^2}{(z+i)^2[z^2+(a-h)/(a+h)]}\right\} = -\frac{ai}{2h}\left(1+\frac{a}{h}\right)$$

and

$$\text{Res}[z=i\sqrt{(a-h)/(a+h)}]$$

$$= \lim_{z\to i\sqrt{(a-h)/(a+h)}}\frac{4z^2}{(z^2+1)^2[z+i\sqrt{(a-h)/(a+h)}]}$$

$$= \frac{i}{2}\left(1+\frac{a}{h}\right)\sqrt{\frac{a^2}{h^2}-1}$$

Transform Methods for Solving Partial Differential Equations

Adding together these two residues and multiplying by $2\pi i$ give the desired result.

13. Because e^{-iz^2} is an entire function,

$$\oint_C e^{-iz^2}\,dz = 0.$$

We now break up the closed line integration into integrations along four legs. Along C_1 from $(0,0)$ to $(\infty,0)$,

$$\int_{C_1} e^{-iz^2}\,dz = \int_0^\infty e^{-i\eta^2}\,d\eta = \frac{\sqrt{\pi}}{2\sqrt{2}}(1-i).$$

Along the leg from $(\infty,0)$ to $(\infty,-i\sqrt{x})$,

$$\left|\int_{C_2} e^{-iz^2}\,dz\right| \le \lim_{R\to\infty}\int_0^{\sqrt{x}} e^{-2R\eta}\,d\eta \to 0.$$

Along the leg from $(\infty,-i\sqrt{x})$ to $(0,-i\sqrt{x})$,

$$\int_{C_3} e^{-iz^2}\,dz = -e^{xi}\int_0^\infty e^{-i\eta^2-2\eta\sqrt{x}}\,d\eta.$$

Finally, along the leg from $(0,-i\sqrt{x})$ to $(0,0)$,

$$\int_{C_4} e^{-iz^2}\,dz = -i\int_{\sqrt{x}}^0 e^{i\eta^2}\,d\eta = \frac{\sqrt{\pi}i}{\sqrt{2}}\left[C\left(\sqrt{\frac{2x}{\pi}}\right) - iS\left(\sqrt{\frac{2x}{\pi}}\right)\right].$$

Solving for

$$\int_0^\infty e^{-i\eta^2-2\eta\sqrt{x}}\,d\eta$$

gives the desired result.

14. The only pole within the closed contour is at $z = \pi i/2$. Therefore,

$$\oint_C \frac{\cosh[(\theta/\pi)z]}{\cosh(z)}\,dz = 2\pi\cos(\theta/2).$$

Integrating along each contour,

$$\int_{-R}^R \frac{\cosh[(\theta/\pi)x]}{\cosh(x)}\,dx + \int_0^\pi \frac{\cosh[(\theta/\pi)(R+iy)]}{\cosh(R+iy)}\,i\,dy$$
$$+ \int_R^{-R} \frac{\cosh[(\theta/\pi)(x+\pi i)]}{\cosh(x+\pi i)}\,dx + \int_\pi^0 \frac{\cosh[(\theta/\pi)(-R+iy)]}{\cosh(-R+iy)}\,i\,dy$$
$$= 2\pi\cos(\theta/2).$$

Taking the real part and letting $R \to \infty$,

$$[1 + \cos(\theta)] \int_{-\infty}^{\infty} \frac{\cosh[(\theta/\pi)x]}{\cosh(x)} \, dx = 2\pi \cos(\theta/2).$$

Finally, we replace $1 + \cos(\theta) = 2\cos^2(\theta/2)$ and change the integration so that it runs from 0 to ∞.

Section 1.6

1. Using the given contour,

$$\oint_C \frac{dz}{z \log(z)} = \int_{C_\infty} + \int_{-\infty+0+i}^{0+i} + \int_{C_\epsilon} + \int_{0-i}^{-\infty+0-i}.$$

Now

$$\int_{C_\epsilon} = \lim_{\epsilon \to 0} \int_{\pi}^{-\pi} \frac{\epsilon i e^{\theta i} d\theta}{\epsilon e^{\theta i}[\ln(\epsilon) + \theta i]} = \lim_{\epsilon \to 0} \log[\ln(\epsilon) + \theta i] \Big|_{\pi}^{-\pi}$$

$$= \lim_{\epsilon \to 0} \left\{ \log[\ln(\epsilon) - \pi i] - \log[\ln(\epsilon) + \pi i] \right\} = -2\pi i$$

because we crossed the branch cut. On the other hand,

$$\int_{C_\infty} = \lim_{r \to \infty} \int_{-\pi}^{\pi} \frac{r i e^{\theta i} d\theta}{r e^{\theta i}[\ln(r) + \theta i]} = \lim_{r \to \infty} \log[\ln(r) + \theta i] \Big|_{-\pi}^{\pi}$$

$$= \lim_{r \to \infty} \left\{ \log[\ln(r) + \pi i] - \log[\ln(r) - \pi i] \right\} = 0$$

because we did not cross the branch cut. Along the branch cut,

$$\int_{-\infty+0+i}^{0+i} = \int_{\infty}^{0} \frac{(-dx)}{(-x)[\ln(x) + \pi i]}$$

and

$$\int_{0-i}^{-\infty+0-i} = \int_{0}^{\infty} \frac{(-dx)}{(-x)[\ln(x) - \pi i]}.$$

Therefore,

$$\int_{0}^{\infty} \frac{dx}{x[\ln(x) - \pi i]} - \int_{0}^{\infty} \frac{dx}{x[\ln(x) + \pi i]} = 2\pi i$$

or

$$\int_{0}^{\infty} \frac{dx}{x[\ln(x)^2 + \pi^2]} = 1.$$

Transform Methods for Solving Partial Differential Equations

2. From the residue theorem,

$$\oint_C \frac{dz}{(z+1)\sqrt{z}} = \int_{C_\infty} + \int_{0+i}^{\infty+0^+i} + \int_{C_\epsilon} + \int_{\infty+0^-i}^{0^-i} = 2\pi.$$

Now

$$\int_{C_\epsilon} = \lim_{\epsilon \to 0} \int_{2\pi}^0 \frac{\epsilon i e^{\theta i} d\theta}{\sqrt{\epsilon e^{\theta i/2}}(1+\epsilon e^{\theta i})} \to 0$$

and

$$\int_{C_\infty} = \lim_{r \to \infty} \int_0^{2\pi} \frac{r i e^{\theta i} d\theta}{\sqrt{r e^{\theta i/2}}(1+r e^{\theta i})} \to 0.$$

Along the branch cut, $z = x e^{0i}$ so that

$$\int_{0+i}^{\infty+0^+i} = \int_0^\infty \frac{dx}{\sqrt{x}(1+x)}$$

while along the other side $z = x e^{2\pi i}$,

$$\int_{\infty+0^-i}^{0^-i} = \int_\infty^0 \frac{dx}{\sqrt{x}(1+x)e^{\pi i}}.$$

Therefore,

$$\int_0^\infty \frac{dx}{\sqrt{x}(1+x)} = \pi.$$

3. From the residue theorem,

$$\oint_C \frac{dz}{z\sqrt{z^2-1}} = \int_{C_\infty} + \int_{Br_1} + \int_{Br_2} + \int_{1+0^+i}^{\infty+0^+i}$$
$$+ \int_{\infty+0^-i}^{1+0^-i} + \int_{-1+0^-i}^{-\infty+0^-i} + \int_{-\infty+0^+i}^{-1+0^+i} = 2\pi.$$

Now

$$\int_{C_\infty} = \lim_{r \to \infty} \int_0^{2\pi} \frac{r i e^{\theta i} d\theta}{r e^{\theta i}\sqrt{r^2 e^{2\theta i}+1}} \to 0.$$

Along the branch cut, $z - 1 = (x-1)e^{0i}$ or $(x-1)e^{2\pi i}$ and $z + 1 = (x+1)e^{0i}$ with $1 < x < \infty$ so that

$$\int_{1+0^+i}^{\infty+0^+i} = \int_1^\infty \frac{dx}{x\sqrt{x^2-1}}$$

422

and

$$\int_{\infty+0-i}^{1+0-i} = -\int_{\infty}^{1} \frac{dx}{x\sqrt{x^2-1}}$$

while along the other side, $z - 1 = (x+1)e^{\pi i}$ and $z + 1 = (x-1)e^{\pm \pi i}$

$$\int_{-1+0-i}^{-\infty+0-i} = \int_{1}^{\infty} \frac{(-dx)}{(-x)\sqrt{x^2-1}}$$

and

$$\int_{-1+0+i}^{-\infty+0+i} = -\int_{\infty}^{1} \frac{(-dx)}{(-x)\sqrt{x^2-1}}.$$

Therefore,

$$\int_{1}^{\infty} \frac{dx}{x\sqrt{x^2-1}} = \frac{\pi}{2}.$$

4. (a) From the residue theorem,

$$\oint_C \frac{dz}{(z^2+1)\sqrt{z^2-1}} = 2\pi i[\mathrm{Res}(z=i) + \mathrm{Res}(z=-i)]$$

where

$$\mathrm{Res}(z=i) = -\frac{1}{2\sqrt{2}}$$

and

$$\mathrm{Res}(z=-i) = -\frac{1}{2\sqrt{2}}.$$

Now, along the top of the branch cut,

$$\int_{top} = \int_{-1}^{1} \frac{dx}{i(x^2+1)\sqrt{1-x^2}}$$

because $z - 1 = (1-x)e^{\pi i}$ and $z + 1 = (1+x)e^{0i}$. Along the bottom, however,

$$\int_{bottom} = \int_{1}^{-1} \frac{dx}{(-i)(x^2+1)\sqrt{1-x^2}}$$

because $z - 1 = (1-x)e^{\pi i}$ and $z + 1 = (1+x)e^{2\pi i}$. Therefore, because the sum of the residues equals the branch cut integrals,

$$\int_{-1}^{1} \frac{dx}{(x^2+1)\sqrt{1-x^2}} = \frac{\pi}{\sqrt{2}}.$$

423

For part (b),

$$\oint_C \frac{dz}{(z^2+1)^2\sqrt{z^2-1}} = 2\pi i[\text{Res}(z=i) + \text{Res}(z=-i)]$$

This time, however, $z = \pm i$ are second-order poles so that

$$\text{Res}(z=i) = \lim_{z \to i} \frac{d}{dz}\left[\frac{1}{(z+i)^2\sqrt{z^2-1}}\right] = -\frac{3}{8\sqrt{2}}$$

and

$$\text{Res}(z=-i) = \lim_{z \to -i} \frac{d}{dz}\left[\frac{1}{(z-i)^2\sqrt{z^2-1}}\right] = -\frac{3}{8\sqrt{2}}.$$

Now, along the top of the branch cut,

$$\int_{top} = \int_{-1}^{1} \frac{dx}{i(x^2+1)^2\sqrt{1-x^2}}$$

because $z - 1 = (1-x)e^{\pi i}$ and $z+1 = (1+x)e^{0i}$. Along the bottom, however,

$$\int_{bottom} = \int_{1}^{-1} \frac{dx}{(-i)(x^2+1)^2\sqrt{1-x^2}}$$

because $z - 1 = (1-x)e^{\pi i}$ and $z+1 = (1+x)e^{2\pi i}$. Therefore, because the sum of the residues equals the branch cut integrals,

$$\int_{-1}^{1} \frac{dx}{(x^2+1)^2\sqrt{1-x^2}} = \frac{3\pi}{4\sqrt{2}}.$$

5. From the residue theorem,

$$\oint_C \frac{1}{z^2}\sqrt{\frac{z-1}{z+1}}\,dz = 2\pi i\,\text{Res}(z=0)$$

where

$$\text{Res}(z=0) = \lim_{z \to 0} \frac{d}{dz}\left(\sqrt{\frac{z-1}{z+1}}\right) = \frac{1}{i}.$$

Now

$$\int_{1+0+i}^{\infty+0+i} \frac{1}{z^2}\sqrt{\frac{z-1}{z+1}}\,dz = \int_{1}^{\infty} \frac{dx}{x^2}\sqrt{\frac{x-1}{x+1}}$$

because $z - 1 = (x-1)e^{0i}$ and $z+1 = (x+1)e^{0i}$.

$$\int_{\infty+0-i}^{1+0^-i} \frac{1}{z^2}\sqrt{\frac{z-1}{z+1}}\,dz = e^{\pi i}\int_{\infty}^{1} \frac{dx}{x^2}\sqrt{\frac{x-1}{x+1}}$$

because $z-1=(x-1)e^{2\pi i}$ and $z+1=(x+1)e^{0i}$.

$$\int_{-\infty+0^+i}^{-1+0^+i} \frac{1}{z^2}\sqrt{\frac{z-1}{z+1}}\,dz = \int_{\infty}^{1} \frac{(-dx)}{x^2}\sqrt{\frac{x+1}{x-1}}$$

because $z-1=(x+1)e^{\pi i}$ and $z+1=(x-1)e^{\pi i}$.

$$\int_{-1+0^-i}^{-\infty+0^-i} \frac{1}{z^2}\sqrt{\frac{z-1}{z+1}}\,dz = e^{\pi i}\int_{1}^{\infty} \frac{(-dx)}{x^2}\sqrt{\frac{x+1}{x-1}}$$

because $z-1=(x+1)e^{\pi i}$ and $z+1=(x-1)e^{-\pi i}$. Adding the branch cut integrals together,

$$\int_1^\infty \frac{dx}{x^2}\sqrt{\frac{x-1}{x+1}} = \pi - \int_1^\infty \frac{dx}{x^2}\sqrt{\frac{x+1}{x-1}}.$$

6. From the residue theorem,

$$\oint_C \frac{dz}{(z^3-z^2)^{1/3}} = 2\pi i.$$

Now, along the top of the branch cut,

$$\int_{top} = e^{-\pi i/3}\int_1^0 \frac{dx}{(x^2-x^3)^{1/3}}$$

because $z=xe^{0i}$ and $z-1=(1-x)e^{\pi i}$. Along the bottom, however,

$$\int_{bottom} = e^{-5\pi i/3}\int_0^1 \frac{dx}{(x^2-x^3)^{1/3}}$$

because $z=xe^{2\pi i}$ and $z-1=(1-x)e^{\pi i}$. Therefore, because the sum of the residues equals the branch cut integrals,

$$\int_0^1 \frac{dx}{(x^2-x^3)^{1/3}} = \frac{2\pi}{\sqrt{3}}.$$

7. The integration comprises four parts: an integration along the top of the branch cut, an integration along the bottom of the branch cut and integrations around the branch points at $z=0$ and $z=-1$. The integrations around the branch points vanish. Along the top of the branch cut, $z=xe^{\pi i}$ and $z+1=(1-x)e^{0i}$ so that

$$\int_{top\ of\ branch\ cut} = -\int_{0+}^{1}\left(\frac{1-x}{x}\right)^{\alpha} e^{-\pi\alpha i}x^{n-1}e^{(n-1)\pi i}dx$$

while along the bottom of the branch cut $z = xe^{-\pi i}$ and $z+1 = (1-x)e^{0i}$ so that

$$\int_{bottom\ of\ branch\ cut} = -\int_1^{0^+} \left(\frac{1-x}{x}\right)^\alpha e^{\pi \alpha i} x^{n-1} e^{-(n-1)\pi i} dx.$$

Combining the two integrals, dividing by $2\pi i$ and simplifying, the desired result is obtained.

8. We have two poles located within the contour $z = \alpha \pm i\sqrt{1-\alpha^2}$. Therefore the residues are

$$\mathrm{Res}(z = \alpha + i\sqrt{1-\alpha^2}) = \lim_{z \to \alpha + i\sqrt{1-\alpha^2}} \left[\frac{z^{p-1}(z - \alpha - i\sqrt{1-\alpha^2})}{z^2 - 2\alpha z + 1}\right]$$

$$= \frac{[\alpha + i\sqrt{1-\alpha^2}]^{p-1}}{2i\sqrt{1-\alpha^2}}.$$

$$\mathrm{Res}(z = \alpha - i\sqrt{1-\alpha^2}) = -\frac{[\alpha - i\sqrt{1-\alpha^2}]^{p-1}}{2i\sqrt{1-\alpha^2}}.$$

The contribution from the integration around the branch point and at infinity vanish. Along the top of the branch cut, $z = xe^{0i}$ and

$$\int_{top\ branch\ cut} = \int_0^\infty \frac{x^{p-1}}{x^2 - 2\alpha x + 1} dx$$

while along the bottom of the branch cut we have $z = xe^{2\pi i}$ and

$$\int_{bottom\ branch\ cut} = e^{2\pi i(p-1)} \int_\infty^0 \frac{x^{p-1}}{x^2 - 2\alpha x + 1} dx.$$

From the residue theorem,

$$\left[1 - e^{2\pi i(p-1)}\right] \int_0^\infty \frac{x^{p-1}}{x^2 - 2\alpha x + 1} dx$$

$$= \frac{\pi}{\sqrt{1-\alpha^2}}\left[(\alpha + i\sqrt{1-\alpha^2})^{p-1} - (\alpha - i\sqrt{1-\alpha^2})^{p-1}\right].$$

or

$$\int_0^\infty \frac{x^{p-1}}{x^2 - 2\alpha x + 1} dx$$

$$= -\frac{\pi}{\sqrt{1-\alpha^2}} \frac{\exp[i(p-1)\cos^{-1}(\alpha)] - \exp[-i(p-1)\cos^{-1}(\alpha)]}{\{\exp[(p-1)\pi i] - \exp[-(p-1)\pi i]\}\exp[(p-1)\pi i]}$$

$$= \frac{\pi}{\sqrt{1-\alpha^2}}\Big\{\cos[(p-1)\cos^{-1}(\alpha)]$$

$$- \cot[(p-1)\pi]\sin[(p-1)\cos^{-1}(\alpha)]\Big\}.$$

Section 2.1

1. From Bromwich's integral,

$$f(t) = \frac{1}{2\pi i} \int_{c-\infty i}^{c+\infty i} \frac{e^{tz}}{z^3(z+1)^2} \, dz = \frac{1}{2\pi i} \oint_C \frac{e^{tz}}{z^3(z+1)^2} \, dz.$$

Computing the residues,

$$\text{Res}(z = 0) = \lim_{z \to 0} \frac{1}{2} \frac{d^2}{dz^2} \left[z^3 \frac{e^{tz}}{z^3(z+1)^2} \right]$$

$$= \lim_{z \to 0} \frac{1}{2} \left[\frac{t^2 e^{tz}}{(z+1)^2} - \frac{4t e^{tz}}{(z+1)^3} + \frac{6 e^{tz}}{(z+1)^4} \right]$$

$$= (t^2 - 4t + 6)/2$$

and

$$\text{Res}(z = -1) = \lim_{z \to -1} \frac{d}{dz} \left[(z+1)^2 \frac{e^{tz}}{z^3(z+1)^2} \right]$$

$$= \lim_{z \to -1} \frac{t e^{tz}}{z^3} - \frac{3 e^{tz}}{z^4} = -t e^{-t} - 3 e^{-t}.$$

Then the inverse is

$$f(t) = \frac{t^2}{2} - 2t + 3 - (t+3)e^{-t}.$$

2. From Bromwich's integral,

$$f(t) = \frac{1}{2\pi i} \int_{c-\infty i}^{c+\infty i} \frac{(z+1)e^{tz}}{(z+2)^2(z+3)} \, dz = \frac{1}{2\pi i} \oint_C \frac{(z+1)e^{tz}}{(z+2)^2(z+3)} \, dz.$$

Computing the residues,

$$\text{Res}(z = -2) = \lim_{z \to -2} \frac{d}{dz} \left[(z+2)^2 \frac{(z+1)e^{tz}}{(z+2)^2(z+3)} \right]$$

$$= \lim_{z \to -2} \left[\frac{t(z+1)e^{tz}}{z+3} + \frac{e^{tz}}{z+3} - \frac{(z+1)e^{tz}}{(z+3)^2} \right]$$

$$= 2e^{-2t} - t e^{-2t}$$

and

$$\text{Res}(z = -3) = \lim_{z \to -3} \left[(z+3) \frac{(z+1)e^{tz}}{(z+2)^2(z+3)} \right] = -2e^{-3t}.$$

427

Then the inverse is

$$f(t) = (2 - t)e^{-2t} - 2e^{-3t}.$$

3. From Bromwich's integral,

$$f(t) = \frac{1}{2\pi i}\int_{c-\infty i}^{c+\infty i}\frac{(z+2)e^{tz}}{z(z-a)(z^2+4)}\,dz = \frac{1}{2\pi i}\oint_C \frac{(z+2)e^{tz}}{z(z-a)(z^2+4)}\,dz.$$

Computing the residues,

$$\text{Res}(z=0) = \lim_{z\to 0}\left[z\frac{(z+2)e^{tz}}{z(z-a)(z^2+4)}\right] = \frac{2}{-4a} = -\frac{1}{2a}$$

$$\text{Res}(z=a) = \lim_{z\to a}\left[(z-a)\frac{(z+2)e^{tz}}{z(z-a)(z^2+4)}\right] = \frac{(a+2)e^{at}}{a(a^2+4)}$$

$$\text{Res}(z=-2i) = \lim_{z\to -2i}\left[(z+2i)\frac{(z+2)e^{tz}}{z(z-a)(z^2+4)}\right]$$

$$= \frac{(2a-4-2ai-4i)e^{-2it}}{8(a^2+4)}$$

and

$$\text{Res}(z=2i) = \lim_{z\to 2i}\left[(z-2i)\frac{(z+2)e^{tz}}{z(z-a)(z^2+4)}\right]$$

$$= \frac{(-2a+4-2ia-4i)e^{2it}}{-8(a^2+4)}.$$

Then the inverse is

$$f(t) = \frac{a+2}{a(a^2+4)}e^{at} - \frac{1}{2a} + \frac{(a-2)\cos(2t) - (a+2)\sin(2t)}{2(a^2+4)}.$$

4. From Bromwich's integral,

$$f(t) = \frac{1}{2\pi i}\int_{c-\infty i}^{c+\infty i}\frac{e^{tz}}{z\sinh(az)}\,dz = \frac{1}{2\pi i}\oint_C \frac{e^{tz}}{z\sinh(az)}\,dz.$$

Some of the poles are where $\sinh(az) = i\sin(-iaz) = 0$ or $z = \pm n\pi i/a$ where $n = 1, 2, 3, \ldots$. On the other hand, at $z = 0$,

$$F(z) = \frac{1}{az^2(1 + \frac{a^2z^2}{3!} + \frac{a^4z^4}{5!} + \cdots)} = \frac{1}{az^2}\left(1 - \frac{a^2z^2}{3!} + \cdots\right)$$

and we have a second-order pole at $z = 0$. Therefore,

$$\text{Res}(z = 0) = \lim_{z \to 0} \frac{d}{dz}\left[\frac{ze^{tz}}{\sinh(az)}\right] = \lim_{z \to 0} \frac{d}{dz}\left[\frac{ze^{tz}}{az(1 + \frac{a^2z^2}{3!} + \frac{a^4z^4}{5!} + \cdots)}\right]$$

$$= \lim_{z \to 0} \frac{t}{a}\left(\frac{e^{tz}}{1 + \frac{a^2z^2}{3!} + \frac{a^4z^4}{5!} + \cdots}\right)$$

$$- \lim_{z \to 0} \frac{e^{tz}\left(\frac{2a^2z}{3!} + \frac{4a^4z^3}{5!} + \cdots\right)}{a(1 + \frac{a^2z^2}{3!} + \frac{a^4z^4}{5!} + \cdots)^2} = \frac{t}{a}.$$

At the remaining poles,

$$\text{Res}(z = n\pi i/a) = \lim_{z \to n\pi i/a}\left[(z - n\pi i/a)\frac{e^{tz}}{z\sinh(az)}\right]$$

$$= \frac{\exp(n\pi it/a)}{n\pi i \cosh(n\pi i)} = \frac{(-1)^n \exp(n\pi it/a)}{n\pi i}.$$

Then the inverse is

$$f(t) = \frac{t}{a} + \frac{1}{\pi}\sum_{\substack{n=-\infty \\ n \neq 0}}^{\infty} \frac{(-1)^n \exp(n\pi it/a)}{ni} = \frac{t}{a} + \frac{2}{\pi}\sum_{n=1}^{\infty}\frac{(-1)^n}{n}\sin\left(\frac{n\pi t}{a}\right).$$

5. From Bromwich's integral,

$$f(t) = \frac{1}{2\pi i}\int_{c-\infty i}^{c+\infty i}\frac{\sinh(az/2)e^{tz}}{z\cosh(az/2)}\,dz = \frac{1}{2\pi i}\oint_C \frac{\sinh(az/2)e^{tz}}{z\cosh(az/2)}\,dz.$$

Some of the poles are where $\cosh(az/2) = \cos(-iaz/2) = 0$ or $z_n = \pm(2n-1)\pi i/a$ where $n = 1, 2, 3, \ldots$. On the other hand, at $z = 0$,

$$F(z) = \frac{\frac{az}{2}(1 + \frac{a^2z^2}{3!4} + \frac{a^4z^4}{5!16} + \cdots)}{z(1 + \frac{a^2z^2}{2!4} + \frac{a^4z^4}{4!16} + \cdots)} = \frac{a}{2}\left(1 + \frac{a^2z^2}{3!4} + \cdots\right)\left(1 - \frac{a^2z^2}{2!4} + \cdots\right)$$

and $z = 0$ is a removable singularity. Therefore, at the remaining poles,

$$\text{Res}(z = z_n) = \lim_{z \to z_n}\left[(z - z_n)\frac{\sinh(az/2)e^{tz}}{z\cosh(az/2)}\right] = \pm\frac{2\exp[\pm(2n-1)\pi it/a]}{(2n-1)\pi i}.$$

Then the inverse is

$$f(t) = \frac{2}{\pi}\sum_{n=1}^{\infty}\frac{\{\exp[(2n-1)\pi it/a] - \exp[-(2n-1)\pi it/a]\}}{(2n-1)i}$$

$$= \frac{4}{\pi}\sum_{n=1}^{\infty}\frac{1}{2n-1}\sin\left[\frac{(2n-1)\pi t}{a}\right].$$

6. From Bromwich's integral,

$$f(t) = \frac{1}{2\pi i} \int_{c-\infty i}^{c+\infty i} \frac{\sinh(a\sqrt{z})e^{tz}}{z\sinh(\sqrt{z})} \, dz = \frac{1}{2\pi i} \oint_C \frac{\sinh(a\sqrt{z})e^{tz}}{z\sinh(\sqrt{z})} \, dz.$$

Some of the poles are where $\sinh(\sqrt{z}) = i\sin(-i\sqrt{z}) = 0$ or $z_n = -n^2\pi^2$ with $\sqrt{z_n} = n\pi i$ where $n = 1, 2, 3, \ldots$. On the other hand, at $z = 0$,

$$F(z) = \frac{az^{1/2} + \frac{a^3 z^{3/2}}{3!} + \frac{a^5 z^{5/3}}{5!} + \cdots}{z(z^{1/2} + \frac{z^{3/2}}{3!} + \frac{z^{5/2}}{5!} + \cdots)} = \frac{a}{z}\left(1 + \frac{a^2 z}{3!} + \cdots\right)\left(1 - \frac{z}{3!} + \cdots\right)$$

and $z = 0$ is a simple pole. The easiest way to compute the residue at $z = 0$ is by constructing its Laurent expansion:

$$\frac{\sinh(a\sqrt{z})e^{tz}}{z\sinh(\sqrt{z})} = \frac{(az^{1/2} + \frac{a^3 z^{3/2}}{3!} + \frac{a^5 z^{5/2}}{5!} + \cdots)(1 + tz + \cdots)}{z(z^{1/2} + \frac{z^{3/2}}{3!} + \frac{z^{5/2}}{5!} + \cdots)}$$

$$= \frac{a}{z}\left(1 + \frac{a^2 z}{3!} + \cdots\right)\left(1 + tz + \cdots\right)\left(1 - \frac{z}{3!} + \cdots\right)$$

$$= \frac{a}{z}\left(1 + \frac{a^2 z}{3!} + tz - \frac{z}{3!} + \cdots\right)$$

and the residue equals a. The remaining poles give the residues

$$\text{Res}(z = z_n) = \lim_{z \to z_n}\left[(z - z_n)\frac{\sinh(a\sqrt{z})e^{tz}}{z\sinh(\sqrt{z})}\right]$$

$$= \lim_{z \to z_n} \frac{2\sqrt{z}\sinh(a\sqrt{z})e^{tz}}{z\cosh(\sqrt{z})}$$

$$= \frac{2\sinh(n\pi a i)(n\pi i)\exp(-n^2\pi^2 t)}{-n^2\pi^2 \cosh(n\pi i)}$$

$$= \frac{2(-1)^n \sin(n\pi a)\exp(-n^2\pi^2 t)}{n\pi}.$$

Adding the residues gives

$$f(t) = a + \frac{2}{\pi}\sum_{n=1}^{\infty} \frac{(-1)^n}{n}\sin(n\pi a)e^{-n^2\pi^2 t}.$$

7. From Bromwich's integral,

$$f(t) = \frac{1}{2\pi i}\int_{c-\infty i}^{c+\infty i} \frac{\sinh(z)e^{tz}}{z^2\cosh(z)} \, dz = \frac{1}{2\pi i}\oint_C \frac{\sinh(z)e^{tz}}{z^2\cosh(z)} \, dz.$$

Some of the poles are where $\cosh(z) = \cos(-iz) = 0$ or $z_n = \pm(2n - 1)\pi i/2$ where $n = 1, 2, 3, \ldots$. On the other hand, at $z = 0$,

$$F(z) = \frac{\sinh(z)}{z^2 \cosh(z)} = \frac{z(1 + \frac{z^2}{3!} + \frac{z^4}{5!} + \cdots)}{z^2(1 + \frac{z^2}{2!} + \frac{z^4}{4!} + \cdots)}$$

$$= \frac{1}{z}\left(1 + \frac{z^2}{3!} + \cdots\right)\left(1 - \frac{z^2}{2!} + \cdots\right)$$

and $z = 0$ is a simple pole. Therefore the $z = 0$ residue is

$$\text{Res}(z = 0) = \lim_{z \to 0}\left[z\frac{\sinh(z)e^{tz}}{z^2 \cosh(z)}\right] = 1.$$

The remaining poles give the residues

$$\text{Res}(z = z_n) = \lim_{z \to z_n}\left[(z - z_n)\frac{\sinh(z)e^{tz}}{z^2 \cosh(z)}\right] = \lim_{z \to z_n}\frac{e^{tz}}{z^2}$$

$$= -\frac{4\exp[\pm(2n - 1)\pi it/2]}{(2n - 1)^2\pi^2}.$$

Adding the residues gives

$$f(t) = 1 - \frac{4}{\pi^2}\sum_{n=1}^{\infty}\frac{\exp[(2n - 1)\pi it/2] + \exp[-(2n - 1)\pi it/2]}{(2n - 1)^2}$$

or

$$f(t) = 1 - \frac{8}{\pi^2}\sum_{n=1}^{\infty}\frac{\cos[(2n - 1)\pi t/2]}{(2n - 1)^2}.$$

8. From Bromwich's integral,

$$f(t) = \frac{1}{2\pi i}\int_{c-\infty i}^{c+\infty i}\frac{e^{tz}}{Rz + M^2/\alpha}\left[1 - \frac{\cosh(a\sqrt{Rz + M^2/\alpha})}{\cosh(\sqrt{Rz + M^2/\alpha})}\right]dz$$

$$= \frac{1}{2\pi i}\oint_C\frac{e^{tz}}{Rz + M^2/\alpha}\left[1 - \frac{\cosh(a\sqrt{Rz + M^2/\alpha})}{\cosh(\sqrt{Rz + M^2/\alpha})}\right]dz.$$

Some of the poles are where

$$\cosh(\sqrt{Rz + M^2/\alpha}) = \cos(-i\sqrt{Rz + M^2\alpha}) = 0$$

or

$$z_n = -M^2/(\alpha R) - (2n - 1)^2\pi^2/(4R)$$

and

$$\sqrt{Rz + M^2\alpha} = (2n - 1)\pi i/2$$

where $n = 1, 2, 3, \ldots$. On the other hand, at $z = -M^2/(\alpha R)$,

$$\text{Res}[z = -M^2/(\alpha R)] = \lim_{z \to -M^2/(\alpha R)} \left\{ [z + M^2/(\alpha R)] \frac{e^{tz}}{Rz + M^2/\alpha} \right.$$
$$\left. \times \left[1 - \frac{\cosh(a\sqrt{Rz + M^2\alpha})}{\cosh(\sqrt{Rz + M^2/\alpha})} \right] \right\}$$
$$= 0$$

while the remaining poles give the residues

$$\text{Res}(z = z_n) = -\lim_{z \to z_n} \left[(z - z_n) \frac{e^{tz}}{Rz + M^2/\alpha} \right.$$
$$\left. \times \frac{\cosh(a\sqrt{Rz + M^2/\alpha})}{\cosh(\sqrt{Rz + M^2/\alpha})} \right]$$
$$= \frac{\exp[-(\frac{M^2}{\alpha} + \frac{(2n-1)^2\pi^2}{4})\frac{t}{R}]}{R[\frac{M^2}{\alpha R} + \frac{(2n-1)^2\pi^2}{4R} - \frac{M^2}{\alpha R}]}$$
$$\times \frac{2\cosh[a(2n-1)\pi i/2][(2n-1)\pi i/2]}{R\sinh[(2n-1)\pi i/2]}$$
$$= -\frac{4(-1)^n}{R\pi(2n-1)} \exp\left\{ -\left[\frac{M^2}{\alpha} + \frac{(2n-1)^2\pi^2}{4} \right] \frac{t}{R} \right\}$$
$$\times \cos\left[\frac{(2n-1)\pi a}{2} \right].$$

Adding the residues gives

$$f(t) = -\frac{4}{\pi R} \sum_{n=1}^{\infty} \frac{(-1)^n}{(2n-1)} \cos\left[\frac{(2n-1)\pi a}{2} \right]$$
$$\times \exp\left\{ -\left[\frac{(2n-1)^2\pi^2}{4} + \frac{M^2}{\alpha} \right] \frac{t}{R} \right\}.$$

9. From Bromwich's integral,

$$f(t) = \frac{1}{2\pi i} \int_{c-\infty i}^{c+\infty i} \frac{e^{tz}}{2za - \sqrt{z}\tanh(\sqrt{z})} \, dz$$
$$= \frac{1}{2\pi i} \oint_C \frac{e^{tz}}{2za - \sqrt{z}\tanh(\sqrt{z})} \, dz.$$

Some of the poles are where $2a = \tanh(\sqrt{z})/\sqrt{z}$ or $\tan(\alpha_n) = 2a\alpha_n$ with $z_n = -\alpha_n^2$ and $\sqrt{z_n} = \alpha_n i$ where $n = 1, 2, 3, \ldots$. We also have a simple pole at $z = 0$. At that point the residue is

$$\text{Res}(z = 0) = \lim_{z \to 0} \left[z \frac{e^{tz}}{2za - \sqrt{z}\tanh(\sqrt{z})} \right]$$
$$= \lim_{z \to 0} \frac{e^{tz}}{2a - \tanh(\sqrt{z})/\sqrt{z}} = \frac{1}{2a - 1}$$

while the remaining poles give the residues

$$\text{Res}(z = -\alpha_n^2) = \lim_{z \to -\alpha_n^2}\left[(z + \alpha_n^2)\frac{e^{tz}}{2za - \sqrt{z}\tanh(\sqrt{z})}\right]$$

$$= \frac{e^{-\alpha_n^2 t}}{2a - i\tan(\alpha_n)/(2i\alpha_n) - i\alpha_n/[2i\sec^2(\alpha_n)\alpha_n]}$$

$$= \frac{2e^{-\alpha_n^2 t}}{2a - 1 + 4a^2\alpha_n^2}.$$

Adding the residue gives

$$f(t) = \frac{1}{2a - 1} + 2\sum_{n=1}^{\infty}\frac{2e^{-\alpha_n^2 t}}{2a - 1 + 4a^2\alpha_n^2}.$$

10. Using the Taylor expansion for hyperbolic tangent, we see that $F(s)$ is a single-valued function. Clearly there is a singularity at $s = 0$. Expanding about that point,

$$F(z)e^{tz} = \frac{(\alpha - \alpha^3 z/3 + \cdots)(1 + tz + \cdots)}{z(m + \alpha - \alpha^3 z/3 + \cdots)}$$

and we have a simple pole there. The residue equals $\alpha/(m + \alpha)$.

The remaining poles occur where $m\sqrt{z} + \tanh(\alpha\sqrt{z}) = 0$. If $\alpha\sqrt{z} = i\lambda$, then these simple poles occur where $\tan(\lambda) = -b\lambda$ and $z_n = -\lambda_n^2/\alpha^2$. The corresponding residues are

$$\text{Res}(z = z_n) = \lim_{z \to z_n}\frac{(z - z_n)\tanh(\alpha\sqrt{z})e^{tz}}{z[m\sqrt{z} + \tanh(\alpha\sqrt{z})]}$$

$$= \frac{1}{b\lambda_n}\left[\frac{\sin(2\lambda_n)\exp(-\lambda_n^2 t/b^2)}{1/b + \cos^2(\lambda_n)}\right].$$

Summing the residues gives the desired result.

11. From Bromwich's integral

$$f(t) = \frac{1}{2\pi i}\oint_C \frac{I_0(r\sqrt{z/\kappa})e^{tz}}{I_0(a\sqrt{z/\kappa})}\,dz$$

where C includes all of the singularities. Let $\sqrt{z/\kappa} = -\alpha i$. Then, $I_0(-\alpha ai) = J_0(\alpha a) = 0$. Let us denote the zeros by $\alpha_n a$. Then $z_n = -\kappa\alpha_n^2$ and the poles lie along the negative real axis. Using the residue theorem,

$$\text{Res}(z = -\kappa\alpha_n^2) = \lim_{z \to -\kappa\alpha_n^2}\frac{(z + \kappa\alpha_n^2)I_0(r\sqrt{z/\kappa})e^{tz}}{I_0(a\sqrt{z/\kappa})}$$

$$= \frac{I_0(-r\alpha_n i)\exp(-\alpha_n^2\kappa t)}{I_1(-a\alpha_n i)[a/(-2\kappa\alpha_n i)]}$$

$$= \frac{2\kappa}{a}\frac{\alpha_n J_0(r\alpha_n)\exp(-\alpha_n^2\kappa t)}{J_1(\alpha_n a)}.$$

433

Summing the residues,

$$f(t) = \frac{2\kappa}{a} \sum_{n=1}^{\infty} \frac{\alpha_n J_0(r\alpha_n)\exp(-\alpha_n^2\kappa t)}{J_1(\alpha_n a)}$$

where $J_0(\alpha_n a) = 0$ for $n = 1, 2, 3, \ldots$

12. $F(s)$ has simple poles located at $s = 0$, $s = k_1^2/\kappa$ and $\sqrt{s/\kappa} = -\alpha i$ where $I_0(-\alpha a i) = J_0(\alpha) = 0$. Therefore, the last group of poles is at $s_n = -\kappa\alpha_n^2$ where α_n is the nth root of $J_0(\alpha a) = 0$. From Bromwich's integral,

$$f(t) = \frac{1}{2\pi i} \oint_C \frac{I_0(r\sqrt{z/\kappa})e^{tz}}{z(k_1^2/\kappa - z)I_0(a\sqrt{z/\kappa})}\,dz$$

where C includes all of the singularities. Now the residues are

$$\text{Res}(z = 0) = \lim_{z \to 0} \frac{I_0(r\sqrt{z/\kappa})e^{tz}}{(k_1^2/\kappa - z)I_0(a\sqrt{z/\kappa})} = \frac{\kappa}{k_1^2},$$

$$\text{Res}(z = k_1^2/\kappa) = \lim_{z \to k_1^2/\kappa} -\frac{I_0(r\sqrt{z/\kappa})e^{tz}}{zI_0(a\sqrt{z/\kappa})} = -\frac{\kappa}{k_1^2}\frac{I_0(k_1 r/\kappa)}{I_0(k_1 a/\kappa)}\exp(k_1^2 t/\kappa)$$

and

$$\text{Res}(z = -\kappa\alpha_n^2) = \lim_{z \to -\kappa\alpha_n^2} \frac{(z + \kappa\alpha_n^2)I_0(r\sqrt{z/\kappa})e^{tz}}{z(k_1^2/\kappa - z)I_0(a\sqrt{z/\kappa})}$$

$$= \frac{I_0(-r\alpha_n i)\exp(-\alpha_n^2\kappa t)}{I_1(-a\alpha_n i)(-a\alpha_n i/2)(k_1^2/\kappa + \kappa\alpha_n^2)}$$

$$= \frac{2\kappa}{a}\frac{J_0(r\alpha_n)\exp(-\alpha_n^2\kappa t)}{\alpha_n(k_1^2 + \kappa^2\alpha_n^2)J_1(\alpha_n a)}.$$

Summing the residues,

$$f(t) = \frac{\kappa}{k_1^2} - \frac{\kappa}{k_1^2}\frac{I_0(k_1 r/\kappa)}{I_0(k_1 a/\kappa)}\exp(k_1^2 t/\kappa) + \frac{2\kappa}{a}\sum_{n=1}^{\infty} \frac{J_0(r\alpha_n)\exp(-\alpha_n^2\kappa t)}{\alpha_n(k_1^2 + \kappa^2\alpha_n^2)J_1(\alpha_n a)}$$

where $J_0(\alpha_n a) = 0$ for $n = 1, 2, 3, \ldots$

13. Because

$$I_0(z) = 1 + \frac{z^2}{4} + \frac{z^4}{64} + \cdots$$

and

$$I_1(z) = \frac{z}{2} + \frac{z^3}{16} + \frac{z^5}{384} + \cdots,$$

then

$$F(s) = \frac{I_0(r\sqrt{s})}{s^{3/2}I_1(a\sqrt{s})}$$

is single-valued and has a second-order pole at $s = 0$. Furthermore, if $a\sqrt{s} = -i\alpha$, then $I_1(-\alpha i) = iJ_1(\alpha) = 0$ where $J_1(\ )$ is the first-order Bessel function of the first kind. The residue for the $s = 0$ is

$$\text{Res}(z = 0) = 1 - \lim_{z \to 0} \frac{d}{dz} \left[\frac{(1 + \frac{r^2 z}{4} + \frac{r^4 z^2}{64} + \cdots)(1 + tz + \frac{t^2 z^2}{2} + \cdots)}{(a/2)(1 + \frac{a^2 z}{8} + \frac{a^4 z^2}{192} + \cdots)} \right]$$

$$= 1 - \frac{2}{a}\left(t + \frac{r^2}{4} - \frac{a^2}{8}\right).$$

On the other hand, the residue at $z_n = -\alpha_n^2/a^2$ is

$$\text{Res}(z = -\alpha_n^2/a^2) = \lim_{z \to -\alpha_n^2/a^2} \frac{(z + \alpha_n^2/a^2)I_0(r\sqrt{z})e^{tz}}{z^{3/2}I_1(a\sqrt{z})}$$

$$= \frac{I_0(-i\alpha_n r/a)\exp(-\alpha_n^2 t/a^2)}{I_1'(-i\alpha_n)(a/2)(-\alpha_n^2/a^2)}$$

$$= 2a\frac{J_0(\alpha_n r/a)\exp(-\alpha_n^2 t/a^2)}{\alpha_n^2 J_0(\alpha_n)}$$

because $2I_1'(z) = I_0(z) + I_2(z)$, $I_0(-i\alpha_n) = J_0(\alpha_n)$ and $I_2(-i\alpha_n) = -J_2(\alpha_n)$. Finally, $J_2(\alpha_n) = J_0(\alpha_n)$ because $J_2(\alpha_n) + J_0(\alpha_n) = 2J_1(\alpha_n)/\alpha_n = 0$. Therefore, summing the residues gives

$$f(t) = 1 - a\left[\frac{2t}{a^2} + \frac{r^2}{2a^2} - \frac{1}{4} - 2\sum_{n=1}^{\infty} \frac{J_0(\alpha_n r/a)}{\alpha_n^2 J_0(\alpha_n)}e^{-\alpha_n^2 t/a^2}\right].$$

14. From Bromwich's integral,

$$f(t) = \frac{1}{2\pi i}\int_{c-\infty i}^{c+\infty i} \frac{I_n(a\sqrt{z})}{\sqrt{z}I_{n+1}(b\sqrt{z})}e^{zt}\,dz.$$

Closing the line integral with a semi-circle of infinite radius in the left side of the z-plane, there are simple poles at $z = 0$ and $z_m = -\alpha_m^2/b^2$ with $\sqrt{z_m} = -i\alpha_m/b$ where $J_{n+1}(\alpha_m) = 0$. The residue at $z = 0$ is

$$\text{Res}(z = 0) = \lim_{z \to 0} \frac{\sqrt{z}I_n(a\sqrt{z})}{I_{n+1}(b\sqrt{z})}e^{zt} = 2(n+1)\frac{a^n}{b^{n+1}}$$

because $I_n(z) \sim (\tfrac{1}{2}z)^n/n!$ as $z \to 0$. On the other hand,

$$
\text{Res}(z = z_m) = \lim_{z \to z_m} \frac{(z - z_m)I_n(a\sqrt{z})}{\sqrt{z}I_{n+1}(b\sqrt{z})}e^{zt}
$$

$$
= \frac{4e^{-n\pi i/2}J_n(a\alpha_m/b)}{b[e^{-n\pi i/2}J_n(\alpha_m) + e^{(n+2)\pi i/2}J_{n+2}(\alpha_m)]}e^{-\alpha_m^2 t/b^2}.
$$

Summing the residues, the inverse is

$$
f(t) = 2(n+1)\frac{a^n}{b^{n+1}} + \frac{4}{b}\sum_{m=1}^{\infty}\frac{J_n(a\alpha_m/b)}{J_n(\alpha_m) - J_{n+2}(\alpha_m)}e^{-\alpha_m^2 t/b^2}.
$$

15. Deforming Bromwich's integral along the imaginary axis,

$$
f(t) = \frac{1}{2\pi i}\left[\int_{-\infty i}^{-0i} + \int_{C_\epsilon} + \int_{+0i}^{\infty i}\right]
$$

where C_ϵ is a very small semi-circle passing to the right of the singularity at $z = 0$. From a straightforward integration with $z = \epsilon e^{\theta i}$ with $\epsilon \to 0$ and $-\pi/2 < \theta < \pi/2$,

$$
\int_{C_\epsilon} = \pi i.
$$

Along the imaginary axis, we have $s = \eta^2 e^{\pm \pi i/2}/2$ and

$$
\tanh(\sqrt{s}) = \frac{\sinh(\eta) \pm i\sin(\eta)}{\cosh(\eta) + \cos(\eta)}.
$$

Then

$$
\int_{-\infty i}^{-0i} = \int_{\infty}^{0+}\frac{\eta\,d\eta}{\eta^2/2}\exp\left[-\frac{r\eta}{2}(1-i)\frac{\sinh(\eta) - i\sin(\eta)}{\cosh(\eta) + \cos(\eta)} - i\frac{\eta^2 t}{2}\right]
$$

and

$$
\int_{+0i}^{\infty i} = \int_{0+}^{\infty}\frac{\eta\,d\eta}{\eta^2/2}\exp\left[-\frac{r\eta}{2}(1+i)\frac{\sinh(\eta) + i\sin(\eta)}{\cosh(\eta) + \cos(\eta)} + i\frac{\eta^2 t}{2}\right]
$$

so that

$$
\int_{-\infty i}^{-0i} + \int_{+0i}^{\infty i} = 4i\int_{0+}^{\infty}\frac{d\eta}{\eta}\exp\left[-\frac{r\eta}{2}\frac{\sinh(\eta) - \sin(\eta)}{\cosh(\eta) + \cos(\eta)}\right]
$$

$$
\times \sin\left[\frac{\eta^2 t}{2} - \frac{r\eta}{2}\frac{\sinh(\eta) + \sin(\eta)}{\cosh(\eta) + \cos(\eta)}\right].
$$

Adding the above result to πi gives the Bromwich integral after dividing the sum by $2\pi i$.

Section 2.2

1. The poles are located at $w = 0$ and $w = a^2 i$. Therefore, the inversion integral is

$$f(t) = \frac{1}{2\pi} \int_{-\infty-\epsilon i}^{\infty-\epsilon i} \frac{e^{iwt}}{w(w-a^2 i)} \, dw.$$

Because there are no singularities in the lower half-plane, $f(t) = 0$ for $t < 0$. In the limit of $\epsilon \to 0$,

$$\text{Res}(w = 0) = \frac{1}{-a^2 i} \quad \text{and} \quad \text{Res}(w = a^2 i) = \frac{e^{-a^2 t}}{a^2 i}.$$

Thus, the residue theorem gives

$$f(t) = \frac{1}{a^2}\left(e^{-a^2 t} - 1\right) H(t).$$

2. The inversion formula is

$$f(t) = \frac{1}{2\pi} \int_{-\infty}^{\infty} \frac{e^{iwt}}{w^2 + a^2} \, dw.$$

The poles are located at $w = \pm ai$. For $t < 0$, we take the pole in the lower half-plane and

$$f(t) = \frac{1}{2\pi}\left[-2\pi i \, \text{Res}(w = -ai)\right] = -i \lim_{w \to -ia} \frac{(w+ai)e^{itw}}{(w+ai)(w-ai)} = \frac{e^{at}}{2a}.$$

The negative sign comes from taking the contour in the negative sense. For $t > 0$, we use the pole $w = ai$ and

$$f(t) = \frac{1}{2\pi}\left[2\pi i \, \text{Res}(w = ai)\right] = i \lim_{w \to ia} \frac{(w-ai)e^{itw}}{(w+ai)(w-ai)} = \frac{e^{-at}}{2a}.$$

Therefore, the total answer, using the absolute value sign, is

$$f(t) = \frac{e^{-a|t|}}{2a}.$$

3. The inversion formula is

$$f(t) = \frac{1}{2\pi} \int_{-\infty}^{\infty} \frac{w e^{iwt}}{w^2 + a^2} \, dw.$$

The poles are located at $\omega = \pm ai$. For $t < 0$, we take the pole in the lower half-plane and

$$f(t) = \frac{1}{2\pi}\left[-2\pi i \; \text{Res}(\omega = -ai)\right] = -i \lim_{\omega \to -ia} \frac{(\omega + ai)\omega e^{it\omega}}{(\omega + ai)(\omega - ai)} = \frac{-ie^{at}}{2}.$$

The negative sign comes from taking the contour in the negative sense. For $t > 0$, we use the pole $\omega = ai$ and

$$f(t) = \frac{1}{2\pi}\left[2\pi i \; \text{Res}(\omega = ai)\right] = i \lim_{\omega \to ia} \frac{(\omega - ai)\omega e^{it\omega}}{(\omega + ai)(\omega - ai)} = \frac{ie^{-at}}{2}.$$

Therefore, the total answer, using the absolute value sign, is

$$f(t) = \frac{i}{2}\text{sgn}(t)e^{-a|t|}$$

where

$$\text{sgn}(t) = \begin{cases} 1 & \text{if } t > 0 \\ -1 & \text{if } t < 0. \end{cases}$$

4. The inversion formula is

$$f(t) = \frac{1}{2\pi}\int_{-\infty}^{\infty} \frac{\omega e^{i\omega t}}{(\omega^2 + a^2)^2}\, d\omega.$$

The poles are located at $\omega = \pm ai$ and are second order. For $t < 0$, we take the pole in the lower half-plane and

$$f(t) = \frac{1}{2\pi}\left[-2\pi i \; \text{Res}(\omega = -ai)\right] = -i \lim_{\omega \to -ia} \frac{d}{d\omega}\left[\frac{(\omega + ai)^2 \omega e^{it\omega}}{(\omega + ai)^2(\omega - ai)^2}\right]$$

$$= -i \lim_{\omega \to -ia}\left[\frac{e^{it\omega}}{(\omega - ai)^2} + \frac{it\omega e^{it\omega}}{(\omega - ai)^2} - \frac{2\omega e^{it\omega}}{(\omega - ai)^3}\right]$$

$$= -i\left(-\frac{e^{at}}{4a^2} - \frac{te^{at}}{4a} + \frac{e^{at}}{4a^2}\right) = \frac{ite^{at}}{4a}.$$

The negative sign comes from taking the contour in the negative sense. For $t > 0$, we use the pole $\omega = ai$ and

$$f(t) = \frac{1}{2\pi}\left[2\pi i \; \text{Res}(\omega = ai)\right] = i \lim_{\omega \to ia} \frac{d}{d\omega}\left[\frac{(\omega - ai)^2 \omega e^{it\omega}}{(\omega + ai)^2(\omega - ai)^2}\right]$$

$$= i \lim_{\omega \to -ia}\left[\frac{e^{it\omega}}{(\omega + ai)^2} + \frac{it\omega e^{it\omega}}{(\omega + ai)^2} - \frac{2\omega e^{it\omega}}{(\omega + ai)^3}\right]$$

$$= i\left(-\frac{e^{-at}}{4a^2} + \frac{te^{-at}}{4a} + \frac{e^{-at}}{4a^2}\right) = \frac{ite^{-at}}{4a}.$$

Therefore, the total answer, using the absolute value sign, is

$$f(t) = \frac{it}{4a} e^{-a|t|}.$$

5. The inversion formula is

$$f(t) = \frac{1}{2\pi} \int_{-\infty}^{\infty} \frac{\omega^2 e^{i\omega t}}{(\omega^2 + a^2)^2} \, d\omega.$$

The poles are located at $\omega = \pm ai$ and are second order. For $t < 0$, we take the pole in the lower half-plane and

$$f(t) = \frac{1}{2\pi} \left[-2\pi i \, \mathrm{Res}(\omega = -ai) \right] = -i \lim_{\omega \to -ia} \frac{d}{d\omega} \left[\frac{(\omega + ai)^2 \omega^2 e^{i\omega t}}{(\omega + ai)^2 (\omega - ai)^2} \right]$$

$$= -i \lim_{\omega \to -ia} \left[\frac{2\omega e^{i\omega t}}{(\omega - ai)^2} + \frac{it\omega^2 e^{i\omega t}}{(\omega - ai)^2} - \frac{2\omega^2 e^{i\omega t}}{(\omega - ai)^3} \right]$$

$$= \frac{e^{at}}{2a} + \frac{te^{at}}{4} - \frac{e^{at}}{4a} = \frac{ate^{at}}{4a} + \frac{e^{at}}{4a}.$$

The negative sign comes from taking the contour in the negative sense. For $t > 0$, we use the pole $\omega = ai$ and

$$f(t) = \frac{1}{2\pi} \left[2\pi i \, \mathrm{Res}(\omega = ai) \right] = i \lim_{\omega \to ia} \frac{d}{d\omega} \left[\frac{(\omega - ai)^2 \omega^2 e^{i\omega t}}{(\omega + ai)^2 (\omega - ai)^2} \right]$$

$$= i \lim_{\omega \to ia} \left[\frac{2\omega e^{i\omega t}}{(\omega + ai)^2} + \frac{it\omega^2 e^{i\omega t}}{(\omega + ai)^2} - \frac{2\omega^2 e^{i\omega t}}{(\omega + ai)^3} \right]$$

$$= \frac{e^{-at}}{2a} - \frac{te^{-at}}{4} - \frac{e^{-at}}{4a} = \frac{e^{-at}}{4a} - \frac{ate^{-at}}{4a}.$$

Therefore, the total answer, using the absolute value sign, is

$$f(t) = \frac{1}{4a}(1 - a|t|)e^{-a|t|}.$$

6. The inversion formula is

$$f(t) = \frac{1}{2\pi} \int_{-\infty}^{\infty} \frac{e^{i\omega t}}{\omega^2 - 3i\omega - 3} \, d\omega.$$

The simple poles are located at $\omega = (\pm\sqrt{3} + 3i)/2$. Because there are

Transform Methods for Solving Partial Differential Equations

no singularities in the lower half-plane, $f(t) = 0$ for $t < 0$. For $t > 0$,

$$f(t) = \frac{1}{2\pi} \left\{ 2\pi i \text{ Res}[\omega = (\sqrt{3} + 3i)/2] \right.$$

$$\left. + 2\pi i \text{ Res}[\omega = (-\sqrt{3} + 3i)/2] \right\}$$

$$= i \left\{ \lim_{\omega \to (\sqrt{3}+3i)/2} \frac{[\omega - (\sqrt{3} + 3i)/2] e^{i\omega t}}{\omega^2 - 3i\omega - 3} \right.$$

$$\left. + \lim_{\omega \to (-\sqrt{3}+3i)/2} \frac{[\omega - (-\sqrt{3} + 3i)/2] e^{i\omega t}}{\omega^2 - 3i\omega - 3} \right\}$$

$$= \frac{ie^{-3t/2}[\cos(\sqrt{3}t/2) + i\sin(\sqrt{3}t/2)]}{2(\sqrt{3}/2 + 3i/2) - 3i}$$

$$+ \frac{ie^{-3t/2}[\cos(\sqrt{3}t/2) - i\sin(\sqrt{3}t/2)]}{2(-\sqrt{3}/2 + 3i/2) - 3i}$$

$$= -\frac{2}{\sqrt{3}} e^{-3t/2} \sin\left(\frac{\sqrt{3}t}{2}\right).$$

Therefore, the total answer is

$$f(t) = -\frac{2}{\sqrt{3}} e^{-3t/2} \sin(\sqrt{3}t/2) H(t).$$

7. The inversion formula is

$$f(t) = \frac{1}{2\pi} \int_{-\infty}^{\infty} \frac{e^{i\omega t}}{(\omega - ai)^{2n+2}} \, d\omega.$$

The pole is located at $\omega = ai$ and it is a $(2n + 2)$th order pole. Because there are no singularities in the lower half-plane, $f(t) = 0$ for $t < 0$. For $t > 0$,

$$f(t) = \frac{1}{2\pi} 2\pi i \text{ Res}(\omega = ai)$$

$$= \lim_{\omega \to ai} \left\{ \frac{1}{(2n + 1)!} \frac{d^{2n+1}}{d\omega^{2n+1}} \left[\frac{(\omega - ai)^{2n+2} e^{i\omega t}}{(\omega - ai)^{2n+2}} \right] \right\}$$

$$= i\frac{1}{(2n + 1)!} (i)^{2n+2} t^{2n+1} e^{-at} = \frac{(-1)^{n+1}}{(2n + 1)!} t^{2n+1} e^{-at}.$$

Therefore, the total answer is

$$f(t) = \frac{(-1)^{n+1}}{(2n + 1)!} t^{2n+1} e^{-at} H(t).$$

8. The inversion formula is

$$f(t) = \frac{1}{2\pi} \int_{-\infty}^{\infty} \frac{2i \sin(\omega h/2) e^{i\omega t}}{\omega^2 + a^2} \, d\omega$$

$$= \frac{1}{2\pi} \int_{-\infty}^{\infty} \frac{e^{i(t+h/2)\omega} - e^{i(t-h/2)\omega}}{\omega^2 + a^2} \, d\omega$$

$$= \frac{1}{2\pi} \int_{-\infty}^{\infty} \frac{e^{i(t+h/2)\omega}}{\omega^2 + a^2} \, d\omega - \frac{1}{2\pi} \int_{-\infty}^{\infty} \frac{e^{i(t-h/2)\omega}}{\omega^2 + a^2} \, d\omega.$$

The simple poles are located at $\omega = \pm ai$.

$$\text{Res}(\omega = ai) \text{ of first integral} = \frac{e^{-a(t+h/2)}}{2ia}.$$

$$\text{Res}(\omega = ai) \text{ of second integral} = \frac{e^{-a(t-h/2)}}{2ia}.$$

$$\text{Res}(\omega = -ai) \text{ of first integral} = \frac{e^{a(t+h/2)}}{-2ia}.$$

$$\text{Res}(\omega = -ai) \text{ of second integral} = \frac{e^{a(t-h/2)}}{-2ia}.$$

If $t > h/2$,

$$f(t) = \frac{1}{2\pi} \left\{ 2\pi i [\text{Res}(\omega = ai) \text{ of first integral}] \right.$$

$$\left. - 2\pi i [\text{Res}(\omega = ai) \text{ of second integral}] \right\}$$

$$= \frac{1}{2a} \left[e^{-a(t+h/2)} - e^{-a(t-h/2)} \right] = -\frac{e^{-at}}{a} \sinh\left(\frac{ah}{2}\right).$$

If $-h/2 < t < h/2$,

$$f(t) = \frac{1}{2\pi} \left\{ 2\pi i [\text{Res}(\omega = ai) \text{ of first integral}] \right.$$

$$\left. + 2\pi i [\text{Res}(\omega = -ai) \text{ of second integral}] \right\}$$

$$= \frac{1}{2a} \left[e^{-a(t+h/2)} - e^{-a(t-h/2)} \right] = -\frac{e^{-ah/2}}{a} \sinh(at).$$

If $t < -h/2$,

$$f(t) = \frac{1}{2\pi} \left\{ -2\pi i [\text{Res}(\omega = -ai) \text{ of first integral}] \right.$$

$$\left. - 2\pi i [\text{Res}(\omega = -ai) \text{ of second integral}] \right\}$$

$$= \frac{1}{2a} \left[e^{a(t+h/2)} - e^{a(t-h/2)} \right] = \frac{e^{at}}{a} \sinh\left(\frac{ah}{2}\right).$$

Therefore, the total answer is

$$f(t) = \begin{cases} -e^{-at}\sinh(ah/2)/a & \text{if } t > h/2 \\ -e^{-ah/2}\sinh(at)/a & \text{if } -h/2 < t < h/2 \\ e^{at}\sinh(ah/2)/a & \text{if } t < -h/2. \end{cases}$$

9. The inversion formula is

$$f(t) = \frac{1}{2\pi}\int_{-\infty}^{\infty}\frac{\omega^2 e^{i\omega t}}{(\omega^2 - 1)^2 + 4a^2\omega^2}\,d\omega.$$

The simple poles are located at $\omega = \pm\sqrt{1 - a^2} \pm ai$. Let us assume that $a < 1$. Then

$$\begin{aligned}
\mathrm{Res}(\omega = \sqrt{1 - a^2} + ai) &= \lim_{\omega\to\sqrt{1-a^2}+ai}\left[\frac{(\omega - \sqrt{1 - a^2} - ai)\omega^2 e^{i\omega t}}{(\omega^2 - 1)^2 + 4a^2\omega^2}\right] \\
&= \lim_{\omega\to\sqrt{1-a^2}+ai}\frac{\omega e^{i\omega t}}{4(\omega^2 - 1) + 8a^2} \\
&= \frac{e^{it\sqrt{1-a^2}}e^{-at}}{8ia\sqrt{1 - a^2}}(\sqrt{1 - a^2} + ia)
\end{aligned}$$

and

$$\begin{aligned}
\mathrm{Res}(\omega = -\sqrt{1 - a^2} + ai) &= \lim_{\omega\to-\sqrt{1-a^2}+ai}\left[\frac{(\omega + \sqrt{1 - a^2} - ai)\omega^2 e^{i\omega t}}{(\omega^2 - 1)^2 + 4a^2\omega^2}\right] \\
&= \lim_{\omega\to-\sqrt{1-a^2}+ai}\frac{\omega e^{i\omega t}}{4(\omega^2 - 1) + 8a^2} \\
&= \frac{e^{-it\sqrt{1-a^2}}e^{-at}}{-8ia\sqrt{1 - a^2}}(-\sqrt{1 - a^2} + ia).
\end{aligned}$$

Therefore, for $t > 0$, the residue theorem gives

$$\begin{aligned}
f(t) &= \frac{e^{-at}}{8a\sqrt{1 - a^2}}\Big[\sqrt{1 - a^2}\cos(t\sqrt{1 - a^2}) - a\sin(t\sqrt{1 - a^2}) \\
&\qquad + ia\cos(t\sqrt{1 - a^2}) + i\sqrt{1 - a^2}\sin(t\sqrt{1 - a^2}) \\
&\qquad + \sqrt{1 - a^2}\cos(t\sqrt{1 - a^2}) - a\sin(t\sqrt{1 - a^2}) \\
&\qquad - ia\cos(t\sqrt{1 - a^2}) - i\sqrt{1 - a^2}\sin(t\sqrt{1 - a^2})\Big] \\
&= \frac{e^{-at}}{4a}\cos(t\sqrt{1 - a^2}) - \frac{e^{-at}}{4\sqrt{1 - a^2}}\sin(t\sqrt{1 - a^2}).
\end{aligned}$$

Similarly, for the lower half-plane,

$$\mathrm{Res}(\omega = \sqrt{1-a^2} - ai) = \lim_{\omega \to \sqrt{1-a^2}-ai} \left[\frac{(\omega - \sqrt{1-a^2} + ai)\omega^2 e^{i\omega t}}{(\omega^2 - 1)^2 + 4a^2\omega^2} \right]$$

$$= \lim_{\omega \to \sqrt{1-a^2}-ai} \frac{\omega e^{i\omega t}}{4(\omega^2 - 1) + 8a^2}$$

$$= \frac{e^{it\sqrt{1-a^2}} e^{at}}{-8ia\sqrt{1-a^2}} (\sqrt{1-a^2} - ia)$$

and

$$\mathrm{Res}(\omega = -\sqrt{1-a^2} - ai) = \lim_{\omega \to -\sqrt{1-a^2}-ai} \left[\frac{(\omega + \sqrt{1-a^2} + ai)\omega^2 e^{i\omega t}}{(\omega^2 - 1)^2 + 4a^2\omega^2} \right]$$

$$= \lim_{\omega \to -\sqrt{1-a^2}-ai} \frac{\omega e^{i\omega t}}{4(\omega^2 - 1) + 8a^2}$$

$$= \frac{e^{-it\sqrt{1-a^2}} e^{at}}{8ia\sqrt{1-a^2}} (-\sqrt{1-a^2} - ia).$$

Therefore, for $t < 0$, the residue theorem gives

$$f(t) = -\frac{e^{at}}{8a\sqrt{1-a^2}}$$

$$\times \left[-\sqrt{1-a^2} \cos(t\sqrt{1-a^2}) - a\sin(t\sqrt{1-a^2}) \right.$$

$$- ia\cos(t\sqrt{1-a^2}) + i\sqrt{1-a^2} \sin(t\sqrt{1-a^2})$$

$$- \sqrt{1-a^2} \cos(t\sqrt{1-a^2}) - a\sin(t\sqrt{1-a^2})$$

$$\left. + ia\cos(t\sqrt{1-a^2}) - i\sqrt{1-a^2} \sin(t\sqrt{1-a^2}) \right]$$

$$= \frac{e^{at}}{4a} \cos(t\sqrt{1-a^2}) + \frac{e^{at}}{4\sqrt{1-a^2}} \sin(t\sqrt{1-a^2}).$$

Now, let us consider $a > 1$. The easiest method is to use the above results with $\sqrt{1-a^2} = i\sqrt{a^2 - 1}$ or $-i\sqrt{a^2 - 1}$. Therefore, the total answer is

$$f(t) = \begin{cases} \frac{e^{-a|t|} \cosh(\sqrt{a^2-1}|t|)}{4a} - \frac{e^{-a|t|} \sinh(\sqrt{a^2-1}|t|)}{4\sqrt{a^2-1}} & \text{if } a > 1 \\ \frac{e^{-a|t|} \cos(\sqrt{1-a^2}|t|)}{4a} - \frac{e^{-a|t|} \sin(\sqrt{1-a^2}|t|)}{4\sqrt{a^2-1}} & \text{if } 0 < a < 1. \end{cases}$$

10. The inversion formula is

$$f(t) = \frac{1}{2\pi} \int_{-\infty}^{\infty} \frac{e^{i\omega t}}{I_0(\omega)} d\omega.$$

Because $I_0(\omega) = J_0(\omega i)$ if $-\pi < \arg(\omega) \le \pi/2$, the the poles are located at $\omega = \pm i\alpha_n$ where $J_0(\alpha_n) = 0$ and $n = 1, 2, 3, \dots$ For $t < 0$,

$$f(t) = \frac{1}{2\pi}(-2\pi i)\left[\sum_{n=1}^{\infty} \text{Res}(\omega = -\alpha_n i)\right]$$

$$= -i\sum_{n=1}^{\infty} \lim_{\omega \to -\alpha_n i}\left[\frac{(\omega + \alpha_n i)e^{it\omega}}{I_0(z)}\right]$$

$$= -i\sum_{n=1}^{\infty} \frac{e^{\alpha_n t}}{I_0'(-i\alpha_n)}.$$

The negative sign comes from taking the contour in the negative sense. Now $I_0'(-i\alpha_n) = I_1(-i\alpha_n) = -iJ_1(\alpha_n)$, so

$$f(t) = \sum_{n=1}^{\infty} \frac{e^{\alpha_n t}}{J_1(\alpha_n)}.$$

For $t > 0$,

$$f(t) = \frac{1}{2\pi}(2\pi i)\left[\sum_{n=1}^{\infty} \text{Res}(\omega = \alpha_n i)\right] = i\sum_{n=1}^{\infty} \lim_{\omega \to \alpha_n i}\left[\frac{(\omega - \alpha_n i)e^{it\omega}}{I_0(z)}\right]$$

$$= i\sum_{n=1}^{\infty} \frac{e^{-\alpha_n t}}{I_0'(i\alpha_n)}.$$

Now $I_0'(i\alpha_n) = -iJ_1(-\alpha_n)$, so

$$f(t) = -\sum_{n=1}^{\infty} \frac{e^{-\alpha_n t}}{J_1(-\alpha_n)} = \sum_{n=1}^{\infty} \frac{e^{-\alpha_n t}}{J_1(\alpha_n)}.$$

Therefore, the total answer, using absolute value signs, is

$$f(t) = \sum_{n=1}^{\infty} \frac{\exp[-z_n|t|]}{J_1(z_n)} \quad \text{where} \quad J_0(z_n) = 0, n = 1, 2, 3, \dots$$

11. The poles are located at $\sinh(\omega h) = -i\sin(i\omega h) = 0$ or $\omega = \pm n\pi i/h$ with $n = 0, 1, 2, \dots$ Therefore, the inversion integral is

$$f(t) = \frac{1}{2\pi}\int_{-\infty-\epsilon i}^{\infty-\epsilon i} \frac{\cosh[\omega(x-h)]\cosh(\omega a)e^{i\omega t}}{i\sinh(\omega h)}\, d\omega.$$

In the limit of $\epsilon \to 0$,

$$\text{Res}(\omega = 0) = \lim_{z \to 0}\left\{z\frac{\cosh[z(x-h)]\cosh(za)e^{izt}}{i\sinh(zh)}\right\} = \frac{1}{ih}$$

and

$$\text{Res}(\omega = n\pi i/h) = \lim_{z \to n\pi i/h} \left\{ (z - n\pi i/h) \frac{\cosh[z(x-h)]\cosh(za)e^{izt}}{i\sinh(zh)} \right\}$$

$$= \frac{\cosh[n\pi(x-h)i/h]\cosh[n\pi ai/h]\exp[-n\pi t/h]}{ih\cosh(n\pi i)}$$

$$= \frac{\cos[n\pi(x-h)/h]\cos(n\pi a/h)\exp(-n\pi t/h)}{ih(-1)^n}.$$

Therefore, $t > 0$,

$$f(t) = \frac{1}{h} + \sum_{n=1}^{\infty} \frac{1}{h}\cos(n\pi x/h)\cos(n\pi a/h)\exp(-n\pi t/h).$$

On the other hand, $t < 0$

$$f(t) = -\sum_{n=1}^{\infty} \frac{1}{h}\cos(n\pi x/h)\cos(n\pi a/h)\exp(n\pi t/h).$$

Thus, the most general form of the inverse is

$$f(t) = \frac{H(t)}{h} + \frac{1}{h}\text{sgn}(t)\sum_{n=1}^{\infty}\cos\left(\frac{n\pi x}{h}\right)\cos\left(\frac{n\pi a}{h}\right)\exp\left(-\frac{n\pi|t|}{h}\right).$$

12. The poles are located at $\cos(\omega h/a) = 0$ or $\omega = \pm(2n-1)\pi a/(2h)$ with $n = 1, 2, 3, \ldots$ The pole $\omega = 0$ is removable. Therefore, the inversion integral is

$$f(t) = \frac{1}{2\pi}\int_{-\infty-\epsilon i}^{\infty-\epsilon i} \frac{\sin(\omega x/a)e^{i\omega t}}{\omega\cos(\omega h/a)}\,d\omega.$$

Because there are no singularities in the lower half of the ω-plane, $f(t) = 0$ for $t < 0$. On the other hand, in the limit of $\epsilon \to 0$,

$$\text{Res}[\omega = (2n-1)\pi a/(2h)]$$

$$= \lim_{z \to (2n-1)\pi a/h}\left\{[z - (2n-1)\pi a/(2h)]\frac{\sin(zx/a)e^{izt}}{z\cosh(zh/a)}\right\}$$

$$= \frac{\sin[(2n-1)\pi x/(2h)]}{[(2n-1)\pi/2]}\frac{\exp[(2n-1)\pi ati/(2h)]}{\{-\sin[(2n-1)\pi/2]\}}$$

$$= (-1)^{n+1}\sin[(2n-1)\pi x/(2h)]\frac{\exp[(2n-1)\pi ati/(2h)]}{[(2n-1)\pi/2]}.$$

A similar result is found for $\omega = -(2n+1)\pi a/(2h)$:

$$\text{Res}[\omega = -(2n-1)\pi a/(2h)] = (-1)^{n+1} \sin[(2n-1)\pi x/(2h)]$$
$$\times \frac{\exp[-(2n-1)\pi a t i/(2h)]}{[(2n-1)\pi/2]}.$$

Therefore, $t > 0$,

$$f(t) = \frac{4}{\pi} \sum_{n=1}^{\infty} (-1)^{n+1} \frac{\sin[(2n-1)\pi x/(2h)] \sin[(2n-1)\pi a t/(2h)]}{(2n-1)}.$$

Thus, the most general form of the inverse is

$$f(t) = \frac{4}{\pi} \sum_{n=1}^{\infty} \frac{(-1)^{n+1}}{2n-1} \cos\left[\frac{(2n-1)\pi x}{2h}\right] \sin\left[\frac{(2n-1)\pi a t}{2h}\right] H(t).$$

13. The poles are located at $\cosh(\omega a) = 0$ or $\omega = \pm(2n-1)\pi i/(2a)$ with $n = 1, 2, 3, \ldots$ The pole $\omega = 0$ is removable. There are also simple poles at $\omega = \pm m\pi$. Therefore, the inversion integral is

$$f(t) = \frac{1}{2\pi} \int_{-\infty-\epsilon i}^{\infty-\epsilon i} \frac{m\pi \sinh(\omega a) e^{i\omega t}}{2\omega(\omega^2 - m^2\pi^2)\cosh(\omega a)}$$
$$\times [1 + (-1)^m \cos(\omega) - i(-1)^m \sin(\omega)]\, d\omega.$$

In the limit of $\epsilon \to 0$,

$$\text{Res}(\omega = m\pi) = \frac{m\pi \sinh(m\pi a) e^{im\pi t}}{2m\pi(2m\pi)\cosh(m\pi a)}$$
$$\times [1 + (-1)^m \cos(m\pi) - i(-1)^m \sin(m\pi)]$$
$$= \frac{\tanh(m\pi a)}{2m\pi} e^{im\pi t},$$

$$\text{Res}(\omega = -m\pi) = \frac{m\pi \sinh(-m\pi a) e^{-im\pi t}}{-2m\pi(-2m\pi)\cosh(-m\pi a)}$$
$$\times [1 + (-1)^m \cos(-m\pi) - i(-1)^m \sin(-m\pi)]$$
$$= -\frac{\tanh(m\pi a)}{2m\pi} e^{-im\pi t},$$

$$\text{Res}[\omega = (2n-1)\pi i/2a] =$$

$$\frac{m\pi \sinh[(2n-1)\pi i/2] \exp[-(2n-1)\pi t/2a]}{2a \sinh[(2n-1)\pi i/2][(2n-1)\pi i/2a][-m^2\pi^2 - (2n-1)^2\pi^2/4a^2]}$$
$$\times \{1 + (-1)^m \cos[(2n-1)\pi i/2a] - i(-1)^m \sin[(2n-1)\pi i/2a)]\}$$

$$= -\frac{m\exp[-(2n-1)\pi t/2a]}{[m^2\pi^2 + (2n-1)^2\pi^2/4a^2](2n-1)i}$$
$$\times \{1 + (-1)^m \cosh[(2n-1)\pi i/2a] + (-1)^m \sinh[(2n-1)\pi i/2a)]\}$$

and

$$\text{Res}[\omega = -(2n-1)\pi i/2a] =$$

$$\frac{m\pi \sinh[-(2n-1)\pi i/2]\exp[(2n-1)\pi t/2a]}{2a\sinh[-(2n-1)\pi i/2][-(2n-1)\pi i/2a][-m^2\pi^2 - (2n-1)^2\pi^2/4a^2]}$$

$$\times\{1 + (-1)^m\cos[-(2n-1)\pi i/2a] - i(-1)^m\sin[-(2n-1)\pi i/2a)]\}$$

$$= \frac{m\exp[(2n-1)\pi t/2a]}{[m^2\pi^2 + (2n-1)^2\pi^2/4a^2](2n-1)i}$$

$$\times\{1 + (-1)^m\cosh[(2n-1)\pi i/2a] - (-1)^m\sinh[(2n-1)\pi i/2a)]\}.$$

Thus, the most general form of the inverse is

$$f(t) = -\frac{\tanh(m\pi)}{m\pi}H(t)$$

$$-\sum_{n=1}^{\infty}\frac{m\exp\left[-\frac{(2n-1)\pi|t|}{2a}\right]\left\{1 - (-1)^m\exp\left[\frac{(2n-1)\pi\,\text{sgn}(t)}{2a}\right]\right\}}{(2n-1)\left[m^2\pi^2 + \frac{(2n-1)^2\pi^2}{4a^2}\right]}.$$

14. The poles are located at $\omega = 0$, $\omega = \omega_n = \pm(2n-1)\pi i/2a$ and $\omega = -\omega_n$ with $n = 1, 2, 3, \ldots$ They are all simple poles. Therefore, the inversion integral is

$$f(t) = \frac{1}{2\pi i}\int_{-\infty-\epsilon i}^{\infty-\epsilon i}\left[\frac{e^{i(t+x)\omega}}{\omega\cosh(\omega)} + \frac{e^{i(t-x)\omega}}{2\omega\cosh(\omega)} - \frac{e^{it\omega}}{\omega}\right]d\omega.$$

For $0 \le x < t$, we close all of the contours in the upper half plane. For $x > t > 0$, however, we must evaluate the second term with a contour in the lower half plane. Now, for the third term,

$$\text{Res}(\omega = 0) = -1.$$

For the first term,
$$\text{Res}(\omega = 0) = 1$$

and

$$\text{Res}(\omega = \omega_n) = \lim_{\omega\to\omega_n}\frac{(\omega - \omega_n)e^{i(t+x)\omega}}{\omega\cosh(\omega)}$$

$$= \frac{2a\exp[-(2n-1)\pi(x+t)/2a]}{(2n-1)\pi ai\sinh[(2n-1)\pi i/2]}$$

$$= -\frac{2(-1)^n}{(2n-1)\pi}\exp[-(2n-1)\pi(x+t)/2a].$$

For the second term,

$$\text{Res}(\omega = 0) = -1,$$

$$\text{Res}(\omega = \omega_n) = -\lim_{\omega \to \omega_n} \frac{(\omega - \omega_n)e^{i(t-x)\omega}}{\omega \cosh(\omega)}$$

$$= \frac{2a \exp[-(2n-1)\pi(t-x)/2a]}{(2n-1)\pi a i \sinh[(2n-1)\pi i/2]}$$

$$= \frac{2(-1)^n}{(2n-1)\pi} \exp[-(2n-1)\pi(t-x)/2a].$$

and

$$\text{Res}(\omega = -\omega_n) = -\lim_{\omega \to \omega_n} \frac{(\omega - \omega_n)e^{i(t-x)\omega}}{\omega \cosh(\omega)}$$

$$= -\frac{2a \exp[(2n-1)\pi(t-x)/2a]}{(2n-1)\pi a i \sinh[(2n-1)\pi i/2]}$$

$$= -\frac{2(-1)^n}{(2n-1)\pi} \exp[(2n-1)\pi(t-x)/2a].$$

Thus, for $0 \le x < t$,

$$f(t) = -1 + g(-x,t) - g(x,t)$$

while for $x > t > 0$

$$f(t) = -g(x,t) + g(x,-t)$$

where

$$g(x,t) = \frac{2}{\pi} \sum_{n=1}^{\infty} \frac{(-1)^n}{2n-1} \exp\left[-\frac{(2n-1)\pi t}{2a}\right] \exp\left[-\frac{(2n-1)\pi x}{2a}\right].$$

15. From the inversion formula,

$$f(t) = -\frac{Pa}{2\pi r \rho c} \int_{-\infty}^{\infty} \frac{\exp[-\omega^2 b^2 (r-a)/(2c^3) + i\omega t]}{(\omega + B - Ai)(\omega - B - Ai)} \, d\omega.$$

The poles are located at $\omega = \pm B + Ai$. Therefore, for $t < 0$, $f(t) = 0$. For $t > 0$,

$$f(t) = -\frac{Pai}{r\rho c}\left[\text{Res}(\omega = B + Ai) + \text{Res}(\omega = -B + Ai)\right].$$

Now

$$\text{Res}(\omega = B + Ai)$$
$$= \frac{\exp[-b^2(B^2 - A^2 + 2iAB)(r - a)/(2c^3) + iBt - At]}{2B}$$

and

$$\text{Res}(\omega = -B + Ai)$$
$$= \frac{\exp[-b^2(B^2 - A^2 - 2iAB)(r - a)/(2c^3) - iBt - At]}{-2B}.$$

Adding the residues and simplifying give

$$f(t) = \frac{Pa}{r\rho cB} \exp\left[-b^2(B^2 - A^2)(r - a)/(2c^3) - At\right]$$
$$\times \sin\left[Bt - \frac{ABb^2}{c^3}(r - a)\right].$$

16. Because P_n is a real positive number, the poles at $\omega_n = i\bar{\omega}_n = i(P_n^2 + \lambda^2)$ lie along the positive imaginary axis. Therefore, $f(t)$ and $g(t)$ are zero for $t < 0$. For $t > 0$, the residue for computing $f(t)$ is

$$\text{Res}(\omega = \omega_n) = \frac{[P_n \cot(P_n) + \lambda]e^{i\omega_n t}}{\{2\lambda \cot(P_n) - 2\lambda P_n[1 + \cot^2(P_n)] - 2P_n\}\{-i/2P_n\}}$$
$$= \frac{[(2P_n^2 - \lambda^2)/2\lambda + \lambda]e^{-\bar{\omega}_n t}}{[(P_n^2 - \lambda^2)/P_n - \bar{\omega}_n^2/2\lambda P_n - 2P_n][-i/2P_n]}$$
$$= \frac{2(\bar{\omega}_n - \lambda^2)\exp(-\bar{\omega}_n t)}{i(\bar{\omega}_n + 2\lambda)}$$

because $\cot(P_n) = (P_n^2 - \lambda^2)/2\lambda_n P_n$, $\cot^2(P_n) = (P_n^2 - \lambda^2)^2/4\lambda^2 P_n^2$ and $1 + \cot^2(P_n) = \bar{\omega}_n^2/4\lambda^2 P_n^2$. Therefore,

$$f(t) = 2\sum_{n=1}^{\infty} \frac{\bar{\omega}_n - \lambda^2}{\bar{\omega}_n + 2\lambda} e^{-\bar{\omega}_n t}.$$

On the other hand, the residue used in computing $g(t)$ is

$$\text{Res}(\omega = \omega_n) = \frac{e^{i\omega_n t}}{\{\cot(P_n) - P_n[1 + \cot^2(P_n)]\}(-i/2P_n)}$$
$$= \frac{2(\bar{\omega}_n - \lambda^2)\exp(-\bar{\omega}_n t)}{i(\bar{\omega}_n + \lambda)}$$

Transform Methods for Solving Partial Differential Equations

because $\cot(P_n) = -\lambda/P_n$, $\cot^2(P_n) = \lambda^2/P_n^2$ and $1 + \cot^2(P_n) = \bar{\omega}_n/P_n^2$. Therefore,

$$g(t) = 2 \sum_{n=1}^{\infty} \frac{\bar{\omega}_n - \lambda^2}{\bar{\omega}_n + \lambda} e^{-\bar{\omega}_n t}.$$

17. The poles of the transform are located at $\zeta = 0$, a second-order pole, and $\zeta_n = n\pi i$ where $n = 1, 2, 3, \ldots$ To express these poles in terms of ω, we solve for ω and find that

$$\omega_n^{\pm} = \pm\sqrt{k^2 + \zeta_n^2} + i\delta = \pm\sqrt{k^2 - n^2\pi^2} + i\delta.$$

For sufficiently large n, say $N + 1$, ω_n becomes purely imaginary. Consequently there are a number of poles that lie along and above the real axis, i.e., $n = 1$ to N; the vast majority lie along the imaginary axis.

To find the inverse for $t > 0$, we convert the line integral into a closed contour by adding an infinite semicircle in the upper half-plane. Therefore,

$$f(t) = \frac{1}{2\pi} \int_{-\infty}^{\infty} \frac{\cosh(\zeta)}{\zeta \sinh(\zeta)} e^{i\omega t} d\omega = \frac{1}{2\pi} \oint_C \frac{\cosh(\zeta)}{\zeta \sinh(\zeta)} e^{izt} dz.$$

Upon applying the residue theorem,

$$\begin{aligned}
f(t) &= \frac{1}{2} \lim_{z \to k+i\delta} \frac{d}{dz} \left[\frac{(z - k - \delta i)^2 \cosh(\zeta) e^{itz}}{\zeta \sinh(\zeta)} \right] \\
&+ \frac{1}{2} \lim_{z \to -k+i\delta} \frac{d}{dz} \left[\frac{(z + k - \delta i)^2 \cosh(\zeta) e^{itz}}{\zeta \sinh(\zeta)} \right] \\
&+ i \sum_{n=0}^{N} \frac{\sinh(\zeta) e^{-\delta t}}{\frac{d}{d\zeta}[\zeta \cosh(\zeta)] \frac{d\zeta}{dz}|_{z=\omega_n^+}} \exp\left[it\sqrt{k^2 - n^2\pi^2} \right] \\
&+ i \sum_{n=0}^{N} \frac{\sinh(\zeta) e^{-\delta t}}{\frac{d}{d\zeta}[\zeta \cosh(\zeta)] \frac{d\zeta}{dz}|_{z=\omega_n^-}} \exp\left[-it\sqrt{k^2 - n^2\pi^2} \right] \\
&+ i \sum_{n=N+1}^{\infty} \frac{\sinh(\zeta) e^{-\delta t}}{\frac{d}{d\zeta}[\zeta \cosh(\zeta)] \frac{d\zeta}{dz}|_{z=\omega_n^+}} \exp\left[-t\sqrt{n^2\pi^2 - k^2} \right]
\end{aligned}$$

or

$$f(t) = -e^{-\delta t} \sin(kt)/2 - 2 \sum_{n=0}^{N} \frac{e^{-\delta t} \sin\left[t\sqrt{k^2 - n^2\pi^2} \right]}{\sqrt{k^2 - n^2\pi^2}}$$

$$+ \sum_{n=N+1}^{\infty} \frac{\exp\left[-\delta t - t\sqrt{n^2\pi^2 - k^2} \right]}{\sqrt{n^2\pi^2 - k^2}}.$$

18. The transform has singularities where

$$\sinh(d\sqrt{\omega i}) = i\sin(-id\sqrt{\omega i}) = 0$$

or $\sqrt{\omega_n i} = n\pi i/d$ and $\omega_n = n^2\pi^2 i/d^2$ with $n = 1, 2, 3, \ldots$ The singularity $\omega = 0$ is a removable singularity. From the Taylor expansion of hyperbolic sine, we find that $F(\omega)$ is a single-valued function. Therefore,

$$f(t) = \frac{1}{2\pi}\oint_C \frac{\sqrt{zi}}{\sinh(d\sqrt{zi})}e^{itz}\,dz$$

where C is a semicircle of infinite radius in the upper half-plane if $t > 0$; a semicircle of infinite radius in the lower half-plane if $t < 0$. Because there are no singularities in the lower half-plane, $f(t) = 0$. In the upper half-plane,

$$\text{Res}(z = z_n) = \lim_{z \to z_n} \frac{(z - z_n)\sqrt{zi}}{\sinh(d\sqrt{zi})}e^{itz} = -\frac{2n^2\pi^2}{d^3 i}e^{-n^2\pi^2 t/d^2}.$$

Multiplying by $2\pi i$ and summing the residues, we obtain the answer for $t > 0$.

19. (a) Poles are at

$$(U\omega - i\delta)^2 \cosh(\omega h) - g\omega \sinh(\omega h) = 0.$$

Let

$$\omega = \pm\omega_0 + \delta\omega_1 + \delta^2\omega_2 + \cdots.$$

Taking the positive sign

$$[U\omega_0 + (U\omega_1 - 1)\delta + \omega_2\delta^2 + \cdots]\cosh[h(\omega_0 + \delta\omega_1 + \delta^2\omega_2 + \cdots)]$$
$$- g(\omega_0 + \delta\omega_1 + \delta^2\omega_2 + \cdots)\sinh[h(\omega_0 + \delta\omega_1 + \delta^2\omega_2 + \cdots)] = 0.$$

Equating powers of δ, then $O(1)$ terms give

$$\tanh(\omega_0 h) = U^2\omega_0/g \quad \text{and} \quad \omega_0 = 0.$$

We get the same results when we take $-\omega_0$. Taking the $O(\delta)$ terms,

$$[2U^2\omega_0\cosh(\omega_0 h) + U_0^2\omega_0^2 h\sinh(\omega_0 h)$$
$$- g\sinh(\omega_0 h) - g\omega_0 h\cosh(\omega_0 h)]\omega_1 = 2iU\omega_0\cosh(\omega_0 h).$$

Using $\tanh(\omega_0 h) = U^2\omega_0/g$,

$$\omega_1 = -\frac{2iU\omega_0}{g[\omega_0 h\,\text{sech}^2(\omega_0 h) - \tanh(\omega_0 h)]}.$$

Transform Methods for Solving Partial Differential Equations

We get the same result for $-\omega_0$. Now, the $O(\delta)$ terms do not give us any information when $\omega_0 = 0$ and we must go the $O(\delta^2)$ terms. The $O(\delta^2)$ terms are

$$[2\omega_0\omega_2 U + (U\omega_1 - i)^2 + U^2\omega_0^2\omega_1^2/2 - gh\omega_0\omega_2 - gh\omega_1^2]\cosh(\omega_0 h)$$
$$+ [\omega_0^2\omega_2 U^2 + 2\omega_0\omega_1 U(U\omega_1 - i) - gh\omega_0\omega_1^2/2 - gh\omega_2]\sinh(\omega_0 h) = 0.$$

For $\omega_0 = 0$,

$$\omega_1 = \frac{i}{U \pm \sqrt{gh}}.$$

(b)

$$f(t) = \frac{1}{2\pi} \int_{-\infty}^{\infty} \frac{\cosh(\omega h)e^{it\omega}}{(U\omega - i\delta)^2 \cosh(\omega h) - g\omega\sinh(\omega)} \, d\omega.$$

Now,

$$\text{Res}(\omega = \omega_n) = \lim_{\omega\to\omega_n} \frac{(\omega - \omega_n)\cosh(\omega h)e^{it\omega}}{(U\omega - i\delta)^2 \cosh(\omega h) - g\omega\sinh(\omega)}$$
$$= \frac{\cosh(\omega_n h)e^{it\omega_n}}{D(\omega_n)}$$

where

$$D(\omega) = 2U(U\omega - i\delta)\cosh(\omega h) + (U\omega - i\delta)^2 h\sinh(\omega h)$$
$$- g\sinh(\omega h) - gh\omega\cosh(\omega h).$$

For $t < 0$, $\omega_I = i\delta/[U - \sqrt{gh}]$ is the only pole. Therefore,

$$\text{Res}(\omega_I) = \frac{\exp[-\delta t/(U - \sqrt{gh})]}{2Ui\delta[U/(U - \sqrt{gh}) - 1] - 2i\delta gh/(U - \sqrt{gh})}$$
$$= \frac{1}{2i\delta\sqrt{gh}} - \frac{t}{2i\sqrt{gh}(U - \sqrt{gh})} + O(\delta).$$

Multiplying by i and noting that the contour is taken in the negative sense

$$f(t) = -\frac{1}{2\delta\sqrt{gh}} - \frac{t}{2i\sqrt{gh}(\sqrt{gh} - U)} + O(\delta).$$

For $t > 0$,

$$\text{Res}(\omega_{II}) = -\frac{1}{2i\delta\sqrt{gh}} + \frac{t}{2i\sqrt{gh}(U + \sqrt{gh})} + O(\delta),$$

$$\text{Res}(\omega_{III}) = \frac{\cosh(\omega_0 h)e^{it\omega_0}}{D(\omega_0)}$$

and

$$\text{Res}(\omega_{IV}) = -\frac{\cosh(\omega_0 h) e^{it\omega_0}}{D(\omega_0)}$$

where

$$\begin{aligned}
D(\omega_0) &= 2U^2\omega_0 \cosh(\omega_0 h) + U^2\omega_0^2 h \sinh(\omega_0 h) \\
&\quad - g\sinh(\omega_0 h) - gh\omega_0 \cosh(\omega_0 h).
\end{aligned}$$

Therefore,

$$\begin{aligned}
i[\text{Res}(\omega = \omega_{III}) &+ \text{Res}(\omega = \omega_{IV})] \\
&= -\frac{2\cosh(\omega_0 h)\sin(\omega_0 t)}{D(\omega_0)} + O(\delta) \\
&= \frac{2\cosh^2(\omega_0 h)\sin(\omega_0 t)}{g\tanh(\omega_0 h)[gh/U^2 - \cosh^2(\omega_0 h)]} + O(\delta).
\end{aligned}$$

In summary, then

$$f(t) = \begin{cases}
-\dfrac{1}{2\delta\sqrt{gh}} - \dfrac{t}{2\sqrt{gh}(\sqrt{gh}-U)} + O(\delta) & \text{if } t < 0 \\[3mm]
-\dfrac{1}{2\delta\sqrt{gh}} + \dfrac{t}{2\sqrt{gh}(\sqrt{gh}+U)} & \\[2mm]
\quad + \dfrac{2\cosh^2(\omega_0 h)\sin(\omega_0 t)}{g\tanh(\omega_0 h)(gh/U^2 - \cosh^2(\omega_0 h))} + O(\delta) & \text{if } t > 0.
\end{cases}$$

20. (a) Poles are at

$$(U\omega - i\epsilon)^2[\omega^2 + (nN/U\kappa)^2] - N^2\omega^2 = 0.$$

Let

$$\omega = \pm\omega_0 + \epsilon\omega_1 + \epsilon^2\omega_2 + \cdots.$$

Taking the positive sign

$$\begin{aligned}
[U\omega_0 + (U\omega_1 - i)\epsilon &+ U\omega_2\epsilon^2 + \cdots]^2[(\omega_0 + \epsilon\omega_1 + \epsilon^2\omega_2 + \cdots)^2 \\
&+ (nN/U\kappa)^2] - N^2(\omega_0 + \epsilon\omega_1 + \epsilon^2\omega_2 + \cdots)^2 = 0.
\end{aligned}$$

Equating powers of ϵ, then $O(1)$ terms give

$$\omega_0 = N\sqrt{1 - (n/\kappa)^2}/U \quad \text{and} \quad \omega_0 = 0.$$

We get the same results when we take $-\omega_0$. Taking the $O(\epsilon)$ terms,

$$2U\omega_0(U\omega_1 - i)[\omega_0^2 + (nN/U\kappa)^2] + 2\omega_1\omega_0^3 U^2 - 2\omega_0\omega_1 N^2 = 0.$$

Simplifying,

$$\omega_1 = \frac{i}{U[1 - (n/\kappa)^2]}.$$

Transform Methods for Solving Partial Differential Equations

We get the same result for $-\omega_0$. Now, the $O(\epsilon)$ terms do not give us any information when $\omega_0 = 0$ and we must go the $O(\epsilon^2)$ terms. For $\omega_0 = 0$, the $O(\epsilon^2)$ terms are

$$(U\omega_1 - i)^2(nN/U\kappa)^2 - N^2\omega_1^2 = 0.$$

or

$$\omega_1 = \frac{ni/U\kappa}{n/\kappa \pm 1}.$$

(b)

$$f(t) = -\frac{1}{2\pi}\int_{-\infty}^{\infty}\frac{iU\omega + \epsilon}{(U\omega - i\epsilon)^2[\omega^2 + (nN/U\kappa)^2] - N^2\omega^2}d\omega.$$

Now, for $n/\kappa < 1$, we have three poles in the upper half plane:

$$\omega_I = \omega_0 + \frac{\epsilon i}{U(1 - n^2/\kappa^2)} + O(\epsilon^2)$$

$$\omega_{II} = -\omega_0 + \frac{\epsilon i}{U(1 - n^2/\kappa^2)} + O(\epsilon^2)$$

$$\omega_{III} = \frac{n\epsilon i/U\kappa}{n/\kappa + 1} + O(\epsilon^2)$$

and one pole in the lower half plane:

$$\omega_{IV} = \frac{n\epsilon i/U\kappa}{n/\kappa - 1} + O(\epsilon^2)$$

where $\omega_0 = N\sqrt{1 - (n/\kappa)^2}/U$. These poles are all simple and the corresponding residues are

$$\mathrm{Res}(\omega_I) = -\frac{iU\exp[itN\sqrt{1 - (n/\kappa)^2}/U]}{2N^2[1 - (n/\kappa)^2]},$$

$$\mathrm{Res}(\omega_{II}) = -\frac{iU\exp[-itN\sqrt{1 - (n/\kappa)^2}/U]}{2N^2[1 - (n/\kappa)^2]},$$

$$\mathrm{Res}(\omega_{III}) = \frac{U}{2N^2i(n/\kappa)(1 + n/\kappa)}$$

and

$$\mathrm{Res}(\omega_{IV}) = \frac{U}{2N^2i(n/\kappa)(n/\kappa - 1)}.$$

Using the residue theorem and recalling that for $t < 0$ we have a contour in the negative sense,

$$f(t) = \frac{U}{2N^2} \left\{ \frac{1}{(n/\kappa)(1 + n/\kappa)} + \frac{2\cos[Nt\sqrt{1 - (n/\kappa)^2}/U]}{1 - (n/\kappa)^2} \right\}$$

for $t > 0$ and

$$f(t) = \frac{U}{2N^2(n/\kappa)(1 - n/\kappa)}$$

for $t < 0$. On the other hand, for $n/\kappa > 1$, we have three poles in the upper half plane:

$$\omega_I = i\omega_0 - \frac{\epsilon i}{U(n^2/\kappa^2 - 1)} + O(\epsilon^2)$$

$$\omega_{II} = \frac{n\epsilon i/U\kappa}{n/\kappa + 1} + O(\epsilon^2)$$

$$\omega_{III} = \frac{n\epsilon i/U\kappa}{n/\kappa - 1} + O(\epsilon^2)$$

and one pole in the lower half plane:

$$\omega_{IV} = -i\omega_0 - \frac{\epsilon i}{U(n^2/\kappa^2 - 1)} + O(\epsilon^2)$$

where $\omega_0 = N\sqrt{(n/\kappa)^2 - 1}/U$. These poles are all simple and the corresponding residues are

$$\text{Res}(\omega_I) = -\frac{iU\exp[-tN\sqrt{(n/\kappa)^2 - 1}/U]}{2N^2[1 - (n/\kappa)^2]},$$

$$\text{Res}(\omega_{II}) = \frac{U}{2N^2 i(n/\kappa)(1 + n/\kappa)}$$

$$\text{Res}(\omega_{III}) = \frac{U}{2N^2 i(n/\kappa)(n/\kappa - 1)}.$$

and

$$\text{Res}(\omega_{IV}) = -\frac{iU\exp[tN\sqrt{(n/\kappa)^2 - 1}/U]}{2N^2[1 - (n/\kappa)^2]},$$

Using the residue theorem and recalling that for $t < 0$ we have a contour in the negative sense,

$$f(t) = \frac{U}{2N^2} \left\{ \frac{2 - \exp[-Nt\sqrt{(n/\kappa)^2 - 1}/U]}{(n/\kappa)^2 - 1} \right\}$$

for $t < 0$ and

$$f(t) = \frac{U}{2N^2} \frac{\exp[Nt\sqrt{(n/\kappa)^2 - 1}/U]}{(n/\kappa)^2 - 1}.$$

Section 2.5

$$\mathcal{H}[\delta(r)] = \int_0^\infty \delta(r)J_0(kr)r \, dr$$

$$= \frac{1}{2\pi} \int_{-\infty}^\infty \delta(x)\left[\int_{-\infty}^\infty \delta(y)J_0(k\sqrt{x^2 + y^2}) \, dy\right] dx$$

$$= \frac{1}{2\pi} \int_{-\infty}^\infty \delta(x)J_0(kx) \, dx = \frac{1}{2\pi}.$$

Section 3.1

1. With the branch cut taken along the negative real axis, Bromwich's integral reduces to residues from the poles $s = \pm\omega i$ plus a contour integral along two semi-circles of infinite radius in the second and third quadrants and a line integration along the branch cut. From Jordan's lemma, the contribution from the arcs vanish. Furthermore, the integration around the branch point vanishes. Computing the residues first,

$$\text{Res}(s = \omega i) = \frac{1}{2i} \exp[i\omega t - xe^{\pi i/4}\sqrt{\omega/\kappa}]$$

$$= \frac{1}{2i} \exp(-x\sqrt{\omega/2\kappa}) \exp[i(\omega t - x\sqrt{\omega/2\kappa})]$$

and

$$\text{Res}(s = -\omega i) = -\frac{1}{2i} \exp[-i\omega t - xe^{-\pi i/4}\sqrt{\omega/\kappa}]$$

$$= -\frac{1}{2i} \exp(-x\sqrt{\omega/2\kappa}) \exp[-i(\omega t - x\sqrt{\omega/2\kappa})].$$

The sum of the residues equals

$$\exp(-x\sqrt{\omega/2\kappa}) \sin(\omega t - x\sqrt{\omega/2\kappa}).$$

Along the top of the branch cut $s = ue^{\pi i}$ where $0 < u < \infty$. Therefore,

$$\int_{top} = -\int_\infty^0 \frac{\omega}{\omega^2 + u^2} e^{-ut - xi\sqrt{u/\kappa}} du.$$

Along the bottom of the branch cut $s = ue^{-\pi i}$ where $0 < u < \infty$. Therefore,

$$\int_{bottom} = -\int_0^\infty \frac{\omega}{\omega^2 + u^2} e^{-ut + xi\sqrt{u/\kappa}} \, du.$$

The contribution to Bromwich's integral equals the sum of these two integrals divided by $2\pi i$ or

$$-\frac{1}{\pi} \int_0^\infty \frac{\omega}{\omega^2 + u^2} e^{-ut} \sin(x\sqrt{u/\kappa}) \, du.$$

Introducing $u = \kappa\eta^2$, this integral becomes

$$-\frac{2\kappa}{\pi} \int_0^\infty \frac{\omega}{\omega^2 + \kappa^2\eta^4} e^{-\kappa t\eta^2} \sin(x\eta)\eta \, d\eta.$$

2. With the branch cut taken along the negative real axis, Bromwich's integral reduces to residues from the poles $s = \pm i$ plus the branch cut integral. Computing the residues first,

$$\text{Res}(s = i) = \frac{1}{2i} \exp(it - ix/\sqrt{1+i})$$

$$= \frac{1}{2i} \exp\{it - ix[\cos(\pi/8) - i\sin(\pi/8)]/\sqrt{2}\}$$

$$= \frac{1}{2i} \exp\{i[t - x\cos(\pi/8)/\sqrt{2}]\} \exp[-x\sin(\pi/8)/\sqrt{2}]$$

and

$$\text{Res}(s = -i) = -\frac{1}{2i} \exp(-it + ix/\sqrt{1-i})$$

$$= -\frac{1}{2i} \exp\{-it + ix[\cos(\pi/8) + i\sin(\pi/8)]/\sqrt{2}\}$$

$$= -\frac{1}{2i} \exp\{-i[t - x\cos(\pi/8)/\sqrt{2}]\} \exp[-x\sin(\pi/8)/\sqrt{2}].$$

The sum of the residues equals

$$\exp[-x\sin(\pi/8)/\sqrt{2}] \sin[t - x\cos(\pi/8)/\sqrt{2}].$$

Along the top of the branch cut $s + 1 = ue^{\pi i}$ where $0 < u < \infty$. Therefore,

$$\int_{top} = \int_\infty^0 \frac{-du}{1 + (-1 - u)^2} \exp\left[-\frac{(-1 - u)x}{i\sqrt{u}} + t(-1 - u)\right].$$

Along the bottom of the branch cut $s + 1 = ue^{-\pi i}$ where $0 < u < \infty$. Therefore,

$$\int_{bottom} = \int_0^\infty \frac{-du}{1 + (-1 - u)^2} \exp\left[-\frac{(-1 - u)x}{-i\sqrt{u}} + t(-1 - u)\right].$$

The contribution to Bromwich's integral equals the sum of these two integrals divided by $2\pi i$ or

$$-\frac{1}{\pi} \int_0^\infty \frac{\exp[-t(1 + u)]}{u^2 + 2u + 2} \sin\left[\frac{(1 + u)x}{\sqrt{u}}\right] du.$$

3. We may close Bromwich's integral with a semicircle in the left half of the s-plane with a cut to exclude the branch cut from $s = -c$ to $s = 0$. Because the arcs at infinity vanish,

$$f(t) = \text{Res}(s = s_0) - \frac{1}{2\pi i} \int_{branch\ cut}$$

where $a + b \log(1 + c/s_0) = 0$ or

$$s_0 = -\frac{c}{1 - e^{-a/b}}.$$

The residue at $s = s_0$ is

$$\text{Res}(s = s_0) = \lim_{s \to s_0} \frac{(s - s_0)e^{st}}{s[a + b \log(1 + c/s)]} = \frac{\exp[-ct/(1 - e^{-a/b})]}{b(e^{a/b} - 1)}.$$

There is no contribution from integrations around the branch points. Along the top of the branch cut, $s = cue^{\pi i}$ and $s + c = c(1 - u)e^{0i}$ so that

$$\int_{top} = \int_1^0 \frac{e^{-cut}}{(-cu)[a + b \ln(1/u - 1) - b\pi i]}(-c\,du)$$

while along the bottom of the branch cut, $s = cue^{\pi i}$ and $s + c = c(1 - u)e^{2\pi i}$ so that

$$\int_{bottom} = \int_0^1 \frac{e^{-cut}}{(-cu)[a + b \ln(1/u - 1) + b\pi i]}(-c\,du).$$

Therefore, the branch cut equals

$$\int_{branch\ cut} = -2b\pi i \int_0^1 \frac{e^{-uct}}{u\{[a + b \ln(1/u - 1)]^2 + b^2\pi^2\}} du.$$

Substitution of the residue and branch cut integral into our first equation gives the final result.

4. With contour along the imaginary s-axis, except for a small semicircle to the right of the singularity at $s = 0$,

$$f(t) = \frac{1}{2\pi i} \int_{-\infty i}^{\infty i} \frac{1}{s} \exp\left[st - r\sqrt{\frac{s(1+s)}{(as+1)}}\right] ds$$

$$= \frac{1}{2\pi i} \left\{ \int_{-\infty i}^{-0^+ i} + \int_{C_\epsilon} + \int_{0+i}^{\infty i} \right\} \frac{1}{s} \exp\left[st - r\sqrt{\frac{s(1+s)}{(as+1)}}\right] ds.$$

Now

$$\int_{C_\epsilon} = \lim_{\epsilon \to 0} \int_{-\pi/2}^{\pi/2} \frac{\exp[-r\sqrt{\epsilon e^{\theta i}(1 + \epsilon e^{\theta i})/(1 + a\epsilon e^{\theta i})}]}{\epsilon e^{\theta i}} i\epsilon e^{\theta i} d\theta = \pi i.$$

Along the contour from $(-\infty i, -0^+ i]$,

$$\int_{-\infty i}^{-0^+ i} = \int_{\infty}^{0^+} \frac{du}{u} \exp\{-iut - r\sqrt{u/2}\, M(1-i)[\cos(\theta) - i\sin(\theta)]\}$$

$$= -\int_{0+}^{\infty} \frac{du}{u} \exp\{-iut - r\sqrt{u/2}\, M$$

$$\times [\cos(\theta) - i\cos(\theta) - i\sin(\theta) - \sin(\theta)]\}$$

$$= -\int_{0+}^{\infty} \frac{du}{u} \exp\{-r\sqrt{u/2}\, M[\cos(\theta) - \sin(\theta)]\}$$

$$\times \exp\{-iut + ir\sqrt{u/2}\, M[\cos(\theta) + \sin(\theta)]\}$$

because $s = -ui$ with $0^+ < u < \infty$, $s + 1 = \sqrt{1 + u^2}\, e^{-2\theta_1 i}$ where $2\theta_1 = \tan^{-1}(u)$, $1 + as = \sqrt{1 + a^2 u^2}\, e^{-2\theta_2 i}$ where $2\theta_2 = \tan^{-1}(au)$, $\theta = \theta_1 - \theta_2$ and $M = \sqrt{(1 + u^2)/(1 + a^2 u^2)}$. Similarly, along $[0^+ i, \infty i)$,

$$\int_{0+i}^{\infty i} = \int_{0+}^{\infty} \frac{du}{u} \exp\{-r\sqrt{u/2}\, M[\cos(\theta) - \sin(\theta)]\}$$

$$\times \exp\{iut - ir\sqrt{u/2}\, M[\cos(\theta) + \sin(\theta)]\}.$$

Therefore,

$$\frac{1}{2\pi i} \left\{ \int_{-\infty i}^{-0^+ i} + \int_{0+i}^{\infty i} \right\} = \frac{1}{\pi} \int_{0+}^{\infty} \frac{du}{u} \exp\{-r\sqrt{u/2}\, M[\cos(\theta) - \sin(\theta)]\}$$

$$\times \sin\{ut - r\sqrt{u/2}\, M[\cos(\theta) + \sin(\theta)]\}.$$

Transform Methods for Solving Partial Differential Equations

5. The contour used to invert the transform is similar to Fig. 3.1.1. The arcs at infinity vanish by Jordan's lemma. Therefore,

$$f(t) = \text{Res}(s = 0) - \frac{1}{2\pi i} \int_{branch\ cut} .$$

Now the residue at $s = 0$ is

$$\text{Res}(s = 0) = \lim_{s \to 0} s \frac{K_0(r\sqrt{s/\nu + 1/b^2})}{s K_0(a\sqrt{s/\nu + 1/b^2})} = \frac{K_0(r/b)}{K_0(a/b)}.$$

The branch cut consists of two parts. Along the top, $s + \nu/b^2 = \nu \chi^2 e^{\pi i}$, $0 < \chi < \infty$. Therefore,

$$\int_{top} = 2 \int_\infty^0 \frac{K_0(ir\chi) \exp(-\nu t/b^2 - \nu t \chi^2)}{(\chi^2 + 1/b^2) K_0(ia\chi)} \chi \, d\chi$$

$$= -2 e^{-\nu t/b^2} \int_0^\infty \frac{\exp(-\nu t \chi^2)[J_0(r\chi) - iY_0(r\chi)]}{(\chi^2 + 1/b^2)[J_0(a\chi) - iY_0(a\chi)]} \chi \, d\chi$$

because $K_0(ir\chi) = -\pi i[J_0(r\chi) - iY_0(r\chi)]/2$. Along the bottom of the branch cut, $s + \nu/b^2 = \nu \chi^2 e^{-\pi i}$, $0 < \chi < \infty$. Therefore,

$$\int_{bottom} = 2 e^{-\nu t/b^2} \int_0^\infty \frac{\exp(-\nu t \chi^2)[J_0(r\chi) + iY_0(r\chi)]}{(\chi^2 + 1/b^2)[J_0(a\chi) + iY_0(a\chi)]} \chi \, d\chi.$$

Therefore,

$$\int_{branch\ cut} = 4ie^{-\nu t/b^2} \int_0^\infty \frac{\exp(-\nu t \chi^2)}{(\chi^2 + 1/b^2)} \chi \, d\chi$$

$$\times \frac{Y_0(r\chi) J_0(a\chi) - J_0(r\chi) Y_0(a\chi)}{J_0^2(a\chi) + Y_0^2(a\chi)}.$$

If we now let $\chi = a\eta$, the inverse becomes

$$f(t) = \frac{K_0(r/b)}{K_0(a/b)} - \frac{2}{\pi} e^{-\nu t/b^2} \int_0^\infty \frac{\exp(-\nu t \eta^2/a^2)}{[\eta^2 + (a/b)^2]} \eta \, d\eta$$

$$\times \frac{Y_0(r\eta/a) J_0(\eta) - J_0(r\eta/a) Y_0(\eta)}{J_0^2(\eta) + Y_0^2(\eta)}.$$

6. There are no singularities, just the branch cut. Therefore, the inverse equals the negative of the branch cut integrals:

$$f(t) = \frac{1}{2\pi i} \int_\infty^0 \frac{K_0(i\eta r)}{(i\eta) K_1(i\eta a)} e^{-\eta^2 t} (2\eta \, d\eta)$$

$$+ \frac{1}{2\pi i} \int_0^\infty \frac{K_0(-i\eta r)}{(-i\eta) K_1(-i\eta a)} e^{-\eta^2 t} (2\eta \, d\eta)$$

$$= \frac{1}{\pi i} \int_0^\infty \frac{J_0(\eta r) - iY_0(\eta r)}{J_1(\eta a) - iY_1(\eta a)} e^{-\eta^2 t} d\eta$$

$$- \frac{1}{\pi i} \int_0^\infty \frac{J_0(\eta r) + iY_0(\eta r)}{J_1(\eta a) + iY_1(\eta a)} e^{-\eta^2 t} d\eta$$

because $K_0(iz) = -\pi i[J_0(z) - iY_0(z)]/2$, $K_0(-iz) = \pi i[J_0(z) + iY_0(z)]/2$, $K_1(iz) = -\pi[J_1(z) - iY_1(z)]/2$, $K_1(-iz) = -\pi[J_1(z) + iY_1(z)]/2$ and $s = -\eta^2$. The first integral comes from the top of the branch cut while the second integral comes from the bottom. After a little bit of complex algebra, we obtain the desired answer.

7. From Bromwich's integral,

$$f(t) = \frac{K_0(r)}{K_0(1)} - \frac{1}{2\pi i} \int_{branch\ cut}$$

where the first term on the right side is from the residue at $z = 0$. Now the branch cut integral is

$$\int_{branch\ cut} = -\int_\infty^0 \frac{K_0(rui)}{(-1 - u^2)K_0(ui)} e^{-(1+u^2)t}(2u\,du)$$
$$- \int_0^\infty \frac{K_0(-rui)}{(-1 - u^2)K_0(-ui)} e^{-(1+u^2)t}(2u\,du)$$
$$= -2\int_0^\infty \frac{J_0(ru) - iY_0(ru)}{(1 + u^2)[J_0(u) - iY_0(u)]} e^{-(1+u^2)t} u\,du$$
$$+ 2\int_0^\infty \frac{J_0(ru) + iY_0(ru)}{(1 + u^2)[J_0(u) + iY_0(u)]} e^{-(1+u^2)t} u\,du$$
$$= 4i\int_0^\infty \frac{Y_0(ru)J_0(u) - J_0(ru)Y_0(u)}{J_0^2(u) + Y_0^2(u)} e^{-(1+u^2)t} \frac{u\,du}{1 + u^2}$$

because $s + 1 = u^2 e^{\pm\pi i}$, $K_0(iz) = -\pi i[J_0(z) - iY_0(z)]/2$ and $K_0(-iz) = \pi i[J_0(z) + iY_0(z)]/2$. After substituting the branch cut integral into the first equation, we obtain the desired result.

8. We have simple poles at $s = \pm 2\pi i$. Therefore, the residues are

$$\text{Res}(s = 2\pi i) = \frac{1}{2i} \exp\left[2\pi t i - x\sqrt{\frac{2\pi i(1 + 2\pi ai)}{1 + 2\pi bi}}\right]$$
$$= \frac{1}{2i} \exp[2\pi t i - x\sqrt{d_3(d_2 + id_1)}]$$
$$= \frac{1}{2i} \exp[2\pi t i - x\Delta e^{\psi i/2}]$$

and

$$\text{Res}(s = -2\pi i) = -\frac{1}{2i} \exp[-2\pi t i - x\Delta e^{-\psi i/2}].$$

Thus, the sum of the residues is

$$\exp[-x\Delta\cos(\psi/2)]\sin[2\pi t - x\Delta\sin(\psi/2)].$$

Transform Methods for Solving Partial Differential Equations

We can break the integration along the branch cuts into three parts above and below the negative real axis. Above the negative real axis from $-1/b$ to 0,

$$s = \eta e^{\pi i}, \quad s + 1/b = (1/b - \eta)e^{0i}, \quad s + 1/a = (1/a - \eta)e^{0i}$$

for $0 < \eta < 1/b$. Below the negative real axis, the only change is $s = \eta e^{-\pi i}$. Above the negative real axis from $-1/a$ to $-1/b$,

$$s = \eta e^{\pi i}, \quad s + 1/b = (\eta - 1/b)e^{\pi i}, \quad s + 1/a = (1/a - \eta)e^{0i}$$

for $1/a < \eta < 1/b$. Below the negative real axis, we have two changes:

$$s = \eta e^{-\pi i} \quad \text{and} \quad s + 1/b = (\eta - 1/b)e^{-\pi i}.$$

Finally, from $-\infty$ to $-1/a$ above the negative real axis,

$$s = \eta e^{\pi i}, \quad s + 1/b = (\eta - 1/b)e^{\pi i}, \quad s + 1/a = (\eta - 1/a)e^{\pi i}$$

for $1/b < \eta < \infty$ while below the negative real axis,

$$s = \eta e^{-\pi i}, \quad s + 1/b = (\eta - 1/b)e^{-\pi i}, \quad s + 1/a = (\eta - 1/a)e^{-\pi i}.$$

Therefore, direct substitution yields the branch cut integral

$$
\begin{aligned}
\int_{branch\ cut} = & -\int_{\infty}^{1/a} \frac{2\pi}{4\pi^2 + \eta^2} \exp\left[-\eta t - xi\sqrt{\frac{\eta(a\eta - 1)}{b\eta - 1}}\right] d\eta \\
& -\int_{1/a}^{\infty} \frac{2\pi}{4\pi^2 + \eta^2} \exp\left[-\eta t + xi\sqrt{\frac{\eta(a\eta - 1)}{b\eta - 1}}\right] d\eta \\
& -\int_{1/a}^{1/b} \frac{2\pi}{4\pi^2 + \eta^2} \exp\left[-\eta t - x\sqrt{\frac{\eta(a\eta - 1)}{b\eta - 1}}\right] d\eta \\
& -\int_{1/b}^{1/a} \frac{2\pi}{4\pi^2 + \eta^2} \exp\left[-\eta t - x\sqrt{\frac{\eta(a\eta - 1)}{b\eta - 1}}\right] d\eta \\
& -\int_{1/b}^{0} \frac{2\pi}{4\pi^2 + \eta^2} \exp\left[-\eta t - xi\sqrt{\frac{\eta(a\eta - 1)}{b\eta - 1}}\right] d\eta \\
& -\int_{0}^{1/b} \frac{2\pi}{4\pi^2 + \eta^2} \exp\left[-\eta t + xi\sqrt{\frac{\eta(a\eta - 1)}{b\eta - 1}}\right] d\eta \\
= & -4\pi i \int_{0}^{1/b} F(x, t, \eta)\, d\eta - 4\pi i \int_{1/a}^{\infty} F(x, t, \eta)\, d\eta
\end{aligned}
$$

where

$$F(x,t,\eta) = \frac{\exp(-\eta t)}{4\pi^2 + \eta^2} \sin\left[x\sqrt{\frac{\eta(1-a\eta)}{1-b\eta}}\right].$$

The final answer is

$$f(t) = \exp[-x\Delta\cos(\psi/2)]\sin[2\pi t - x\Delta\sin(\psi/2)] - \frac{1}{2\pi i}\int_{branch\ cut}.$$

9. The closed Bromwich integral is similar to Fig. 3.1.2. With the contour we have singularities at $s = \pm\omega i$. The residue are

$$\text{Res}(s = \omega i) = \frac{1}{2i}\frac{K_1(r\sqrt{\omega i})}{K_1(\sqrt{\omega i})}e^{i\omega t}$$

and

$$\text{Res}(s = -\omega i) = -\frac{1}{2i}\frac{K_1(r\sqrt{-\omega i})}{K_1(\sqrt{-\omega i})}e^{-i\omega t}.$$

Because

$$K_1(ze^{\pm\pi i/4})e^{\mp\pi i/2} = \text{ker}_1(z) \pm i\text{kei}_1(z)$$

and defining

$$\text{ker}_1(z) + i\text{kei}_1(z) = N_1(z)e^{-i\phi_1(z)},$$

the sum of the residue gives

$$\frac{N_1(r\sqrt{\omega})}{N_1(\sqrt{\omega})}\sin[\omega t + \phi_1(r\sqrt{\omega}) - \phi_1(\sqrt{\omega})].$$

Along the top of the branch cut, $s = \eta^2 e^{\pi i}$ and

$$\int_{top} = -2\int_{\infty}^{0}\frac{\omega}{\omega^2 + \eta^4}\frac{K_1(r\eta i)}{K_1(\eta i)}e^{-\eta^2 t}\eta\,d\eta$$

while along the bottom of the branch cut, $s = \eta^2 e^{-\pi i}$ and

$$\int_{bottom} = -2\int_{0}^{\infty}\frac{\omega}{\omega^2 + \eta^4}\frac{K_1(-r\eta i)}{K_1(-\eta i)}e^{-\eta^2 t}\eta\,d\eta.$$

Using the relationship

$$K_1(ze^{\pm i/2}) = \pm\pi i e^{\mp i/2}[-J_1(z) \pm iY_1(z)]/2,$$

the sum of the two branch cut integrals reduces to

$$4i\int_{0}^{\infty}\frac{\omega\eta}{\omega^2 + \eta^4}e^{-\eta^2 t}\frac{J_1(r\eta)Y_1(\eta) - Y_1(r\eta)J_1(\eta)}{J_1^2(\eta) + Y_1^2(\eta)}\,d\eta.$$

The final solution $f(t)$ consists of the residues minus the branch cut integrals after we divide them by $2\pi i$.

10. From Bromwich's integral, the inversion consists of two parts. From the integration around the branch point $s = 0$, we have $-2\pi i$. The second part comes from the integration along the branch cut. Along the top of the branch cut, $s = \eta^2 e^{\pi i}$ and $\sqrt{s} = i\eta$ while along the bottom of the branch cut, $s = \eta^2 e^{-\pi i}$ and $\sqrt{s} = -i\eta$. Therefore the branch cut integration is

$$
\int_{branch\ cut} = -\int_{\infty}^{0} \frac{1}{-\eta^2} \frac{1 + r\eta i}{1 + a\eta i} e^{-(r-a)\eta i} e^{-\eta^2 t} 2\eta \, d\eta
$$
$$
- \int_{0}^{\infty} \frac{1}{-\eta^2} \frac{1 - r\eta i}{1 - a\eta i} e^{(r-a)\eta i} e^{-\eta^2 t} 2\eta \, d\eta
$$
$$
= 4i \int_{0}^{\infty} \left\{ \frac{1 + ar\eta^2}{1 + a^2\eta^2} \sin[(r-a)\eta] \right.
$$
$$
\left. - \frac{r\eta - a\eta}{1 + a^2\eta^2} \cos[(r-a)\eta] \right\} e^{-\eta^2 t} \frac{d\eta}{\eta}.
$$

Thus the inverse equals the negative of the integration around the branch point and the branch cut integral after we divide them by $2\pi i$.

11. First, we note that

$$
\frac{\partial K_1(r\sqrt{z})}{\partial r} \bigg|_{r=a} = -\frac{\sqrt{z}}{2} \left[K_0(a\sqrt{z}) + K_2(a\sqrt{z}) \right].
$$

The contour integral is identical to Fig. 3.1.2. The integration around the branch point is

$$
\int_{branch\ point} = -\frac{2\pi M i}{4\pi \mu r}
$$

where we have used the asymptotic forms of $K_0(\ )$, $K_1(\ )$ and $K_2(\ )$ as $a\sqrt{z} \to 0$.

Turning to the branch cut integrals, $z = \eta^2 e^{\pi i}$,

$$
K_0(a\sqrt{z}) = -\pi i [J_0(a\eta) - iY_0(a\eta)]/2
$$
$$
K_1(a\sqrt{z}) = -\pi [J_1(a\eta) - iY_1(a\eta)]/2
$$

and

$$
K_2(a\sqrt{z}) = \pi i [J_2(a\eta) - iY_2(a\eta)]/2
$$

along the top of the branch cut so that

$$
\int_{top} = \frac{2aM}{I} \int_{0}^{\infty} \frac{[J_1(r\eta) - iY_1(r\eta)] \exp(-\eta^2 t)}{[\eta J_1(a\eta) - \chi J_2(a\eta)] - i[\eta Y_1(a\eta) - \chi Y_2(a\eta)]} \frac{d\eta}{\eta^2}.
$$

Along the bottom of the branch cut, $z = \eta^2 e^{-\pi i}$,

$$K_0(a\sqrt{z}) = \pi i[J_0(a\eta) + iY_0(a\eta)]/2$$

$$K_1(a\sqrt{z}) = -\pi[J_1(a\eta) + iY_1(a\eta)]/2$$

and

$$K_2(a\sqrt{z}) = -\pi i[J_2(a\eta) + iY_2(a\eta)]/2$$

so that

$$\int_{bottom} = -\frac{2aM}{I} \int_0^\infty \frac{[J_1(r\eta) + iY_1(r\eta)]\exp(-\eta^2 t)}{[\eta J_1(a\eta) - \chi J_2(a\eta)] + i[\eta Y_1(a\eta) - \chi Y_2(a\eta)]} \frac{d\eta}{\eta^2}.$$

We have used the relationship that

$$J_0(a\eta) = \frac{2}{a\eta} J_1(a\eta) - J_2(a\eta)$$

and

$$Y_0(a\eta) = \frac{2}{a\eta} Y_1(a\eta) - Y_2(a\eta)$$

to eliminate $J_0(\ )$ and $Y_0(\ )$ from the integrals. The inverse equals the negative of the branch cut integrals after we divide them by $2\pi i$.

12. We close Bromwich's integral with an infinite semi-circle in the left side of the s-plane, except for a cut running along the negative real axis. By Jordan's lemma the arcs at infinity vanish. Along the cut from $-\infty$ to $-a$, both $\sqrt{s}$ and $\sqrt{s+a}$ have the same arguments π or $-\pi$ along the top and bottom of the branch cut, respectively. The integral along the top of the branch cut will cancel the integral along the bottom of the branch cut. Consequently the only contribution comes from that portion of the branch cut integral from the segment from $-a$ to 0 along the top and bottom of the branch cut as well as integrations around the branch points $s = 0$ and $s = -a$. The contribution from the branch point $s = 0$ is $2\pi i$ while the other branch point gives zero. Along the top of the branch cut $s = \eta e^{\pi i}$ and $s + a = (a - \eta)e^{0i}$. Therefore,

$$\int_{top} = \int_a^0 \frac{d\eta}{\eta} \exp\left(-t\eta - xi\sqrt{\frac{\eta}{a-\eta}}\right).$$

Along the bottom of the branch cut $s = \eta e^{-\pi i}$ and $s + a = (a - \eta)e^{0i}$. Therefore,

$$\int_{bottom} = \int_0^a \frac{d\eta}{\eta} \exp\left(-t\eta + xi\sqrt{\frac{\eta}{a-\eta}}\right).$$

Combining the integrals yields

$$f(t) = 1 - \frac{1}{\pi} \int_0^a e^{-t\eta} \sin\left(x\sqrt{\frac{\eta}{a-\eta}}\right) \frac{d\eta}{\eta}.$$

Finally, we substitute $\eta = a\sigma^2/(1+\sigma^2)$ to obtain the final result.

13. We close Bromwich's contour in a manner *similar* to Fig. 3.1.2. If we take the branch cuts along the negative real axis of the s-plane, then for $-\infty < s < -b^2$ the phase of $\sqrt{s+a^2}$ and $\sqrt{s+b^2}$ equals π just above the branch cut and equals $-\pi$ just below the branch cut. Therefore, the argument of the modified Bessel function is the same on either side of the cut and the value of the integration along the top of the branch cut equals the negative of the integration just below the branch cut. Thus there is no contribution from the integration between $s = -\infty$ and $s = -b^2$.

However, between $s = -b^2$ and $s = -a^2$, $s = -\rho$, $\sqrt{s+a^2} = (\rho - a^2)e^{\pm\pi i}$ and $\sqrt{s+a^2} = (b^2 - \rho)e^{0i}$. Therefore,

$$f(t) = \frac{1}{2\pi i} \int_{b^2}^{a^2} \frac{1}{b^2 - \rho} K_0\left(rci\sqrt{\frac{\rho - a^2}{b^2 - \rho}}\right) e^{-\rho t} d\rho$$

$$- \frac{1}{2\pi i} \int_{a^2}^{b^2} \frac{1}{b^2 - \rho} K_0\left(-rci\sqrt{\frac{\rho - a^2}{b^2 - \rho}}\right) e^{-\rho t} d\rho.$$

Because $K_0(\pm xi) = \mp\pi i[J_0(x) - iY_0(x)]/2$,

$$f(t) = \int_{a^2}^{b^2} \frac{1}{b^2 - \rho} J_0\left(rc\sqrt{\frac{\rho - a^2}{b^2 - \rho}}\right) e^{-\rho t} d\rho.$$

14. For $t < r/c$ we close the contour on the right side of the plane by Jordan's lemma. Because there are no singularities, $f(t) = 0$. For $t > r/c$ we close it on the left side. Then with the branch cut along the negative real axis, Fig. 3.1.2 shows the contour for the inversion. If $sa/c = \eta e^{\pm\pi i}$, it follow that

$$f(t) = -\frac{1}{2\pi i} \int_\infty^0 \frac{\exp(-\eta\tau)}{(-\eta)[K_0(\eta) + i\pi I_0(\eta)]} d\eta$$

$$- \frac{1}{2\pi i} \int_0^\infty \frac{\exp(-\eta\tau)}{(-\eta)[K_0(\eta) - i\pi I_0(\eta)]} d\eta$$

$$= \int_0^\infty \frac{I_0(\eta)}{K_0^2(\eta) + \pi^2 I_0^2(\eta)} e^{-\eta\tau} \frac{d\eta}{\eta}.$$

15. Closing Bromwich's integral as shown in Fig. 3.1.2,

$$f(t) = \frac{1}{2\pi i} \oint_C \frac{K_1(az)e^{tz}}{z\, K_1(bz)} \, dz.$$

There are no singularities inside the contour. The integration around the branch point yields

$$\int_{branch\ point} = -2\pi i.$$

Along the top of the branch cut, $z = \eta e^{\pi i}$ and $K_1(\eta e^{\pi i}) = K_1(\eta) - \pi i I_1(\eta)$ while along the bottom of the branch cut $z = \eta e^{-\pi i}$ and $K_1(\eta e^{-\pi i}) = K_1(\eta) + \pi i I_1(\eta)$. Therefore,

$$\int_{top\ branch\ cut} = \int_\infty^0 e^{-\eta t} \left[\frac{K_1(a\eta) + \pi i I_1(a\eta)}{K_1(b\eta) + \pi i I_1(b\eta)} \right] \frac{d\eta}{\eta},$$

$$\int_{bottom\ branch\ cut} = \int_0^\infty e^{-\eta t} \left[\frac{K_1(a\eta) - \pi i I_1(a\eta)}{K_1(b\eta) - \pi i I_1(b\eta)} \right] \frac{d\eta}{\eta}$$

and

$$\int_{branch\ cut} = -2\pi i \int_0^\infty e^{-\eta t} \left[\frac{I_1(a\eta)K_1(b\eta) - K_1(a\eta)I_1(b\eta)}{K_1^2(b\eta) + \pi^2 I_1^2(b\eta)} \right] \frac{d\eta}{\eta}.$$

The final answer equals the negative of the branch cut and branch point integrals after we divide them by $2\pi i$.

16. The closed contour for the inversion is similar to Fig. 3.1.2 where the point $z = k$ lies on the positive real axis between the origin and the Bromwich contour. The arcs at infinity vanish by Jordan's lemma as does the integration around the branch point at $z = 0$. Therefore,

$$f(t) = \frac{\exp(kt - \alpha\sqrt{k})}{\sqrt{k}} - \frac{1}{2\pi i} \int_{branch\ cut}$$

where the first term on the right side comes from the residue at $z = k$. Along the branch cut $z = \eta^2 e^{\pm\pi i}$. Therefore,

$$\int_{branch\ cut} = -\int_\infty^0 \frac{\exp(-\alpha\eta i - \eta^2 t)}{(i\eta)(-k - \eta^2)} 2\eta\, d\eta - \int_0^\infty \frac{\exp(\alpha\eta i - \eta^2 t)}{(-i\eta)(-k - \eta^2)} 2\eta\, d\eta$$

$$= 4i \int_0^\infty \frac{\cos(\alpha\eta)}{k + \eta^2} e^{-\eta^2 t} d\eta$$

Substitution into the first equation gives the final answer.

17. Closing Bromwich's integral as shown in Fig. 3.1.2,

$$f(t) = \frac{1}{2\pi i} \oint_C \frac{\sqrt{z}\,e^{tz}}{z\sqrt{z} + a^3}\,dz.$$

Inside the contour there are simple poles at $z_{1,2} = a^2 \exp(\pm 2\pi i/3)$ with $\sqrt{z_{1,2}} = a\exp(\pm\pi i/3)$. Therefore,

$$\mathrm{Res}(z = s_1) = \lim_{z \to s_1} \frac{(z - s_1)\sqrt{z}\,e^{tz}}{z\sqrt{z} + a^3} = 2e^{-a^2 t/2}\exp(ia^2 t\sqrt{3}/2)/3$$

and

$$\mathrm{Res}(z = s_2) = \lim_{z \to s_2} \frac{(z - s_2)\sqrt{z}\,e^{tz}}{z\sqrt{z} + a^3} = 2e^{-a^2 t/2}\exp(-ia^2 t\sqrt{3}/2)/3.$$

Therefore,

$$\mathrm{Res}(z = s_1) + \mathrm{Res}(z = s_2) = 4e^{-a^2 t/2}\cos(a^2 t\sqrt{3}/2)/3.$$

Along the top of the branch cut, $z = x^2 e^{\pi i}$ and $\sqrt{z} = x e^{\pi i/2}$ so that

$$\int_{top} = -\int_\infty^0 \frac{xi\,e^{-x^2 t}}{a^3 - ix^3}(2x\,dx)$$

while along the bottom of the branch cut $z = x^2 e^{-\pi i}$ and $\sqrt{z} = x e^{-\pi i/2}$ so that

$$\int_{bottom} = \int_0^\infty \frac{xi\,e^{-x^2 t}}{a^3 + ix^3}(2x\,dx).$$

Consequently, the contribution from the branch cut is

$$\int_{branch\ cut} = 4ia^3 \int_0^\infty \frac{x^2}{x^6 + a^6}\,dx.$$

The final answer equals the sum of the residues minus the branch cut integral after we divide them by $2\pi i$.

18. Let us first find the inverse of

$$G(s) = \frac{K_1(a\sqrt{s}\,)}{bK_1(\sqrt{s}\,) + \sqrt{s}K_0(\sqrt{s}\,)}.$$

This transform has no poles, so

$$\int_{branch\ cut} = -\int_{\infty}^{0} \frac{K_1(\eta ai)}{bK_1(\eta i) + \eta iK_0(\eta i)} e^{-\eta^2 t}(2\eta\,d\eta)$$

$$- \int_{0}^{\infty} \frac{K_1(-\eta ai)}{bK_1(-\eta i) - \eta iK_0(-\eta i)} e^{-\eta^2 t}(2\eta\,d\eta)$$

$$= 2\int_{0}^{\infty} \frac{[J_1(\eta a) - iY_1(\eta a)]e^{-\eta^2 t}}{b[J_1(\eta) - iY_1(\eta)] - \eta[J_0(\eta) - iY_0(\eta)]}\eta\,d\eta$$

$$- 2\int_{0}^{\infty} \frac{[J_1(\eta a) + iY_1(\eta a)]e^{-\eta^2 t}}{b[J_1(\eta) + iY_1(\eta)] - \eta[J_0(\eta) + iY_0(\eta)]}\eta\,d\eta$$

$$= 4i\int_{0}^{\infty} \eta e^{-\eta^2 t}d\eta$$

$$\times \frac{[bY_0(\eta) - \eta Y_0(\eta)]J_1(\eta a) + Y_1(\eta a)[bJ_0(\eta) - \eta J_0(\eta)]}{[bJ_1(\eta) - \eta J_0(\eta)]^2 + [bY_1(\eta) - \eta Y_0(\eta)]^2}$$

because $s = \eta^2 e^{\pm\pi i}$, $K_0(zi) = -\pi i[J_0(z) - iY_0(z)]/2$, $K_0(-zi) = \pi i$ $[J_0(z) + iY_0(z)]/2$, $K_1(zi) = -\pi[J_1(z) - iY_1(z)]/2$ and $K_1(-zi) = -\pi$ $[J_1(z) + iY_1(z)]/2$. Therefore,

$$g(t) = \frac{2}{\pi}\int_{0}^{\infty} e^{-\eta^2 t}d\eta$$

$$\times \frac{[bY_0(\eta) - \eta Y_0(\eta)]J_1(\eta a) + Y_1(\eta a)[bJ_1(\eta) - \eta J_0(\eta)]}{[bJ_1(\eta) - \eta J_0(\eta)]^2 + [bY_1(\eta) - \eta Y_0(\eta)]^2}.$$

Finally,

$$f(t) = \int_{0}^{\infty} g(\tau)d\tau$$

$$= \frac{2}{\pi}\int_{0}^{\infty} \left(1 - e^{-\eta^2 t}\right)\frac{d\eta}{\eta}$$

$$\times \frac{[bY_0(\eta) - \eta Y_0(\eta)]J_1(\eta a) + Y_1(\eta a)[bJ_1(\eta) - \eta J_0(\eta)]}{[bJ_1(\eta) - \eta J_0(\eta)]^2 + [bY_1(\eta) - \eta Y_0(\eta)]^2}.$$

19. Because there are no poles, the only contribution comes from the branch cut integral. Along the top of the negative real axis, there are three regions. In region 1, $-\infty < \eta e^{\pi i} < -b$, $\sqrt{s} = i\sqrt{\eta}$, $\sqrt{s+a} = i\sqrt{\eta - a}$ and $\sqrt{s+b} = i\sqrt{\eta - b}$ so that

$$K_0\left[r\sqrt{\frac{s(s+b)}{c(s+a)}}\right] = -\frac{\pi i}{2}\left\{J_0\left[r\sqrt{\frac{\eta(\eta - b)}{c(\eta - a)}}\right] - iY_0\left[r\sqrt{\frac{\eta(\eta - b)}{c(\eta - a)}}\right]\right\}.$$

Transform Methods for Solving Partial Differential Equations

In region 2, $-b < \eta e^{\pi i} < -a$, $\sqrt{s} = i\sqrt{\eta}$, $\sqrt{s+a} = i\sqrt{\eta - a}$ and $\sqrt{s+b} = \sqrt{b-\eta}$ so that

$$K_0\left[r\sqrt{\frac{s(s+b)}{c(s+a)}}\right] = K_0\left[r\sqrt{\frac{\eta(b-\eta)}{c(\eta-a)}}\right].$$

In region 3, $-a < \eta e^{\pi i} < 0$, $\sqrt{s} = i\sqrt{\eta}$, $\sqrt{s+a} = \sqrt{a-\eta}$ and $\sqrt{s+b} = \sqrt{b-\eta}$ so that

$$K_0\left[r\sqrt{\frac{s(s+b)}{c(s+a)}}\right] = -\frac{\pi i}{2}\left\{J_0\left[r\sqrt{\frac{\eta(b-\eta)}{c(a-\eta)}}\right] - iY_0\left[r\sqrt{\frac{\eta(b-\eta)}{c(a-\eta)}}\right]\right\}.$$

Similarly, below the branch cut, in region 1, $-\infty < \eta e^{-\pi i} < -b$, $\sqrt{s} = -i\sqrt{\eta}$, $\sqrt{s+a} = -i\sqrt{\eta - a}$ and $\sqrt{s+b} = -i\sqrt{\eta - b}$ so that

$$K_0\left[r\sqrt{\frac{s(s+b)}{c(s+a)}}\right] = \frac{\pi i}{2}\left\{J_0\left[r\sqrt{\frac{\eta(\eta-b)}{c(\eta-a)}}\right] + iY_0\left[r\sqrt{\frac{\eta(\eta-b)}{c(\eta-a)}}\right]\right\}.$$

In region 2, $-b < \eta e^{-\pi i} < -a$, $\sqrt{s} = -i\sqrt{\eta}$, $\sqrt{s+a} = -i\sqrt{\eta - a}$ and $\sqrt{s+b} = \sqrt{b-\eta}$ so that

$$K_0\left[r\sqrt{\frac{s(s+b)}{c(s+a)}}\right] = K_0\left[r\sqrt{\frac{\eta(b-\eta)}{c(\eta-a)}}\right].$$

In region 3, $-a < \eta e^{-\pi i} < 0$, $\sqrt{s} = -i\sqrt{\eta}$, $\sqrt{s+a} = \sqrt{a-\eta}$ and $\sqrt{s+b} = \sqrt{b-\eta}$ so that

$$K_0\left[r\sqrt{\frac{s(s+b)}{c(s+a)}}\right] = \frac{\pi i}{2}\left\{J_0\left[r\sqrt{\frac{\eta(b-\eta)}{c(a-\eta)}}\right] + iY_0\left[r\sqrt{\frac{\eta(b-\eta)}{c(a-\eta)}}\right]\right\}.$$

Now

$$f(t) = \frac{1}{2\pi i}\left[\int_{-\infty-0i}^{-b-0i} + \int_{-b-0i}^{-a-0i} + \int_{-a-0i}^{0-0i}\right.$$
$$\left. + \int_{0+0i}^{-a+0i} + \int_{-a+0i}^{-b+0i} + \int_{-b+0i}^{-\infty+0i}\right].$$

However, the second and fifth integrals cancel, leaving

$$f(t) = \frac{1}{2}\int_0^a J_0\left[r\sqrt{\frac{\eta(b-\eta)}{c(a-\eta)}}\right]e^{-\eta t}d\eta + \frac{1}{2}\int_b^\infty J_0\left[r\sqrt{\frac{\eta(\eta-b)}{c(\eta-a)}}\right]e^{-\eta t}d\eta$$

after substitution.

20. From Bromwich's integral,

$$f(t) = \frac{1}{2\pi i} \oint_C \frac{e^{zt}}{1 + a^p z^p} dz$$

where our closed contour is the same as Fig. 3.1.2. Along the top of the branch cut,

$$\int_{top} = -\int_\infty^0 \frac{e^{-t\eta}}{1 + a^p \eta^p e^{p\pi i}} d\eta$$

while along the bottom of the branch cut,

$$\int_{bottom} = -\int_0^\infty \frac{e^{-t\eta}}{1 + a^p \eta^p e^{-p\pi i}} d\eta.$$

Thus, the total branch cut contribution is

$$\int_{branch\ cut} = -2i a^p \sin(p\pi) \int_0^\infty \frac{\eta^p e^{-t\eta}}{1 + a^{2p}\eta^{2p} + 2a^p \eta^p \cos(p\pi)} d\eta.$$

The final answer equals the negative of the branch cut integral after we divide it by $2\pi i$. Note that integrand does have simple poles located at

$$s_{1,2} = [\cos(\pi/p) \pm i \sin(\pi/p)]/a.$$

However, they are located on different Riemann surfaces.

21. From Bromwich's integral,

$$f(t) = \frac{1}{2\pi i} \int_{c-\infty i}^{c+\infty i} \frac{a}{z(z^{1/n} + a)} e^{zt} dz.$$

If we take the branch cut along the negative real axis, we close Bromwich's integral as shown in Fig. 3.1.2. The integral has poles at $z = a^n e^{2n(2m-1)\pi i}$ where m is an integer. Because we restrict our phase between $-\pi$ to π, these poles lie on another Riemann surface and we may exclude them from further consideration. Integration around the branch point $z = 0$ yields

$$\int_{branch\ point} = -2\pi i.$$

Along the top of the branch cut $z = a^n \eta^n e^{\pi i}$ so that

$$\int_{top} = n \int_\infty^0 \frac{\exp(-a^n \eta^n t)}{1 + \eta \cos(\pi/n) + i\eta \sin(\pi/n)} \frac{d\eta}{\eta}$$

where along the bottom of the branch cut $z = a^n \eta^n e^{-\pi i}$ so that

$$\int_{bottom} = n \int_0^\infty \frac{\exp(-a^n \eta^n t)}{1 + \eta \cos(\pi/n) - i\eta \sin(\pi/n)} \frac{d\eta}{\eta}.$$

Combining the two branch cuts integrals,

$$\int_{branch\ cut} = 2in \sin\left(\frac{\pi}{n}\right) \int_0^\infty \frac{\exp(-a^n \eta^n t)}{1 + \eta^2 + 2\eta \cos(\pi/n)} \frac{d\eta}{\eta}.$$

The inverse equals the negative of the branch cut and branch point integrals after dividing the sum by $2\pi i$.

22. Fig. 3.1.2 shows the closed Bromwich integral. The integration around the branch point yields $-2\pi i$. Along the top of the branch cut $s = \eta e^{\pi i}$ so that

$$\int_{top} = \int_\infty^0 \exp\left[-t\eta - \lambda\sqrt{\frac{\eta^\alpha e^{\pi \alpha i}}{\eta^\alpha e^{\pi \alpha i} + a}}\right] \frac{d\eta}{\eta}.$$

Now, $\eta^\alpha e^{\pi \alpha i} + a = \rho e^{\phi i}$. Therefore,

$$\int_{top} = -\int_0^\infty \exp\left[-t\eta - \lambda\sqrt{\frac{\eta^\alpha}{\rho}} e^{(\pi\alpha - \phi)i/2}\right] \frac{d\eta}{\eta}$$

$$= -\int_0^\infty e^{-t\eta} \exp\left[-\lambda\sqrt{\frac{\eta^\alpha}{\rho}} \cos\left(\frac{\pi\alpha - \phi}{2}\right)\right]$$

$$\times \exp\left[-i\lambda\sqrt{\frac{\eta^\alpha}{\rho}} \sin\left(\frac{\pi\alpha - \phi}{2}\right)\right] \frac{d\eta}{\eta}.$$

Along the bottom of the branch cut, $s = \eta e^{-\pi i}$ and

$$\int_{bottom} = \int_0^\infty \exp\left[-t\eta - \lambda\sqrt{\frac{\eta^\alpha}{\rho}} e^{-(\pi\alpha - \phi)i/2}\right] \frac{d\eta}{\eta}$$

$$= \int_0^\infty e^{-t\eta} \exp\left[-\lambda\sqrt{\frac{\eta^\alpha}{\rho}} \cos\left(\frac{\pi\alpha - \phi}{2}\right)\right]$$

$$\times \exp\left[i\lambda\sqrt{\frac{\eta^\alpha}{\rho}} \sin\left(\frac{\pi\alpha - \phi}{2}\right)\right] \frac{d\eta}{\eta}.$$

Consequently, the branch cut integral is

$$\int_{branch\ cut} = -2\pi i + 2i \int_0^\infty e^{-t\eta} \exp\left[-\lambda\sqrt{\frac{\eta^\alpha}{\rho}} \cos\left(\frac{\pi\alpha - \phi}{2}\right)\right]$$

$$\times \sin\left[\lambda\sqrt{\frac{\eta^\alpha}{\rho}} \sin\left(\frac{\pi\alpha - \phi}{2}\right)\right] \frac{d\eta}{\eta}.$$

The final answer equals the negative of the branch cut integral after we divide it by $2\pi i$.

23. We close Bromwich's integral in a manner similar to Fig. 3.1.10. From the expansion about $s = 0$, the residue at $s = 0$ equals $k^2/[2(k^2 - 1)]$. At $s = i/\gamma$,

$$\text{Res}(s = i/\gamma) = \lim_{s \to i/\gamma} \frac{(s - i/\gamma)s\sqrt{1 + s^2/k^2}}{g(s)} e^{ts}$$

$$= \frac{(i/\gamma)\sqrt{1 - 1/(k\gamma)^2}}{g'(i/\gamma)} e^{it/\gamma}.$$

Similarly, at $s = -i/\gamma$,

$$\text{Res}(s = -i/\gamma) = \lim_{s \to -i/\gamma} \frac{(s + i/\gamma)s\sqrt{1 + s^2/k^2}}{g(s)} e^{ts}$$

$$= \frac{(-i/\gamma)\sqrt{1 - 1/(k\gamma)^2}}{g'(-i/\gamma)} e^{-it/\gamma}.$$

Because $g'(i/\gamma) = -g'(-i/\gamma)$, the sum of the residues equals

$$\frac{k^2}{2(k^2 - 1)} + \frac{2\sqrt{1 - 1/(k\gamma)^2}}{[-i\gamma g'(-i/\gamma)]} \cos(t/\gamma)$$

which is purely real.

The integrations around the branch points equal zero. From $s = i$ to ki,

$$s - i = (\eta - 1)e^{\pi i/2} \text{ or } (\eta - 1)e^{-3\pi i/2}, \quad s + i = (\eta + 1)e^{\pi i/2}$$

$$s - ki = (k - \eta)e^{-\pi i/2}, \qquad s + ki = (k + \eta)e^{\pi i/2}$$

and

$$\int_{C_1} = \int_k^1 \frac{i\eta\sqrt{k^2 - \eta^2}}{k(2 - \eta^2)^2 - 4i\sqrt{(k^2 - \eta^2)(\eta^2 - 1)}} e^{it\eta}\, i\, d\eta$$

$$+ \int_1^k \frac{i\eta\sqrt{k^2 - \eta^2}}{k(2 - \eta^2)^2 + 4i\sqrt{(k^2 - \eta^2)(\eta^2 - 1)}} e^{it\eta}\, i\, d\eta.$$

From $s = -i$ to $-ki$,

$$s - i = (\eta - 1)e^{-\pi i/2}, \quad s + i = (\eta - 1)e^{-\pi i/2} \text{ or } (\eta - 1)e^{3\pi i/2}$$

$$s - ki = (k + \eta)e^{-\pi i/2}, \qquad s + ki = (k - \eta)e^{\pi i/2}$$

and

$$\int_{C_2} = \int_k^1 \frac{-i\eta\sqrt{k^2 - \eta^2}}{k(2 - \eta^2)^2 - 4i\sqrt{(k^2 - \eta^2)(\eta^2 - 1)}} e^{-it\eta} \, (-i \, d\eta)$$

$$+ \int_1^k \frac{-i\eta\sqrt{k^2 - \eta^2}}{k(2 - \eta^2)^2 + 4i\sqrt{(k^2 - \eta^2)(\eta^2 - 1)}} e^{-it\eta} \, (-i \, d\eta).$$

Combining the integrals and simplifying,

$$\int_{C_1} + \int_{C_2} = 16i \int_1^k \frac{\eta(1 - \eta^2/k^2)\sqrt{\eta^2 - 1}}{(2 - \eta^2)^4 + 16(1 - \eta^2/k^2)(\eta^2 - 1)} \cos(\eta t) \, d\eta.$$

From $s = ki$ to ∞i,

$$s - i = (\eta - 1)e^{\pi i/2} \text{ or } (\eta - 1)e^{-3\pi i/2}, \quad s + i = (\eta + 1)e^{\pi i/2}$$

$$s - ki = (\eta - k)e^{\pi i/2} \text{ or } (\eta - k)e^{-3\pi i/2}, \quad s + ki = (\eta + k)e^{\pi i/2}$$

and

$$\int_{C_3} = \int_\infty^k \frac{i\eta(i\sqrt{\eta^2 - k^2})}{k(2 - \eta^2)^2 - 4(i\sqrt{\eta^2 - k^2})(i\sqrt{\eta^2 - 1})} e^{it\eta} \, i \, d\eta$$

$$+ \int_k^\infty \frac{i\eta(-i\sqrt{k^2 - \eta^2})}{k(2 - \eta^2)^2 - 4(-i\sqrt{\eta^2 - k^2})(-i\sqrt{\eta^2 - 1})} e^{it\eta} \, i \, d\eta.$$

From $s = -ki$ to $-\infty i$,

$$s - i = (\eta - 1)e^{-\pi i/2}, \quad s + i = (\eta - 1)e^{-\pi i/2} \text{ or } (\eta - 1)e^{3\pi i/2}$$

$$s - ki = (k + \eta)e^{-\pi i/2}, \quad s + ki = (k - \eta)e^{-\pi i/2} \text{ or } (k - \eta)e^{3\pi i/2}$$

and

$$\int_{C_4} = \int_\infty^k \frac{(-i\eta)(i\sqrt{\eta^2 - k^2})}{k(2 - \eta^2)^2 - 4(i\sqrt{\eta^2 - k^2})(i\sqrt{\eta^2 - 1})} e^{-it\eta} \, (-i \, d\eta)$$

$$+ \int_k^\infty \frac{-i\eta(-i\sqrt{k^2 - \eta^2})}{k(2 - \eta^2)^2 - 4(-i\sqrt{\eta^2 - k^2})(-i\sqrt{\eta^2 - 1})} e^{-it\eta} (-i \, d\eta).$$

Combining the integrals and simplifying,

$$\int_{C_3} + \int_{C_4} = 4i \int_k^\infty \frac{\eta\sqrt{\eta^2/k^2 - 1}}{(2 - \eta^2)^2 + 4\sqrt{(\eta^2/k^2 - 1)(\eta^2 - 1)}} \cos(\eta t) \, d\eta.$$

The final answer consists of the sum of the residues minus the branch cut integrals after we divide the sum by $2\pi i$.

474

24. From Bromwich's integral,

$$f(t) = \frac{a}{2\pi i} \int_{c-\infty i}^{c+\infty i} \frac{\exp(zt - x\sqrt{z}\,)}{z[a + (2-a)\sqrt{\pi z/2}\,]}\,dz.$$

Closing the line integral as shown in Fig. 3.1.2, we only have an integration along the branch cut. The integration around the branch point $z = 0$ yields

$$\int_{branch\ point} = -2\pi i.$$

Along the branch cut we have $z = \eta^2 e^{\pm\pi i}$. Therefore, along the top of the branch cut,

$$\int_{top} = a \int_{\infty}^{0} \frac{\exp(-\eta^2 t - x\eta i)}{a + (2-a)\eta i\sqrt{\pi/2}} \frac{2\eta\,d\eta}{\eta^2}$$

$$= -2a \int_{0}^{\infty} e^{-\eta^2 t} \frac{(a - c\eta i)e^{-x\eta i}}{a^2 + \eta^2 c^2} \frac{d\eta}{\eta}$$

while along the bottom of the branch cut

$$\int_{bottom} = 2a \int_{0}^{\infty} \frac{\exp(-\eta^2 t + x\eta i)}{a - c\eta i} \frac{d\eta}{\eta}$$

$$= 2a \int_{0}^{\infty} e^{-\eta^2 t} \frac{(a + c\eta i)e^{-x\eta i}}{a^2 + \eta^2 c^2} \frac{d\eta}{\eta}.$$

The inverse equals the negative of the sum of the branch cut and branch point integrals after dividing the sum by $2\pi i$ and replacing the complex exponentials with sines and cosines.

25. From Bromwich's integral

$$f(t) = \frac{a}{2\pi i} \int_{c-\infty i}^{c+\infty i} \frac{\sqrt{z}\exp[tz - x\sqrt{z(z+1)}\,]}{z[a\sqrt{z+1} + (2-a)\sqrt{\pi z/2}\,]}\,dz.$$

By Jordan's lemma we close the contour in the right half-plane if $t < x$; in the left half-plane, if $t > x$. If we take the branch cuts associated with $\sqrt{z}$ and $\sqrt{z+1}$ along the negative real axis so that the phase lies between $-\pi$ and π, the integration along the top of the branch cut cancels the integration along the bottom of the branch cut in the interval $[-\infty, 1)$. Consequently the branch cut for $\sqrt{z(z+1)}$ results in a dumbbell shaped contour along the negative real axis from -1 to 0.

Integration around the branch point $z = 0$ and $z = -1$ equal zero. Along the branch cut, $z = \eta^2 e^{\pm \pi i}$ and $z+1 = (1-\eta^2)e^{0i}$ with $0 < \eta < 1$. Along the top of the branch cut,

$$\int_{top} = \int_1^0 \left(\frac{2a}{\eta}\right) \frac{\eta i \exp(-\eta^2 t - ix\eta\sqrt{1-\eta^2})}{a\sqrt{1-\eta^2} + (2-a)i\eta\sqrt{\pi/2}}\, d\eta$$

$$= -2ai \int_0^1 e^{-\eta^2 t} \frac{[a\sqrt{1-\eta^2} + (2-a)i\eta\sqrt{\pi/2}\,]}{a^2 + b^2\eta^2}$$

$$\times \exp(-ix\eta\sqrt{1-\eta^2})\, d\eta$$

while along the bottom of the branch cut,

$$\int_{bottom} = \int_0^1 \left(\frac{2a}{\eta}\right) \frac{-\eta i \exp(-\eta^2 t + ix\eta\sqrt{1-\eta^2})}{a\sqrt{1-\eta^2} - (2-a)i\eta\sqrt{\pi/2}}\, d\eta$$

$$= -2ai \int_0^1 e^{-\eta^2 t} \frac{[a\sqrt{1-\eta^2} + (2-a)i\eta\sqrt{\pi/2}\,]}{a^2 + b^2\eta^2}$$

$$\times \exp(ix\eta\sqrt{1-\eta^2})\, d\eta.$$

The inverse equals the negative of the sum of the two branch cut integrals after dividing it by $2\pi i$ and replacing the complex exponentials by sine and cosine.

26. The inversion contour is identical to Fig. 3.1.2. The contribution from the integration around the branch point at $z = 0$ is $2\pi i$. Along CD, $z = \eta e^{\pi i}$, $\sqrt{z} = i\sqrt{\eta}$ so that

$$\int_{CD} = \int_0^\infty \exp[-\eta t - ai\sqrt{\eta}\,(1 - ce^{-bi\sqrt{\eta}})/(1 + ce^{-bi\sqrt{\eta}})] \frac{d\eta}{\eta}.$$

Along FG, $z = \eta e^{-\pi i}$, $\sqrt{z} = -i\sqrt{\eta}$ so that

$$\int_{FG} = \int_\infty^0 \exp[-\eta t + ai\sqrt{\eta}\,(1 - ce^{bi\sqrt{\eta}})/(1 + ce^{bi\sqrt{\eta}})] \frac{d\eta}{\eta}.$$

Using Euler's formula,

$$\frac{1 - c\exp(-bi\sqrt{\eta})}{1 + c\exp(-bi\sqrt{\eta})} = \frac{1 - c^2 + 2ic\sin(b\sqrt{\eta})}{1 + c^2 + 2c\cos(b\sqrt{\eta})} = v(\eta) + iu(\eta)$$

and

$$\frac{1 - c\exp(bi\sqrt{\eta})}{1 + c\exp(bi\sqrt{\eta})} = \frac{1 - c^2 - 2ic\sin(b\sqrt{\eta})}{1 + c^2 + 2c\cos(b\sqrt{\eta})} = v(\eta) - iu(\eta).$$

Therefore,

$$
\int_{CD} = \int_0^\infty \exp\{-\eta t - a\sqrt{\eta}\,[iv(\eta) - u(\eta)]\}\frac{d\eta}{\eta}
$$

$$
= \int_0^\infty e^{-\eta t + a\sqrt{\eta}\,u(\eta)}e^{-ai\sqrt{\eta}\,v(\eta)}\frac{d\eta}{\eta}
$$

and

$$
\int_{FG} = -\int_0^\infty e^{-\eta t + a\sqrt{\eta}\,u(\eta)}e^{ai\sqrt{\eta}\,v(\eta)}\frac{d\eta}{\eta}.
$$

Finally,

$$
f(t) = 1 + \frac{1}{2\pi i}\int_{FG} + \frac{1}{2\pi i}\int_{CD}
$$

$$
= 1 - \frac{1}{\pi}\int_0^\infty e^{-\eta t + a\sqrt{\eta}\,u(\eta)}\sin[a\sqrt{\eta}\,v(\eta)]\frac{d\eta}{\eta}.
$$

27. From the simple pole at $z = 0$,

$$
\mathrm{Res}(z = 0) = \sqrt{\frac{v\beta}{c}}.
$$

Let us rewrite the transform as

$$
F(s) = \frac{1}{s}\sqrt{\frac{v}{c}\left(s + \frac{\beta}{2}\right) + \sqrt{(s - \alpha_1)(s - \alpha_2)}}.
$$

In the interval from α_2 to α_1,

$$
s - \alpha_1 = \eta e^{\pm\pi i}, s - \alpha_2 = (\beta\sqrt{1 - v^2/c^2} - \eta)e^{0i}
$$

and

$$
\sqrt{(s - \alpha_1)(s - \alpha_2)} = \sqrt{\eta(\beta\sqrt{1 - v^2/c^2} - \eta)}\;e^{\pm\pi i/2}.
$$

Therefore,

$$
\sqrt{\frac{v}{c}\left(s + \frac{\beta}{2}\right) + \sqrt{(s - \alpha_1)(s - \alpha_2)}}
$$

$$
= \sqrt{\frac{v}{c}\left(\frac{\beta}{2}\sqrt{1 - \frac{v^2}{c^2}} - \eta\right) \pm i\sqrt{\eta(\beta\sqrt{1 - v^2/c^2} - \eta)}}
$$

$$
= \sqrt{\frac{r + x}{2} \pm i\sqrt{\frac{r - x}{2}}}
$$

where $0 < \eta < \beta\sqrt{1 - v^2/c^2}$. Combining both sides of the branch cut integral,

$$I_1 = 2i \int_0^{\beta\sqrt{1-v^2/c^2}} \frac{e^{(\alpha_1 - \eta)t}}{\alpha_1 - \eta} \sqrt{\frac{r - x}{2}} \, d\eta.$$

In the interval from $-\infty$ to α_2, $s - \alpha_1 = \eta e^{\pm\pi i}$, $s - \alpha_2 = (\eta - \beta\sqrt{1 - v^2/c^2})e^{\pm\pi i}$ and

$$\sqrt{(s - \alpha_1)(s - \alpha_2)} = \sqrt{\eta(\eta - \beta\sqrt{1 - v^2/c^2})} \, e^{\pm\pi i}.$$

Therefore,

$$\sqrt{\frac{v}{c}\left(s + \frac{\beta}{2}\right)} + \sqrt{(s - \alpha_1)(s - \alpha_2)}$$

$$= \sqrt{\frac{v}{c}\left(\eta - \frac{\beta}{2}\sqrt{1 - \frac{v^2}{c^2}}\right) + \sqrt{\eta(\eta - \beta\sqrt{1 - v^2/c^2})}} \, e^{\pm\pi i/2}.$$

where $\beta\sqrt{1 - v^2/c^2} < \eta < \infty$. Thus,

$$I_2 = 2i \int_{\beta\sqrt{1-v^2/c^2}}^{\infty} \frac{e^{(\alpha_1 - \eta)t}}{\alpha_1 - \eta} \, d\eta$$

$$\times \sqrt{\frac{v}{c}\left(\eta - \frac{\beta}{2}\sqrt{1 - \frac{v^2}{c^2}}\right) + \sqrt{\eta(\eta - \beta\sqrt{1 - v^2/c^2})}}.$$

The inverse equals the residue minus the sum of the branch cut integrals after we divide them by $2\pi i$.

Section 3.2

1. For $t > t_0$ we close the contour in the upper half-plane while for $t < t_0$ we close the contour in the lower half-plane. Because the only singularities are in the upper half-plane, $f(t) = 0$ if $t < t_0$. Along the branch cut on the positive real axis, $\omega - a = (u - a)e^{\theta i}$ where $\theta = 0$ or 2π and $\omega + a = (u + a)e^{\pi i}$. Along the branch cut on the negative real axis, $\omega - a = (u + a)e^{\pi i}$ and $\omega + a = (u - a)e^{\theta i}$ where $\theta = -\pi$ or π. The nonvanishing portions of the line integrals come from the branch cut edge below the positive real axis, the branch cut edge above the positive real axis, the branch cut edge above the negative real axis and the branch cut edge below the negative real axis. In that order,

$$f(t) = \frac{-i}{2\pi} \int_\infty^a \frac{e^{iut} \exp(it_0\sqrt{u^2 - a^2})}{-\sqrt{u^2 - a^2}} \, du$$

$$+ \frac{-i}{2\pi} \int_a^\infty \frac{e^{iut} \exp(-it_0\sqrt{u^2 - a^2})}{\sqrt{u^2 - a^2}} \, du$$

$$+ \frac{-i}{2\pi} \int_\infty^a \frac{e^{-iut} \exp(it_0\sqrt{u^2 - a^2})}{-\sqrt{u^2 - a^2}} \, (-du)$$

$$+ \frac{-i}{2\pi} \int_a^\infty \frac{e^{-iut} \exp(-it_0\sqrt{u^2 - a^2})}{\sqrt{u^2 - a^2}} \, (-du)$$

$$= \frac{2}{\pi} \int_a^\infty \frac{\cos(ut)\cos(t_0\sqrt{u^2 - a^2})}{\sqrt{u^2 - a^2}} \, du$$

$$= J_0\left(a\sqrt{t^2 - t_0^2}\right).$$

2. We can rewrite

$$F(\omega) = \frac{\exp[-\omega i\sqrt{r^2 + (z + a/\omega i)^2}]}{\sqrt{r^2 + (z + a/\omega i)^2}}$$

as

$$F(\omega) = \omega i \frac{\exp[-\sqrt{(r\omega i)^2 + (z\omega i + a)^2}]}{\sqrt{(r\omega i)^2 + (z\omega i + a)^2}} = \omega i G(\omega).$$

Therefore,

$$f(t) = \frac{dg(t)}{dt}.$$

Now, we can write

$$G(\omega) = \frac{\exp[-ir_i\sqrt{(\omega - \omega_1)(\omega - \omega_2)}]}{ir_i\sqrt{(\omega - \omega_1)(\omega - \omega_2)}}$$

where $\omega_1 = \beta + \alpha i$, $\omega_2 = -\beta + \alpha i$, $\alpha = az/r_i^2$, $\beta = ar/r_i^2$ and $r_i^2 = r^2 + z^2$. In order to satisfy the conditions on the square root,

$$w - \omega_{1,2} = re^{\theta i}, \qquad 2\pi < \theta < 4\pi.$$

The integration is similar to Fig. 3.2.2; the only nonzero contribution comes from the integration along the branch cuts. Therefore,

$$g(t) = \frac{1}{2\pi} \int_{-\infty}^{\infty} G(\omega)e^{i\omega t} d\omega$$

$$= \frac{e^{-\alpha t}}{2\pi} \int_{-\infty-i\epsilon}^{\infty-i\epsilon} \frac{e^{-ir_i\sqrt{(\chi+\beta)(\chi-\beta)}}e^{i\chi t}}{ir_i\sqrt{(\chi+\beta)(\chi-\beta)}} d\chi$$

$$= \frac{e^{-\alpha t}}{2\pi r_i}\left\{ \int_0^\beta \frac{e^{r_i\sqrt{\beta^2-\eta^2}}e^{i\eta t}}{\sqrt{\beta^2-\eta^2}} d\eta - \int_\beta^0 \frac{e^{-r_i\sqrt{\beta^2-\eta^2}}e^{i\eta t}}{\sqrt{\beta^2-\eta^2}} d\eta \right.$$

$$\left. + \int_0^\beta \frac{e^{-r_i\sqrt{\beta^2-\eta^2}}e^{-i\eta t}}{\sqrt{\beta^2-\eta^2}} d\eta - \int_\beta^0 \frac{e^{r_i\sqrt{\beta^2-\eta^2}}e^{-i\eta t}}{\sqrt{\beta^2-\eta^2}} d\eta \right\}$$

$$= \frac{e^{-\alpha t}}{\pi r_i}\left\{ \int_0^\beta \frac{e^{r_i\sqrt{\beta^2-\eta^2}}\cos(\eta t)}{\sqrt{\beta^2-\eta^2}} d\eta + \int_0^\beta \frac{e^{-r_i\sqrt{\beta^2-\eta^2}}\cos(\eta t)}{\sqrt{\beta^2-\eta^2}} d\eta \right\}$$

$$= \frac{2e^{-\alpha t}}{\pi r_i} \int_0^\beta \frac{\cosh(r_i\sqrt{\beta^2-\eta^2})\cos(\eta t)}{\sqrt{\beta^2-\eta^2}} d\eta$$

$$= \frac{2e^{-\alpha t}}{\pi r_i} J_0(\beta\sqrt{t^2-r_i^2})H(t-r_i).$$

3. For $t > 0$, we close *both* integrals with a semi-circle of infinite radius in the upper half of the ω-plane. Within one of these contours we have a second-order poles at $\omega = 0$,

$$\text{Res}(\omega = 0) = \lim_{\omega \to 0} \frac{d}{d\omega}\left\{ \frac{\omega^2 \exp[z\sqrt{\omega^2+k^2}+i\omega t]}{k-\sqrt{\omega^2+k^2}} \right\}$$

$$= -2ikte^{kz}.$$

From our choice of branch cuts $\omega - ki = re^{\theta i}$ where $-3\pi/2 < \theta < \pi/2$ and $\omega + ki = re^{\theta i}$ where $-\pi/2 < \theta < 3\pi/2$. Therefore, the integration along the branch cut in the upper half plane is

$$\int_{branch\ cut} = \int_{\infty}^{k} \frac{\exp(ia\sqrt{u^2-k^2}-ut)}{k-i\sqrt{u^2-k^2}} i\,du$$

$$+ \int_{k}^{\infty} \frac{\exp(-ia\sqrt{u^2-k^2}-ut)}{k+i\sqrt{u^2-k^2}} i\,du$$

$$= 2\int_{k}^{\infty} \frac{e^{-ut}}{u^2}[k\sin(a\sqrt{u^2-k^2})$$

$$+ \sqrt{u^2-k^2}\cos(a\sqrt{u^2-k^2})]\,du.$$

Thus,

$$\frac{1}{2\pi} \int_{-\infty-\epsilon i}^{\infty-\epsilon i} = 2kte^{kz} - \frac{1}{2\pi} \int_{branch\ cut}.$$

On the other hand,

$$\frac{1}{2\pi} \int_{-\infty+\epsilon i}^{\infty+\epsilon i} = -\frac{1}{2\pi} \int_{branch\ cut}.$$

We now substitute these results into the formula.

For $t < 0$, we close both integrals in the lower half plane and repeat the previous analysis. Remember that the contour is now in the negative sense so we must take the negative of the residue.

4.
$$f(t) = -\frac{1}{2\pi} \int_{-\infty}^{\infty} \frac{|\omega| \exp[-|\omega|s + i\omega t]}{|\omega| - a_1} \, d\omega$$

where we pass over the singularities at $\omega = \pm a_1$. The residues for these singularities are
$$\text{Res}(\omega = a_1) = a_1 e^{-a_1 s + i a_1 t}$$

and
$$\text{Res}(\omega = -a_1) = -a_1 e^{-a_1 s - i a_1 t}.$$

Therefore, the sum of the residues is

$$2 i a_1 e^{-a_1 s} \sin(a_1 t).$$

For $t < 0$, we close the contour in the lower half-plane. This includes are two simple poles. Recalling that the closed contour in the lower half-plane is in the negative sense,

$$f(t) = i[2 i a_1 e^{-a_1 s} \sin(a_1 t)] - \frac{1}{2\pi} \int_{branch\ cut}.$$

Now

$$\int_{branch\ cut} = -\int_{\infty}^{0} \frac{(-i\eta) \exp(i\eta s + \eta t)}{-i\eta - a_1} (-i \, d\eta)$$
$$- \int_{0}^{\infty} \frac{(i\eta) \exp(-i\eta s + \eta t)}{i\eta - a_1} (-i \, d\eta)$$
$$= 2 \int_{0}^{\infty} \frac{\eta \exp(t\eta)}{\eta^2 + a_1^2} [a_1 \cos(s\eta) + \eta \sin(s\eta)] \, d\eta.$$

For $t > 0$, we close the contour in the upper half-plane. There are no poles in this case and

$$f(t) = -\frac{1}{2\pi} \int_{branch\ cut}$$

where

$$
\int_{branch\ cut} = -\int_{\infty}^{0} \frac{(i\eta)\exp(-i\eta s - \eta t)}{i\eta - a_1}(i\,d\eta)
$$
$$
-\int_{0}^{\infty} \frac{(-i\eta)\exp(i\eta s - \eta t)}{-i\eta - a_1}(i\,d\eta)
$$
$$
= 2\int_{0}^{\infty} \frac{\eta\exp(-t\eta)}{\eta^2 + a_1^2}[a_1\cos(s\eta) + \eta\sin(s\eta)]\,d\eta.
$$

5. The inverse is

$$
f(t) = \frac{1}{2\pi}\int_{-\infty-\epsilon i}^{\infty-\epsilon i}\left\{\frac{V(z)i}{z[1 - cV(z)]} - \frac{V(z)zi}{(z^2 + a^2)[1 - cV(z)]}\right\}e^{izt}dz.
$$

For $t > 0$, we close the contour with an infinite semi-circle in the upper half plane of the z plane, except for a branch cut along the imaginary axis from i to ∞i. Within the closed contour there are three poles $z = 0$, $z = \alpha i$ and $z = \omega_0 i$ where $cV(\omega_0 i) = 1$. The corresponding residues are

$$
\text{Res}(z = 0) = \frac{i}{1 - c}
$$

$$
\text{Res}(z = \alpha i) = -\frac{iV(\alpha i)}{2[1 - cV(\alpha i)]}e^{-\alpha t}
$$

and

$$
\text{Res}(z = \omega_0 i) = \frac{i\alpha^2(1 - \omega_0^2)}{c(\alpha^2 - \omega_0^2)(\omega_0^2 + c - 1)}e^{-\omega_0 t}.
$$

Along the branch cut we have $\omega - i = (\eta-1)e^{\pi i/2}$ or $\omega - i = (\eta-1)e^{-3\pi i/2}$ and $\omega + i = (\eta + 1)e^{\pi i/2}$. Then

$$
\int_{cut} = \frac{1}{2\pi}\int_{\infty}^{1}\left\{\frac{V_+(\eta)}{\eta[1 - cV_+(\eta)]} - \frac{V_+(\eta)(-\eta)}{(\alpha^2 - \eta^2)[1 - cV_+(\eta)]}\right\}e^{-\eta t}i\,d\eta
$$
$$
+ \frac{1}{2\pi}\int_{1}^{\infty}\left\{\frac{V_-(\eta)}{\eta[1 - cV_-(\eta)]} - \frac{V_-(\eta)(-\eta)}{(\alpha^2 - \eta^2)[1 - cV_-(\eta)]}\right\}e^{-\eta t}i\,d\eta
$$

where

$$
V_+(\eta) = -\frac{1}{2\eta}\ln\left(\frac{\eta - 1}{\eta + 1}\right) - \frac{\pi i}{2\eta}
$$

and

$$
V_-(\eta) = -\frac{1}{2\eta}\ln\left(\frac{\eta - 1}{\eta + 1}\right) + \frac{\pi i}{2\eta}.
$$

A little algebra gives

$$\int_{branch\ cut} = \frac{\pi \alpha^2}{2\pi} \int_1^\infty \left\{ \left[1 + \frac{c}{2\eta} \ln\left(\frac{\eta-1}{\eta+1}\right) \right]^2 + \frac{\pi^2 c^2}{4\eta^2} \right\}^{-1} \frac{e^{-\eta t}\ d\eta}{\eta^2(\eta^2 - \alpha^2)}.$$

We obtain the final result by summing the residue times $2\pi i$ and subtracting off the branch cut integral. Note that we have made the substitution $x = 1/\eta$ in the branch cut integral.

6. From the definition of the generalized Fourier transform, the inverse is

$$f(t) = \frac{1}{2\pi} \int_{-\infty-\epsilon i}^{\infty-\epsilon i} \frac{\log(i\omega\tau)}{(\lambda + \sqrt{\lambda^2 + \omega i})^2} e^{i\omega t} d\omega$$

where $\epsilon > 0$. For $t > 0$ we close the contour with a semi-circle in the upper half of the ω-plane, except for a cut along the imaginary axis. For $t < 0$ we close the contour with a semi-circle in the lower half of the ω-plane. Within the closed contours there are no singularities. Therefore $f(t) = 0$ if $t < 0$. For $t > 0$ we have four line integrals that comprise the branch cut integration: (1) from $0^+ + \infty i$ to $0^+ + \lambda^2 i$, (2) from $0^+ + \lambda^2 i$ to $0^+ + 0^+ i$, (3) from $0^- + 0^+ i$ to $0^- + \lambda^2 i$ and (4) from $0^- + \lambda^2 i$ to $0^- + \infty i$. The integrations around the branch points equal zero.

For the first contour, $\omega = \lambda^2 \eta i$, $\omega - \lambda^2 i = \lambda^2(\eta - 1)e^{\pi i/2}$ and $\log(i\omega\tau) = \ln(\lambda^2\tau\eta) + \pi i$ so that

$$\int_{C_1} = \frac{1}{2\pi} \int_\infty^1 \frac{[\ln(\lambda^2\tau\eta) + \pi i]}{(\lambda + \lambda i\sqrt{\eta-1})^2} e^{-\lambda^2 t\eta} i\lambda^2\ d\eta$$

while along the fourth contour $\omega = \lambda^2 \eta i$, $\omega - \lambda^2 i = \lambda^2(\eta - 1)e^{-3\pi i/2}$ and $\log(i\omega\tau) = \ln(\lambda^2\tau\eta) - \pi i$ so that

$$\int_{C_4} = \frac{1}{2\pi} \int_1^\infty \frac{[\ln(\lambda^2\tau\eta) - \pi i]}{(\lambda - \lambda i\sqrt{\eta-1})^2} e^{-\lambda^2 t\eta} i\lambda^2\ d\eta.$$

Consequently, the sum of the integrals equals

$$\int_{C_1} + \int_{C_4} = -\frac{1}{2\pi} \int_1^\infty [2\sqrt{\eta-1}\ \ln(\lambda^2\tau\eta) - \pi(2 - \eta)]e^{-\lambda^2 t\eta} \frac{d\eta}{\eta^2}.$$

Along the second contour, $\omega = \lambda^2 \eta i$, $\omega - \lambda^2 i = \lambda^2(1 - \eta)e^{-\pi i/2}$ and $\log(i\omega\tau) = \ln(\lambda^2\tau\eta) + \pi i$ so that

$$\int_{C_2} = \frac{1}{2\pi} \int_1^0 \frac{[\ln(\lambda^2\tau\eta) + \pi i]}{\lambda^2(1 + \sqrt{1-\eta})^2} e^{-\lambda^2 t\eta} i\lambda^2\ d\eta$$

while along the third contour $\omega = \lambda^2 \eta i$, $\omega - \lambda^2 i = \lambda^2(1-\eta)e^{-\pi i/2}$ and $\log(i\omega\tau) = \ln(\lambda^2 \tau \eta) - \pi i$ so that

$$\int_{C_3} = \frac{1}{2\pi} \int_0^1 \frac{[\ln(\lambda^2 \tau \eta) - \pi i]}{\lambda^2(1 + \sqrt{1-\eta})^2} e^{-\lambda^2 t \eta_i \lambda^2} \, d\eta.$$

Consequently, the sum of the integrals equals

$$\int_{C_2} + \int_{C_3} = \int_0^1 \frac{1}{(1 + \sqrt{1-\eta})^2} e^{-\lambda^2 t \eta} \, d\eta.$$

The final answer is the negative of the sum of all of the integrals.

8. From the inversion integral,

$$f(x, y) = \frac{1}{2\pi} \int_{-\infty}^{\infty} \frac{\exp(ikx - y\sqrt{k^2 - ikV/c})}{\sqrt{k^2 - ikV/c}} \, dk$$

$$= \frac{1}{2\pi} \int_{-\infty}^{\infty} \frac{\exp[ikx - y\sqrt{(k - iV/2c)^2 + V^2/4c^2}]}{\sqrt{(k - iV/2c)^2 + V^2/4c^2}} \, dk$$

$$= \frac{e^{-Vx/2c}}{2\pi} \int_{-\infty - Vi/2c}^{\infty - Vi/2c} \frac{\exp[i\alpha x - y\sqrt{\alpha^2 + V^2/4c^2}]}{\sqrt{\alpha^2 + V^2/4c^2}} \, d\alpha$$

if $\alpha = k - Vi/2c$. Let

$$i\alpha x - y\sqrt{\alpha^2 + V^2/4c^2} = -s,$$

so that

$$\alpha_\pm = \frac{ixs}{r^2} \pm \frac{z}{r^2} \sqrt{s^2 - (rV/2c)^2}$$

where α_- gives the contour in the second (third) quadrant and α_+ is the contour in the first (fourth) quadrant if $x > 0$ ($x < 0$). We can deform our contour because there are no singularities between the original contour and the new contour. Straightforward substitution and algebra leads to

$$f(x, y) = \frac{e^{-Vx/2c}}{\pi} \int_{rV/2c}^{\infty} \frac{e^{-s}}{\sqrt{s^2 - (rV/2c)^2}} \, ds.$$

Now, let $s = rV\eta/2c$ and

$$f(x, y) = \frac{e^{-Vx/2c}}{\pi} \int_1^{\infty} \frac{e^{-rV\eta/2c}}{\sqrt{\eta^2 - 1}} \, d\eta.$$

Section 3.3

1. See Section 3.1, problem 2.

2. See Section 3.1, problem 4.

3. See Section 3.1, problem 5.

4. See Section 3.1, problem 6.

5. See Section 3.1, problem 7.

6. See Section 3.1, problem 8.

7. See Section 3.1, problem 9.

8. See Section 3.1, problem 10.

9. See Section 3.1, problem 11.

10. See Section 3.1, problem 12.

11. See Section 3.1, problem 13.

Section 3.4

1. See Section 3.2, problem 4.

Index